UNIVERSITY OF STRATHCLYDE
30125 00464762 3

KV-038-794
WITHDRAWN
FROM
LIBRARY
STOCK
UNIVERSITY OF STRATHCLYDE

BIOLOGICAL DEGRADATION OF WASTES

Other titles in the Elsevier Applied Biotechnology Series

K. Carr-Brion (ed.). *Measurement and Control in Bioprocessing*

M. Y. Chisti. *Airlift Bioreactors*

W. M. Fogarty/C. T. Kelly (eds). *Microbial Enzymes and Biotechnology, 2nd Edition*

T. J. R. Harris (ed.). *Protein Production by Biotechnology*

R. Isaacson (ed.). *Methane from Community Wastes*

A. M. Martin (ed.). *Bioconversion of Waste Materials to Industrial Products*

E. J. Vandamme (ed.). *Biotechnology of Vitamins, Pigments and Growth Factors*

BIOLOGICAL DEGRADATION OF WASTES

Edited by

A. M. MARTIN

Department of Biochemistry, Memorial University of Newfoundland, St John's, Newfoundland, Canada A1B 3X9

ELSEVIER APPLIED SCIENCE
LONDON and NEW YORK

ELSEVIER SCIENCE PUBLISHERS LTD
Crown House, Linton Road, Barking, Essex IG11 8JU, England

Sole distributor in the USA and Canada
ELSEVIER SCIENCE PUBLISHING CO., INC
655 Avenue of the Americas, New York, NY 10010, USA

WITH 73 TABLES AND 134 ILLUSTRATIONS

British Library Cataloguing in Publication Data

Biological degradation of wastes.
1. Waste materials. Disposal
I. Martin, A. M. (Antonio M.)
628.4
ISBN 1-85166-635-4

Library of Congress Cataloging-in-Publication Data

Biological degradation of wastes/edited by A. M. Martin.
p. cm.
Includes bibliographical references and index.
ISBN 1-85166-635-4
1. Refuse and refuse disposal—Biodegradation. I. Martin, A. M. (Antonio M.)
TD796.5.B564 1991
628.4′45—dc20 91-15471
CIP

Printed in Great Britain by Galliard (Printers) Ltd, Great Yarmouth

PREFACE

The cumulative effects of pollution have led, in recent years, to increased public concern, which is resulting in stricter legislation on the discharge of wastes in whatever state they are present: gaseous, liquid or solid. The treatment and disposal of wastes has become one of the most important problems facing mankind. This is a problem which will not disappear, and could even worsen, if it is not faced with resolution by all the main parties involved: consumers, governments, producers and scientists.

Some wastes could be reused, producing some economic return which could pay for the waste-treatment process. In the best of cases, this could become an economically attractive recycling operation. However, in many situations, waste treatment is considered to be an unproductive process which entails additional costs to an otherwise productive operation.

Methods for the removal and purification of wastes (including those considered to be 'toxic wastes', the most dreaded form of pollution), if developed at all, suffer from serious limitations. Two of these are the high energy input into the process and, after the contaminants have been removed, the lingering problem of what to do with them, as they will then exist as some kind of concentrate. The ideal solution is none other than a natural, biological process to degrade wastes. Fortunately, mankind is increasingly choosing that option, as exemplified by the general acceptance of the role of biotechnology in modern society.

This book deals with precisely those biological processes, mostly derived from the action of enzymes and microorganisms, used for the degradation of wastes. Although many publications are available dealing with biological treatments as they apply to liquid effluents (and more recently the biodegradation of organic compounds and toxic wastes has been dealt with in specialized publications), this book is intended to present a comprehensive coverage of biodegradation, not only of liquid but also of solid wastes and gaseous effluents. It was also considered necessary that topics important to the future of biodegradation, such as the application of computers to the operations involved in biological pollution control, be incorporated into the book. This book is not intended to be a study of the processes of the bioconversion of wastes to useful products, as this is the subject of a recently published volume—*Bioconversion of Waste Materials to Industrial Products*—by the same editor (Elsevier Applied Science, 1991).

The contents of the chapters in the book have been grouped into two parts: General Topics, and Applications. This has been done only to provide a preliminary organization of the materials presented and does not imply an absolute division of them into those two categories, as many of the chapters and topics are interrelated. Indeed, some of the chapters have characteristics which could justify their placement in either section.

Although it is not possible to present in one volume all the advances occurring in this very dynamic field, this book is intended to illustrate, by reporting specific biodegradation processes, the usefulness and potential of biological processes in pollution control. The diversity of expertise of the authors indicates the mixed population of scientists and engineers to whom this book is addressed. Only a combined effort, incorporating the latest advances in biological sciences and in processing plant design, will enable mankind to apply biological degradation processes to fulfil the public hope of protecting the environment.

ANTONIO M. MARTIN

CONTENTS

LIST OF CONTRIBUTORS

K. Ahmed
Institute of Public Health Engineering and Research, University of Engineering and Technology, Lahore 54890, Pakistan

Y. Argaman
Department of Civil Engineering, Technion, Israel Institute of Technology, Haifa, Israel 32000

N. Athanasopoulos
Perivallontiki-Energiaki Ltd, 106B Lontou Street, 26224 Patras, Greece

M. I. D. Chughtai
Division of Biochemistry, Institute of Chemistry, University of the Punjab, Lahore 54590, Pakistan

D. Couillard
Institut National de la Recherche Scientifique (INRS-Eau), Université du Québec, 2700 rue Einstein, CP 7500, Sainte-Foy, Québec, Canada G1V 4C7

C. Durán de Bazúa
Departamento de Alimentos y Biotecnología, División de Ingeniería, Facultad de Química, Universidad Nacional Autónoma de México, 04510 México DF

H. Endo
Research Center for Advanced Science and Technology, University of Tokyo, 4-6-1 Kamaba, Meguro-ku, Tokyo 153, Japan

H. J. GIJZEN
Department of Microbiology, Faculty of Science, University of Nijmegen, Toernooiveld, NL-6525 ED Nijmegen, The Netherlands
Present address: Applied Microbiology Unit, Department of Botany, University of Dar-es-Salaam, PO Box 35060, Dar-es-Salaam, Tanzania

D. C. HEMPEL
Department of Technical Chemistry and Chemical Engineering, University of Paderborn, Warburger Str. 100, 4790 Paderborn, Federal Republic of Germany

G. L. JONES
4 Grovelands Avenue, Hitchin, Hertfordshire SG4 0QT, UK

I. KARUBE
Research Center for Advanced Science and Technology, University of Tokyo, 4-6-1 Kamaba, Meguro-ku, Tokyo 153, Japan

A. M. MARTIN
Department of Biochemistry, Memorial University of Newfoundland, St John's, Newfoundland, Canada A1B 3X9

F. C. MILLER
Department of Microbiology, La Trobe University, Bundoora, Victoria 8083, Australia
Present address: Director of Research, Sylvan Foods Inc., One Moonlight Drive, Worthington, Pennsylvania 16262, USA

G. MORIN
Institut National de la Recherche Scientifique (INRS-Eau), Université du Québec, 2700 rue Einstein, CP 7500, Sainte-Foy, Québec, Canada G1V 4C7

B. NÖRTEMANN
Department of Technical Chemistry and Chemical Engineering, University of Paderborn, Warburger Str. 100, 4790 Paderborn, Federal Republic of Germany

A. NOYOLA
Coordinación de Ingeniería Ambiental, Instituto de Ingeniería, Universidad Nacional Autónoma de México, 04510 México DF

H. J. M. OP DEN CAMP
Department of Microbiology, Faculty of Science, University of Nijmegen, Toernooiveld, NL-6525 ED Nijmegen, The Netherlands

H. POGGI
Departamento de Biotecnología y Bioingeniería, Centro de Investigación y Estudios Avanzados, Instituto Politécnico Nacional, 07300 México DF

R. J. PORTIER
Aquatic and Industrial Toxicology Laboratory, Institute for Environmental Studies, Louisiana State University, Baton Rouge, Louisiana 70803, USA

M. SHODA
Research Laboratory of Resources Utilization, Tokyo Institute of Technology, 4259 Nagatsuta, Midori-ku, Yokohama 227, Japan

D. THIRUMURTHI
Department of Civil Engineering, Technical University of Nova Scotia, PO Box 1000, Halifax, Nova Scotia, Canada B3J 2X4

R. D. TYAGI
Institut National de la Recherche Scientifique (INRS-Eau), Université du Québec, 2700 rue Einstein, CP 7500, Sainte-Foy, Québec, Canada G1V 4C7

L. P. WACKETT
Department of Biochemistry and Gray Freshwater Biological Institute, College of Biological Sciences, University of Minnesota, PO Box 100, County Road 15 and 19, Navarre, Minnesota 55392, USA

L. E. ZEDILLO
Instituto para el Mejoramiento de la Producción de Azúcar, Grupo de Países de Latinoamérica y el Caribe Exportadores de Azúcar, Tuxpan 2, 12avo. Piso, 06760 México DF

Chapter 1

BIODEGRADATION OF SOLID WASTES BY COMPOSTING

FREDERICK C. MILLER*

Department of Microbiology, La Trobe University, Bundoora, Victoria, Australia

CONTENTS

* Present address: Director of Research, Sylvan Foods, Inc., One Moonlight Drive, Worthington, Pennsylvania 16262, USA.

1 INTRODUCTION

1.1 General Introduction

Composting has become increasingly popular in the past decade as an alternative to incineration or tipping of decomposable organic wastes. It can now be considered a useful treatment process for almost every kind of biodegradable waste (US Congress, Office of Technology Assessment, 1989). In the United States in 1989 there were eight operating municipal solid waste (MSW) facilities. In the same year there were also 75 MSW projects under development, up from 42 in 1988 (Goldstein, 1988). Many of these projects are quite large, such as the Delaware Reclamation Project, which receives 1000 tons per day of MSW and composts a degradable fraction of 350 tons per day. In Europe in 1988 there were at least 23 composting plants for refuse, most combining refuse and sewage sludge (Bardos & Lopez-Real, 1988). Sewage sludge composting has become very common since the 1970s, with perhaps 150 facilities planned or in operation in the United States alone. While complete estimates are not available, facilities for composting collected yard waste are becoming so popular that there were 180 permitted facilities in the state of New Jersey alone by the end of 1988 (Glenn, 1988). Yard waste and similar waste streams can easily be treated in fairly simple facilities (Strom & Finstein, 1985).

Interest in solid waste composting for refuse developed in the 1950s, as evident in the early work of McCauley and Shell (1956), Wiley (1956), Wiley and Pearce (1957) and Schulze (1958), among others. Little of this early work was put into practice, partly because of a lack of convincing large-scale experience, and a lack of appropriate technologies for separating compostable wastes from other materials, but mainly because of cheap tipping fees. Some successful projects were carried out in Europe, notable VAM in The Netherlands (Teensma, 1961), but these projects would not be looked on with favour today. Early efforts attempted mass composting, where the entire waste stream was composted, lead to high metal concentrations and other unwanted materials in the final compost. New separation technologies that can economically remove the compostable materials from the refuse stream are making refuse composting feasible.

In the United States in the 1970s restrictions began to be placed on the ocean dumping and the landfilling of sewage sludge. Composting is inexpensive, rapidly implemented, and a publicly acceptable treatment process, and as a result became quickly popular. Almost 200 sludge composting facilities are operating or planned in the United States alone, mostly in the north-eastern part of the country (US Congress, Office of Technology Assessment, 1989). As a category of solid waste, sewage sludge composting has been fairly successful, leading the way in the development of process control strategies.

Major problems still occur at some facilities, such as odours, poor product quality, operational problems including long processing times, and excessive moisture and materials handling problems. Most of these problems are a direct result of poor process management, often arising from facilities designs that leave little opportunity for good process management (Finstein *et al.*, 1987*a,b,c,d*). Composting is a complex biological process. This complexity makes the process not easy to understand, and when not well understood difficult to manage. Most troubled systems evidence detailed consideration of materials handling but ineffective process control. In developing facilities design, serving the needs of the composting ecosystem to achieve the desired processing goals needs to be foremost, and not sacrificed to other design considerations. This chapter will consider in depth composting as a controllable ecosystem. It will emphasise the physical ecology of the composting process, as this marks the boundary between biological control and systems engineering, and makes it possible to accommodate biological control and systems engineering.

1.2 Composting Defined

Composting is an ecosystem which self heats, i.e. temperature within the composting mass rises because heat released metabolically accumulates faster than it is dissipated to the surrounding environment. This self heating tends to increase decomposition rates unless inhibitively high temperatures are reached. Activity is much more rapid and less odorous under fully aerobic conditions.

2 COMPOSTING CONFIGURATIONS

2.1 Description of Some Common Systems

Useful definitions of terms and concepts referred to in composting are given in the reviews of Finstein *et al.* (1985) and Zucconi and De Bertoldi (1987). Important to both basic design configuration and process control is the choice between *batch* and *continuous* processes. Batch systems have advantages in process control because all material in the system is of equal age. Continuous processes can offer mechanical advantages in materials handling. Either system can work well, but within the confines of different design constraints.

Open composting systems are suitable for many applications, with advantages of low capital investment and rapid implementation. A *pile* is a simple arrangement of material into a conical or pyramidal shape, or a sloping sided row. *Static piles* are built over ventilation duct work, and are not turned. *Heap* often means the same as pile, but also refers to a more indiscriminately configured, larger mass. *Turned windrows* are long narrow piles turned using various machinery. Some windrows are both turned and aerated using ventilation duct work. *Bin composting* falls between open and enclosed composting, normally having walls but not fully enclosed. Bin systems can offer much of the simplicity of fully open systems, with some advantage in materials handling.

Enclosed composting is a broad term including all composting not open to the environment or ambient surroundings. *In-vessel* is a popular term in advertisements suggesting an advanced configuration, including silos, trenches and rotating drums, with mixing or not, but 'in vessel' means nothing in reference to actual process. *Silos* are vertical reactors in which material is continuously added to the top of the reactor, with finished material removed from the bottom. Silos are often aerated in an up-flow direction. Good process control is not possible in a silo configuration because ventilation control is independent of compost maturity, and heavy compaction interferes with gas and heat exchange. *Trenches* are reactors in which the compost is placed in long straight-walled trenches, with subfloor ventilation. The physical compartments of trench systems can permit good process control, even in a continuous feed (non-batch) mode if designed properly (Kuter *et al.*, 1985). Trench systems can incorporate mixing during processing, which can promote product uniformity. *Drums* are horizontal tube reactors, which are able to rotate at a slow rate of speed so that materials placed in one end of the drum will after a specific residence time exit from the other end; the best known of this type is the Dano drum, which has had some popularity over the years (BioCycle, 1986). Drums can provide good mixing and reduction of particle size, but process control is not easy, and long curing outside the drum is normally needed. A *Fairfield–Hardy digester* is a unique proprietary system in which material is introduced in the centre of a round tank, which is then moved outward during processing by vertical augers mounted on rotating arms; finished material is removed at the outer edge. Zonal-based ventilation is possible with this design (Schulze, 1965). *Tunnel* is a term referring to a long and low reactor design common to the mushroom industry. This type of batch reactor offers excellent process control and may be adaptable to solid waste composting.

2.2 Other Terms Related to Composting

High rate is a fuzzy but popular term, implying speed and sophistication but meaning little in process terms. *Mechanical composting* is sometimes used, but like high rate is normally restricted to advertising, lacking a real technical meaning.

An important distinction exists between composting (the process) and compost (a product) (Finstein & Miller, 1982). Other terms to be distinguished are process strategy and process configuration (Finstein *et al.*, 1986). Process strategy refers to the means of managing the composting ecosystem to achieve processing goals.

Configuration refers to the physical process implementation. Implementation of the same processing strategy in different configurations will produce similar results. Outdoor composting and enclosure of composting into drums, silos, troughs, etc., are not different processes. Fundamental inadequacies of processing strategy will not be solved by changing configurations, although some configurations preclude reasonable process management. Designers of many failed composting configurations often appear to have failed to adequately consider composting strategy except the more dangerous one: that composting is a 'natural process' and composting will naturally occur regardless of strategy.

As a technical term, *fermentation* refers to substrate-based phosphorylation without an external inorganic terminal electron acceptor (i.e. oxygen, nitrate or sulphate) and produces compounds such as organic acids or alcohols as a result (Brock & Madigan, 1988). 'Fermentation' is sometimes used generally to mean biological decomposition but the more precise technical usage should be encouraged.

3 PHYSICAL FACTORS AFFECTING PROCESSING

3.1 Physical Structure of the Matrix

Unlike most other waste treatment processes which occur in an aqueous phase, composting occurs within a physical matrix. The consequences of this matrix structure are physical and chemical gradients, and site specificity of various factors which influence microbial activity (Miller, 1984, 1989). From an ecological perspective, composting systems are similar to soil systems except that much more substrate is available. Composting matrices are characteristically organic, with a high volumetric substrate density, and potentially high rates of metabolic activity per unit volume. Also, as composting progresses, the structural nature of the matrix changes, softening the texture and losing volume.

Unless mixed, the composting matrix limits the transfer of gases, heat, water, nutrients and microbial populations, causing distinct gradients. A large part of process management consists of overcoming these gradients, and establishing a uniformly favourable environment for composting. Mixing may relieve heterogeneity to varying degrees, but mixing may not be the most economical and practical means of achieving uniformity. An important part of process design is determining the compromise between economics and material uniformity within the matrix.

It has been pointed out that a composting system is a matrix on two levels. On one level it is a stable collection of aggregates, and on the smaller scale it is a matrix at the level of the aggregates themselves. Non-homogeneity of physical conditions within aggregates creates a large variety of nonequivalent microsites which foster greater species diversity (Wimpenny *et al.*, 1984). Composting refuse, Filip (1978) found that the association of microbes with particles allowed them to tolerate greater heat stress. Atkey and Wood (1983), using an electron microscope, showed the heterogeneity in the colonisation of straw particles, and how this affected decomposition. Association of cells with matrices may also induce variations in cellular morphology and biochemistry, including the production of extracellular enzymes (Hubert Lechevalier, personal communication).

3.2 Heat Evolution

Working with stored hay, Browne (1933) was the first to demonstrate that biological activity was the cause of the self heating commonly observed in stored organic materials. Waksman and colleagues (Waksman & Cordon, 1939; Waksman *et al.*, 1939*a,b*) investigated microbial populations and their dynamic changes within composting ecosystems, and determined the effect of various physical factors on activity. Work by a number of investigators (Carlyle & Norman, 1941; Norman *et al.*, 1941; Walker & Harrison, 1960; Rothbaum, 1961, 1963; Dye, 1964; Dye & Rothbaum, 1964; Rothbaum & Dye, 1964) further described the metabolic nature of self heating, with elucidation of the basic thermodynamics and kinetics of biological self heating. Controversy over the contributions of chemical and biological reactions to heat evolution was resolved by Nell and Wiechers (1978), who determined that the heat released during composting was almost exclusively derived from biological reactions. Microbial energetics related to metabolism have been comprehensively reviewed in the recent work of Battley (1987).

Heat evolution and oxygen uptake are proportionally linked during aerobic metabolism, with approximately 14 000 kJ released per kilogram oxygen consumed during the complete oxidation of organic matter (Finstein *et al.*, 1986); this can be used to convert between heat evolution data and oxygen consumption.

Heat evolution from the composting of different substrates can be either calculated based on the extent of decomposition and the heats of combustion for the decomposed fraction (Haug, 1979) or measured directly (Hogan *et al.*, 1989). Heats of combustion are invariably higher than heats of metabolism, but are useful for approximation. By combustion, proteins and carbohydrates will yield around 15–25 kJ per gram, and lipids about 35–45 kJ. Based on heats of combustion, Haug (1979) calculated for sewage sludge a theoretical heat output of 23·0 kJ per gram decomposed. Using direct measurement while composting refuse with a high lipid content (14·7%), Wiley (1956) measured heat outputs of 22·1–28·5 kJ per gram decomposed. Using a scaled heat model composting system, Miller (1984) reported two data sets for primary sewage sludge which averaged 21·8 and 15·2 kJ per gram decomposed. Composting rice hulls and rice flour, Hogan *et al.* (1989) measured heats per gram decomposed of 14·2 and 16·7 kJ in different trials.

Storage of evolved heat within the mass itself leads to the elevated temperatures characteristic of composting; it is also characteristic that without intervention heat accumulation commonly leads to high temperatures, which severely inhibit further microbial activity (Finstein *et al.*, 1983). Designing for maximum heat evolution rates is critical, in that this defines the requirements for heat removal needed to maintain favourable temperatures. Summarised in Table 1 are heat evolution data from a wide variety of materials during composting.

Table 1
Heat Output from Various Composting Situations

Substrate	*Peak heat output (J per g volatile matter per hour)*[a]	*Trial temperature (°C)*	*Average time period (h)*	*Reference*
Straw[b]	48·2	40	1	Carlyle and Norman (1941)
	26·6	60	1	
Refuse	76·7	60	1	Wiley (1957)
Sewage sludge and wood chip	46·7–54·0	50–60	12	Miller (1984)
	23·5	72	12	
	17·8	74	12	
Oak leaves	5·8	50	24	Finstein *et al.* (1986)
Oak leaves[b]	11·6	50	24	
Maple leaves	13·1	50	24	
Mushroom compost, straw-based	8·4	72	12	Miller *et al.* (1989*a*)
	31·4	55	12	
	37·7	45	12	
Rice hulls and rice flour	41·8	40–55	12	Hogan *et al.* (1989)
Mushroom compost, straw-based[c]	66	54	12	Harper, Miller and Macauley, unpublished data

[a] In the reporting of peak heat output, volatile refers to the non-ash dry matter which is equivalent to organic matter. The grams volatile referred to is the initial loading.
[b] Inorganic nutrients added.
[c] Small commercial-scale (10 tonne) reactor, operated under conditions for optimal activity.

3.3 Temperatures as a Selective Factor

Temperatures within composting materials are a function of the rates of heat evolution and heat loss to the environment. In composting systems, temperature is both a result of and a determinant of activity.

Enzyme activity rates generally double with each 10°C rise in temperature, until an inactivation temperature is reached. Elevated temperatures during composting can promote rapid decomposition, but can also lead to thermal death. Thermal killing eliminates pathogens (Burge *et al.*, 1978; Finstein *et al.*, 1982), but can also kill microbes desired for decomposition. Temperature selects for or against populations based on temperature tolerance (Atlas & Bartha, 1981). Microbial activity is limited to a species-specific optimal temperature range. Most species can adapt somewhat to varying temperatures, but the temperature variation common to composting systems is much broader than the broadest capabilities of an individual species. Thermal killing is important during composting, in that if extensive pasteurisation occurs within the composting mass the rate of composting will be greatly retarded, and if reinoculation potential is limited the extent of decomposition will be decreased.

It can be difficult to initiate composting in open system configurations if the temperatures of the composting mass are much below 20°C, and if initiated an appreciable lag period may be evident (Mosher & Anderson, 1977). Mass temperatures above 20°C favour activity, and composting activity rapidly increases with temperature. Mesophilic populations are inhibited by temperatures exceeding the lower 40s (°C), while in the mid 40s conditions become favourable for the growth of thermophiles (Walker & Harrison, 1960; Dye, 1964; Dye & Rothbaum, 1964; Rothbaum & Dye, 1964). Above 60°C the temperature optimum for various thermophiles is exceeded. Composting activity rates decrease at temperatures above 60°C (MacGregor *et al.*, 1981; Finstein *et al.*, 1983, 1985; Bach *et al.*, 1984; McKinley & Vestal, 1984, 1985; Kuter *et al.*, 1985), with optimal decomposition rates in the mid to upper 50s. Nakasaki *et al.* (1987*b*) also found the upper 50s to be most favourable for decomposition, and claimed novelty for such a discovery by misinterpreting earlier work by MacGregor *et al.* (1981). Biologically, the maximum temperature achievable through composting is approximately 82°C, at which point biological activity and metabolic heat evolution cease (Walker & Harrison, 1960; Suler & Finstein, 1977; Nell & Wiechers, 1978; Fermor *et al.*, 1985; Finstein *et al.*, 1986). It has also been observed in the field that 82°C cannot be exceeded in any size mass, no matter how well insulated (Willson *et al.*, 1980; Finstein *et al.*, 1983; Randle, 1983).

Sikora *et al.* (1983) suggested that extreme thermophiles may, after adaptation, have a role in composting at temperatures higher than 70°C, but this has not been demonstrated. Extreme thermophiles have been shown to be important in some geothermally heated aqueous systems, but these are very different ecologically from composting systems (Brock, 1978; Kelly & Deming, 1988). Most extreme thermophiles are inhibited by high substrate concentrations, and aerobic extreme thermophiles can only live in low densities because of oxygen limitations related to temperature–solubility functions (Sundaram, 1986). To date the most extreme thermophile yet recovered and identified from composting systems is *Bacillus stearothermophilus* (Strom, 1985*b*). McKinley and Vestal (1984) found no evidence that extreme thermophiles with temperature optimum above 60°C contributed to decomposition during composting. Some of the heat evolution which occurs at extreme temperatures could be the result of enzyme function by thermally incompetent cells or an uncoupling of ATP production systems (H. A. J. Hoitink, personal communication).

3.4 Heat Flow and Control

Heat flow is of critical importance in the control of composting temperatures. Composting temperatures are determined by the rate of heat evolution, heat storage and the rate of heat loss. Heat loss is a function of conduction, evaporation of water and sensible heating of air. Radiation is not significant (Finstein *et al.*, 1980; Miller, 1984).

3.4.1 Heat Storage

Heat storage affects temperatures, especially when heat evolution rates are low, such as early or late in processing, or with low energy substrates. Water within the composting mass represents most of the heat storage capacity because of its high specific heat and the high moisture content (50–75%) of composted materials. The specific heat of wood is 0·45–0·65 W/m°C (Perry & Chilton, 1973), and composting materials could be assumed to fall into the same range. Excessive wetting of composting substrates, especially at low starting temperatures or with less energetic substrates, could interfere with high temperature achievement.

3.4.2 *Conduction*

In materials commonly composted the potential for conductive heat loss is small. Thermal conductivity of refuse, leaves, sewage sludge and the like is low, and is further reduced by the large volume of air contained in small pores within the composting matrix. Additionally, the potential volumetric rate of heat evolution is quite high compared to the potential for loss. Mears *et al.* (1975) found with swine waste compost that thermal conductivity varied linearly with moisture content, and the dry solids fraction had a thermal conductivity of 0·22 W/m°C.

The rate of conductive heat loss includes a distance function, so that decreasing the path length of heat flow can increase the total flow per unit time. Small composting masses with a large surface area to volume ratio can transfer significant amounts of heat through conduction, but this is not normally factored into bench-scale studies. Composting can be realistically carried out in small systems by controlling conductive heat losses to mimic full-scale systems (Hogan *et al.*, 1989). Conduction is a poor means of temperature control in full-scale systems, and process control by improving conductive loss through the use of heat exchangers would be limited by many problems, including condensation soaking the compost at cooler surfaces.

3.4.3 *Evaporative Cooling*

Evaporation of water has a great potential to remove heat from composting masses because of the great heat required for a phase change from liquid to vapour, and the high moisture content of the composting mass. Turning compost can remove large amounts of heat through this mechanism as the hot compost is moved through the air (Miller *et al.*, 1989*b*). Quite practical is the use of forced pressure (or less efficient vacuum) ventilation to promote evaporative heat loss (Miller *et al.*, 1982).

Air-related heat transfer mechanisms, i.e. evaporative cooling and sensible heating of dry air, are important in composting systems (Finstein *et al.*, 1985; Kuter *et al.*, 1985). Moving air has the potential to remove heat in a much more uniform manner than other mechanisms. Evaporation removes large amounts of heat because of the high heat of vaporisation of water. In an analysis of composting where ventilative removal of heat was used to control temperature, MacGregor *et al.* (1981) calculated that approximately 90% of the heat removed through ventilation was through evaporative cooling, and only 10% was due to sensible heating of the dry air. This was based on air inlet conditions of 60% RH (relative humidity) at 25°C and exit conditions of 100% RH at 50–70°C.

3.4.4 *Convective Transfer*

Convection is driven by buoyancy differences related to temperature differences in a fluid medium such as air. In many compost pile configurations like the pyramidal piles common in waste composting, significant convection does not occur (Finstein *et al.*, 1980). Convective heat transfer can be significant in the straight-sided (rectangular cross-section) stacks used in preparing mushroom growing compost (Gerrits, 1972; Tschierpe, 1972; Randle & Flegg, 1978). High oxygen concentrations are often observed in mushroom composting stacks but may be as much a function of activity inhibition because of high temperature inhibition of biological activity as of convective mass transfer (Miller *et al.*, 1989*b*).

3.4.5 *Heat Control*

Removal of heat for temperature control can be achieved through numerous strategies. Heat removal can be increased by controlling the configuration alone of the composting mass, i.e. size and shape. For example, controlling heat flow through configuration can be quite suitable for materials that can be composted slowly, such as leaves, wood wastes and other plant materials (Strom & Finstein, 1985). Temperature distribution can be improved, at the expense of more effort, using turning machines with drums and flails which throw the compost through the air, promoting significant heat removal. Stack turning, which is still popular in waste treatment applications (Goldstein, 1988), promotes significant cooling during turning events (Randle & Flegg, 1978; Miller *et al.*, 1989*b*). Temperature control is most effective in systems that remove heat through temperature feedback controlled ventilation (De Bertoldi *et al.*, 1982*b*; Kuter *et al.*, 1985; Finstein *et al.*, 1986). Systems that use active ventilation for the purpose of supplying oxygen without the temperature feedback control characteristics (i.e. the earlier and still common Beltsville Process; Willson *et al.*, 1980) should not be confused with the use of temperature feedback controlled ventilation.

De Bertoldi *et al.* (1988) proposed the use of oxygen feedback control as a means to control ventilation. Oxygen consumption and heat evolution are strongly correlated (Cooney *et al.*, 1968), but it takes nine times more ventilation to remove heat as to supply oxygen at a compost exit temperature of 60°C (Finstein *et al.*, 1986). Oxygen control is a less sensitive measure for controlling heat evolving activity than temperature, and under normal composting conditions temperature range is a more critical selective factor than oxygen concentration. In some situations, however, such as enclosed reactors, control based on exit oxygen levels might have some operational advantage.

3.5 Temperature Dynamics in Batch Processing

In batch processing, heat flow and temperature change can be described as a series of different stages. Aspects of this have been discussed by Haug (1979) and Finstein *et al.* (1985, 1986).

3.5.1 *Temperature Rising Stage*

In most processes, materials at the start of processing will be close to the ambient temperature, perhaps 15–25°C. At this stage there can be some lag in activity, because of the lower activity at such temperatures, and a lag in the growth of the appropriate biomass. Minimal heat removal is desired at this stage, but ventilation is still required to maintain oxygen concentrations that will provide optimal activity. This requires ventilation to supply optimal oxygen with minimal heat loss. Finstein *et al.* (1986) recommended as an approximation 9 m^3 per tonne per hour. Harper, Miller and Macauley (unpublished data) found that maintaining a 14% interstitial oxygen concentration promoted the most rapid temperature increase. As the temperature reaches the upper 30s, the rate of temperature increase can rise sharply.

3.5.2 Constant Temperature Stage

The constant temperature stage represents that period where heat output equals heat loss. In a temperature controlled composting situation, heat removal must equal heat production to maintain optimal temperatures. That is, the total heat production must be balanced by the sum of the losses by conduction and ventilation; ventilation is further divided into evaporative and sensible heat loss. Most heat is removed through ventilation, conductive losses being small (Finstein *et al.*, 1980). Ventilative heat removal is a function of the heat gain of the ventilative air times its mass. Because the amount of heat removed per unit mass of air cannot be changed without changing the exit air temperature under the influence of the composting mass (and therefore the temperature of the composting mass), process control is maintained by controlling the mass of ventilative air. Under isothermal composting conditions, the mass of air required will vary with and be proportional to the rate of heat evolution.

Interstitial oxygen concentration is not limiting to microbial activity during the constant temperature phase because of the ratio of heat evolved per unit oxygen consumed. Assuming 14 000 kJ/kg oxygen consumed, inlet air at 20°C and 50% RH, and exit air at 60°C and 100% RH, it would take 38·7 kg dry air to remove the heat but only 4·31 kg to replenish the oxygen, giving a ratio of 8·98 (Finstein *et al.*, 1986).

3.5.3 Temperature Decline Stage

Substrate energy will eventually become exhausted within the composting mass, causing a concomitant decrease in heat evolution. At some point even the very small heat losses of conduction and convection will become sufficient to remove heat faster than it can be liberated, and temperatures will decline. Sometimes in high energy substrates sufficient water will be lost during processing that water limitations will become rate limiting, and addition of water will renew activity (MacGregor *et al.*, 1981).

3.6 Water

Water is a critical factor in composting systems. Microbial cells have physiological needs for water not considered in depth herein. Microbial cells can also be physically affected by the solution of substrates and salts (i.e. nutrient availability), the effect of percent water content on gas exchange, and the role of water as a medium for bacterial colonisation (Griffin, 1981*b*; Harris, 1981). Generally, optimum activity rates can be achieved by reaching the maximum water content that does not restrict oxygen utilisation.

Water content must be high enough at the beginning of composting so that acceptable rates of activity can be realised, without producing an excessively wet final product or interfering with materials handling. For waste treatment a dry final compost is generally desirable because it decreases weight and bulk, and improves handling, storage and transport (Finstein *et al.*, 1986). With high energy substrates like sewage sludge, the heat released through composting can drive considerable water removal, and final moisture contents as low as 20–30% can be achieved (Miller *et al.*, 1980; De Bertoldi *et al.*, 1982*b*; Finstein *et al.*, 1983; Miller & Finstein, 1985).

As water content increases, the rate of gas transfer decreases. As the rate of oxygen transfer becomes insufficient to meet the metabolic demand, the composting system will become restricted in activity, and eventually anaerobic. The coefficient of oxygen diffusion through gas-filled space is 0·189 cm^2/s (Letey *et al.*, 1967) but only $2{\cdot}56 \times 10^{-5}$ cm^2/s through water (Baver *et al.*, 1972), or almost 10 000 times slower. According to Campbell (1985), the gravimetric (%) water content is a useful predictor of gas diffusion limitations, as the apparent gas diffusion coefficient is proportional to the square of the gas-filled porosity. Water tends to both decrease pore continuity and fill up available pore space. For waste composting systems requirements for free air space have been proposed for the maintenance of sufficient gas diffusion (Schulze, 1961; Jeris & Regan, 1973*a*; Haug, 1978). Wiley (1957) observed that in materials with a large amount of lipids that are liquid at composting temperatures, the lipid fraction can be considered part of the water fraction as gas exchange will be determined by the entire liquid fraction.

Water limitation is complex, and the mechanisms of water limitation have received limited investigation. Microbiologists are familiar with water activity (a measure based on vapour pressure depression) but this system is not useful in systems with a high amount of readily available water. Finstein and Morris (1975) proposed the use of water activity for composting investigations, but water activity is much too insensitive to be of much use. Water potential, a system based on water energetics, is the only practical approach for investigating water limitations caused by dryness (Miller, 1989). Water potential is the system familiar to soil physics and soil microbiology (Griffin, 1981*a*,*b*; Harris, 1981; Campbell, 1985). Matric water potential is especially useful, in that it describes the lowering of the free energy of water through the more thermodynamically stable associations of water with surfaces and capillaries. Matric potential is expressed as a negative pressure in units of pascals (Pa). The relationship between matric potential and the radius of a water-filled capillary can be described thus

$$r = -2\sigma/\psi_m \doteqdot -0{\cdot}147/\psi_m$$

where ψ_m = matric water potential
r = capillary radius (metres)
σ = surface tension in newton metres (N m)

This relationship permits the conversion of a matric water potential measure into an estimation of the size distribution of water-filled capillaries and surface film thickness, which are ecologically of great significance. For specific waste mixtures the relationship between matric potential and gravimetric water potential can be determined. For a sewage sludge–wood chip mixture Miller (1989) found the empirical relationship to be a second-degree polynomial with $Y = 64{\cdot}049 - 0{\cdot}0142X$ with a correlation of $r = 0{\cdot}95$, where X and Y are the line intercepts and r is the Pearson r product moment correlation coefficient.

In composting systems, which are generally quite wet, water activity (as measured by relative humidity) is too insensitive to be of use because the relationship of water content to vapour pressure is logarithmic. A matric water potential of −1380 kPa, which is the wilting point of vascular plants, would be equivalent to a water activity of 0·990 (Griffin, 1981*b*). Water limitations would severely retard composting activity at a potential of −100 kPa (Miller, 1989), yet from a purely practical perspective instrumentation for measuring water activity (relative humidity) cannot distinguish between −100 and −5 kPa; −5 kPa would indicate a system too wet for good gas transfer. Water activity also fails to distinguish between matric and osmotic water potentials which affect microorganisms differently (Griffin, 1981*a*).

Drying inhibits composting activity, in that as matric potentials become more negative than −70 kPa colonisation by bacteria becomes strongly inhibited because of a lack of adequate water for cell transport (Miller, 1989). This colonisation limitation related to matric water potential was earlier demonstrated in a soil matrix by Griffin and Quail (1968). Composting ecosystems are sensitive to water transport limitations because at the very active high temperature stage *Bacillus* spp. are predominant (Strom, 1985*a*), and these cells tend to be quite large (Gordon *et al.*, 1973). Optimal bacterial colonisation is achieved at matric potentials above approximately −20 kPa. Fungi and actinomycetes are less affected by matric potential because they colonise via mycelium. Fungi and actinomycetes, however, tend to become insignificant decomposers as temperatures exceed 50°C, and most waste composting is carried out at temperatures well above 50°C.

Osmotic potential refers to the lowering in free energy of water because of an association with dissolved salts and other materials. In sewage sludge composting, osmotic potentials are non-limiting (approximately −25 kPa), even to very sensitive organisms (Miller, 1989), and osmotic potential would not be expected to limit waste composting systems.

Most waste composting is carried out at 50–75% water content, depending on the material and the specific process. The lower moisture content for successful composting with materials such as sludge cake–bulking agent mixtures or refuse is about −50 to −60 kPa matric potential, or approximately 45–40% moisture (Miller, 1989). With the initiation of vigorous composting, activity can continue under conditions of increasing drying to moisture contents as low as 22% (Finstein *et al.*, 1983). This can result from extensive substrate colonisation early in composting when matric potentials are still favourable, so that the substrate is already well colonised before dryness becomes limiting (Miller, 1989). Colonisation limitations are partly mitigated by a high degree of continuous mixing, explaining the high activity rates with very dry substrate reported by Shell and Boyd (1969).

Metabolism produces water, which should be considered in composting modelling and management. Early mass balance studies in garbage composting by Wiley and Pearce (1957) found 0·63 g water produced per gram decomposed. In a system investigating fungal decomposition, Griffin (1977) found 0·55 g produced per gram cellulose decomposed. In a bench-scale composting system designed for heat modelling studies, Hogan *et al.* (1989) recovered 0·50–0·53 g of water produced per gram of rice hulls and rice flour decomposed. Composting straw and poultry manure, Harper *et al.* (1992) found water production of 0·50–0·60 g per gram decomposed. Haug (1979), using a theoretical model, calculated that 0·72 g of water should be produced for every gram of sewage sludge decomposed. Combustion models and actual results differ because of biomass production and its compositional difference from the original substrate.

3.7 Available Energy Density

Available energy density (available substrate per unit volume) considers the potential for activity on a volumetric basis. Energy density is a function of substrate availability and bulk density. Heat removal and oxygen supply requirements need to be considered on a volumetric basis because heat and mass transfer are rate functions of unit distance per time. Composting management that works well for leaves (low available energy density) would work poorly for primary sewage sludge (high available energy density). Available energy density as related to bulk density can be greatly affected by the preparation of the initial ingredients (O'Dogherty & Gilbertson, 1988).

In composting systems which favour high rates of decomposition, using a bulking agent to decrease available energy density can be an aid in controlling the distribution of both oxygen and temperature. In forced pressure ventilated static piles, the Rutgers Strategy (Finstein *et al.*, 1986) promotes high rates of activity through temperature optimisation (50–60°C). In temperature controlled static piles, vertical temperature gradient and energy density are positively related. Limiting pile heights to less than 2 m in primary sludge–wood chip systems is necessary if excessively high or low temperatures are to be avoided (Finstein *et al.*, 1986). Materials such as straw, leaves or wood wastes do not need bulking agents because of their low energy density and high natural porosity.

3.8 Availability of Substrate

Substrate availability is related to available energy density, in that the nutritional availability per unit mass proportionally increases the unit volume available energy. Biochemical availability is important, but this is a characteristic of the waste being decomposed and is not readily changed by processing before composting. Substrate availability can also vary greatly because of physical structure, making the composting of wood wastes very slow (months) as wood is extensively polymerised, while food wastes and organic sludges contain large porportions of soluble materials which are readily available. Substrate availability can be increased by size reduction through grinding or shredding (Gray & Sherman, 1969); this also increases bulk density. Other than particle surface area available for attack, substrate availability can also be affected by physical factors related to water (see Section 3.6).

4 CHEMICAL FACTORS AFFECTING PROCESSING

4.1 Interstitial Oxygen Concentration

The fact that 'anaerobic' and 'aerobic' are only general regions on a gradient of redox potential values is often ignored. Like soils (Greenwood, 1961; Foster, 1988), composting matrices can contain aerobic and anaerobic microenvironments coexisting within close proximity, or larger zones can be aerobic or anaerobic (Eicker, 1981; Miller *et al.*, 1989*b*). Products of anaerobic metabolism (such as H_2S or CH_4) can be recovered from areas containing interstitial oxygen (Op den Camp, 1989).

Maintenance of oxygen in the interstitial atmosphere of a composting mass is determined by the rate of utilisation and the rate of supply. Potentially rapid utilisation rates are possible in composting materials because of high substrate density and ready availability, these rates often being well beyond potential rates of passive replenishment. Composting is greatly retarded in the absence of oxygen; this is a common reason for composting failures (Gray *et al.*, 1971; Poincelot, 1975). Walker and Harrison (1960) found the rate of heat evolution was decreased over 100-fold by the exclusion of oxygen.

Oxygen supply rates are determined by diffusion potential and process management. Diffusion rates are determined by the gas concentration gradient and the resistance to flow. In turn, flow resistance is a function of pore size, pore continuity and moisture content (Section 3.6). Flow potential is best determined by the cross-sectional void space (Papendick & Campbell, 1981). Diffusion rates have been measured for refuse (McCauley & Shell, 1956) and for sewage sludge cakes (Nakasaki *et al.*, 1987*a*). Diffusion alone is much too slow to supply a large composting mass with sufficient oxygen.

Large heaps or windrows are often used in leaf composting, with oxygen supply a function of gas diffusion rates; in these masses an oxygen front moves slowly inward over many months as outer material is stabilised (Strom *et al.*, 1980). Straight-sided rectangular stacks, common to mushroom compost preparation, are constructed so that both diffusion and convection driven mass flow are promoted (Gerrits, 1972; Randle & Flegg, 1978). Stack oxygen utilisation potential exceeds supply until extreme thermophilic temperatures are reached; core areas are anaerobic at lower temperatures and aerobic at high temperatures (70–80°C) (Randle & Flegg, 1978; Miller *et al.*, 1989*b*). Waste composting usually occurs in windrows of pyramidal cross-section that are resistant to convective mass transfer (chimney effect) because the shape causes excessively long path lengths. Diffusion limitations have been greatly overcome by constructing composting systems which use forced ventilation to supply oxygen (Goldstein, 1988).

Diffusion at the small particle level is also driven by concentration gradients. Some investigators have suggested that aerobic conditions can be maintained with interstitial oxygen concentrations of 5% (Schulze, 1962; Parr *et al.*, 1978; Willson *et al.*, 1980) or 10% (Suler & Finstein, 1977). Activity can be retarded by oxygen limitations, even with seemingly high interstitial oxygen concentrations. Using an enclosed composting system, Miller (1984) found that in trials where interstitial oxygen concentrations were maintained above 18%, but ventilation velocity was doubled to improve diffusion by increasing turbulence, decomposition rate was increased by 25%. Miller *et al.* (1990) found rates of heat evolution decreased when interstitial oxygen concentrations dropped below 12–14%. De Bertoldi *et al.* (1988) reported work in which activity rates were highest when interstitial oxygen concentrations were maintained between 15% and 20%.

If composting is allowed to become predominantly anaerobic, volatile organic acids, volatile sulphur and nitrogen compounds, and other compounds can be produced which can cause severe odour problems (Miller & Macauley, 1988, 1989). Under anaerobic conditions up to 15% of the total organic carbon content can be converted in the compost to volatile organic acids (Chanyasak *et al.*, 1980). These organic acids are toxic to higher plants and such composts can remain phytotoxic for years (Hoitink, 1980). Composted materials that have been processed mostly anaerobically can be very difficult to dispose of in any manner because of odours and poor handling characteristics, such as loss of structure.

4.2 Effect of pH

In most waste materials the initial pH is rarely extreme enough to cause a processing failure. Both fairly acidic and basic materials can be successfully composted and lead to a product near neutrality. Rates of decomposition during composting increase with increasing pH in the range of 6–9, and low pH early in processing can retard subsequent processing (Wiley & Pearce, 1957; Schulze, 1962; Jeris & Regan, 1973*b*).

During composting pH change is quite predictable (Wiley & Pearce, 1957; MacGregor *et al.*, 1981; Pereira-Neto *et al.*, 1987). Fermentation-caused oxygen limitation can drop pH slightly early in processing. With rapid early activity, pH can rise up to approximately 8·5 because of ammonification. At the completion of ammonification pH will decrease to about 7·5–8·0. pH change can be used in many facilities as a rough gauge of composting progress.

4.3 Effect of Ammonia

The effects of ammonia concentrations achieved during composting are rarely considered in waste treatment composting, except for nuisance problems of equipment corrosion or air pollution. Ammonification is frequently a short process in systems that promote high activity rates, completed in less than a week. Ammonia concentrations have been studied during the preparation of mushroom or horticultural composts, where final nitrogen contents are important. Ammonification is temperature dependent, with the quickest rate of completion between 40 and 50°C (Ross & Harris, 1982) but the greatest conservation of ammonia nitrogen between 50 and 55°C (Burrows, 1951). Ammonia concentrations can peak at 1000 ppm or more in high protein substrates.

Ion equilibrium between ammonia and ammonium is determined by pH; at pH 7 and lower the ammonium ion is almost exclusively present, while at pH 9 and above free ammonia almost exclusively predominantes (Koster, 1986). Free ammonia can form stable reaction products with organic matter (Nommik, 1965) by reacting with sugars (Sharon, 1965) and carboxyls, carbonyls, enolic, phenolic or quinone hydroxyls, or unsaturated carbons near their functional groups (Mortland & Wolcott, 1965). Rates for these reactions also increase with temperature (Nommik, 1970). The presence of oxygen permits much more ammonia to react with organic matter (Mortland & Wolcott, 1965). The conversion of ammonia into stable forms which resist further ammonification is important to compost product end usage.

Free ammonia is very reactive and in soils has been observed to dissolve organic matter and make it susceptible to further decomposition (Myers & Thien, 1988). High ammonia concentrations can inhibit methanogenesis under anaerobic conditions, causing volatile fatty acids to accumulate (Koster, 1986). Free ammonia can change population structures because of its toxicity to many microbial populations.

4.4 Redox Potential

The effect of redox potential is not well understood at this time because so little consideration has been given to it in composting systems. Redox potential may not greatly affect composting activity, but it could certainly affect the quality of the final compost product. As a novel topic in composting, redox potential is at least worthy of consideration as a processing indicator. It is probably the only means of directly evaluating the oxidation stage of a composting material. Redox potential has real advantage over interstitial oxygen concentration measurement, in that the redox potential represents the actual historical availability of oxygen within the composting mass. Redox potential is an important determinant of various chemical reactions (Ponnamperuma, 1981), and a significant ecological selective factor (Postgate, 1984). Redox potential can indicate those conditions under which objectionable odours may form. Redox potential might prove to be a useful measure of compost maturity, as the organic matter remaining becomes more oxidised.

Basic research in a mushroom composting system (Miller *et al.*, 1991) revealed that even under aerobic conditions composting systems tend to be reduced (−100 to 100 mV) but redox potentials rise with maturity, to approximately 0 mV after a few weeks, 100 mV soon after the addition of mushroom spawn, 250–450 mV late in spawn run and 300–400 mV during cropping. Leachate samples from anaerobic composting heaps are characteristically −400 to −450 mV. While not measuring redox potential, Garcia *et al.* (1989) found that there was an increase in carboxyl and carbonyl group content which would be consistent with a more oxidised material.

5 MICROBIOLOGY OF COMPOSTING

5.1 Broad Consideration of Populations

Populations of bacteria, actinomycetes and fungi have been investigated in various self heating systems, including mushroom composts (Waksman & Cordon, 1939; Waksman *et al.*, 1939*a,b*; Hayes, 1968; Chanter & Spencer, 1974; Fermor *et al.*, 1979), straw (Carlyle & Norman, 1941; Chang, 1967; Chang & Hudson, 1967), wool (Walker & Harrison, 1960; Rothbaum, 1961; Dye, 1964), tree bark (Bagstam, 1978, 1979), sludges (De Bertoldi *et al.*, 1980, 1982*b*) and refuse (Strom, 1985*a,b*). Results of these studies vary based on substrate, investigative interest, microbial recovery techniques and means of expression, and especially on composting process management. These variations preclude making a clean comparison of the results of these different studies. Especially for fungi, the enumeration technique can cause significant variation in the reported results. Reducing the above reports into general trends, peak bacterial counts per substrate gram can range from 10^8 to 10^{12} at temperatures of 55–65°C. At mesophilic temperatures bacterial counts can be an order of magnitude higher. Populations of thermophilic actinomycetes peak later than those of bacteria, often achieving counts in the 10^7–10^9 range. Below 65°C total counts of bacteria and actinomycetes are commonly linked with actinomycete populations about one order of magnitude less than the bacterial total count. Thermophilic fungal populations peak much later than the bacteria, normally in the declining activity stage as temperatures drop into the lower 50s and lower; fungal colony forming units can be found in the 10^5–10^8 range. At temperatures above 60°C fungi are normally absent, and above 70°C actinomycetes are generally excluded.

5.2 Actinomycetes

The role of actinomycetes in composting was established early, to a large extent because of Waksman's great interest in this group (Waksman & Hutchings, 1937; Waksman & Cordon, 1939; Waksman *et al.*, 1939*a,b*). There are many thermophilic actinomycetes which can tolerate composting temperatures in the 50s and a smaller number of species into the mid 60s (Lacey, 1973). Actinomycetes prefer moist but aerobic conditions (Chen & Griffin, 1966) with neutral or slightly alkaline pH (Lacey, 1973). Complex organic materials including polymers are substrates utilised by these slower growing filamentous bacteria. Actinomycetes tend to be common in the later stages of composting and can exhibit extensive growth.

Genera commonly found in composting materials are *Streptomyces*, *Thermoactinomyces* and *Thermomonospora*. Specific species of importance would include *Micropolyspora faeni*, *S. rectus*, *S. thermovulgaris* and *Thermomonospora viridis* (Henssen, 1957; Gregory *et al.*, 1963; Hayes, 1968; Fergus, 1971; Stanek, 1971; Lacey, 1973; Fermor *et al.*, 1979; Makawi, 1980).

5.3 Bacteria

Bacteria are by far the most important decomposers during the most active stages of composting, partly because of their ability to grow rapidly on soluble proteins, and other readily available substrates, and partly because they are the most tolerant of high temperatures. Waste composting is usually managed to achieve processing temperatures above 55°C to ensure pathogen destruction, restricting activity almost exclusively to thermophilic bacteria (Strom, 1985*a*). This bacterial decomposition can be rapid; in sewage sludge composting under favourable temperatures (less than 60°C) over 40% of sludge volatile solids can be decomposed in the first 7 days (Miller & Finstein, 1985). Reports of thermophilic obligate anaerobic bacteria are lacking, but this is because efforts have not been made to isolate them; there is evidence for their existence, such as Eicker's (1981) findings of sulphate reduction occurring under thermophilic conditions.

In windrows (Finstein & Morris, 1975) or Beltsville Process aerated static piles (Willson *et al.*, 1980; Finstein *et al.*, 1987*c*), temperatures in the range of 60–80°C are commonly achieved during the first few weeks of composting. In controlled temperature waste composting a range of 50–65°C is commonly maintained (Finstein *et al.*, 1986). Such high temperatures are selective for bacteria, and especially for the genus *Bacillus* (Hayes, 1968; Fermor *et al.*, 1979). Strom (1985*a,b*) found species of *Bacillus* made up the majority of isolates from materials composting in the upper 50s and low 60s (°C), and that above 65°C sampling provided almost monocultures of *B. stearothermophilus*.

Bacterial genera reported commonly during composting include *Bacillus*, *Clostridium* and *Pseudomonas*. Most commonly reported species would include *B. coagulans*, *B. licheniformis*, *B. spharicus*, *B. stearothermophilus* and *B. subtilis*; *Pseudomonas* have been reported by a few authors but without any species identified (Henssen, 1957; Niese, 1959; Rothbaum, 1961; Gregory *et al.*, 1963; Okafor, 1966; Hayes, 1968; Fermor *et al.*, 1979; Strom, 1985*a,b*).

5.4 Fungi

Fungi have a limited role in waste composting except during curing, as they are excluded by the temperature ranges common to waste composting. Most fungi are eliminated above 50°C; only a few have been recovered that can grow at all up to 62°C, and their optimal temperatures are lower (Brock, 1978). During composting, temperatures above 55°C discourage fungal growth (Fermor *et al.*, 1979). Fungi are excluded from waste composting during the earlier high temperature stages (Kane & Mullins, 1973). Fungi are commonly recovered from composting materials later in processing when temperatures are more moderate, and remaining substrates are predominantly cellulose and lignins (Eastwood, 1952; Chang, 1967; Chang & Hudson, 1967; De Bertoldi *et al.*, 1983).

Large numbers of different species of fungi have been reported from composted materials, especially common being *Mucor pusillus*, *Chaetomium thermophilum*,

Talaromyces (*Penicillium*) *dupontii* and *T. thermophilus*, *Thermoascus aurantiacus*, *Aspergillus fumigatus* and *Humicola grisea* (Cooney & Emerson, 1964; Chang & Hudson, 1967; Hayes, 1968; Stutzenberger, 1971; Tansey, 1981; Millner *et al.*, 1977; Fermor *et al.*, 1979; De Bertoldi *et al.*, 1983).

5.5 Interaction Between Population and Nutritional Factors

Bacteria, actinomycetes and fungi assimilate carbon and nitrogen differently. For mixed populations 5–10% of substrate carbon is assimilated by bacteria, 15–30% by actinomycetes and 30–40% by fungi. Both bacteria and actinomycetes have a 5:1 protoplasmic C:N ratio, while fungi have a 10:1 ratio (Alexander, 1977). As a result of assimilation of carbon and nitrogen, bacteria would need 1–2% nitrogen to degrade a unit of carbon, while actinomycetes would need 3–6% and fungi 3–4%. The early attack by thermophilic bacteria on initially highly proteinaceous substrate frees nitrogen through ammonification and makes it available for subsequent populations.

Differing nutrients available during composting will preferentially favour different populations. Bacteria can utilise narrow carbon:nitrogen ratios of 10–20:1, while fungi can utilise wide ratios of 150–200:1 or even much higher for wood decay fungi (Griffin, 1985). Waste substrates that are rich in protein (animal wastes, sewage sludges, fermentation wastes) would initially favour bacteria because of the high levels of organic nitrogen. After consumption of most of the readily available protein, fungi and actinomycetes would be better adapted to consume the remaining complex carbohydrates (cellulose, hemicellulose, lignin).

5.6 Population Dynamics

Maximal specific growth rates and respiration rates of fungi are generally about an order of magnitude lower than those of bacteria (Griffin, 1985). This gives bacteria an advantage over fungi in the early utilisation of composting substrates. Most fungi are aerobes (Griffin, 1985). Fungi are disadvantaged under anaerobic conditions, and such conditions can often exist within aggregates of material even when the air-filled pore space contains some oxygen (Miller, 1984; Op den Camp, 1989). Drier substrates favour the filamentous growth of fungi and actinomycetes which can bridge air gaps that are effective barriers to colonisation by non-filamentous bacteria (Miller, 1989).

Population succession occurs rapidly during composting. The initial mesophilic population is later supplanted by a thermophilic one as temperatures increase. A drop in heat output can be observed in the 44–52°C range during the shift from mesophiles to thermophiles (Walker & Harrison, 1960). Peak temperature achievement lasts days or weeks, depending on substrate and processing strategy, but as the substrate becomes exhausted temperatures decline (Finstein *et al.*, 1986), leading to a reappearance of mesophiles and moderate thermophiles, especially fungi (Chang & Hudson, 1967).

5.7 Interactions—Synergisms and Antagonisms

Waksman and colleagues (Waksman & Hutchings, 1937) pioneered the investigation of microbial interactions, observing that actinomycetes could not attack corn stalks, but if preceded by the thermophilic fungus *Humicola* they grew extensively. Previous decomposition can be favourable or unfavourable to subsequent microorganisms based on what was made nutritionally available or unavailable, and the production of various co-factors and intermediate products of decomposition. Mixed populations achieve greater decomposition than pure cultures because of greater metabolic diversity and nutritional synergisms.

Increasingly important in the horticultural usage of composts is the establishment of populations within the compost which can suppress the growth of plant pathogens (Hoitink, 1980). Production of large populations of *Trichoderma hamatum* and *T. harzianum* in composted hardwood bark is strongly suppressive against *Rhizoctonia* damping-off (Kuter *et al.*, 1983; Nelson *et al.*, 1983). Sewage sludge compost can be made to be suppressive to a variety of plant pathogens (Lumsden *et al.*, 1983). Agricultural wastes can be composted and made suppressive to *Pythium* damping-off (Mandelbaum *et al.*, 1988). Hoitink and Fahy (1986) have thoroughly reviewed the control of plant diseases based on biological suppression.

An especially successful example of developing populations that determine subsequent populations is found in the production of compost for the cultivation of the common mushroom *Agaricus brunnescens* (Peck). Mushroom compost is selective biologically and nutritionally for the growth of mushrooms while suppressive to other possible competitors. Temperatures above 60°C and antimicrobial chemicals can destroy this selectivity (Ross & Harris, 1983*a*). Mushroom mycelial growth is strongly stimulated by the previous colonisation of the compost with the thermophilic fungus *Scytalidium thermophilum* (formerly *Torula thermophila*). Ross and Harris (1983*b*) proposed that *S. thermophilum* (*T. thermophila* in their papers) was particularly good in changing the compost substrate to product a nutritionally optimal substrate, while Straatsma *et al.* (1989) have proposed that *S. thermophilum* protects against negative effects of compost bacteria on the mycelial growth of *A. brunnescens*. In mushroom composts, weed moulds can be inhibited by preparing a suitably selective compost (Fermor *et al.*, 1985). *Chaetomium olivaceum*, which is associated with secondary ammonification (ammonia is toxic to mushrooms), can in turn be controlled using antagonistic thermophilic *Bacillus* spp. (Tautorus & Townsley, 1983).

6 HEALTH RISKS FROM PATHOGENS

6.1 Pathogen Reduction

Pathogen reduction can be effectively achieved through composting (Wiley & Westerberg, 1969; Burge *et al.*, 1978; Finstein *et al.*, 1982; Pereira-Neto *et al.*, 1986), primarily because of thermal killing but also because of biological antagonisms.

Microbial antagonisms during composting reduce pathogen populations more rapidly than would be predicted from time–temperature kill relationships alone (Banse & Strauch, 1966).

Composting has proved to be an effective process for the destruction of pathogens found in solid waste materials. Two mechanisms of destruction are thermal killing (Wiley & Westerberg, 1969; Finstein *et al.*, 1982) and biological antagonism (Makawi, 1980; Millner *et al.*, 1987). Sewage sludge normally contains pathogenic bacteria, viruses, helminthic ova and protozoan cysts, but all of the pathogens in these groups can be greatly reduced or eliminated by composting (Burge *et al.*, 1978; Pereira-Neto *et al.*, 1986; Hirotani *et al.*, 1988).

Recent United States Environmental Protection Agency (EPA) studies have shown that sludge-derived composts are generally free of most pathogens (Goldstein *et al.*, 1988). Process control is a factor in pathogen reduction; turned windrow systems are not as good as temperature feedback control static pile systems in the rapid elimination of pathogens (De Bertoldi *et al.*, 1982*a*). Current EPA guidance requires that for 'further pathogen reduction' (the most stringent classification) composting materials must be '... maintained at operating condition of 55°C for three days' (Federal Register, 1979). EPA guidance related to pathogen kill has been reviewed by Finstein *et al.* (1987*c*), who feel that some revision is in order based on newer information.

6.2 Pathogen Risks

Salmonella spp. have been of concern because of their common occurrence in wastes of faecal origin. *Salmonella* can certainly be thermally killed during composting but has an unusual ability to regrow to potentially hazardous levels in some composts if the remaining substrate is nutritionally suitable and microbial antagonists are absent (Burge *et al.*, 1987; Millner *et al.*, 1987). In well-composted materials *Salmonella* growth is greatly suppressed through the growth of antagonistic bacteria and actinomycetes (Makawi, 1980; Millner *et al.*, 1987). Suppression of *Salmonella* is very strong in composts produced at 55°C (well stabilised with ecologically complex populations) but not at all in composts produced above 70°C (less organic decomposition and simple populations because of thermal stress) (Millner *et al.*, 1987). *Shigella* can also be suppressed by populations of thermophilic actinomycetes (Makawi, 1980).

A potential pathogen that grows well in composting materials is *Aspergillus fumigatus*, a thermophilic fungus that can tolerate temperatures into the 50s (°C) and is one of the most common organisms recoverable from composting yards (Millner *et al.*, 1977). It can cause opportunistic infection, but normally only in immune suppressed individuals, and poses very little risk to compost workers (Clark *et al.*, 1984). *Thermoactinomyces vulgaris* and other thermophilic actinomycetes can occasionally cause an allergic alveolitis known as farmer's or mushroom worker's lung (Festenstein *et al.*, 1965; Blyth, 1973). Studies of workers at waste composting facilities have shown that health risks related to biological agents are extremely small (Cookson *et al.*, 1983; Clark *et al.*, 1984).

7 COMPOSTING ODOURS AND THEIR MANAGEMENT

7.1 Odour Sources

Odours have been one of the main difficulties encountered during composting, often leading to additional expensive air treatment facilities, replacement of facilities by more expensive new facilities, or even site closure or movement to new sites. Everywhere the public is becoming increasingly less tolerant of composting odours (Anon., 1986). Two important factors related to odours are the precursors to odoriferous compounds in the starting substrate and the processing environment of the composting material.

Compounds implicated in composting odours include organic acids, amines, aldehydes, sulphides, thiols and ammonia (Miller & Macauley, 1988). Of these, the organic acids, sulphides and thiols are normally the worst offenders. Proteins are often associated with odour problems because they are highly susceptible to bacterial decomposition, and the amine nitrogen can be a precursor of odorous nitrogen compounds. Of the proteins, the sulphur containing cystine and methionine are troublesome, with methionine by far the worst. Banwart and Bremner (1975) found that the sulphur in methionine was rapidly decomposed, producing various volatile sulphur compounds, and that methanethiol could be produced even under fully aerobic conditions. Organic acids can also produce bad odours and can be produced from a large variety of materials under conditions of fermentation.

Odours can also be related to general material characteristics. For example, grass clippings can cause odour problems because of their high moisture content, high nitrogen levels and readily available organic matter (Finstein & Strom, 1985). Materials lacking in structure, such as sewage sludges, often produce odours.

Environmental conditions which lead to odour problems can be obvious, such as anaerobic conditions. It must be remembered, however, that oxygen in the interstitial matrix of a composting material does not preclude odours arising from anaerobic zones within particles (see Section 4). Moreover, the degree of anaerobic condition (redox potential) will determine the types of odorous compounds formed (Beard & Guenzi, 1983), and the most anaerobic conditions might not be the worst for odour production. High temperature achievement has also been linked to worst odours, with a number of factors contributing (MacGregor *et al.*, 1981).

7.2 Odour Management

Odours can be treated or prevented. Odour treatment consists of some means of containing and treating the air after it has left the composting mass. Completely enclosed composting systems are becoming more commonplace because of the advantages of air containment and treatment (Walker *et al.*, 1986). Popular treatment methods can include spent compost scrubber piles and scrubbers using sulphuric acid, sodium hypochlorite, permanganate or ozone (Murray & Thompson, 1986), but the chemical treatment systems can be quite expensive.

Biofiltration can be effective in removing odours, while at the same time cost-effective and with minimal environmental impact (Van Der Hoek & Oosthoek, 1985).

Odour prevention, or at least significant reduction, can be achieved through careful process management. Lower temperature achievement (less than 60°C) and uniformly well oxygenated conditions that most favour decomposition also produce less odour (MacGregor *et al.*, 1981; Finstein *et al.*, 1986). Optimising process control can control odours and do it much more cost-effectively than treatment systems (Cerenzio, 1987).

8 HAZARDOUS WASTES

Some research has been carried out specifically to assess the ability of composting to break down hazardous materials. Specific compounds observed to decompose include diazinon, parathion and dieldrin (Rose & Mercer, 1968), trinitrotoluene (TNT) (Osmon *et al.*, 1978), and crude oil, refinery sludge, crankcase oil and polychlorinated biphenyls (PCBs) (Hunter *et al.*, 1981). In the TNT study, breakdown through composting was complete, while landfarming left numerous residual chemicals. In the investigation of PCB decomposition, the less highly chlorinated PCBs were the likely ones degraded, while the highly chlorinated ones remained. Vallini *et al.* (1989) used a temperature-controlled composting process to detoxify vegetable–tannery sludges that contained mixed polyphenols and sulphides. These materials were rapidly broken down and oxidised during composting to become non-toxic.

Hogan *et al.* (1988) composted a variety of hazardous wastes with a mixture of wastewater sludge cake, sludge compost and an acclimatised inoculum for 35 days. With accounting and correction for volatile losses and extraction methods, the following had truly disappeared at 35°C (in %): 1-octadecene (98·0); 2,6,10,15,19,23-hexamethyltetracosane (75·1); phenanthrene (99·7); fluoranthrene (94·4); and pyrene (93·0). At a composting temperature of 50°C disappearance rates for the above compounds were 95·1, 50·9, 97·0, 90·0 and 87·1%. This indicates that not only can these compounds be decomposed but that there may be advantages of mesophilic composting over thermophilic composting for many recalcitrant compounds. In the same study, 81·4% of Aroclor 1232 disappeared at 50°C. This study indicates that composting can achieve decompositions very comparable to that achieved by the white rot fungi under soil decomposition conditions (John Bumpus, personal communication).

Racke and Frink (1989) investigated the fate of ^{14}C-labelled phenanthrene and carbaryl at low concentrations (1·3–2·2 ppm) under composting conditions. Penenanthrene showed little decomposition over 21 days. Maximum temperature achievement was 69, 71 and 55°C; these temperatures would be too hot for *Pseudomonas* and *Aeromonas* that have been demonstrated to be able to decompose phenanthrene (Gibson & Subramanian, 1984). Carbaryl was composted under very high temperature conditions (69–71°C), and showed little decomposition (1·6–4·9%)

but a high level (92–96%) of binding to compost organic matter. Comparison between these results and those of Hogan *et al.* (1988) above indicates that temperature may be a critical factor in decomposing hazardous materials, as a result of temperature being a factor selecting for different populations.

Composting may offer advantages over other biological treatments for hazardous wastes because of the combination of high unit volume activity and robust mixed populations. Solid wastes often contain some amount of hazardous material that has the potential for being biologically destroyed. Further investigation is needed on this potential, and the most favourable composting conditions for the destruction of various hazardous materials.

REFERENCES

Alexander, M. (1977). *Introduction to Soil Microbiology*. John Wiley, New York.

Anon. (1986). *BioCycle*, **27**(6), 21.

Atkey, P. T. & Wood, D. A. (1983). *Journal of Applied Bacteriology*, **55**, 293.

Atlas, R. M. & Bartha, R. (1981), *Microbial Ecology: Fundamentals and Applications*. Addison-Wesley, Reading, MA.

Bach, P. D., Shoda, M. & Kubota, H. (1984). *Journal of Fermentation Technology*, **62**(3), 285.

Bagstam, G. (1978). *European Journal of Applied Microbiology & Biotechnology*, **5**, 315.

Bagstam, G. (1979). *European Journal of Applied Microbiology & Biotechnology*, **6**, 279.

Banse, H. J. & Strauch, D. (1966). *Compost Science*, **7**(3), 17.

Banwart, W. L. & Bremner, J. M. (1975). *Journal of Environmental Quality*, **4**, 363.

Bardos, R. P. & Lopez-Real, J. M. (1988). *EC Workshop on Compost Process in Waste Management*, September 1988, Monastery of Neresheim, FRG.

Battley, E. H. (1987). *Energetics of Microbial Growth*. John Wiley, New York.

Baver, C. A., Gardner, W. H. & Gardner, W. R. (1972). *Soil Physics*, 4th edn. John Wiley, New York.

Beard, W. E. & Guenzi, W. D. (1983). *Journal of Environmental Quality*, **12**, 113.

BioCycle (1986). *The BioCycle Guide to In-Vessel Composting*. JG Press, Emmaus, PA.

Blyth, W. (1973). In *Actinomycetales: Characteristics and Practical Importance*, eds G. Sykes & F. A. Skinner. Academic Press, London, p. 261.

Brock, T. D. (1978). *Thermophilic Microorganisms and Life at High Temperatures*. Springer-Verlag, New York.

Brock, T. D. & Madigan, M. T. (1988). *Biology of Microorganisms*, 5th edn. Prentice-Hall, Englewood Cliffs, NJ.

Browne, C. A. (1933). *Science*, **77**, 223.

Burge, W. D., Cramer, W. N. & Epstein, E. (1978). *Transactions of the ASAE—1978*, 510.

Burge, W. D., Enkiri, N. K. & Hussong, D. (1987). *Microbial Ecology*, **14**, 243.

Burrows, S. (1951). *Journal of the Science of Food and Agriculture*, **2**, 403.

Campbell, G. S. (1985). *Soil Physics with Basic Transport Models for Soil–Plant Systems*. Elsevier Science Publishers, Amsterdam.

Carlyle, R. E. & Norman, A. G. (1941). *Journal of Bacteriology*, **41**, 699.

Cerenzio, P. F. (1987). *BioCycle*, **28**(4), 26.

Chang, Y. (1967). *Transactions of the British Mycological Society*, **50**, 667.

Chang, Y. & Hudson, H. J. (1967). *Transactions of the British Mycological Society*, **50**, 649.

Chanter, D. P. & Spencer, D. M. (1974). *Scientia Horticulturae*, **2**, 249.

Chanyasak, V., Yoshida, T. & Kubota, H. (1980). *Journal of Fermentation Technology*, **58**, 533.

Chen, W. A. C. & Griffin, D. M. (1966). *Transactions of the British Mycological Society*, **49**, 419.

Clark, C. S., Bjornson, H. S., Schwartz-Fulton, J., Holland, J. W. & Gartside, P. S. (1984). *Journal of the Water Pollution Control Federation*, **56**(12), 1269.
Cookson, J. T., Smith, R. B. & Deugwillo, K. R. (1983). *Proceedings of the International Conference on Composting of Solid Wastes and Slurries*, ed. E. I. Stentiford. University of Leeds, Leeds, UK, 148.
Cooney, C. L., Wang, D. I. C. & Mateles, R. L. (1968). *Biotechnology and Bioengineering*, **11**, 269.
Cooney, D. G. & Emerson, E. (1964). *Thermophilic Fungi: an Account of Their Biology, Activities, and Classification.* W. H. Freeman, San Francisco.
De Bertoldi, M., Citernesi, U. & Griselli, M. (1980). *Compost Science*, **21**(1), 32.
De Bertoldi, M., Coppola, S. & Spinosa, L. (1982*a*). In *Disinfection of Sewage Sludge: Technical, Economic and Microbiological Aspects.* D. Reidel, Dordrecht, The Netherlands, p. 165.
De Bertoldi, M., Vallini, G., Pera, A. & Zucconi, F. (1982*b*). *BioCycle*, **23**(2), 45.
De Bertoldi, M., Vallini, G. & Pera, A. (1983). *Waste Management & Research*, **1**, 157.
De Bertoldi, M., Rutili, A., Citterio, B. & Civillini, M. (1988). *Waste Management & Research*, **6**, 239.
Dye, M. H. (1964). *New Zealand Journal of Science*, **7**, 87.
Dye, M. H. & Rothbaum, H. P. (1964). *New Zealand Journal of Science*, **7**, 97.
Eastwood, D. J. (1952). *Transactions of the British Mycological Society*, **35**(3), 215.
Eicker, A. (1981). *Mushroom Science*, **11**, 27.
Federal Register (1979). 40 CFR Part 257, EPA Part IX. Vol. 44(179):53438–53468 (13 September 1979).
Fergus, C. L. (1971). *Mycologia*, **63**, 426.
Fermor, T. R., Smith, J. F. & Spencer, D. M. (1979). *Journal of Horticultural Science*, **54**(2), 137.
Fermor, T. R., Randle, P. E. & Smith, J. F. (1985). In *The Biology and Technology of the Cultivated Mushroom*, eds P. B. Flegg, D. M. Spencer & D. A. Wood. Wiley, Chichester, UK, p. 81.
Festenstein, G. N., Lacey, J., Skinner, F. A., Jenkins, P. A. & Pepys, J. (1965). *Journal of General Microbiology*, **41**, 389.
Filip, Z. (1978). *Applied Microbiology and Biotechnology*, **6**, 87.
Finstein, M. S. & Miller, F. C. (1982). *BioCycle*, **23**(6), 56.
Finstein, M. S. & Morris, M. L. (1975). *Advances in Applied Microbiology*, **19**, 113.
Finstein, M. S. & Strom, P. F. (1985). *Leaf Composting Manual for New Jersey Municipalities.* New Jersey Department of Energy, Office of Recycling, Newark, NJ.
Finstein, M. S., Cirello, J., MacGregor, S. T., Miller, F. C. & Psarianos, K. M. (1980). *Sludge composting and utilization: rational approach to process control.* US EPA project no. C-340-678-01-1, Accession no. PB82-13623. National Technical Information Service, Springfield, VA.
Finstein, M. S., Lin, K. W. & Fischler, G. E. (1982). *Sludge composting and utilization: review of the literature on the temperature inactivation of pathogens.* Report of New Jersey Agricultural Experiment Station, Project No. 03543, New Brunswick, NJ.
Finstein, M. S., Miller, F. C., Strom, P. F., MacGregor, S. T. & Psarianos, K. M. (1983). *Bio/Technology*, **1**, 347.
Finstein, M. S., Miller, F. C., MacGregor, S. T. & Psarianos, K. M. (1985). *The Rutgers strategy for composting: process control and design and control.* EPA/600/2-85/059, Accession no. PB85-207538. National Technical Information Service, Springfield, VA.
Finstein, M. S., Miller, F. C. & Strom, P. F. (1986). In *Biotechnology*, Vol. 8, eds H. J. Rehm & G. Reed. VCH Verlagsgesellschaft, Weinheim, FRG, p. 363.
Finstein, M. S., Miller, F. C., Hogan, J. A. & Strom, P. F. (1987*a*). *BioCycle*, **28**(1), 20.
Finstein, M. S., Miller, F. C., Hogan, J. A. & Strom, P. F. (1987*b*). *BioCycle*, **28**(2), 42.
Finstein, M. S., Miller, F. C., Hogan, J. A. & Strom, P. F. (1987*c*). *BioCycle*, **28**(3), 38.
Finstein, M. S., Miller, F. C., Hogan, J. A. & Strom, P. F. (1987*d*). *BioCycle*, **28**(4), 56.
Foster, R. C. (1988). Biology and Fertility of Soils, **6**(3), 189.

Garcia, C., Hernandez, T., Costa, F. & del Rio, J. C. (1989). *The Science of the Total Environment*, **81/82**, 551.
Gerrits, J. P. G. (1972). *Mushroom Science*, **9**, 43.
Gibson, D. T. & Subramanian, V. (1984). In *Microbial Degradation of Organic Compounds*, ed. D. T. Gibson. Marcel Dekker, New York, p. 181.
Glenn, J. (1988). *BioCycle*, **29**(7), 49.
Goldstein, N. (1988). *BioCycle*, **29**(10), 27.
Goldstein, N., Yanko, W. A., Walker, J. M. & Jakubowski, W. (1988). *BioCycle*, **29**(5), 44.
Gordon, R. E., Haynes, W. C. & Hor-Nay Pang, C. (1973). *The Genus Bacillus*. Agricultural Handbook No. 427, Agricultural Research Service, USDA, Washington, DC.
Gray, K. R. & Sherman, K. (1969). *Birmingham University Chemical Engineering*, **20**(3), 64.
Gray, K. R., Sherman, K. & Biddlestone, A. J. (1971). *Process Biochemistry*, **6**(10), 22.
Greenwood, D. J. (1961). *Plant & Soil*, **14**, 360.
Gregory, P. H., Lacey, M. E., Festenstein, G. N. & Skinner, F. A. (1963). *Journal of General Microbiology*, **33**, 147.
Griffin, D. M. (1977). *Annual Reviews of Phytopathology*, **15**, 319.
Griffin, D. M. (1981*a*). *Advances in Microbial Ecology*, **5**, 91.
Griffin, D. M. (1981*b*). In *Water Potential Relations in Soil Microbiology*, SSSA special publication no. 9, eds J. F. Parr, W. R. Gardner & L. F. Elliott. Soil Science Society of America, Madison, WI, p. 141.
Griffin, D. M. (1985). In *Bacteria in Nature*, Vol. I, eds E. R. Leadbetter & J. S. Poindexter. Plenum, London, p. 221.
Griffin, D. M. & Quail, G. (1968). *Australian Journal of Biological Science*, **21**, 579.
Harper, E. R., Miller, F. C. & Macauley, B. J. (1992). *Australian Journal of Experimental Agriculture*, in press.
Harris, R. F. (1981). In *Water Potential Relations in Soil Microbiology*, SSSA special publication no. 9, eds J. F. Parr, W. R. Gardner & L. F. Elliott. Soil Science Society of America, Madison, WI, p. 23.
Haug, R. T. (1978). *Proceedings—National Conference on Design of Municipal Sludge Composting Facilities*, Chicago, IL, p. 27.
Haug, R. T. (1979). *Journal of the Water Pollution Control Federation*, **51**, 2189.
Hayes, W. A. (1968). *Mushroom Science*, **7**, 173.
Henssen, A. (1957). *Archive für Mikrobiologie*, **27**, 63.
Hirotani, H., Suzuki, M., Kobayashi, M. & Takahashi, E. (1988). *Soil Science and Plant Nutrition*, **34**(3), 467.
Hogan, J. A., Toffoli, G. R., Miller, F. C., Hunter, J. V. & Finstein, M. S. (1988). In *International Conference on Physiological and Biological Detoxification of Hazardous Wastes*, ed. Y. C. Wu. Technomic Publishing Co., Lancaster, PA, p. 742.
Hogan, J. A., Miller, F. C. & Finstein, M. S. (1989). *Applied and Environmental Microbiology*, **55**(5), 1082.
Hoitink, H. A. J. (1980). *Plant Diseases*, **64**, 142.
Hoitink, H. A. J. & Fahy, P. C. (1986). *Annual Reviews of Phytopathology*, **24**, 93.
Hunter, J. V., Finstein, M. S., Suler, D. J. & Bobal, R. R. (1981). *Fate of concentrated industrial wastes during laboratory scale composting of sewage sludge*. Report of New Jersey Agricultural Experiment Station, New Brunswick, NJ.
Jeris, J. S. & Regan, W. R. (1973*a*). *Compost Science*, **14**(2), 8.
Jeris, J. S. & Regan, W. R. (1973*b*). *Compost Science*, **14**(3), 16.
Kane, B. E. & Mullins, J. T. (1973). *Mycologia*, **65**, 1087.
Kelly, R. M. & Deming, J. W. (1988). *Biotechnology Progress*, **4**(2), 47.
Koster, I. W. (1986). *Journal of Chemical Technology and Biotechnology*, **36**, 445.
Kuter, G. A., Nelson, E. B., Hoitink, H. A. J. & Madden, L. V. (1983). *Phytopathology*, **73**(10), 1450.
Kuter, G. A., Hoitink, H. A. J. & Rossman, L. A. (1985). *Journal of the Water Pollution Control Federation*, **57**, 309.

Lacey, J. (1973). In *Actinomycetales: Characteristics and Practical Importance*, eds G. Sykes & F. A. Skinner. Academic Press, London, p. 231.
Letey, J. Jr, Stolzy, L. H. & Kemper, W. D. (1967). In *Irrigation of Agricultural Lands, Agronomy 11*, eds R. M. Hagan, H. R. Haise & T. W. Edminster. Academic Press, New York, 941.
Lumsden, R. D., Lewis, A. J. & Millner, P. D. (1983). *Phytopathology*, **73**, 1543.
MacGregor, S. T., Miller, F. C., Psarianos, K. M. & Finstein, M. S. (1981). *Applied and Environmental Microbiology*, **41**, 1321.
Makawi, A. A. M. (1980). *Zeitblatt Bakteriologie*, **135**, 12.
Mandelbaum, R., Hadar, Y. & Chen, Y. (1988). *Biological Wastes*, **26**, 261.
McCauley, R. F. & Shell, B. J. (1956). *Proceedings of the Purdue Industrial Waste Conference*, **11**, 436.
McKinley, V. L. & Vestal, J. R. (1984). *Applied and Environmental Microbiology*, **47**, 933.
McKinley, V. L. & Vestal, J. R. (1985). *Canadian Journal of Microbiology*, **31**, 919.
Mears, D. R., Singley, M. E., Ali, C. & Rupp, F. (1975). In *Energy, Agriculture and Waste Management*, ed. W. J. Jewell. Ann Arbor Science Publishers, Ann Arbor, MI, p. 83.
Miller, F. C. (1984). *Thermodynamic and matric water potential analysis in field and laboratory scale composting ecosystems*. PhD dissertation, Rutgers University. University Microfilms, Ann Arbor, MI.
Miller, F. C. (1989). *Microbial Ecology*, **18**, 59.
Miller, F. C. & Finstein, M. S. (1985). *Journal of the Water Pollution Control Federation*, **57**(2), 122.
Miller, F. C. & Macauley, B. J. (1988). *Australian Journal of Experimental Agriculture*, **28**, 553.
Miller, F. C. & Macauley, B. J. (1989). *Australian Journal of Experimental Agriculture*, **29**, 119.
Miller, F. C., MacGregor, S. T., Finstein, M. S. & Cirello, J. (1980). *Proceedings of the ASCE Environmental Engineering Division Specialty Conference*, American Society of Civil Engineers, New York, p. 40.
Miller, F. C., MacGregor, S. T., Psarianos, K. M., Cirello, J. & Finstein, M. S. (1982). *Journal of the Water Pollution Control Federation*, **54**(1), 111.
Miller, F. C., Hogan, J. A. & Macauley, B. J. (1989*a*). *Abstracts of Papers, 5th International Symposium on Microbial Ecology*, Kyoto, Japan, Abstract 0-9-7.
Miller, F. C., Harper, E. R. & Macauley, B. J. (1989*b*). *Australian Journal of Experimental Agriculture*, **29**, 741.
Miller, F. C., Harper, E. H., Macauley, B. J. & Gulliver, A. (1990). *Australian Journal of Experimental Agriculture*, **30**, 287.
Miller, F. C., Macauley, B. J. & Harper, E. R. (1991). *Australian Journal of Experimental Agriculture*, **31**, 415.
Millner, P. D., Marsh, P. B., Snowden, R. B. & Parr, J. F. (1977). *Applied and Environmental Microbiology*, **34**, 765.
Millner, P. D., Powers, K. E., Enkiri, N. K. & Burge, W. D. (1987). *Microbial Ecology*, **14**, 255.
Mortland, M. M. & Wolcott, A. R. (1965). In *Soil Nitrogen*, eds W. V. Bartholomew & F. E. Clark. American Society of Agronomy, Madison, WI, p. 151.
Mosher, D. & Anderson, R. K. (1977). *Composting sewage sludge by high-rate suction aeration techniques—the process as conducted at Bangor, ME, and some guides of general applicability*. Interim report SW-614d. US Government Printing Office, Washington, DC.
Murray, C. M. & Thompson, J. L. (1986). *BioCycle*, **27**(4), 22.
Myers, R. G. & Thien, S. J. (1988). *Soil Science Society of America Journal*, **52**, 516.
Nakasaki, K., Nakano, Y., Akiyama, T., Shoda, M. & Kubota, H. (1987*a*). *Journal of Fermentation Technology*, **65**(1), 43.
Nakasaki, K., Kato, J., Akiyama, T. & Kubota, H. (1987*b*). *Journal of Fermentation Technology*, **65**(4), 441.
Nell, J. H. & Wiechers, S. G. (1978). *Water South Africa*, **4**(4), 203.
Nelson, E. B., Kuter, G. A. & Hoitink, H. A. J. (1983). *Phytopathology*, **73**(10), 1457.

Niese, G. (1959). *Archive für Mikrobiologie*, **34**, 285.

Nommik, H. (1965). In *Soil Nitrogen*, eds W. V. Bartholomew & F. E. Clark. American Society of Agronomy, Madison, WI, p. 198.

Nommik, H. (1970). *Plant & Soil*, **33**, 581.

Norman, A. G., Richards, L. A. & Carlyle, R. E. (1941). *Journal of Bacteriology*, **41**, 689.

O'Dogherty, M. J. & Gilbertson, H. G. (1988). *Journal of Agricultural Engineering Research*, **40**, 245.

Okafor, N. (1966). *Nature*, **210**, 220.

Op den Camp, H. J. M. (1989). *De Champignonculture*, **31**(10), 513.

Osmon, J. L., Andrews, C. C. & Tatyreh, A. (1978). *The biodegradation of TNT in enhanced soil and compost systems*. Technical report ARLCD-TR-77032, US Army Armament Research and Development Command, Large Caliber Weapons Station, Dover, NJ.

Papendick, R. I. & Campbell, G. S. (1981). In *Water Potential Relations in Soil Microbiology*, SSSA special publication no. 9, eds J. F. Parr, W. R. Gardner & L. F. Elliott. Soil Science Society of America, Madison, WI, p. 1.

Parr, J. F., Epstein, E. & Willson, G. B. (1978). *Agriculture and Environment*, **4**, 123.

Pereira-Neto, J. T., Stentiford, E. I. & Smith, D. V. (1986). *Waste Management & Research*, **4**, 397.

Pereira-Neto, J. T., Stentiford, E. I. & Mara, D. D. (1987). In *Compost: Production Quality and Use*, eds M. De Bertoldi, M. P. Ferranti, P. L'Hermite & F. Zucconi. Elsevier Applied Science, London, p. 276.

Perry, R. H. & Chilton, C. H. (eds) (1973). *Chemical Engineers Handbook*, 5th edn. McGraw-Hill, New York.

Poincelot, R. P. (1975). *The Biochemistry and Methodology of Composting*. Bulletin 754, Connecticut Agricultural Experiment Station, New Haven, CT.

Ponnamperuma, F. N. (1981). In *Proceedings of Symposium on Paddy Soil*, Institute of Soil Science, Academia Sinica. Science Press, Beijing, p. 59.

Postgate, J. R. (1984). *The Sulphate-reducing Bacteria*, 2nd edn. Cambridge University Press, Cambridge, UK.

Racke, K. D. & Frink, C. R. (1989). *Bulletin of Environmental Contamination and Toxicology*, **42**, 526.

Randle, P. (1983). *Scientia Horticulturae*, **20**, 53.

Randle, P. & Flegg, P. B. (1978). *Scientia Horticulturae*, **8**, 315.

Rose, W. W. & Mercer, W. A. (1968). *Fate of Insecticides in Composted Agricultural Wastes*. National Canners Association, Washington, DC.

Ross, R. C. & Harris, P. J. (1982). *Scientia Horticulturae*, **17**, 223.

Ross, R. C. & Harris, P. J. (1983*a*). *Scientia Horticulturae*, **19**, 55.

Ross, R. C. & Harris, P. J. (1983*b*). *Scientia Horticulturae*, **20**, 61.

Rothbaum, H. P. (1961). *Journal of Bacteriology*, **81**(2), 165.

Rothbaum, H. P. (1963). *Applied Chemistry*, **13**, 291.

Rothbaum, H. P. & Dye, M. H. (1964). *New Zealand Journal of Science*, **7**, 119.

Schulze, K. L. (1958). *Proceedings of the 13th Industrial Waste Conference* (Purdue), **13**, 541.

Schulze, K. L. (1961). *Compost Science*, **2**(2), 32.

Schulze, K. L. (1962). *Applied Microbiology*, **10**(2), 108.

Schulze, K. L. (1965). *Compost Science*, **2**(4), 31.

Sharon, N. (1965). In *The Amino Sugars—The Chemistry and Biology of Compounds Containing Amino Sugars*, Vol. IIA, eds E. A. Balaz & R. W. Jeanloz. Academic Press, New York, p. 1.

Shell, G. L. & Boyd, J. L. (1969). *Composting dewatered sewage sludge*. Report SW-12c, USHDEW/Public Health Service, Environmental Health Services Environmental Control Administration, Bureau of Solid Waste Management, Washington, DC.

Sikora, L. J., Ramirez, M. A. & Troeschel, T. A. (1983). *Journal of Environmental Quality*, **12**(2), 219.

Stanek, M. (1971). *Mushroom Science*, **8**, 797.

Straatsma, G., Gerrits, J. P. G., Augustijn, M. P. A. M., Op den Camp, H. J. M., Vogels, G. D. & Van Griensven, L. J. L. D. (1989). *Journal of General Microbiology*, **135**, 751.
Strom, P. F. (1985*a*). *Applied and Environmental Microbiology*, **50**(4), 899.
Strom, P. F. (1985*b*). *Applied and Environmental Microbiology*, **50**(4), 906.
Strom, P. F. & Finstein, M. S. (1985). *Leaf Composting Manual for New Jersey Municipalities*. Office of Recycling, New Jersey Departments of Energy and Environmental Protection, Newark, NJ.
Strom, P. F., Morris, M. L. & Finstein, M. S. (1980). *Compost Science*, **21**(6), 44.
Stutzenberger, F. J. (1971). *Applied Microbiology*, **22**, 147.
Suler, D. J. & Finstein, M. S. (1977). *Applied and Environmental Microbiology*, **32**, 345.
Sundaram, T. K. (1986). In *Thermophiles—General, Molecular, and Applied Microbiology*, ed. T. D. Brock. John Wiley, New York, p. 75.
Tansey, M. R. (1981). *Mycologia*, **58**(3), 537.
Tautorus, T. E. & Townsley, P. M. (1983). *Applied and Environmental Microbiology*, **45**, 511.
Teensma, B. (1961). *Compost Science*, **1**(4), 11.
Tschierpe, H. J. (1972). *Mushroom Science*, **8**, 553.
US Congress, Office of Technology Assessment (1989). *Facing America's Trash: What Next for Municipal Solid Waste*. OTA-O-424, US Government Printing Office, Washington, DC.
Vallini, G., Pera, A., Cecchi, F., Briglia, M. & Perghem, F. (1989). *Waste Management & Research*, **7**, 277.
Van Der Hoek, K. W. & Oosthoek, J. (1985). In *Composting of Agricultural and Other Wastes*, ed. J. K. R. Gasser. Elsevier Applied Science, Amsterdam, p. 271.
Waksman, S. A. & Cordon, T. C. (1939). *Soil Science*, **47**, 217.
Waksman, S. A. & Hutchings, I. J. (1937). *Soil Science*, **43**, 77.
Waksman, S. A., Cordon, T. C. & Hulpoi, N. (1939*a*). *Soil Science*, **47**, 83.
Waksman, S. A., Umbreit, W. W. & Cordon, T. C. (1939*b*). *Soil Science*, **47**, 37.
Walker, I. K. & Harrison, W. J. (1960). *New Zealand Journal of Agricultural Research*, **3**, 861.
Walker, J., Goldstein, N. & Chen, B. (1986). *BioCycle*, **27**(4), 22.
Wiley, B. B. & Westerberg, S. C. (1969). *Applied Microbiology*, **18**, 994.
Wiley, J. S. (1956). *Proceedings of the Purdue Industrial Waste Conference*, **12**, 596.
Wiley, J. S. (1957). *Proceedings of the ASCE, Sanitary Engineering Division*, Paper 1411.
Wiley, J. S. & Pearce, G. W. (1957). *Proceedings of the Purdue Industrial Waste Conference*, **12**, 596.
Willson, G. B., Parr, J. F., Epstein, E., Marsh, P. B., Chaney, R. L., Colcicco, W. D., Burge, W. D., Sikora, L. J., Tester, C. F. & Hornic, S. (1980). *Manual for Composting Sewage Sludge by the Beltsville Aerated Pile Method*. USEPA, USDA, US Government Printing Office, Washington, DC.
Wimpenny, J. W. T., Coombs, P. J. & Lovitt, R. W. (1984). In *Current Perspectives in Microbial Ecology—Proceedings of the Third International Symposium Microbial Ecology*, American Society for Microbiology, Washington, DC, p. 291.
Zucconi, F. & De Bertoldi, M. (1987). In *Compost: Production Quality and Use*, eds M. De Bertoldi, M. P. Ferranti, P. L'Hermite & F. Zucconi. Elsevier Applied Science, London, p. 30.

Chapter 2

METHODS FOR THE BIOLOGICAL TREATMENT OF EXHAUST GASES

MAKOTO SHODA

Research Laboratory of Resources Utilization, Tokyo Institute of Technology, Yokohama, Japan

CONTENTS

1 INTRODUCTION

Offensively odorous substances emitted from various industrial areas and water treatment plants are causing many problems. In Japan complaints about malodorous gases are numerically second among seven environmental problems regulated by the laws. For treatment of exhaust gases, physical and/or chemical methods (combustion, activated carbon adsorption, acid–alkali treatment, etc.) have been popularly used and their efficiency is generally satisfactory. However, maintenance and operation costs of these methods are high and thus biological methods have attracted attention as a more economical alternative. A review on exhaust gas purification was published in 1986 (Ottengraf, 1986) and has been discussed intensively from the point of view of chemical engineering. Here recent information and advanced technology on biological degradation methods of exhaust gas treatment will be described.

2 OUTLINES OF METHODS OF BIOLOGICAL TREATMENT OF EXHAUST GASES

Biological methods to treat exhaust gases will be categorized as follows:

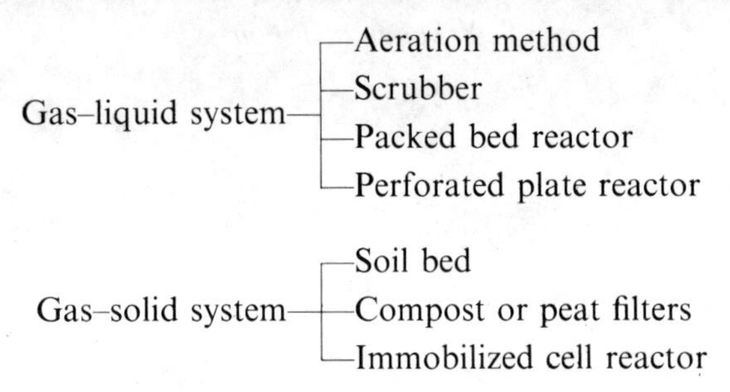

Some of these methods will be outlined here.

2.1 Gas–Liquid System

2.1.1 Aeration Method

This method consists of supplying the exhaust gas directly to the aeration tank in an activated sludge treatment plant and degrading the gas components using the microorganisms in the sludge. If the oxygen concentration in the exhaust gas is high enough, the gas supply will not adversely affect the activity of the activated sludge to decompose the organic substances in the waste water. Thus this is an energy-saving method for the treatment of exhaust gases, but some chemical or physical treatment facilities have to be constructed, depending on the concentration and flow rate of the exhaust gas. Such plants are in practical operation in Japan in the following fields: night soil treatment plant, sewage sludge tank, composting plant, rendering

Table 1
Treatment of Exhaust Gas from Night Soil Treatment Plant by Aeration Method (The Japanese Society of Industrial Machinery Manufacturers, 1987)

	Before treatment	*After treatment*
Odor concentration[a]	10 420	43
NH_3 (μl/liter)	13	0·5
H_2S (μl/liter)	20	0·000 2
Methanethiol (MT) (μl/liter)	0·25	0·000 1
Dimethyl sulfide (DMS) (μl/liter)	0·09	0·001 9
Dimethyl disulfide (DMDS) (μl/liter)	0·002	0·000 1

[a] Measured by a triangle bag method (a standard sensory method in Japan).

factories, kraft-pulp factories, chicken farming, paint manufacturing, printing factory, incineration of sewage sludge, etc. An example of treatment efficiency for high concentrations of exhaust gas from a night soil treatment plant is shown in Table 1.

The features of this method are as follows: (1) operation is simple; (2) it is applicable to various gases such as sulfur-containing gases, amines, aldehydes, hydrocarbons, etc.; (3) neither additional chemicals nor fuel is needed; and (4) costs of construction and maintenance are low (Fukuyama *et al.*, 1981; Fukuyama & Honda, 1988).

2.1.2 Scrubber Method

The bioscrubber is a tower reactor where exhaust gas is adsorbed into circulating activated sludge and the adsorbed gas is metabolized by microbes in the sludge. As the contact time of the gas with the liquid sludge is relatively short, high efficiency of gas–liquid contact is needed. One example of the scrubber method is shown in Fig. 1. This system is designed to degrade the exhaust gas completely via a circulation tank, a re-aeration tank to supply additional oxygen, a sedimentation tank to remove excess sludge and digested products, and a nutrient reservoir for supply of nitrogen or phosphate, if necessary. To avoid clogging of sludge inside the scrubber reactor several modifications are proposed. Figure 2 shows a typical perforated plate tower reactor (Moretana reactor). The plates placed inside have holes (diameter 2–3 mm) to achieve efficient gas–liquid contact by a simple structure. As operating conditions, a gas flow rate of 1–4 m/s, a liquid/gas ratio of 2–10 liter/m^3, a pH of 6–7, a dissolved oxygen concentration of 1–3 mg/liter, and a mixed liquor suspended solid (MLSS) of 700–10 000 mg/liter are recommended.

2.2 Gas–Solid System

2.2.1 Soil Filter (Soil Bed)

Soil treatment of exhaust gas is a traditional method utilizing microbial activities

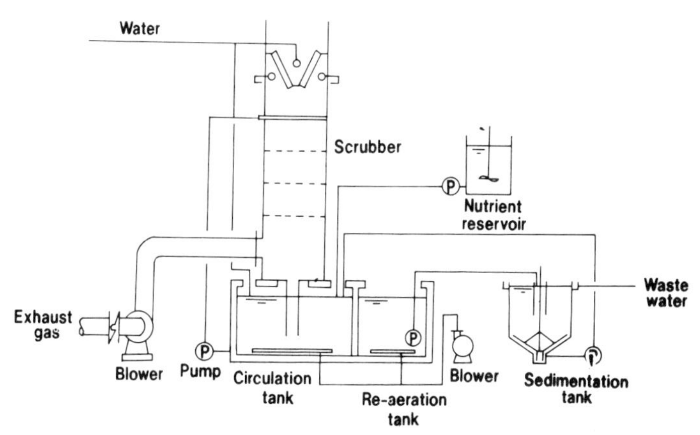

Fig. 1. Scrubber method by using activated sludge (The Japanese Society of Industrial Machinery Manufacturers, 1987).

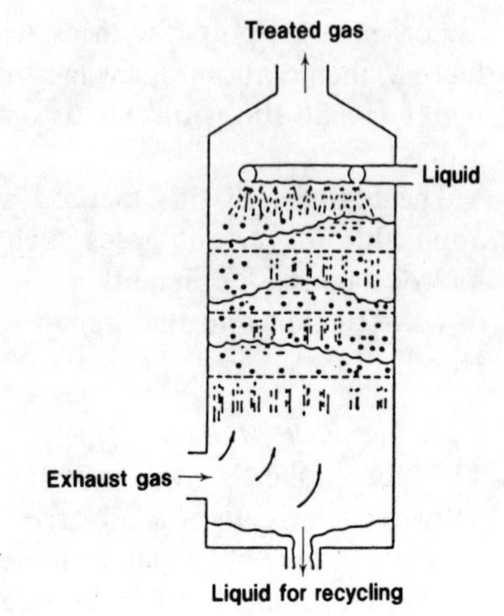

Fig. 2. Perforated plate reactor (The Japanese Society of Industrial Machinery Manufacturers, 1987).

inhabiting soil in cooperation with the chemical and physical adsorption capacity of the soil, and with solubilization of the gas components into the water phase in the soil.

When the structure of the soil filter is properly designed to maximize the activity of microorganisms in the soil and operating conditions are properly selected, maintenance and running costs of this method are significantly lower compared with other methods.

The structure of the soil filter generally consists of a diffusion layer, a homogeneous layer, and a soil layer. The diffusion layer, which is the bottom layer of the filter, is of low resistance to flow rate and to dispersion of the gas supplied from outside. The homogeneous layer is located above the diffusion layer; its function is to keep the gas pressure in this layer homogeneous before the gas enters the soil layer.

Standard specifications for designing the soil filter are as follows:

—Soil: of high air permeability and water-holding capacity.
—Linear velocity of gas: 0·5–1 cm/s.
—Static pressure of gas: less than 250 mm Hg.
—Humidity of gas: non-saturated with water.
—Temperature of gas: 5–40°C.

2.2.1.1 MOISTURE CONTENT OF SOIL

Microbial activities in soil are optimal in the range of 40–70% of the maximum holding capacity of water. As gas supply generally causes a decrease in the moisture content in soil, a supply of water is essential by any means available, for example by spraying water onto the surface of the soil filter.

2.2.1.2 TEMPERATURE OF SOIL

The optimal temperature for microbial activity in soil is between 20 and 37°C, although some microbes are active even below 5°C. As it is difficult to control the temperature of soil, cooling or heating of the inflow gas should be considered if necessary.

2.2.1.3 pH

Although some autotrophic sulfur bacteria grow in the acidic range of pH, most soil microorganisms have an optimal pH of about 7. Acidic soil should be neutralized by treatment with $CaCO_3$ or $Ca(OH)_2$.

2.2.1.4 PERMEABILITY OF AIR IN SOIL

Homogeneous distribution of air into the soil layer is important for stimulating activity of aerobic microorganisms which are responsible for degradation of organic or inorganic components in exhaust gas. Heavy rainfall often causes packing of soil which leads to an increase in the pressure drop of soil and thus to channeling of air through the soil layer. The permeability of air in soil depends mostly on properties of the soil. Therefore selection of soil is a key point for efficient removal of gas. The soil filter is widely applied in food manufacturing, wood processing, pulping factories, the chemical industry, night soil treatment plants, etc. A defect of the soil filter is the high pressure drop caused by a thick soil layer. This means that a wide area must be prepared for a high flow rate or high concentration of exhaust gas. A tower-type soil filter is not practical due to its weight and pressure drop. Thus peat or compost have been tried as a carrier of microorganisms instead of soil. These show a lower pressure drop than soil when a tower-type reactor is designed for treatment of exhaust gas. Figure 3 shows pressure drop against moisture content for various materials. The pressure drop of soil (not shown) is much larger than that of garbage compost shown in the figure. Biological treatment methods using compost or peat as a carrier of microorganisms are often referred to as compost biofilters or peat biofilters.

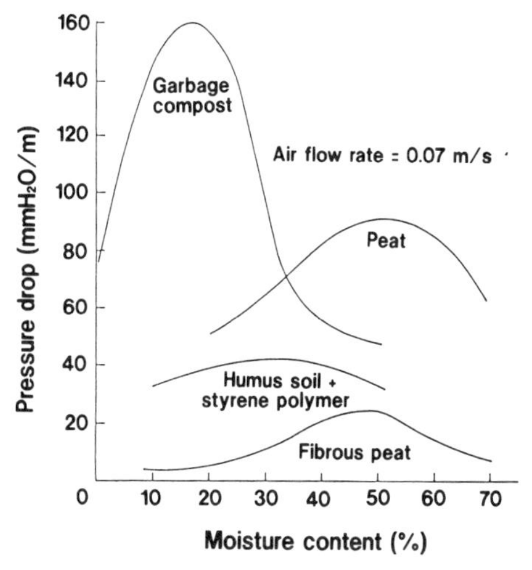

Fig. 3. Pressure drop of various materials vs moisture content (Zeisig *et al.*, 1977).

2.2.2 Compost or Peat Biofilters

Although compost is defined as a solid organic waste which has been decomposed and stabilized aerobically by microbial degradation, immature composts often evolve malodorous components. Thus, as a carrier of microorganisms, well-matured compost should be selected.

Treatment of exhaust gas by peat biofilter has been carried out mainly in Europe. As shown in Fig. 3, fibrous peat has a relatively lower pressure drop. However, the peat biofilter in the vertical tower type is not popular, mainly because basic information on its design and control is lacking, and because the biological phenomena inside the peat during the treatment of exhaust gas have not been fully analyzed. In this circumstance pilot-scale peat biofilters have been constructed and their operational characteristics intensively investigated to assess the data for a practical tower-type biofilter in Japan.

The removal characteristics of a pilot-scale peat biofilter (400 mm in diameter,

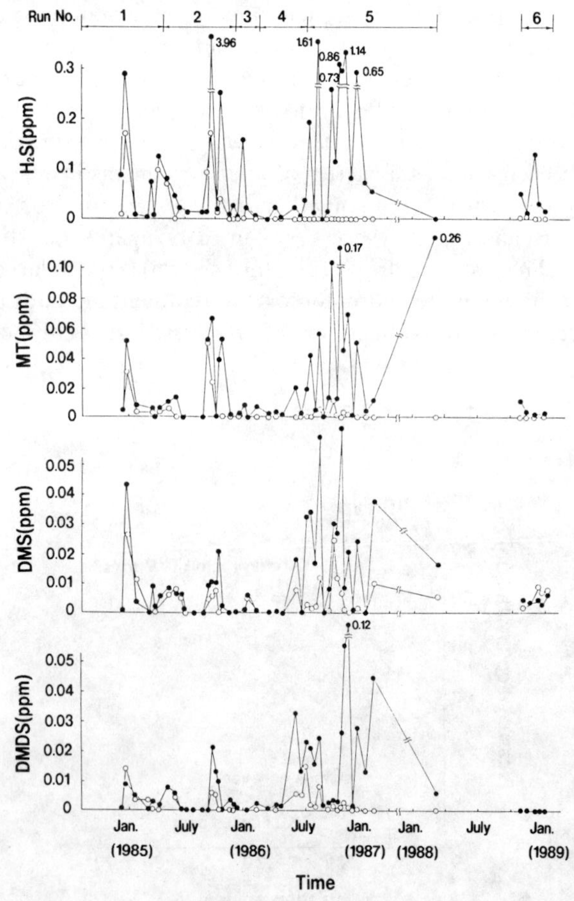

Fig. 4. Change of concentrations of sulfur-containing compounds at inlet (●) and outlet (○) in a pilot-scale peat biofilter constructed in a waste water treatment plant (Liang *et al.*, 1990).

Table 2
Estimated Operating Costs (1000 Yen/Year) (Liang, 1991)

	Peat biofilter	*Chemical treatment*
Electric power	3 630	11 226
Chemicals	0	801
Total	3 630	12 027

3660 mm in height) in a long-term operation are shown in Fig. 4. This filter was constructed in a waste water treatment plant to treat the gases from both aeration and sludge sedimentation tanks. The main components of the exhaust gas were H_2S, methanethiol (MT), dimethyl sulfide (DMS), dimethyl disulfide (DMDS) and NH_3. For $4\frac{1}{2}$ years of operation H_2S, MT and DMDS were effectively removed. Ammonia was taken up biologically and also chemically reacted with sulfate, which was a final product of the oxidation process of sulfur-containing gases by bacteria inhabiting the peat. DMS, however, was less liable to be degraded biologically. In long-term operation the clogging or packing phenomenon in peat was less serious than in soil and only a 20% increase of pressure drop was observed after $4\frac{1}{2}$ years of operation (Liang *et al.*, 1990). The estimated operating costs for the peat biofilter and the present chemical treatment are compared in Table 2, based on the same treatment efficiency to clear the values regulated by the Odor Control Law in Japan. Only one-third of the present cost is needed for the biological treatment process. In another pilot plant test in a waste water treatment plant more than 200 ppm of H_2S was effectively degraded by a peat biofilter, where pH was varied in the range of 2–4. This reflected the active oxidation of H_2O by acidophilic bacteria inhabiting the peat (Liang, 1991).

3 BASIC ANALYSIS FOR EACH GAS

3.1 NH_3

When air containing NH_3 was supplied to the peat biofilter, the change of NH_3 concentration at the outlet is shown in Fig. 5 on different NH_3 loads. After complete removal of NH_3 a breakthrough was observed, indicating that NH_3 gas was mainly removed by adsorption to peat or absorption into water in peat. Proposed NH_3 trapping schemes with the functional groups of the peat are shown in Table 3 (Togashi *et al.*, 1986). When chemical adsorption is dominant, NH_3 will be saturated and eventually the breakthrough will occur. This means that microbial contribution to the removal of NH_3 was significantly small. In order to enhance removal efficiency of NH_3 by biological reaction and to maintain steady removal of NH_3, addition of nitrification activity by seeding nitrifying bacteria is a possible means, as

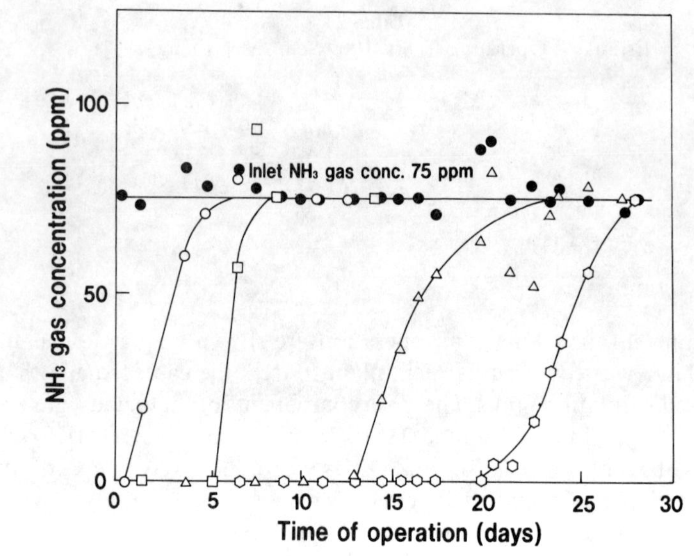

Fig. 5. NH_3 removal in peat biofilter at different loadings of NH_3. Ammonia loading (g N/day/kg dry peat): ⬡, 0·7; △, 1·6; □, 3·1; ○, 7·1 (Togashi *et al.*, 1986).

Table 3
Proposed NH_3 Trapping Schemes with the Functional Groups of the Peat (Togashi *et al.*, 1986)

Neutralizing trap	*Hydrogen bonding trap*
Aromatic carboxylic acid	Carbonyl group
$ArCOOH + NH_3 \rightarrow ArCO\overset{-}{O}\overset{+}{N}H_4$	$>C{=}O \cdots H{-}N(H)_2$
Aliphatic carboxylic acid	Methoxyl group
$RCOOH + NH_3 \rightarrow RCO\overset{-}{O}\overset{+}{N}H_4$	$-O(Me) \cdots H{-}N(H)_2$
Aromatic alcohol	Alcoholic OH
$ArOH + NH_3 \rightarrow Ar\overset{-}{O}\overset{+}{N}H_4$	$-O{-}H \cdots N(H)_2{-}H$ or $-O(H) \cdots H{-}N(H)_2$
	Amino group (less contribution)
	$>N{-}H \cdots N(H)_2{-}H$ or $>N \cdots H{-}N(H)_2$

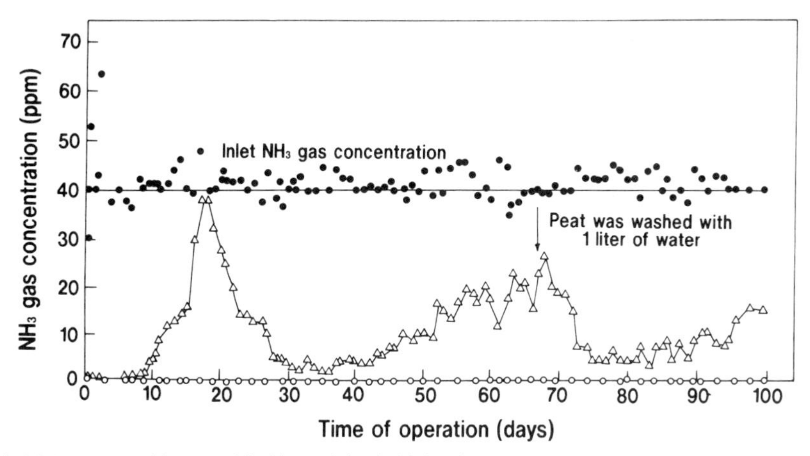

Fig. 6. NH_3 removal by peat biofilter with nitrifying bacteria. Ammonia loading (g N/day/kg dry peat): ○, 0·16; △, 0·32 (Togashi *et al.*, 1986).

shown in Fig. 6. When nitrifying sludge in the nitrification plant was inoculated onto peat, constant removal of NH_3 was seen at an NH_3 loading of 0·16 (g N/day/kg dry peat), while at the higher loading of 0·32 NH_3 was detected at the outlet. In practical operation of the peat biofilter in a waste water treatment plant, as shown in Fig. 4, no NH_3 was detected during the period of the experiment, mainly because sulfate produced by oxidation of sulfur-containing gases (H_2S, MT, DMS, DMDS) by sulfur-oxidizing bacteria neutralized NH_3, and thus the maximum loading of NH_3 was five times larger than the result obtained in Fig. 6.

3.2 Trimethyl Amine (TMA)

When TMA was supplied to the compost biofilter, it was degraded into NH_3 in the scheme shown in Fig. 7, indicating that breakthrough will be observed after a certain period of time, just as with NH_3. This also means that the introduction of nitrification capacity into the biofilter is essential for stable treatment of TMA if TMA is a main component in the exhaust gas.

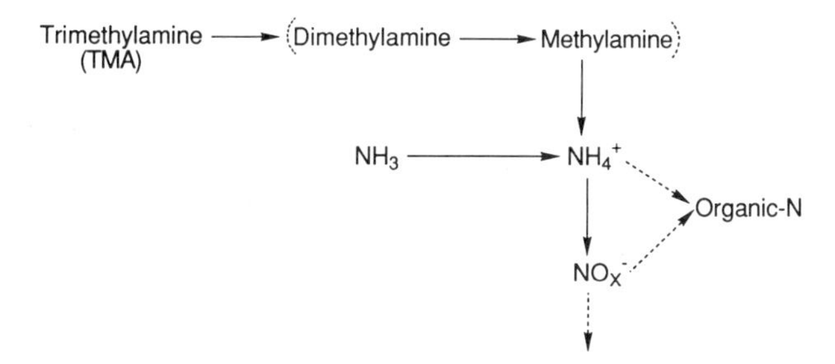

Fig. 7. Biological change of trimethyl amine (TMA) and NH_3.

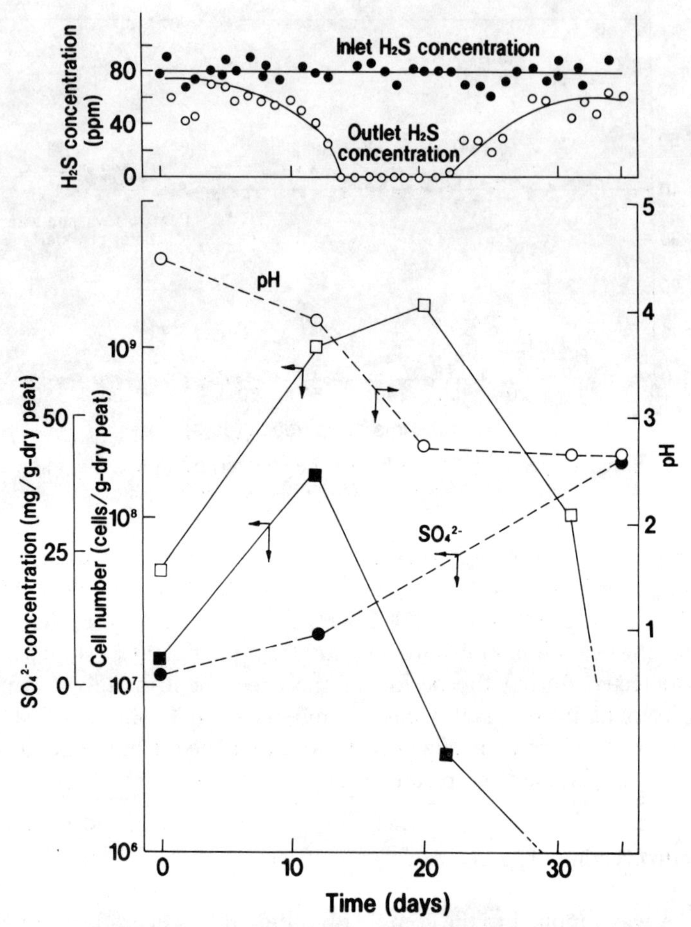

Fig. 8. Time course of H_2S removal by peat biofilter at inlet H_2S concentration of about 80 ppm and air flow rate of 3 liter/min. ○, pH; ●, SO_4^{2-} concentration; □, cell number counted on modified Waksman medium; ■, cell number counted on Trypticase-soy medium (Wada *et al.*, 1986).

3.3 Hydrogen Sulfide (H_2S)

For different loadings of H_2S supplied to the peat packed bed biofilter the H_2S concentrations in the effluent gas from the bed are shown in Fig. 8 (Wada *et al.*, 1986). The H_2S removal pattern showed a transition phenomenon, indicating that removal of H_2S was accelerated due to acclimation and increase of population of H_2S-oxidizing bacteria inhabiting the peat. No H_2S removal was observed when the peat was sterilized with gamma-ray irradiation. Kinetic analysis is as follows (Furusawa *et al.*, 1984). Assuming plug flow of air in the packed bed, the removal rate, *r*, is given by

$$r = -F\,dC/dW = k \quad (1)$$

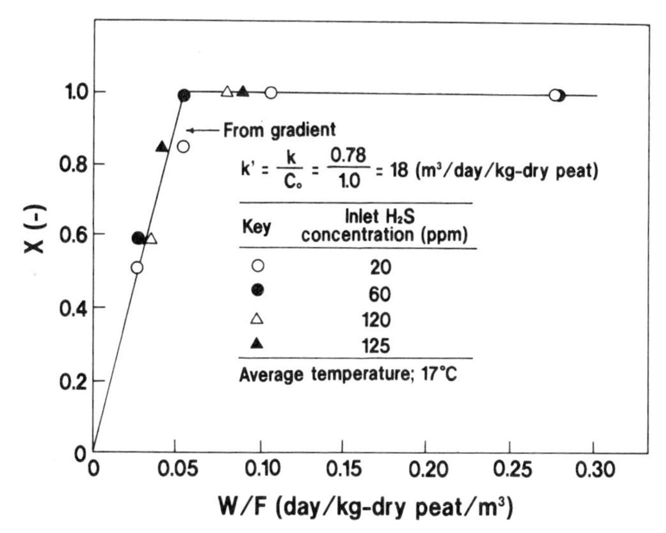

Fig. 9. Relationship between conversion of H_2S, x and W/F (Furusawa *et al.*, 1984).

where F is the flow rate of air, C is the H_2S concentration in the peat, W is the dry weight of peat, and k is the reaction rate constant. Integration of eqn (1) under an inlet concentration of $C = C_0$ at $W = 0$, and introduction of $x = (C_0 - C)/C_0$, $k' = k/C_0$, give

$$x = k'W/F \qquad (x < 1) \tag{2}$$

Figure 9 shows the relationship of eqn (2) with the experimental data. The rate constant k is proportional to C_0, indicating that the number of microorganisms inhabiting the peat that are responsible for oxidation of H_2S is proportional to C_0. During the acclimation period in the initial stage of gas supply the cell number appearing on modified Waksman medium increased while pH decreased (Fig. 8). Once pH went down to below 3, the cell number decreased and breakthrough of H_2S occurred. However, when pH was controlled no breakthrough appeared. One autotrophic bacterium, *Thiobacillus intermedius*, was found to be responsible for the oxidation of H_2S in peat (Wada *et al.*, 1986). As the relationship between the cell number of the bacterium and the inlet concentration of H_2S was linear, the result based on the kinetic analysis above was verified.

3.4 Other Sulfur-containing Compounds

The peat inoculated with aerobically digested sludge removed methanethiol (MT), dimethyl sulfide (DMS) and dimethyl disulfide (DMDS), as well as H_2S. The maximum removal rate, V_m, of the peat was determined by the following equation (based on the Michaelis–Menten equation):

$$C_{ln}/R = K_s/V_m + C_{ln}/V_m \tag{3}$$

Table 4
Kinetic Parameters for H_2S, DMS, MT and DMDS by Peat Biofilter (Cho *et al.*, 1991*a*)

	V_m (*g S/kg dry peat/day*)	K_s (*ppm*)
H_2S	5·0	55
MT	0·9	10
DMS	0·38	10
DMDS	0·68	1

where $R = SV(C_0 - C_e)/a$ is the removal rate, $C_{1n} = (C_0 - C_e)/\ln(C_0/C_e)$ is the log mean concentration, SV is the space velocity, K_s is a constant, a is a conversion factor, C_0 is the inlet concentration, and C_e is the outlet concentration.

The linear relationship between C_{1n}/R and C_{1n} gives V_m and K_s, as shown in Table 4. Apparently V_m for DMDS, MT and DMS are significantly smaller than that for H_2S, reflecting the removability of these sulfur compounds by peat in the following order:

$$H_2S > MT > DMDS > DMS$$

Microbial analysis of peat before and after the supply of organosulfur compounds, MT, DMS and DMDS showed that the number of microorganisms which grew on thiosulfate–agar medium (pH = 7) increased 1000-fold, suggesting that chemolithotrophic sulfur-oxidizing bacteria like *Thiobacillus* sp. were responsible for oxidation of these gases (Cho *et al.*, 1991*a*).

4 FUTURE ASPECTS

In order to enhance the microbial removal rate for exhaust gas and to develop a new microbial process, there are several methods to be introduced in the future.

4.1 Seeding of Useful Bacteria

Inoculation of some microorganisms which are associated with treatment of exhaust gas will be effective (1) to enhance the degradation rate of gases, (2) to shorten the start-up time or acclimation time, and (3) to resume normal conditions quickly when microbial activity deteriorates during operation.

Thiobacillus sp. which was isolated from H_2S-acclimated peat as H_2S-oxidizing bacteria was inoculated into fresh peat and the maximum removal rate of H_2S, V_m, was enhanced almost six times, from 5 to 33 (g S/kg dry peat/day) (Cho *et al.*, 1991*b*).

DMS is one of the most difficult substances to degrade biologically, especially in mixed gas conditions. As DMS-degrading bacteria, *Thiobacillus thioparus* (Kanagawa & Mikami, 1989; Tanji *et al.*, 1989), *Hyphomicrobium* S (de Bont *et al.*,

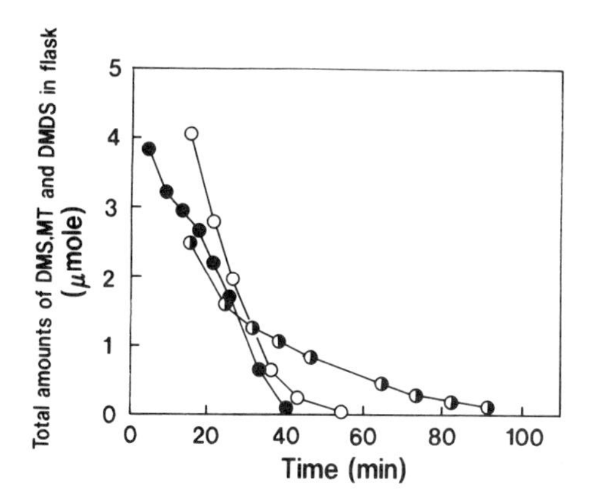

Fig. 10. Removal of DMS (○), MT (●) and DMS (◑) by *Hyphomicrobium* sp. I55 in batch culture (Liang, 1991).

1981) and *Hyphomicrobium* EG (Suylen & Kuenen, 1986) have been reported. The characteristics of *Hyphomicrobium* sp. I55 for removal of DMS, MT and DMDS are shown in Fig. 10. This bacterium degrades these gases at almost the same rate in single gas supply. However, when three mixed gases (H_2S, MT and DMS) were introduced, these gases were co-oxidized but the removal efficiency of each gas was varied; in particular, DMS removal was decreased. When *Pseudomonas acidovorance*, which degrades DMS to dimethyl sulfoxide (DMSO), was seeded with *Hyphomicrobium* sp. I55 into peat, DMS removal was significantly enhanced in cooperative degradation of two bacteria for DMS (Liang, 1991).

4.2 Immobilization of Microorganisms and Selection of Carriers

The immobilization of useful microorganisms on appropriate carriers is one possible way to increase removal rate. Phenol-degrading bacteria were immobilized on alginate and polyacrylamide-hydrazide (Bettmann & Rehm, 1984), and on activated carbon (Ehrhardt & Rehm, 1985). Industrially significant pollutants like propionaldehyde, ethylacetate, butanol and acetone were effectively degraded by bacteria immobilized on activated carbon (Kirchner *et al.*, 1987). Polyvinyl alcohol gel coated with powdered activated carbon gave satisfactory conditions for the buffering effect of pH and high moisture content in the removal of H_2S and MT (Fujie *et al.*, 1990). In practical reactors, fibrous materials as a carrier of immobilized cells have several advantages over other materials in that they are light, flexible, lower in pressure drop, less microbially degradable and easy to handle. Table 5 shows the measured kinetic parameters, V_m and K_s, of DMS removal in different fabrics which were used to immobilize sludge as a source of microorganisms (Tiwaree, 1990). Less biodegradable DMS was removed at the largest removal rate for activated carbon among these fabrics. This fabric also demonstrated better efficiency than peat for the removal of MT, with a maximum removal rate of 0·69 (g S/kg dry fabrics/day), as shown in Fig. 11 (Lee & Shoda, 1989).

Table 5
Kinetic Parameters of DMS Removal for Different Fabrics (Tiwaree, 1990)

Material	V_m (*g S/kg dry fabrics/day*)	K_s (*ppm*)
Polypropylene	0·34	3·4
Nylon	0·47	4·1
Rayon	0·50	3·7
Acrylic	0·45	3·1
Polyester	0·83	1·8
Activated carbon 1	1·3	8·7
Activated carbon 2	2·3	6·3
Peat	0·38	10

4.3 Application of New Microorganisms

More efficient microorganisms which are appropriate for the treatment of gas should be explored. During treatment of biodegradable substances, increase of cell mass is rapid and removal of the increased cell mass is a critical problem. Microorganisms with a smaller yield of cell mass against exhaust gas should be used, based on the investigation of metabolic pathways for the gas components. In this sense, strict autotrophy is better, to simplify nutritional requirements with smaller cell yields. Facultative anaerobes will be better because they are active under either aerobic or anaerobic conditions.

To develop microbial desulfurization, *Thiobacillus denitrificans* was intensively investigated. This bacterium can grow anaerobically on H_2S as an energy source if H_2S is a growth-limiting factor. Maximum loadings for H_2S oxidation under anaerobic and aerobic conditions were 5·4–7·7 and 15·1–20·9 mmol/h/g biomass,

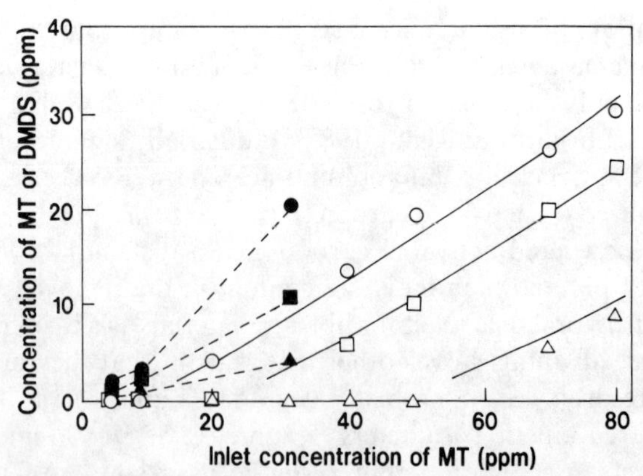

Fig. 11. Relationship between inlet MM concentration and outlet DMDS concentration for activated carbon fabric and outlet MM concentration for peat. Activated carbon fabrics: △, space velocity = 50 h^{-1}; □, space velocity = 100 h^{-1}; ○, space velocity = 200 h^{-1}. Peat: ▲, space velocity = 50 h^{-1}; ■, space velocity = 100 h^{-1}; ●, space velocity = 200 h^{-1} (Lee & Shoda, 1989).

respectively (Sublette, 1987). The biomass yield was lower in aerobic conditions than in anaerobic conditions, mainly due to inhibition of growth by oxygen. In both cases H_2S was oxidized to sulfate stoichiometrically. Heterotrophic contamination, as of aerobic *Pseudomonas* spp., gave a negligible effect on the growth of this bacterium on H_2S (Sublette & Sylvester, 1987*a*,*b*,*c*). However, use of this bacterium is limited, mainly because the H_2S removal rate was totally inhibited by CH_3SH and CS_2.

In the microbial oxidation of sulfur-containing substances, decline of pH is inevitable and this causes the deterioration of biological activities. Selection of acid-tolerant microorganisms is essential, especially for treatment of high concentrations of sulfur gases.

An anaerobic autotrophic bacterium, *Chlorobium thiosulfatophilum*, converted 67·1% of 27 mmol H_2S to insoluble elemental sulfur and 28·3% of 26 mmol CO_2 to organic carbon, respectively. This indicates that the stoichiometric ratio of elemental sulfur to organic carbon is approximately 1:1 (Cork *et al.*, 1983). However, how to conquer the disadvantage of requiring solar or artificial energy is a key point in applying phototrophic microorganisms.

For future potential the use of fungi (Ishikawa *et al.*, 1980), actinomycetes (Ohta & Ikeda, 1978; Hayashi *et al.*, 1980; Ishizaki *et al.*, 1987) or thermophilic bacteria will be promising, although only few examples have been reported. Actinomycetes were found to be useful, especially for removal of DMS, DMDS and volatile fatty acids.

Genetic manipulation of microorganisms, which is already being practically applied in the production of biologically active products such as amino acids, enzymes and medicines, will also be a prominent tool in the breeding of a new and active microorganism applicable to the treatment of exhaust gases.

REFERENCES

Bettmann, H. & Rehm, H. J. (1984). *Applied Microbiology and Biotechnology*, **20**, 285.
Cho, K. S., Hirai, M. & Shoda, M. (1991*a*). *J. Ferment. Bioeng.*, **71**, 289.
Cho, K. S., Liang, Z., Hirai, M. & Shoda, M. (1991*b*). *J. Ferment. Bioeng.*, **71**, 44.
Cork, D. J., Garunas, R. & Sajjad, A. (1983). *Applied and Environmental Microbiology*, **45**, 913.
de Bont, J. A. M., van Dijken, J. P. & Harder, W. (1981). *Journal of General Microbiology*, **127**, 315.
Ehrhardt, H. M. & Rehm, H. J. (1985). *Applied Microbiology and Biotechnology*, **21**, 32.
Fujie, K., Urano, K., Nogi, K., Shimoura, K. & Funahashi, E. (1990). *J. Odor Research Eng.*, **21**, 299.
Fukuyama, J. & Honda, A. (1988). *J. Odor Research Eng.*, **19**, 227.
Fukuyama, J., Honda, A. & Ose, Y. (1981). *J. Jpn. Soc. Air Pollut.*, **14**, 422.
Furusawa, N., Togashi, I., Hirai, M., Shoda, M. & Kubota, H. (1984). *Journal of Fermentation Technology*, **62**, 589.
Hayashi, S., Fujio, Y. & Ueda, S. (1980). *Journal of Fermentation Technology*, **58**, 197.
Ishikawa, H., Kita, Y. & Horikoshi, K. (1980). US Patent 4,225,381.
Ishizaki, A., Singh, H., Lim, C.-L. & Lim, W. H. (1987). *Agricultural and Biological Chemistry*, **51**, 1155.
Kanagawa, T. & Mikami, E. (1989). *Applied and Environmental Microbiology*, **55**, 555.
Kirchner, K., Hauk, G. & Rehm, H. J. (1987). *Applied Microbiology and Biotechnology*, **26**, 579.

Lee, S. K. & Shoda, M. (1989). *J. Ferment. Bioeng.*, **68**, 437.
Liang, Z. (1991). *Research on practical application of peat biofilter in fields and microbial analysis inhabiting peat.* PhD thesis, Tokyo Institute of Technology, Tokyo.
Liang, Z., Suzuki, M., Terasawa, M., Hirai, M. & Shoda, M. (1990). *J. Odor Research Eng.*, **21**, 1.
Ohta, Y. & Ikeda, M. (1978). *Applied and Environmental Microbiology*, **36**, 487.
Ottengraf, S. P. P. (1986). In *Biotechnology*, Vol. 8, eds H. J. Rehm & G. Reed. VCH Verlagsgesellschaft, Weinheim, FRG, p. 425.
Sublette, K. L. (1987). *Biotechnology and Bioengineering*, **29**, 690.
Sublette, K. L. & Sylvester, N. D. (1987*a*). *Biotechnology and Bioengineering*, **29**, 249.
Sublette, K. L. & Sylvester, N. D. (1987*b*). *Biotechnology and Bioengineering*, **29**, 753.
Sublette, K. L. & Sylvester, N. D. (1987*c*). *Biotechnology and Bioengineering*, **29**, 759.
Suylen, G. M. H. & Kuenen, J. G. (1986). *Antonie van Leeuwenhoek*, **52**, 281.
Tanji, Y., Kanagawa, T. & Mikami, E. (1989). *J. Ferment. Bioeng.*, **67**, 280.
The Japanese Society of Industrial Machinery Manufacturers (1987). *Research report on standardization of maintenance and management methods for deodorization facilities.*
Tiwaree, R. S. (1990). *Biological deodorization of sulfur compound (dimethyl sulfide) using different fabrics as the carriers of microorganisms.* Dissertation for Postgraduate University Course of UNSCO, Tokyo Institute of Technology, Tokyo.
Togashi, I., Suzuki, M., Hirai, M., Shoda, M. & Kubota, H. (1986). *Journal of Fermentation Technology*, **64**, 425.
Wada, A., Shoda, M., Kubota, H., Kobayashi, T., Katayama, Y. & Kuraishi, H. (1986). *Journal of Fermentation Technology*, **64**, 161.
Zeisig, H. D., Kreitmeier, J. & Franzspeck, J. (1977). *Schriftenreihe der Landtechnik Weihenstephan.* Bayerische Landesanstalt für Landtechnik, Freising-Weihenstephan, FRG, p. 5.

Chapter 3

BIOLOGICAL TREATMENT OF LIQUID EFFLUENTS

M. I. D. CHUGHTAI
Division of Biochemistry, Institute of Chemistry, University of the Punjab, Lahore, Pakistan

&

KHURSHID AHMED
Institute of Public Health Engineering and Research, University of Engineering and Technology, Lahore, Pakistan

CONTENTS

1 INTRODUCTION

In the developing countries rapid urbanization and industrialization is making the satisfactory collection, treatment, and disposal of liquid effluents a formidable problem with serious implications for public health.

The total use of water for municipal, agricultural, and industrial purposes is increasing considerably, with a resultant increase not only in the volume of wastewater, but also in the concentration of the pollutants it contains. The increase in discharge of these wastewaters into streams that are already diminishing in flow, because of increased withdrawals, makes it no longer possible to rely as much as in the past on the self purifying capacity of receiving bodies of water. The decreasing available dilution for wastes, the increasing need for reuse of water, and increasing public interest in the maintenance of clean streams compound the problem of liquid effluent treatment and disposal. Thus new and improved processes must be developed so that a higher percentage of the pollutants can be removed.

In addition to the increased production of waste due to rapid growth in urban population in the developing countries, the per capita contribution of wastewater is also increasing. In many cities the latter is as much as 600 litres per day and no limit is in sight.

Such increases in per capita production of wastewater result from the increased availability of piped water supplies and improved standards of living. Water-using labour-saving devices such as washing machines and garbage grinders have a significant effect on household water consumption.

Not only is the volume of wastewater very large but its content of organic and mineral pollutants is also large, amounting to about 50 kg per capita of dry solids annually, not including industrial wastes. Handled expeditiously and with full use of modern technology, these putrescible pollutants need not pose an aesthetic or public health problem in urban centres.

In treating liquid effluents, the concentration of polluting substances and organisms must be reduced. A wide variety of processes for achieving such reduction of pollution are available. The process that is to be selected must be appropriate to the pollution situation; however, conditions in developing countries frequently impose limitations on the choice of such treatments.

2 WASTEWATER TREATMENT OBJECTIVES

Liquid effluent treatment methods were developed in response to the concern for public health and the adverse conditions caused by the discharge of liquid effluents to the environment. The purpose of treatment was to accelerate the forces of nature under controlled conditions in such facilities of comparatively small size.

There is a wide range of biological treatment systems in current use for the purification of liquid effluents based on the apparently simple processes by which mixed populations of micro-organisms break down organic material, using it as a source of nutrients. In short, biological liquid effluent treatment has three vital objectives:

(1) The destruction of the causative agents of those water-related diseases which are associated with domestic wastewater. This is particularly important in areas where the major cause of morbidity and mortality is the improper disposal of human faeces.
(2) To convert the wastewater into a readily reusable resource and so conserve both water and nutrients.
(3) To prevent the pollution of any body of water (groundwater or surface water) to which the effluent escapes after reuse or into which it is discharged without reuse.

3 BACTERIAL PROCESSES AND GROWTH KINETICS

Bacteria may be strictly aerobic, strictly anaerobic or facultative types, the latter existing and growing in either condition. They are generally divided into groups depending on their optimum growth temperature range. Thus psychrophiles (sometimes cryophiles) grow from 0 to 30°C, mesophiles from 15 to 45°C and thermophiles from 45 to 60°C. A simple grouping can be made on the basis of substrate usage. Autotrophs utilize carbon dioxide in solution and can synthesize their requirements from inorganic materials—e.g., the chemosynthetic autotrophic bacterium *Nitrosomonas europaea* converts ammonium ion to nitrite ion. The largest group are heterotrophs, which require an organic carbon source (e.g. sugars, alcohols, etc.). They may be further subdivided into three groups, the following two being most important in biological wastewater treatment.

Non-exacting heterotrophs will grow on a wide variety of substrates. They are often cultured in the laboratory on agar and their flexible growth requirements are useful not only in wastewater treatment, but also in pollution tracing techniques—e.g., *Escherichia coli* will grow on a wide variety of substrates and are therefore often used as an indicator of faecal pollution.

Exacting heterotrophs require specific substrates to grow well—e.g., the amino acid tryptophan is needed for *Salmonella typhosa*.

4 LIQUID EFFLUENT CHARACTERISTICS

Sewage or municipal wastewater refers to liquid discharges from residences, business buildings and institutions. Industrial waste is discharged from manufacturing plants. The volume of wastewater from residential areas varies from 200 to 400 litres per person per day depending on the type of dwellings. Largest flows come from single-family houses that have several bathrooms, automatic washing machines, and other water-using appliances.

Sewage or municipal wastewater is a complex mixture of minerals and organic matter in many forms, including (a) large and small particles of solid matter floating and in suspension, (b) colloidal and pseudo-colloidal dispersion, and (c) true solution. Sewage also contains living matter, especially bacteria, viruses and protozoa; it is an excellent medium for the development of bacteria, some of which may be pathogenic.

The water content of sewage is very high (99·9% or more), which means that the total solid matter is only 0·1% or less. Among the organic substances present in sewage are carbohydrates, lignins, fats, soaps, synthetic detergents and proteins and their decomposition products. Ammonia and ammonium salts are always present, being produced by the decomposition of complex nitrogenous organic matter. The objectionable character of sewage is due mainly to the presence of nitrogenous, sulphur and phosphorus containing organic matter which readily undergoes putrefaction by anaerobic bacteria, with the formation of foul smelling compounds; these include malodorous hydrogen sulphide, organic sulphides and mercaptans and also certain organic amines (especially indole and skatole) which impart a characteristic unpleasant faecal odour.

It is customary to express the concentration of organic matter in terms of the oxygen required by the bacteria for oxidation of the wastes. This is the biochemical oxygen demand (BOD) of the waste, measured in milligrams of oxygen per litre of waste. BOD is defined as the amount of oxygen required by the bacteria for stabilizing decomposable organic matter under aerobic conditions. Liquid effluents are normally classified in terms of BOD_5. Those effluents having BOD_5 (BOD over 5 days) more than 750 mg/litre are classified as very strong, those with BOD_5 of about 500 mg/litre are considered to be strong, those with BOD_5 of 350 mg/litre are considered to be medium, and those with BOD_5 of 200 mg/litre or less are considered to be weak (Khurshid, 1979).

The strength of a wastewater depends on two factors, the quantity of organic matter and the volume of water associated with it. The daily per capita output of organic wastes is in the range 30–50 g as BOD, about half of this being associated with faeces and urine and half with sullage.

Typical data on the individual constituents found in domestic wastewater of Lahore, Pakistan, are given in Table 1. In general, the constituents reported in Table 1 are those that were analysed more or less routinely. Apart from the constituents reported in Table 1, sometimes it becomes necessary to analyse additional constituents if biological treatment is being considered. For example, many of the metals are necessary for the growth of micro-organisms, such as calcium, cobalt,

Table 1
Typical Composition of Untreated Domestic Wastewater

Constituents (mg/litre)	*Concentration*		
	Min.	*Max.*	*Av.*[a]
Solids, total	350	720	535
Solids, suspended	100	220	154
Settleable solids[b]	5	10	6
BOD 20°C 5 days	110	220	160
COD	250	500	344
Total organic carbon (TOC)	80	160	116
Nitrogen, total	20	40	33
Nitrogen, organic	8	15	11
Ammonia, free	12	25	17
Phosphorus, total	4	8	6
Chlorides	30	50	39
Oil and grease	50	100	71

[a] Average of the entire data. [b] ml/l.

copper, iron, magnesium, manganese and zinc. The concentration of sulphate should be determined to assess the suitability of anaerobic waste treatment. The presence of filamentous organisms in the wastewater should also be determined.

4.1 Biological Characteristics

Pathogenic organisms found in wastewater may be discharged by human beings who are infected with disease or who are carriers of a particular disease. The usual bacterial pathogenic organisms that may be excreted by man cause diseases of the gastrointestinal tract, such as typhoid and paratyphoid, dysentery, diarrhoea and cholera.

Although bacterial pathogenic organisms are the most numerous, they are by no means the only pathogens in wastewater. Pathogenic organisms that are usually found in wastewater where the climate is hot and humid include *Ascaris* spp., *Bacillus anthracis*, *Brucella* spp., *Entamoeba histolytica*, *Leptospira* spp., *Mycobacterium*, *Salmonella* spp., *Shigella* spp., *Taenia* spp., *Vibrio cholera* and viruses (Khurshid, 1989*a*).

Because the identification of pathogenic organisms in water and wastewater is both extremely time-consuming and difficult, the coliform group of organisms is now used as an indicator of the presence in wastewater of faeces and hence of pathogenic organisms.

Table 2 shows the bacteriological quality of domestic wastewater. The presence of indicators in significant amounts indicated the wastewater to be fresh and faecally polluted. Wastewater also contains significant populations of pathogens and parasites. Usually the presence of faecal streptococci is an indication of recent pollution, and that of *Escherichia coli* of less recent pollution, while the presence of *Clostridium perfringens* may be an indication of remote pollution.

Table 2
Bacteriological Characteristics of Untreated Liquid Effluents[a]

Micro-organisms[b]	*Minimum*	*Maximum*	*Average*[c]
Coliforms (MPN/100 ml)	$0{\cdot}4 \times 10^9$	$3{\cdot}0 \times 10^9$	$1{\cdot}2 \times 10^9$
Escherichia coli (MPN/100 ml)	$0{\cdot}2 \times 10^7$	$2{\cdot}8 \times 10^7$	$1{\cdot}7 \times 10^7$
Streptococci (MPN/100 ml)	$4{\cdot}4 \times 10^6$	$7{\cdot}4 \times 10^6$	$6{\cdot}2 \times 10^6$
Clostridium perfringens (MPN/100 ml)	$4{\cdot}0 \times 10^4$	$9{\cdot}0 \times 10^4$	$8{\cdot}0 \times 10^4$
Salmonella spp. (MPN/100 ml)	$0{\cdot}1 \times 10^3$	$2{\cdot}0 \times 10^3$	$1{\cdot}0 \times 10^3$
Shigella spp. (MPN/100 ml)	$2{\cdot}0 \times 10^2$	$6{\cdot}0 \times 10^2$	$5{\cdot}0 \times 10^2$
Pseudomonas spp. (MPN/100 ml)	$3{\cdot}0 \times 10^3$	$5{\cdot}6 \times 10^3$	$4{\cdot}0 \times 10^3$
Entamoeba histolytica (no./litre)	$8{\cdot}0 \times 10$	$9{\cdot}6 \times 10$	$9{\cdot}0 \times 10$
Ascaris lumbricoides (no./litre)	$2{\cdot}0 \times 10$	$7{\cdot}0 \times 10$	$6{\cdot}0 \times 10$

[a] The values are based on one year's sampling at Lahore.
[b] MPN = most probable number.
[c] Average of the entire data.

4.2 Bacterial Ratio

It has been observed that the quantities of faecal coliforms (FC) and faecal streptococci (FS) that are discharged by human beings are significantly different from the quantities discharged by animals. Therefore it has been suggested that the ratio of the FC count to the FS count in a sample can be used to show whether the suspected contamination is derived from human or animal wastes. The FC/FS ratio for domestic animals is less than 1·0, whereas the ratio for human beings is more than 4·0 (Chughtai & Khurshid, 1988). Use of the FC/FS ratio can be very helpful in establishing the source of pollution.

5 BIOLOGICAL WASTEWATER TREATMENT SYSTEM

Treatment methods in which the removal of contaminants is brought about by biological activity are known as biological treatment processes. Biological treatment is used primarily to remove the biodegradable organic substances (colloidal or dissolved) in wastewater. Basically, these substances are converted into gases that can escape to the atmosphere and into biological cell tissue that can be removed by settling. Biological treatment is also used to remove the nitrogen in wastewater. With proper environmental control, wastewater can be treated biologically in most cases.

The various types of biological treatment processes commonly employed for domestic and industrial effluents are shown in Table 3.

Treatment of wastewaters by biological filtration and biological aeration are the most commonly applied techniques for the treatment of both domestic and industrial effluents. Aerated lagoons, oxidation ditches and sludge digestion are usually reserved for use in large cities. Waste stabilization ponds (WSPs) are suitable

Table 3
Commonly Employed Biological Treatment Processes for Domestic and Industrial Effluents

Type	*Common names*
Combined process	Waste stabilization ponds (WSPs): —anaerobic WSP —facultative WSP —maturation WSP
Biological filtration	Percolating/trickling filter Biological rotating contractor
Biological aeration	Activated sludge Aerated lagoons Oxidation ditches
Anaerobic suspended growth	Anaerobic digestion

for all community sizes, and are considered most suitable for developing countries (both for rural areas and for small urban communities).

6 WASTE STABILIZATION PONDS (WSPs)

Waste stabilization ponds (WSPs), lagoons or oxidation ponds are shallow rectangular lakes in which raw (or screened) wastewater is treated by natural processes based on the activities of both algae and bacteria. They are without doubt the most important effective method of wastewater treatment in hot climates—not only are they the least expensive, but also they are considerably more efficient in destroying pathogenic bacteria and the ova of intestinal parasites. The importance of this latter advantage can hardly be stressed enough in tropical and semi-tropical developing countries where water-borne diseases, especially those associated with the improper disposal of human wastes, are responsible for a high level of both mortality and morbidity. This, combined with their low cost and their extremely simple maintenance requirements, makes WSPs the ideal form of wastewater treatment in hot climates.

The principal types of stabilization ponds that are in common use include anaerobic, facultative and maturation waste stabilization ponds. Strong wastewaters with high BOD_5 will frequently be introduced into a first-stage anaerobic pond which achieves a high volumetric rate of removal. Weaker wastes or, where anaerobic ponds are environmentally unacceptable, even stronger wastes (say up to 1000 mg/litre BOD) may be discharged directly into primary facultative ponds. Effluent from the first-stage anaerobic pond will flow into secondary facultative ponds which comprise the second stage of biological treatment. Following facultative ponds, if further pathogen reduction is necessary, maturation ponds will provide tertiary treatment.

Depending on the type of waste to be treated and the effluent quality to be achieved, there can be many pond combinations. For example, there can be a single anaerobic pond or a facultative pond, there can be one anaerobic pond followed by one facultative pond, or there can be one anaerobic pond followed by a facultative and a maturation pond. Other arrangements with more than one anaerobic, facultative or maturation pond are possible.

6.1 Anaerobic WSPs

Anaerobic WSPs are very cost-effective for the removal of BOD, when it is present in high concentration. Normally a single anaerobic pond is sufficient, if the strength of the influent is less than 1000 mg/litre BOD_5. For higher strength industrial wastewater, up to three anaerobic ponds in series might be justifiable, but the detention time in any of the three ponds should not be less than 1 day (McGarry & Pescod, 1970).

The stabilization process in an anaerobic pond is illustrated in Fig. 1. Anaerobic degradation of organic material is a sequential process involving two distinct groups of bacteria. The facultative organisms hydrolyse or convert complex organic compounds into simpler organic molecules, primarily organic acids, methane and carbon dioxide.

Although the mechanism of anaerobic decomposition is sequential in nature, both processes take place simultaneously and synchronously in a well-buffered system. Temperature affects the rate of acid fermentation; however, these organisms

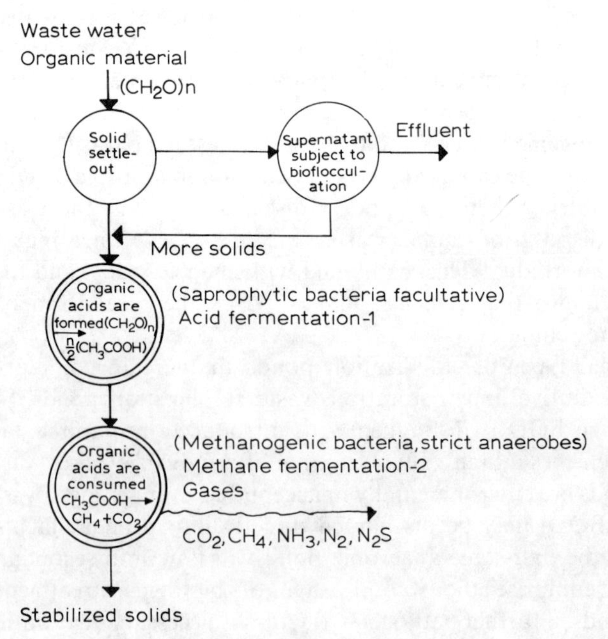

Fig. 1. Stabilization process in an anaerobic waste stabilization pond.

are active over a wide temperature range from approximately 5°C to in excess of 60°C. During acid fermentation almost no net reduction in the BOD, the chemical oxygen demand (COD), or the total organic carbon (TOC) takes place. The conversion of the volatile acids by the methane-forming bacteria results in a marked reduction in the amount of material in the system. The amount of organic material stabilized during methane fermentation is directly proportional to the amount of methane produced (World Health Organization, 1987).

Anaerobic digestion is an extremely temperature-sensitive process. It was established (Khurshid, 1985) that digestion of organic material below 15°C is almost negligible.

During the passage of the liquid effluent through the pond, and after the required detention time (1–5 days), the following changes will take place.

Most of the suspended solids will have settled to the bottom of the pond, and some removal of pathogenic agents will have been achieved. Floating materials will have been carried to the surface, where they will build up in a scum layer. Here they undergo anaerobic decomposition and part-mineralization. The organic material is broken down by anaerobic bacteria. Through the metabolism of these bacteria, part of the organic matter is converted into mineral matter. During this phase, gases are generated, primarily CO_2, CH_4 and H_2S; these gases are dispersed into the atmosphere through the liquid surface (Fig. 1). A portion of the sludge resulting from the settling of solids will be transformed into gas. This reaction, together with sludge thickening, accounts for the very slow build-up of solids in an anaerobic pond. As a result, the accumulated sludge may be estimated at only 40 litres per person per year. The liquid effluent from the anaerobic pond is nearly always transferred to a facultative pond; this effluent will have low levels of suspended and settleable solids and worm eggs. In terms of BOD_5 (BOD over 5 days), the effluent will often have a 40–60% reduction in concentration from that in the raw influent to the primary, anaerobic pond, depending on temperature and detention time. This topic is also dealt with elsewhere in this book.

6.2 Facultative WSPs

The effluent from anaerobic ponds will require some form of aerobic treatment before discharge, and facultative WSPs will often be more appropriate than conventional forms of secondary treatment for application. Solids and excess biomass produced in the pond will settle down, forming a sludge layer at the bottom. This benthic layer will be anaerobic and, as a result of anaerobic breakdown of organics, will release soluble organic products dissolved or suspended which will be metabolized by heterotrophic bacteria with the uptake of oxygen.

However, unlike in the conventional processes, the dissolved oxygen utilized by the bacteria in facultative WSPs is replaced through photosynthetic oxygen production by algae, rather than by aeration equipment. The environment in facultative WSPs is ideal for the proliferation of algae. High temperature and ample sunlight create conditions which encourage algae to utilize CO_2 released by bacteria in breaking down the organic components of the wastewater and take up nutrients

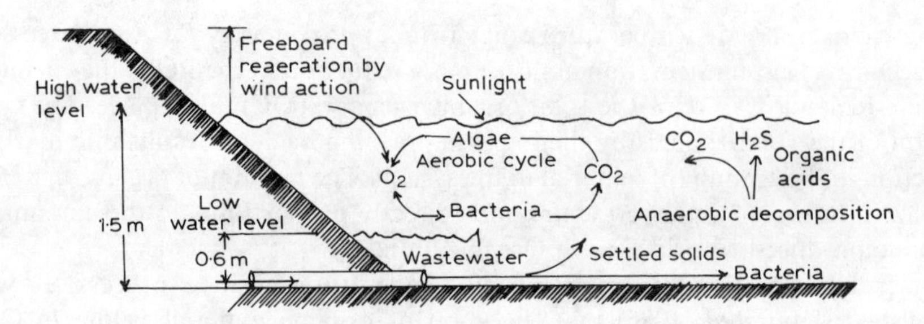

Fig. 2. Schematic diagram of a facultative stabilization pond showing the basic biological reactions of bacteria and algae. Source: Hammer (1977).

contained in the wastewater. This symbiotic relationship contributes to the overall removal of BOD in facultative WSPs shown in Fig. 2.

To maintain the balance necessary to allow this symbiosis to persist, the organic loading in facultative WSPs must be strictly limited. Even under satisfactory operating conditions, the dissolved oxygen concentration (DO) in a facultative pond will vary diurnally as well as over the depth. Maximum DO will occur at the surface of the pond and will usually reach supersaturation at the time of maximum radiation intensity.

From that time until sunrise, DO will decline and may well disappear completely for a short period. For a typical facultative pond depth of 1·5 m the water column will be predominantly aerobic at the time of peak radiation and predominantly anaerobic at sunrise. The pH of the pond content will also vary diurnally as algae utilize CO_2 throughout daylight hours and respire, along with bacteria and other organisms, releasing CO_2 during the night. During the passage of liquid effluent through a facultative pond, the following changes may be observed.

Most of the remaining suspended solids settle to the bottom of the pond, where they develop a layer that works like an anaerobic sludge digester (Section 8). This is the anaerobic zone of a facultative pond. Above the anaerobic sludge layer, an intermediate zone exists in which dissolved oxygen is present some of the time, fed from the upper layer. The upper layer is a natural culture medium for algae and operates as an aeration system, producing oxygen for the aerobic and intermediate zones.

During the stabilization process, much of the biodegradable organic matter is transformed, mainly into living organic matter in the form of algae, bacteria, protozoa, etc.

The effluent of a facultative WSP taken from the surface layer is strongly green-coloured because of the presence of algae. However, there are practically no suspended solids that will settle.

A facultative WSP used as a secondary pond in climates where the average coldest month air and water temperatures do not go below *c.* 10°C and 15°C, respectively, should have a minimum detention time of 5 days for typical domestic wastewater. The upper limit is determined by design parameters and the area of land available, and can extend to 10 days. A facultative pond used as a secondary pond (i.e.

following an anaerobic pond) in climates where the average air and water temperature do not go below 10–15°C, should have a minimum detention time of 5 days for domestic sewage. When facultative ponds are used as primary ponds, the detention time should be 10 days.

This topic is also dealt with elsewhere in this book.

6.3 Energy Requirements

WSPs exert minimum demand on energy resources. According to Gloyna (1976), 1 386 000 kcal are required to produce 907 kg of lime. In comparison, 2520 kcal are required to produce and distribute 1 kWh of electrical energy. For a wastewater treatment plant treating readily biodegradable wastes, equivalent to about 250 mg/litre BOD and flows of 1 000 000 US gal/day (3785 m^3/day), the electrical consumption is about 1000 kWh. For a 10 000 000 gal/day (0·44 m^3/s) plant the energy usage is about 9000 kWh.

6.4 Pond Ecology

A facultative pond typically contains abiotic substances, producer organisms, consumer organisms, and decomposable organisms. The plants, mainly algae, are producer organisms. These microscopic plants are capable of synthesizing organic substrate from complex wastes and inorganic materials. In facultative WSPs, algae are valuable in that they have the ability to produce oxygen through the mechanism of photosynthesis. At night, when light is no longer available for photosynthesis, they use up the oxygen in respiration. The animals, predominantly protozoa, ingest and digest synthesized organic substrate. In fact, the protozoa act as polishers of the final effluent. The decomposer organisms, mainly fungi and bacteria, reduce organic material to basic inorganic constituents by extracellular digestion, through intracellular metabolic pathways. Rotifers are very effective in consuming dispersed and flocculated bacteria and small particles of organic matter. Their presence is indicative of an effluent that is low in organic matter and high in dissolved oxygen.

6.5 Algal Action

The algal forms may be blue-green algae, non-motile green algae, pigmented flagellates and diatoms.

Typical of the green algal species in facultative WSPs are *Chlamydomonas*, *Chlorella* and *Scenedesmus*. Blue-green algal species include *Oscillatoria*, *Pharmidium*, *Anacystis* and *Anabaena* (Tariq & Khurshid, 1980). In the operation of a pond, *Chlamydomonas* and *Chlorella* are usually the first planktonic genera to appear. Algal mats frequently develop in ponds during summer. Problems develop when detached patches of benthic algae begin to accumulate. *Euglena* show a high degree of adaptability to various pond conditions and are present during all seasons and under all climatological conditions. Types of algae along with their numbers, as

Table 4
Algal Species and Numbers Identified from Facultative WSPs at Lahore During the Five Seasons

Season	*Algal species*	*Algal cells (no./ml)*	
		Minimum	*Maximum*
Autumn (mid Sept. to mid Nov.)	*Chlorella*	$5{\cdot}0 \times 10^4$	$12{\cdot}0 \times 10^5$
	Cladophora	$2{\cdot}0 \times 10^4$	$6{\cdot}0 \times 10^5$
	Scenedesmus	$2{\cdot}0 \times 10^4$	$7{\cdot}0 \times 10^5$
	Oscillatoria	$4{\cdot}0 \times 10^5$	$6{\cdot}0 \times 10^5$
	Chlamydomonas	$1{\cdot}0 \times 10^5$	$2{\cdot}0 \times 10^5$
	Euglena	$1{\cdot}0 \times 10^4$	$3{\cdot}0 \times 10^4$
Winter (mid Nov. to mid Feb.)	*Chlorella*	$3{\cdot}0 \times 10^4$	$6{\cdot}0 \times 10^5$
	Euglena	$2{\cdot}0 \times 10^4$	$7{\cdot}0 \times 10^5$
	Scenedesmus	$1{\cdot}0 \times 10^4$	$3{\cdot}0 \times 10^4$
Spring (mid Feb. to mid April)	*Chlorella*	$7{\cdot}0 \times 10^4$	$14{\cdot}0 \times 10^5$
	Chlorococcum	$3{\cdot}0 \times 10^4$	$9{\cdot}0 \times 10^5$
	Cladophora	$3{\cdot}0 \times 10^4$	$7{\cdot}0 \times 10^5$
	Microactium	$2{\cdot}0 \times 10^4$	$6{\cdot}0 \times 10^5$
	Tetrahedron	$2{\cdot}0 \times 10^4$	$6{\cdot}0 \times 10^5$
	Euglena	$3{\cdot}0 \times 10^3$	$5{\cdot}0 \times 10^4$
	Phacus	$1{\cdot}0 \times 10^3$	$3{\cdot}0 \times 10^4$
	Synura	$0{\cdot}0 \times 10^4$	$3{\cdot}0 \times 10^4$
Summer (mid April to mid July)	*Oscillatoria*	$14{\cdot}0 \times 10^5$	$35{\cdot}0 \times 10^6$
	Chlorella	$9{\cdot}0 \times 10^3$	$14{\cdot}0 \times 10^4$
	Euglena	$3{\cdot}0 \times 10^2$	$8{\cdot}0 \times 10^3$
Monsoon (rainy) (mid July to mid Sept.)	*Chlorella*	4.0×10^4	$6{\cdot}0 \times 10^5$
	Oscillatoria	$4{\cdot}0 \times 10^5$	$9{\cdot}0 \times 10^5$
	Euglena	$2{\cdot}0 \times 10^2$	$4{\cdot}0 \times 10^3$

identified in different seasons from facultative WSPs at Lahore, are shown in Table 4. Frequently *Euglena* and *Chlamydomonas* tend to dominate during the cooler weather, while the various chlorococcales are most numerous during summer months. The maximum numbers of algal species were observed during spring and the minimum numbers during the monsoon (rainy season) because of cloud cover.

Animals in facultative WSP systems consist of protozoa, rotifers, worms and larvae. The latter two are found in the bottom muds. The distribution of the animal population may be seasonal and highly subject to a variety of environmental influences (Khurshid & Chughtai, 1986).

6.6 Light Intensity

Algal photosynthesis is directly related to geographical location of the ponds for application of solar radiation intensity and depends to some extent on predation dynamics. The presence of one or more algal species and the depth and density of the algal population present a high degree of uniformity in the efficiency of light energy conversion. This conversion efficiency ranges from 1 to 5% (Chughtai & Khurshid, 1980). The optimum light intensity or saturation values for most algal species found in WSPs range from 400 to 600 foot-candles.

Light intensity inhibition is usually of the order of 2000–4000 foot-candles, again depending on the species. In a typical pond during a bright day, inhibition usually occurs in the first few centimetres, followed by minimum efficiency proceeding to maximum efficiency as the depth increases. The minimum light compensation point for many species is approximately 20 foot-candles.

6.7 Temperature

The overall performance of a facultative WSP system is highly temperature dependent. Sludge deposits will be degraded by anaerobic bacteria; throughout most of the pond depth the soluble BOD will be biodegraded by facultative bacteria. Thermal stratification of the pond liquid is partially responsible for maintaining separate aerobic and anaerobic zones for extended periods of time.

The useful temperature range is from 5°C to about 35°C, the lower limit resulting from retardation of aerobic bacteria and algal activity. Anaerobic bacteria are not very active below 15°C. An upper limit of 35°C is imposed by inactivation of many green algal species.

6.8 Maturation WSPs

Maturation ponds are used mainly for reduction of pathogenic organisms. Beside removing a very high percentage of faecal bacteria, viruses, protozoa and other pathogens, maturation ponds may also remove some algae and nutrients.

The bactericidal effect of maturation ponds is due to several natural factors, including sedimentation, lack of food and nutrients, solar ultraviolet radiation, high temperatures, high pH predators, and the toxins and antibiotics excreted by some organisms (Chughtai & Khurshid, 1982), as well as natural die-off.

It has been established that a series of ponds with an overall detention of 20 days or more will produce an effluent totally free of cysts and ova (Feacham *et al.*, 1983).

Maturation ponds will be aerobic throughout the water column during daylight hours and the pH will rise above 9·0. The algal population of many species of non-flagellate unicellular and colonial forms will be distributed over the full depth of a maturation pond. Large numbers of filamentous algae, particularly blue-greens, will emerge under very low BOD loading conditions. Maturation WSPs should only be used to upgrade the effluent from a facultative WSP, from another type of wastewater treatment plant. They should not receive raw wastewaters or anaerobic WSP effluents. Maturation WSPs are normally designed for a detention time from 3 to 10 days per pond when two or more are in series, while typical depth ranges between 1·0 and 1·5 metres.

6.9 The Pakistani Experience

WSPs are becoming popular in Pakistan because of their low maintenance and operation costs and maximization of the use of local resources. WSPs are particularly suited to the semi-arid climate of Pakistan as they provide an ideal environment for natural treatment of wastewater to the extent of safe reuse of treated effluent for irrigation, etc. WSPs are either in planning stages or under construction in various cities of Pakistan.

To investigate the feasibility of WSP application in Pakistan, pilot-scale facilities were constructed in Lahore near Bund Road about 16 km from the University of Engineering and Technology. This is also the site of one of the major sewage pumping stations built by the Water and Sanitation Agency (WASA). The pumps are meant to convey wastewater from a part of the city across the protective embankment into the River Ravi.

The facility consisted of one anaerobic pond, followed by two facultative ponds operating in parallel, followed by two maturation ponds operating in series. The anaerobic pond was 15·24 m in diameter, having a total depth of 3·65 m and an effective depth of 3·048 m. Both the facultative WSPs were of identical dimensions, measuring 45·72 m by 13·71 m with a total depth of 1·82 m. The dimensions of the maturation WSPs were the same as those of the facultative WSPs. The influent was supplied to the anaerobic pond by a pipe 0·16 m in diameter from a pump of 1 cusec capacity (0·028 32 m^3/s). These facilities are illustrated by a line diagram in Fig. 3.

The influent consisted of 80% domestic sewage, 15% sullage and 5% industrial wastewater. The characteristics of the influent are given along with those of the treated effluent for each type of pond.

In all it took 50 days for the anaerobic pond to develop an environment conducive to methane fermentation. During the acid-phase period the pond emitted a slight odour while undergoing adaptation to methane fermentation. Hence no necessity was felt to apply odour control techniques, as the facility was far away from residential areas.

The performance of the anaerobic WSP at detention times of 1·0, 2·5 and 5·0 days is shown in Table 5. It is observed from the table that with increase in detention time

Table 5
Performance of Anaerobic WSP at Lahore

Detention time (days)	*Volumetric loading (kg $BOD_5/m^3/day$)*	*BOD_5 reduction*[a] *(% av.)*	*COD reduction*[b] *(% av.)*	*SS reduction*[c] *(% av.)*
1·0	0·276	55·80	58·48	55·55
2·5	0·110	68·11	72·92	72·72
5·0	0·055	66·30	68·00	80·55

[a] BOD_5 = biochemical oxygen demand (five days).
[b] COD = chemical oxygen demand.
[c] SS = suspended solids.

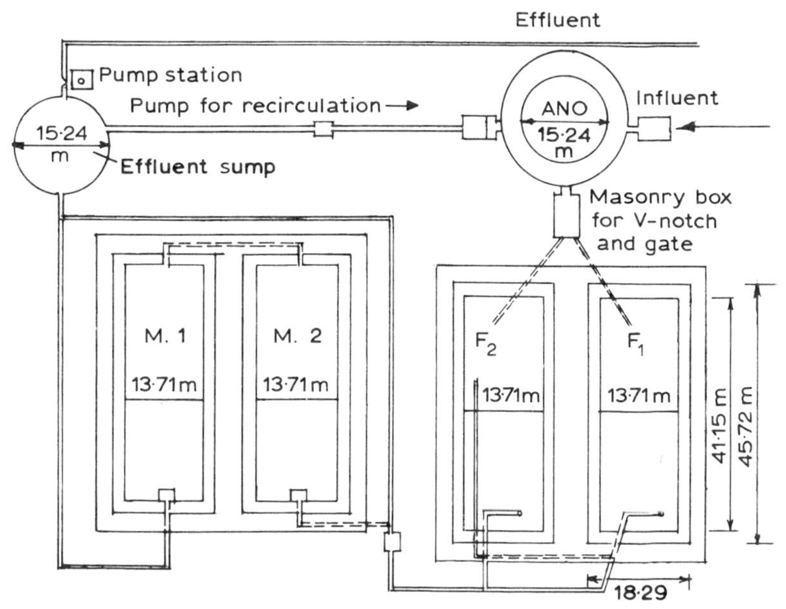

Fig. 3. Layout of waste stabilization pond system consisting of one anaerobic (ANO) and two facultative ($F_1 . F_2$) ponds operating in parallel and two maturation ($M_1 . M_2$) ponds operating in series. They are situated at Bund Road, Lahore. The anaerobic pond has a total depth of 3·65 m, and an effective depth of 3·048 m. Both the facultative ponds have identical dimensions of 45·72 m by 13·71 m with a side slope of 1:1·5 and a total depth of 1·82 m. The dimensions of the maturation ponds are the same as those of the facultative ponds.

there is a substantial increase in the reduction of BOD_5, COD and suspended solids. Bacterial reduction in the anaerobic pond is always less than that in the other types of ponds.

Heavy metals, in particular copper, zinc and nickel, are known to retard the anaerobic process due to the metal toxicity. To be toxic, these metals must be in solution. These metals easily form insoluble salts by reacting with hydroxide, carbonate, sulphide or phosphate. Because these ions are common in an anaerobic system, therefore, metal toxicity was not a problem. The most important anion in the control of metal toxicity is sulphide. Metal sulphide salts have very low solubilities.

In the influent of Lahore, the concentration of heavy metals like chromium, copper and cadmium is quite low (< 1 mg/litre). Although sulphide is present at a concentration of 2·2 mg/litre, because of the low concentration of these metals the possibility of metal toxicity does not exist in the pond.

The effluent from the anaerobic pond was taken into two facultative WSPs. These ponds were operated at loadings of 220, 330 and 440 kg/ha/day. Both the WSPs took 16 days for their surface to show a green colour. The performance of these two facultative WSPs at a loading of 330 kg/ha/day is shown in Table 6. In both the facultative WSPs a BOD_5 removal of more than 70% was achieved. The increase in

Table 6
Performance of Facultative WSPs at Lahore[a]

Parameter	Influent[b]	Pond No. 1		Pond No. 2	
		Value	*% Red.*	*Value*	*% Red.*
pH	8·2	8·6	—	8·5	—
Total solids (mg/litre)	266	138	48·12	120	51·50
Suspended solids (mg/litre)[c]	70	105	—	118	—
BOD_5 (mg/litre)	100	45	71·85	42	73·75
COD (mg/litre)	261	139	46·74	144	44·84
NO_3 (mg/litre)	1·2	Nil	—	Nil	—
Coliforms (MPN/100 ml)[d]	$3·0 \times 10^7$	$4·6 \times 10^5$	98·46	$4·4 \times 10^5$	98·53
Escherichia coli (MPN/100 ml)[d]	$5·6 \times 10^5$	$3·5 \times 10^3$	99·37	$3·9 \times 10^3$	99·30

[a] During the investigation, the ponds were operating in parallel at a loading of 330 kg/ha/day and 7 days' detention time.
[b] Influent represents the effluent from the anaerobic WSP.
[c] Increases in the suspended solids contents were due to increase in the algal population.
[d] MPN = most probable number.

the suspended solids was due to algae. The coliform and *Escherichia coli* removal rates were 98·5% and 99·30%, respectively.

To further improve the quality of effluent in terms of bacterial removal, effluents from both the facultative WSPs were taken into two maturation ponds. Reduction rates of indicators, pathogens and parasites after various detention times are shown in Table 7. After a detention time of 10 days the parasites are totally removed, and the indicators and pathogens are reduced by more than 99·9%.

Table 7
Performance of Maturation WSPs in Removing Micro-organisms at Lahore

Micro-organisms	*Influent*	*Effluent after detention time of*			
		4·0 days	*5·0 days*	*6·0 days*	*10·0 days*
INDICATORS (MPN/100 ml)[a]					
Coliforms	$1·2 \times 10^9$	$3·8 \times 10^6$	$2·0 \times 10^6$	$7·2 \times 10^5$	$1·0 \times 10^5$
Escherichia coli	$1·7 \times 10^7$	$3·3 \times 10^2$	$1·5 \times 10^2$	$4·4 \times 10$	$1·4 \times 10$
Faecal streptococci	$6·5 \times 10^6$	$1·8 \times 10^4$	$6·2 \times 10^3$	$9·6 \times 10^2$	$2·2 \times 10$
Clostridium perfringens	$8·2 \times 10^4$	$5·0 \times 10^2$	$2·9 \times 10^2$	$1·0 \times 10^2$	$0·8 \times 10$
PATHOGENS (MPN/100 ml)[a]					
Salmonella spp.	$1·1 \times 10^3$	$9·0 \times 10$	$3·2 \times 10$	$9·0 \times 10$	$0·7 \times 10$
Shigella spp.	$5·0 \times 10^2$	$5·1 \times 10$	$2·6 \times 10$	$5·1 \times 10$	$0·4 \times 10$
Pseudomonas spp.	$4·0 \times 10^3$	$4·4 \times 10$	$4·0 \times 10$	$4·4 \times 10$	$0·6 \times 10$
PARASITES (number/litre)					
Entamoeba histolytica	90	1	1	0	0
Ascaria lumbricoides	60	1	1	0	0
Hookworms	41	0	0	0	0

[a] MPN = most probable number.

6.10 Bacterial Reduction Model

Reduction of indicator organisms in maturation ponds is generally taken as following first-order kinetics:

$$N_e = \frac{N_i}{1 + K_b t} \tag{1}$$

where N_e = number of FCs per 100 ml of effluent
N_i = number of FCs per 100 ml of influent
K_b = first-order rate constant for FC removal (day^{-1})
t = detention time in any pond (days)

For n ponds in series, eqn (1) becomes

$$N_e = N_i/(1 + K_b t_f)(1 + K_b t_{m_1})\ldots(1 + K_b t_{m_n}) \tag{2}$$

where f indicates facultative WSP
m indicates maturation WSP
t_{m_n} = detention time in the nth maturation pond

The value of K_b is extremely sensitive to temperature and is given by the following equation:

$$K_b(T) = 2{\cdot}6 \times (1{\cdot}19)^{T-20} \tag{3}$$

where $K_b(T)$ = value of K_b at T°C.

By applying the above equation, K values were calculated for the maturation ponds at Lahore (Table 8). It is observed from the table that different rates of removal occur for different organisms. However, it must be kept in mind that the constant $K = 2$ day^{-1} for *E. coli* does not apply to other pathogens. Coliform faecal bacteria are not necessarily pathogenic, and they are considered only as an indicator of the risk of infection. It is therefore incorrect to generalize when considering a K value, and its application to other organisms should be done with caution and understanding of the phenomena involved; however, the equations given are useful for estimating the require volume of maturation ponds.

Table 8
Death Rate Constant (K) for Indicators and Pathogens in Maturation WSPs at Lahore

Micro-organisms	*Temperature (°C)*	*K value (day^{-1})*
Coliforms	17·0–23·5	1·82
Escherichia coli	11·5–23·5	2·00
Faecal streptococci	11·5–23·5	2·20
Clostridium perfringens	11·5–23·5	1·40
Salmonella spp.	15·0–24·0	0·85
Shigella spp.	15·4–24·0	0·72
Pseudomonas spp.	15·0–24·4	0·78

7 CONVENTIONAL BIOLOGICAL TREATMENT

Conventional biological treatment is a combination of physical and biological processes designed to remove organic matter from liquid effluents. The easiest method was plain sedimentation in septic tanks. The final step in the development of primary treatment was complete separation of sedimentation and sludge processing units. Currently, raw sludge is handled independently by biological digestion.

The first major breakthrough in secondary treatment occurred when it was observed that the slow movement of wastewater through a gravel bed resulted in rapid reduction of organic matter. This process, referred to as trickling filtration, was developed for municipal installations starting around 1910.

A second major advance in biological treatment took place when it was observed that biological solids, developed in polluted water, flocculated organic colloids. These masses of micro-organisms, referred to as activated sludge, rapidly metabolized pollutants from solution and could be subsequently removed by gravity settling. In the 1920s the first continuous flow treatment plants were constructed using activated sludge to remove BOD from wastewater. The two conventional biological wastewater treatment processes widely in use depend on biological filtration and biological aeration.

7.1 Biological Filtration

Fixed-growth biological systems are those that contact wastewater with microbial growths attached to the surfaces of supporting media where the wastewater is sprayed over a bed of crushed rock. Such a unit is commonly referred to as a trickling filter. With the development of synthetic media to replace the use of stone, the term 'biological tower' was introduced. These installations are often about 6 m in depth rather than the traditional 1·8 m rock-filled filter. Another type of fixed-growth system is the biological disk (see Section 7.3), where a series of circular plates on a common shaft are slowly rotated while partly submerged in a trough of wastewater. Microbes attached to the disks extract waste organics. Although the physical structures differ, the biological process is essentially the same in all of these fixed-growth systems.

7.2 Trickling Filter or Percolating Filter

Trickling filters (also called percolating filters) were first introduced on a large scale in 1893 at the Salford Sewage Works, Lancashire, UK, by Joseph Corbett. They have since been used all over the world and have taken their place as a popular, rugged and dependable means of sewage purification, especially for strong and difficult industrial effluents.

The trickling filter is an example of a closed heterogeneous continuous culture system. It consists of a cylindrical column with a diameter of up to 30–40 m and a height of 1·5–4·0 m. The column is filled with pieces of rock or plastic with diameters of 5–9 cm, providing a surface of 100–200 m^2/m^3. The settled sewage is evenly

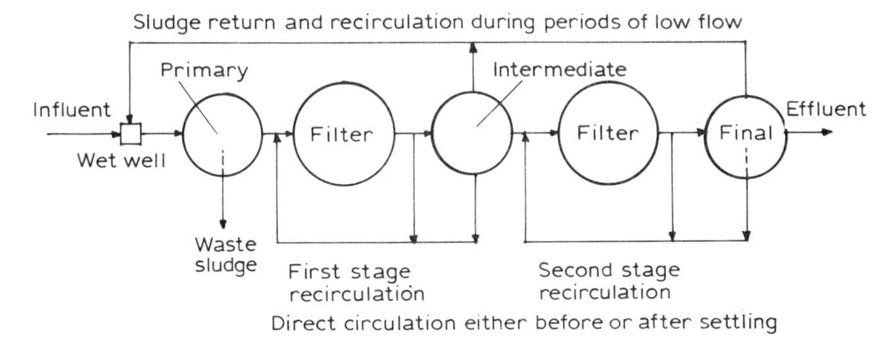

Fig. 4. Typical flow diagram for a two-stage trickling filter plant source (Hammer, 1977).

distributed over the top surface by a sprinkling device and flows downwards over the rock or plastic surface on which a microbial film maintains itself and mineralizes the dissolved organic matter as it passes by. Oxygen supply is ensured by an upward or downward draught of air through the column, which has ventilation holes in the bottom. As a rule no forced ventilation is required. This type of film reactor is heterogeneous, because the microbial composition of the film differs with depth as a result of change of medium composition with progressive mineralization. It is 'closed' because the film cannot keep growing indefinitely without bringing the process to a standstill. In practice, however, the balance of shearing forces and cohesion of the film is such that filters can be kept in operation for indefinite periods of time as sloughing off pieces of film ('humus') keeps pace with its growth. This three-phase system (air, water, microbial film) is then followed by a secondary sedimentation tank. A flow diagram of a trickling filter is shown in Fig. 4.

Different filters can be used in parallel or in series, and recirculation is frequently applied as the hydraulic load is instrumental in keeping the biological film thin enough to prevent clogging. Recirculation is also an instrument to dilute peak loads of toxic substances.

Facultative bacteria are the predominant micro-organisms in the trickling filter. Along with the aerobic and anaerobic bacteria, their role is to decompose the organic material in the wastewater. *Acromobacter*, *Flavobacterium*, *Pseudomonas* and *Alcoligenes* are among the bacterial species commonly associated with trickling filters (Chughtai & Khurshid, 1985). Within the slime layer, where adverse conditions prevail with respect to growth, the filamentous forms *Sphaerotilus natans* and *Beggiatoa* will be found; in the lower reaches of the filter, the nitrifying bacteria *Nitrosomonas* and *Nitrobacter* will be present. Algae can grow only in the upper reaches of the filter where sunlight is available. The protozoa in the filter are predominantly of the Ciliata group, including *Vorticella*, *Opercularia* and *Epistylis*.

In the trickling filter oxygen and organic matter diffuse into the film where oxidation and synthesis occur. End products (CO_2, NO_3, etc.) counter-diffuse back into the flowing liquid and appear in the filter effluent as shown in Fig. 5. Process parameters for trickling filters for domestic sewage treatment are shown in Table 9.

Several mathematical equations have been developed for calculating BOD

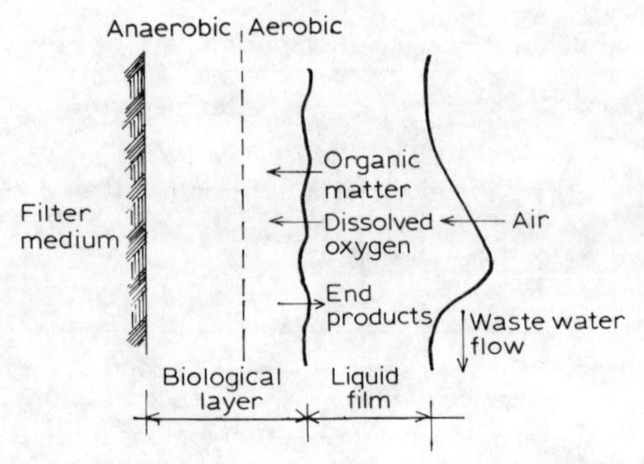

Fig. 5. Sketch illustrating the biological process in a filter bed.

removal efficiency of biological filters based on factors such as depth of bed, type of medium, temperature, recirculation and organic loading. The exactness of these formulae depends on the filter medium having a uniform biological layer and an evenly distributed hydraulic load. Such conditions rarely occur in rock filters which may develop unequal biological growths resulting in short circulating of wastewater through the bed. Therefore, general practice has been to use empirical relationships based on operational data collected from existing treatment plants. The overall treatment plant efficiency of a two-stage filter system can be calculated by the following equation (Schroeder, 1977):

$$E = 100 - 100\left[\left(1 - \frac{35}{100}\right)\left(1 - \frac{E_1}{100}\right)\left(1 - \frac{E_2}{100}\right)\right] \tag{4}$$

where E = treatment plant efficiency (%)
35 = percentage of BOD removed in primary settling
E_1 = BOD efficiency of first-stage filter and intermediate clarifier corrected for temperature (%)
E_2 = BOD efficiency of second-stage filter and final clarifier corrected for temperature (%)

This topic is also dealt with elsewhere in this book.

Table 9
Process Parameters for Trickling Filter for Domestic Sewage Treatment

Parameter	*Value for low rate performance*	*Value for high rate performance*
Hydraulic load ($m^3/m^2/h$)	0·05–0·03	0·6–2·0
BOD_5 load ($kg/m^3/day$)	0·1–0·4	0·6–1·8
Biological film material (kg/m^3)	5·0–7·0	3·0–6·0
BOD load/kg film material per day	0·01–0·08	0·1–0·6
Contact time	1·00–3·00 h	10–30 min
BOD removal efficiency (%)	80–90	65–85

Source: La Riviere (1981).

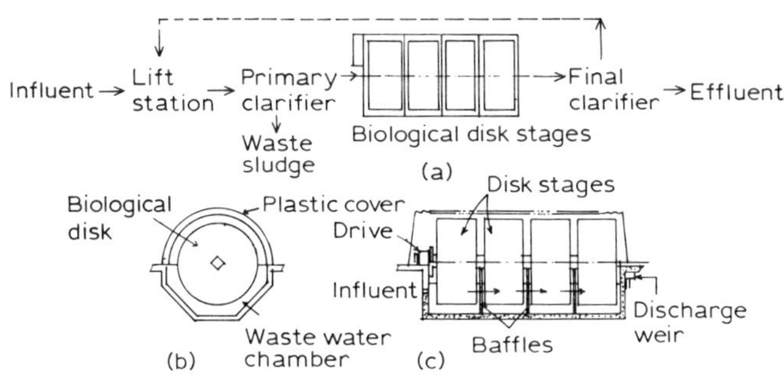

Fig. 6. Wastewater flow pattern and details of a biological disk treatment system: (a) plant flow diagram: (b) cross-section; (c) longitudinal section. Source: Hammer (1977).

7.3 Biological Disks

Biological disks, also known as rotating biological contractors (RBCs), consist of a series of closely spaced circular disks of polystyrene or polyvinyl chloride. A series of these mounted on a horizontal shaft are placed in a contour-bottomed tank and immersed to approximately 40% of their diameter. The disks are spaced so that during submergence wastewater can enter the separation between the corrugated surfaces. When rotated out of the tank, the liquid trickles out of the voids between the plates and is replaced by air. A fixed-film biological growth, similar to that on a trickling filter medium, adheres to the rotating surfaces. Alternating exposure to organics in the wastewater and oxygen in the air during rotation is like the dosing of a trickling filter with a rotating distributor. Excess biomass sloughed from the disks is carried out in the process effluent for gravity separation.

A flow diagram of a biological disk treatment system is illustrated in Fig. 6. After primary settling the wastewater is applied to disk chambers separated by baffles, each containing several stages of media. Organics are extracted by the biological growth as wastewater slowly passes through the disk stages. Solids sloughed from the media are collected in a final clarifier and are recycled to the head of the plant for removal in the primary clarifer.

Three of the design variables in a biological disk system are detention time of the wastewater in the chambers, rotational velocity of the media, and arrangement of the disk stages. Advantages of bio-disk treatment, relative to other biological systems, include ease of operation, high degree of BOD removal achievable, and good settleability of solids flushed from the disk surfaces.

In recent years RBCs have been used for the treatment of various types of industrial effluents (Saw, 1989), though most of the published data and design guidelines available to date are for the treatment of domestic sewage. The RBC system is suitable for small to medium sized factories because, in addition to its operational simplicity and flexibility, it is also a low-energy consuming system.

7.4 Biological Aeration

Understanding of aeration processes in wastewater treatment requires a greater knowledge of biological processes than that required for biological filtration. Raw wastewater flowing into the aeration basin contains organic matter as a food supply. Bacteria metabolize the waste solids, producing new growth while taking in dissolved oxygen and releasing CO_2. Protozoa graze on bacteria for energy to reproduce. Some of the new microbial growth dies, releasing cell contents to the liquid for resynthesis. After the addition of a large population of microorganisms, aerating raw wastewater for a few hours removes organic matter from liquid by synthesis into microbial cells.

The liquid suspension of micro-organisms in an aeration basin is generally referred to as a mixed liquor, and the biological growths are called mixed liquor suspended solids (MLSS).

Biological aeration processes include:

—Activated sludge process
—Aerated lagoons or ponds
—Oxidation ditches

7.5 Activated Sludge Process

In the 1920s, the first continuous flow treatment plants were constructed using activated sludge to remove BOD from wastewater. The name 'activated sludge' originated with reference to the return biological suspension, since these masses of micro-organisms were observed to be very active in removing soluble organic matter from liquid.

A flowsheet of an activated sludge plant of the conventional type is shown in Fig. 7.

Primary treatment involves screening, grit removal and settling, sludge digestion, and drying. Only the settleable and floating solids are removed and about 30–35% reduction in BOD can be expected. If further treatment is required, a biological process is generally adopted in the secondary stage. Changes in the manner of distributing the air supply or raw waste or return sludge over the length of the aeration tank have led to the development of modifications of the original activated sludge process, which are referred to as:

—Tapered aeration
—Step aeration
—Contact stabilization
—Complete mixing activated sludge

Flow diagrams of tapered and step aeration are illustrated in Fig. 8.

It has been frequently observed that removal of BOD from a wastewater by contact with activated sludge occurs in a few minutes, the remainder of the aeration period being required to restore the activity of the sludge. In contact stabilization, advantage is taken of this phenomenon by adding the sewage in effect near the end of the aeration compartment of a plug flow plant so that the remainder of the plant is available for re-aeration of the returned activated sludge. In this way the aeration capacity of the plant is effectively increased.

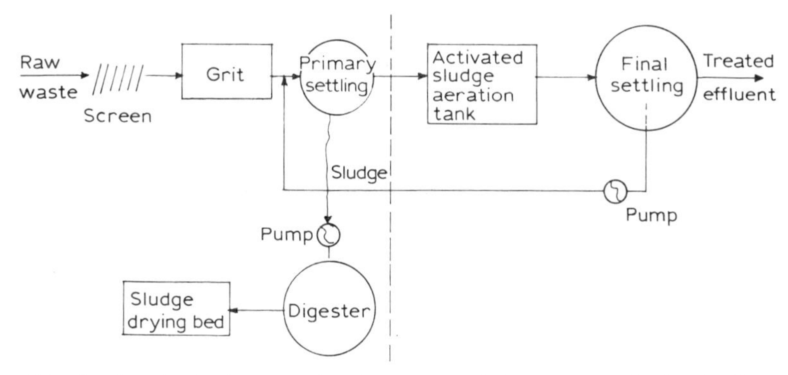

Fig. 7. Flowsheet of conventional activated sludge process (Amin, 1989).

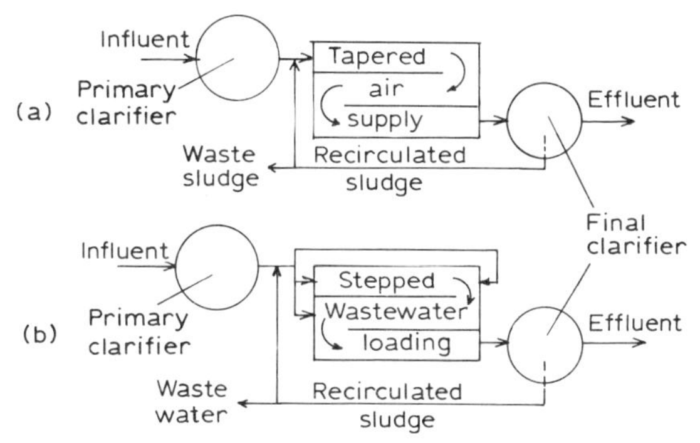

Fig. 8. Flow schemes for conventional and step-aeration activated sludge processes: (a) conventional tapered aeration; (b) step aeration. Source: Hammer (1977).

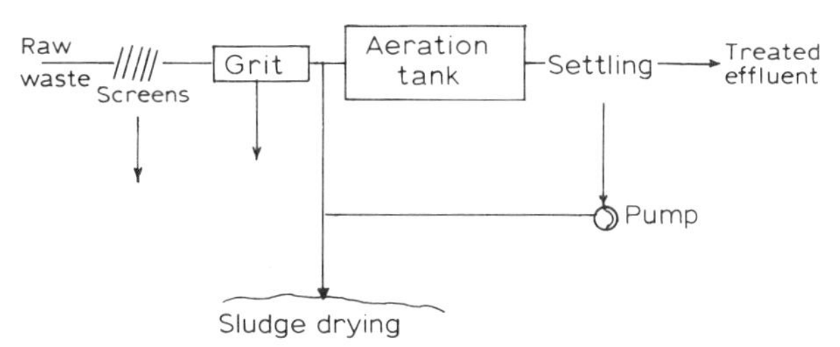

Fig. 9. Flowsheet of extended aeration process as used in Pasveer-type oxidation ditches (Amin, 1989).

Another modification which has gained popularity over the last few years is the extended aeration process of which a common application is to be found in the Pasveer-type oxidation ditches widely used in Europe and other developed countries. This process is simpler to construct and operate than the conventional activated sludge process owing to omission of primary settling and separate sludge digestion (Fig. 9). But this omission is compensated for by a longer aeration time which has earned for the process the name 'extended aeration', with special features that lend the process its name residing in the settleability of the microbial cell mass.

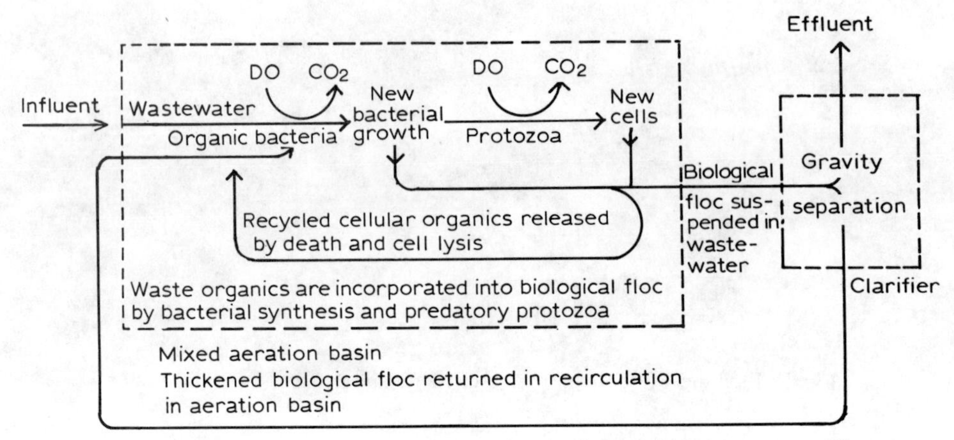

Fig. 10. Generalized biological process in aeration (activated sludge) treatment (Hammer, 1977).

Organic waste is introduced into a reactor where an aerobic bacterial culture is maintained in suspension. The reactor contents are referred to as the mixed liquor. In the reactor, the bacterial culture carries out the conversion in general accordance with the stoichiometry. The aerobic environment in the reactor is achieved by the use of diffused or mechanical aeration, which also serves to maintain the mixed liquor in a completely mixed regime. After a specified period of time, the mixture of new and old cells is passed into a settling tank where the cells are separated from the treated wastewater. A portion of the settled cells is recycled to maintain the desired concentration of organisms in the reactor, and a portion is wasted. The portion wasted corresponds to the new growth of cell tissue associated with the particular wastewater. The generalized biological process in aeration treatment is given in Fig. 10.

The level at which the biological mass in the reactor should be kept depends on the desired treatment efficiency and other considerations related to growth kinetics. The treatment efficiency for the activated sludge plant at Islamabad, Pakistan (Farooqi, 1987), is given in Table 10. It is a conventional-type plant with a design capacity of 5 million gal/day. The air is supplied by diffused air mechanical aerators. The treatment efficiency in terms of BOD is 80–90%.

This topic is also dealt with elsewhere in this book.

7.6 Aerated WSPs

In temperate climates it was found that during winter the algae in WSPs did not produce enough oxygen to satisfy their bacterial demand. Public health engineers therefore provided mechanical oxygenation to supplement the algal oxygen supply. They found, however, that within a few days of the aerators being switched on, the algae in the pond disappeared and the bacterial flora became more flocculent and soon resembled activated sludge. It was thus that aerated WSPs, aerated lagoons or mechanically aerated WSPs became recognized as a distinct biological treatment

Table 10
Performance of Activated Sludge Treatment Plant at Islamabad

Parameter	*Influent*	*Effluent*
Colour	Grey	Light grey
pH	7·2	7·1
Total solids (mg/litre)	600–800	300–450
Total suspended solids (mg/litre)	200–500	20–200
Settleable solids (mg/litre)	5·0–8·0	Nil
Chlorides (mg/litre)	20–100	10–50
Oil and grease (mg/litre)	300–500	200–300
BOD_5 (mg/litre)	200–300	80–90
COD (mg/litre)	300–500	90–100

Source: Farooqi (1987).

process in their own right. In essence aerated WSPs convert sewage into bacterial cells:

$$\text{Wastes} + \text{oxygen} \xrightarrow{\text{bacteria}} \text{Oxidized wastes} + \text{new bacterial cells}$$

Depending on temperature, this process takes some 2–6 days (4 days is the detention time most commonly provided). However, the cells ('sludge') so produced exert a high BOD and need to be removed before the effluent is discharged into a watercourse or used for irrigation. This is most simply achieved in a series of maturation ponds. The first pond in a series acts principally as a settling basin and a detention time of 5–10 days should be provided. The total number of ponds to be provided depends on the required reduction of faecal bacteria.

It is possible to use a conventional secondary sedimentation tank to remove the sludge from aerated pond effluent; in this case, because the cells have been aerated for only 2–6 days, they need further treatment before they can be placed on drying beds without creating an odour problem. This treatment is best done in an aerobic digester which is essentially a small aerated pond with a detention time (DT) of some 10 days. However, all these extra processes drastically increase the maintenance requirements.

With 4 days DT, BOD removals of 85–90% can be readily achieved in aerated

Table 11
Effluent Quality from Aerated WSPs

Type of wastewater	*Det. time (days)*	*Oxygen supply (lb/day)*	*Influent (mg/litre)*		*Effluent (mg/litre)*		*% Removal*	
			BOD_5	*SS*	*BOD_5*	*SS*	*BOD_5*	*SS*
Domestic	8·0	1 430	145	120	30	50	80	58
Chicken processing	18·2	1 300	600	368	85	56	86	85
Potato processing	14·0	2 650	1 200	400	1 100	249	8	38
Fruit canning	38·6	2 800	1 900	236	500	25	74	89

Source: Khurshid (1984).

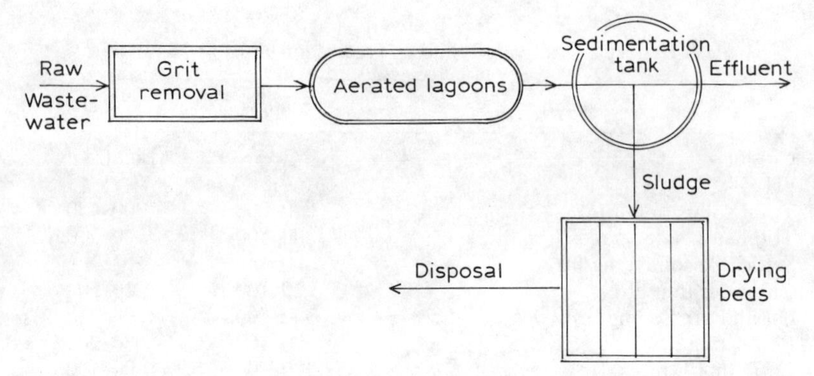

Fig. 11. Schematic diagram of aerated lagoon at Karachi. Capacity 22 750 m^3/day; organic loading 10 465 kg BOD/day; grit chambers, 2 channel type, each 948 m^3/h; aeration units, 4 semi-carousel type, depth 2 m; aerators, cage rotor type, 16 nos with speed 90 rpm.

ponds. The removal of faecal coliforms is, however, poor, only 85–95%; further treatment in maturation ponds is usually necessary. Treatment of domestic and some industrial effluents by aerated WSPs is shown in Table 11: 80% removal was achieved for domestic sewage; 74–88% efficiency in terms of BOD_5 was obtained for chicken processing and fruit canning waste. The efficiency for treating potato processing waste was poor, as only 8·0% reduction in BOD was achieved. A schematic diagram of an aerated WSP located at Karachi, Pakistan, is shown in Fig. 11. This plant has a capacity of 5 million gal/day (22 750 m^3/day), having 16 cage rotors 1·06 m × 4.42 m in size with a speed of 90 rpm, and is being operated at an organic loading of 10 465 kg BOD/ha/day (Khurshid, 1984).

This topic is also dealt with elsewhere in this book.

7.7 Oxidation Ditches

An oxidation ditch is similar to an aerated pond in that the wastewater is oxidized by bacteria in flocculent suspension and that the oxygen required for bio-oxidation is supplied by mechanical aeration. It differs in three main respects:

—Reactor geometry
—Type of aerator
—Sludge recycling

Oxidation ditches are long continuous channels, some 1·5–2·0 m deep and usually oval in shape. Aeration is effected by horizontal cage rotors placed across the channel; these not only provide oxygen but also impart a flow velocity of 0·3–0·4 m/s to the ditch contents, which is sufficient to prevent the bacterial flocs from settling to the bottom of the ditch where they would rapidly decay.

The practice of sludge recycling is the major difference between aerated ponds and oxidation ditches. To recycle the sludge some means of separating the sludge from the ditch effluent must be available. A conventional secondary sedimentation tank is therefore provided. A sludge lifting wheel or screw pump must also be provided to

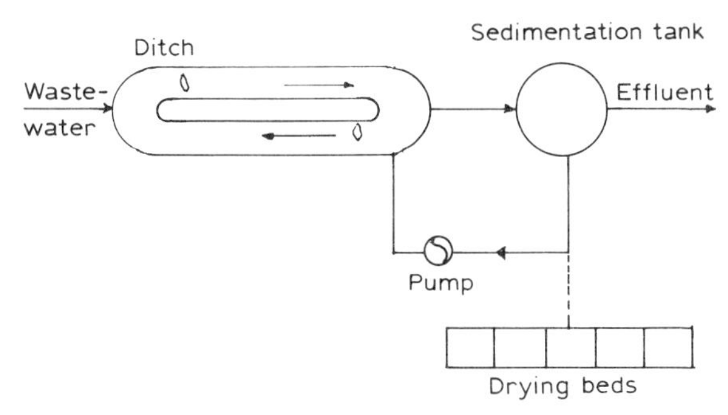

Fig. 12. Flowsheet of a typical oxidation ditch (La Riviere, 1981).

return the sludge back to the ditch. A flowsheet of a typical oxidation ditch is shown in Fig. 12.

The oxidation ditch has been adopted in many urban situations in developing countries for the treatment of organic effluents from small factories. Sometimes, the 'carousel' version of the oxidation ditch, using conventional mechanical surface aerators, has been installed to overcome the high cost of the 'potented cage' or 'manomoth' rotors, which are most satisfactory for the shallow ditch. In some developing countries, locally designed rotors have been manufactured at low cost but rarely have they given reliable service or long life. Oxidation ditches have the advantage of being relatively simple in operation and producing a minimum of excess sludge for disposal, but their long detention time (24–36 h) puts them at a disadvantage compared with the conventional activated sludge process for treating large wastewater flows. Process parameters of an oxidation ditch for domestic sewage are given in Table 12. As noted from the table, the oxidation ditch differs from conventional systems in that it treats both the suspended and the dissolved organic matter in one basin which often also serves as a secondary sedimentation tank.

With the trend towards larger oxidation ditches, the adoption of vertical axis aerators or surface aerators instead of cage rotors with a horizontal axis enables the use of deeper ditches, also known as 'carousels' (2·5–4·0 m deep), thus saving on land

Table 12
Process Parameters for Oxidation Ditch for Domestic Sewage Treatment

Parameter	*Values*
BOD_5 of settled sewage (g/m^3)	200–400
BOD_5 of effluent (g/m^3)	20–40
Treatment efficiency (%)	>95
BOD_5 (kg BOD/m^3 aeration basin/day)	0·1–0·2
Aeration (detention) time (h)	60–70
Mixed liquor suspended solids (MLSS) (kg/m^3)	4·0
Sludge age (days)	25

Source: La Riviere (1981).

requirement. The width of the ditches is also not restricted by the availability of cage rotors or by the need to provide intermediate supports for long horizontal rotor shafts.

8 ANAEROBIC DIGESTION

Among the anaerobic suspended growth type of treatment, anaerobic digestion is most commonly employed. Anaerobic digestion is the stepwise conversion of large molecules of organic compounds into methane and carbon dioxide by bacteria in the absence of free oxygen. Three different but related groups of bacteria are involved, namely hydrolytic fermentative organisms, acetogenic bacteria and methane forming bacteria. The hydrolytic bacteria hydrolyse a wide range of complex polymeric substrates to form simple organic compounds, carbon dioxide and hydrogen; the acetogenic bacteria convert these organic fermentation products into acetate, hydrogen and carbon dioxide. In the methanogenic phase, the acetate and hydrogen will be converted to methane and carbon dioxide. A simplified representation of the processes occurring in an anaerobic system is shown in Fig. 13. In the case of sludge digestion the first stage will involve liquefication of insoluble substances (cellulose, fats, micro-organisms) into simpler organic molecules which are then further degraded into short-chain fatty acids and hydrogen. These materials are converted by different groups of bacteria into methane and carbon dioxide.

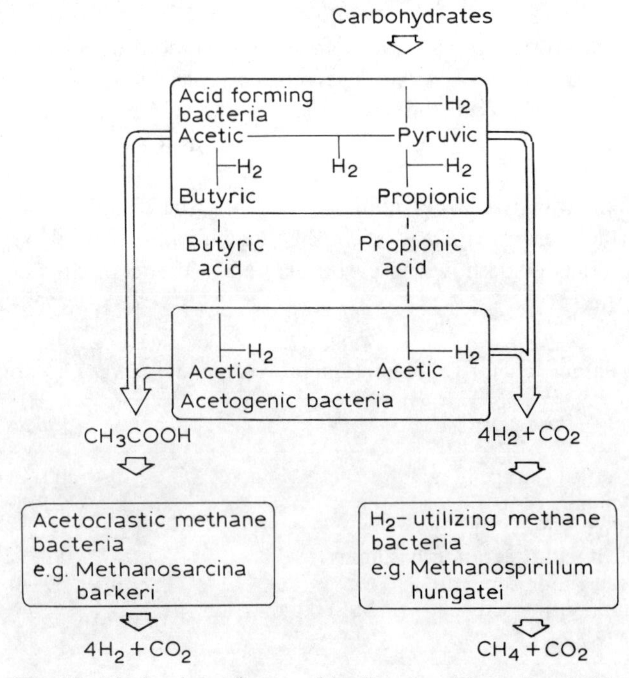

Fig. 13. Mechanism of anaerobic digestion (Downing, 1989).

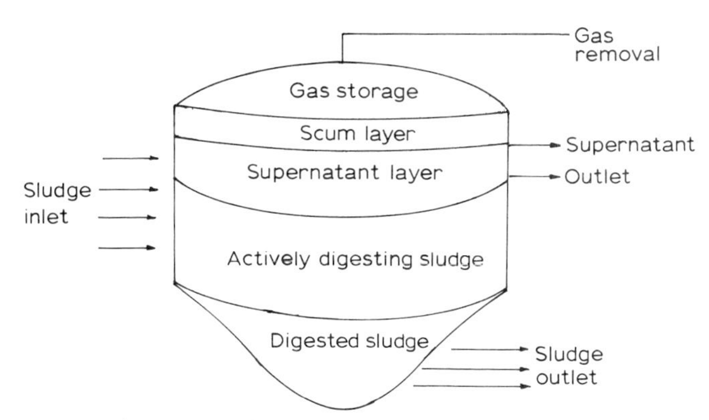

Fig. 14. Typical anaerobic digester (Saw, 1989).

Figure 14 illustrates the anaerobic treatment system (digestion). The anaerobic method of treating high strength industrial effluents offers the following significant advantages over conventional processes:

(1) The process, by its very nature, is totally enclosed and does not produce any environmental nuisance.

(2) Due to the very long solid detention time and consequent low growth rate, the cell yield (i.e. the solid production) is also extremely low; thus most of the carbon in the waste is available for methanogenesis and under normal circumstances the yield of methane would, on average, be 0·33–0·36 m^3 per kg COD utilized at 35°C and atmospheric pressure. The potential for energy recovery (Khurshid, 1988) from the anaerobic process may be summarized as follows:

COD utilized (kg)	1 000
Methane produced (m^3)	340
Energy value of gas (kcal)	2·55 million
Heavy oil equivalent (litres)	260
Gas oil equivalent (litres)	281
Coal equivalent (tonnes)	0·38

Biogas ignites at approximately 690°C compared with 645°C for natural gas and burns with a flame speed of approximately 40 cm/s. Under these circumstances most equipment fitted for natural gas can be operated with biogas after suitable modification. The production of biogas is generally in excess of that needed to operate the anaerobic treatment system, and can be utilized to generate power for other on-site services. Hence, anaerobic digestion is an energy producing process rather than one that demands a regular input of energy as in an aerobic system.

Removal of organic material occurs by microbial conversion, first to hydrogen, carbon dioxide and acetic acid, which then act as precursors to the final by-products—methane and carbon dioxide. Hence it is largely in this last step that the majority of the COD is removed from the raw effluent. Thus gas production and COD removal are inextricably linked in the anaerobic treatment plant.

A major benefit from the use of anaerobic fermentation for the treatment of industrial effluents is the extremely low biomass yield. Volatile organic material in the form of suspended solids is also dealt with more effectively in the anaerobic system which hydrolyses the volatile solids into soluble organics to allow the eventual conversion into methane and carbon dioxide.

With the present state of process technology, anaerobic digestion represents an attractive alternative to aerobic methods for wastewater treatment. It offers the potential for energy savings and an ability to operate at loadings in excess of equivalent aerobic reactor types, which will ultimately result in more economic industrial wastewater treatment.

9 RECENT TRENDS IN LIQUID EFFLUENT TREATMENT

Soil-aquifer treatment systems and the root zone method are the two most recently developed technologies for liquid effluent treatment. They are considered to be cheaper than conventional treatment plants and, at the same time, efficient in removal of pathogens.

9.1 Soil-aquifer Treatment System

Soil-aquifer treatment processes are biological processes, using fixed bacteria. The technique, incorporating infiltration basins, the unsaturated zone, and the aquifer, is adapted to permeable and aquifer soils. The system offers the advantages of low-cost purification and requires much less land (0·5–1 m^2 per person) than WSPs (5–10 m^2 per person).

The principles of treatment are:

—Percolation of effluent through a sufficiently thick layer of soil which acts as a mineral granular medium with numerous pores to harbour an abundant microflora.

—Permanent presence of a bacterial medium; any soil containing an abundant bacterial flora, thus avoiding the necessity of seeding and permitting minimum efficiency right from the beginning of the cycle.

—For aeration, neither equipment nor energy is required. It is carried out naturally by vertical and lateral air penetration during the drying phase.

Continuous or frequent application of biologically treated wastewater on fixed or moving surfaces results in the growth of a bacterial and zoogloeal layer (biocrust). The micro-organisms (bacteria, protozoa and nematodes) convert complex organic compounds in the wastewater into simpler compounds and produce by-products as well as end-products.

The biocrust is constantly in a state of flux that is building and degrading. Initially the biocrust is absent in leaching systems; however, as the wastewater effluent is applied the crust will gradually develop. As the crust develops the acceptance rate of the applied wastewater effluent through the soil surface will decrease. After a period

Table 13
Overall Performance of Soil-aquifer Treatment (SAT)

Parameter	*Raw wastewater*	*Reclaimed water*
pH	7·9	8·0
Suspended solids (mg/litre)	230–250	0
BOD_5 (mg/litre)	220–230	<1
COD (mg/litre)	460–510	10
Ammonia-N (mg/litre)	40–45	0·02
Total N (mg/litre)	60–70	5–8
P (mg/litre)	11–12	<0·05
Total dissolved solids (mg/litre)	800–850	600–650

Source: Amin (1989).

of about 3–6 months, the crust's average liquid acceptance rate stabilizes and reaches its long-term acceptance rate.

Most successful soil treatment systems utilize long distances and long times between application and withdrawal. The long time effectively destroys pathogenic micro-organisms since none of them are adapted to life underground. Also fine pores and biological slimes effectively filter out most micro-organisms. To achieve a very high level of purification and final disinfection of the effluent the thickness of sand layer in the infiltration basin would have to be increased to 1·50 m. BOD removal rates of 90% have been achieved by the process. Overall performance of soil-aquifer treatment for raw sewage is shown in Table 13.

The soil-aquifer treatment system, in addition to its remarkable disinfecting powers, enables the aquifer to be used to store treated water, thus protecting it from evaporation and loss, and permitting natural transfer of this valuable resource to agricultural wells. The development of this low-cost technology can lead to conservation and safe reuse of wastewater for agricultural purposes in developing countries with arid and semi-arid climates.

9.2 The Root Zone Method (RZM)

The root zone method (RZM) is a wetland method of sewage treatment developed over the last 20 years by Professor Dr Kickuth of Kassel University, Germany. It depends upon the flow of sewage through soil in which common reeds (*Phragmites australis*) are growing. The perspective and a typical root zone installation are shown in Fig. 15.

The key features of the RZM process are as follows.

(1) Rhizomes of the reeds grow vertically and horizontally, opening up the soil to provide a 'hydraulic pathway'.
(2) Wastewater is treated by bacterial activity. Aerobic treatment takes place in the rhizosphere with anoxic and anaerobic treatment taking place in the surrounding soil.

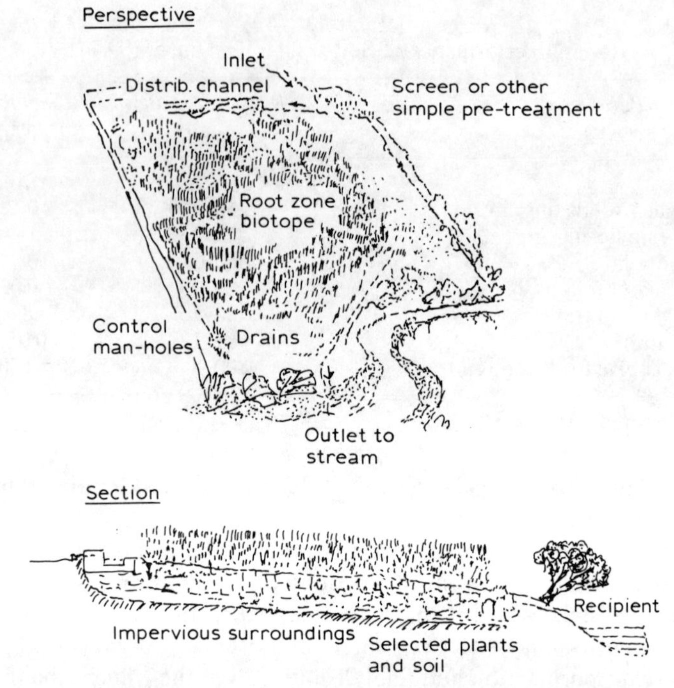

Fig. 15. Perspective and section of typical root zone installation (M. M. Amin, personal communication, 1990).

(3) Oxygen is passed from the atmosphere to the rhizosphere via the leaves and stems of the reeds through the hollow rhizomes and roots.
(4) Suspended solids in the sewage are aerobically composted in the above ground layer of straw formed from dead leaves and stems.

The following advantages have been claimed for RZM:

—Simple construction, no mechanical or electrical equipment
—Low maintenance cost
—Robust process able to withstand wide range of operating conditions
—Consistent effluent quality
—Environmentally acceptable with potential for wildlife conservation

The land area needed for treatment of screened, degritted sewage is about 2–4 m^2 per person to achieve an effluent with BOD less than 20 mg/litre. The method can also be effectively used for tertiary treatment and improving poor quality effluent. The capital cost of the RZM plant is likely to be 25–75% of that of conventional treatment works. Operating costs will depend on the need to remove accumulated composted sludges from the surface of RZM beds (every 25–35 years) but they are likely to be about 10–25% of those of the conventional process.

The basic principles of the RZM are as follows.

(1) The rate of treatment depends on the treatability of the wastewater.

(2) The gradient (artificial or natural) of 1–3% depends on the hydraulic conditions and effluent quality required.
(3) The concentration of BOD at the outlet of the RZM works (C_T) is related to the BOD of the inlet sewage (C_0) and the retention period in the reed bed (T) by the following equation:

$$C_T = C_0\,\mathrm{e}^{-kT} \tag{5}$$

where k is a function of temperature and treatability of sewage. At 8°C the value of k for domestic sewage is equal to 0·032 if the porosity of the soil is 42%.

In soil with a hydraulic conductivity of 10^{-3} m/s the total volume of the voids (due to the rhizosphere) is about 6–7% of that of the total bed. The area (A_h) of the RZM bed (assumed to be 0·6 m deep) can be calculated from the equation

$$A_h = 5{\cdot}2Q_d(\ln C_0 - \ln C_T) \tag{6}$$

where A_h is the area (m^2), and Q_d is the average flow rate of sewage (m^3/day). The cross-sectional area (A_c) can be calculated from the following equation:

$$A_c = \frac{Q_s}{K_f\,\mathrm{d}H/\mathrm{d}s} \tag{7}$$

where A_c = cross-sectional area (m^2)
Q_s = average flow rate of sewage (m^3/s)
K_f = hydraulic conductivity of a fully developed RZM bed (about 3×10^{-3} m^3/m^2 s)
$\mathrm{d}H/\mathrm{d}s$ = slope of the bed (m/m)

A number of RZM systems for sewage treatment have been constructed in Germany and are in operation. The BOD of the effluent from various RZM treatment works is reported to be less than 20 mg/litre.

The RZM technique may be cost-effective and appropriate for developing countries in an irrigation reuse context.

10 EFFLUENT REUSE

Water shortages are becoming a serious problem in the developing countries due to rapid growth of population, urbanization and industrialization coupled with the introduction of modern intensive agricultural techniques causing increasingly heavy demand on water resources. Thus the reuse of wastewater has become an attractive option for increasing water resources.

Effluents from biological wastewater treatment plants may lack the characteristics required for direct reuse. However, the most important uses of the reclaimed water are agricultural irrigation, fish rearing and groundwater recharge.

The design of biological treatment facilities must ensure that the characteristics of the effluent are those required for reuse. Proposed standards (Khurshid, 1989*b*) for

Table 14
Effluent Reuse Standards

Parameter	*Agriculture/ irrigation*	*Fish rearing*	*Recharge*
BOD_5 (mg/litre)	No limit	<10	<5
Suspended solids (mg/litre)	<30	Low	<30
Total dissolved solids (mg/litre)	2 500	<2 000	Low
Total Kjeldahl nitrogen (mg/litre)	No limit	See ammonia	Low
Ammonia-N (mg/litre)	No limit	<0·5	Virtually none
Nitrate-N (mg/litre)	No limit	No limit	<50
Total phosphorus	No limit	No limit	<10
Faecal coliforms (MPN/100 ml)	<1 000	<1 000	<1 000
Cd, Hg, and other heavy metals	Virtually none	Virtually none	<0·001

Source: Khurshid (1989*b*).

these reuses are summarized in Table 14. Effluents will more often than not fail to meet these stringent standards and would require upgrading before reuse. However, it should be noted that, for many places, these standards will be unnecessarily restrictive and they should be relaxed wherever feasible. Moreover, many effluent quality improvement methods are not appropriate for use in developing countries and are too costly.

10.1 Agricultural Irrigation

The reuse of wastewater for agricultural purposes is an age-old and common practice. Shuval (1977) cites many references on the use of sewage effluent as an agricultural water resource. The quality of the reuse water is important both for the health of the workers and for the particular application for which it is used. Trace elements toxic to crops may be a problem; for example, boron, a component of many commercial laundry powders and one that is not removed by conventional treatment processes, is well known as a toxicant in citrus fruit crops.

10.2 Fish Rearing

Effluent from biological treatment plants has been used for rearing of fish. The most popular fish which have been successfully reared in effluents include carp and tilapia. The silver carp and the big head are capable of direct feeding on the plankton. The carp's popularity is largely due to its rapid growth rate and consequently high productivity under pond conditions. Wide ranges of productivity are reported in the literature. In Europe (Pescod & Alka, 1985) carp productivities from sewage-fed ponds are reported to range from 400 to 900 kg/ha/year, whereas through careful feeding and mixing of fish species, productivities as high as 5000–7000 kg/ha/year have been achieved. Pond productivities of about 1000 kg/ha/year are not uncommon in Asia where night soil is applied and the pond is well maintained.

Industrial wastes have also been used in fish culture. Studies using dairy, sugar-mill, abattoir and starch-mill wastes have been conducted in Czechoslovakia and practised by an Austrian starch mill (Overman, 1979).

10.3 Groundwater Recharge

Effluents can also be used for the purposeful recharge of groundwater. Such effluents should be free from heavy metals, with nitrogen and phosphorus contents of less than 50 and 10 mg/litre respectively and faecal coliforms always less than 1000/100 ml.

10.4 Hydroponics and Aquaculture

Domestic wastewater, fully treated in WSPs, has been successfully used in growing vegetables in gravels rather than soil according to the horticultural practice known as hydroponics.

In Papua New Guinea (Ducan, 1985) aquaculture and agriculture based on maturation WSPs have been successfully carried out. While fish and ducks are reared in the fish ponds, the overflow from these ponds is used for hydroponics and finally the overflow from the hydroponics basin is allowed to seep into the soil. Such an arrangement is worth following in rural areas of developing countries.

Hydroponics have a great future in countries like the Gulf states and the Middle East where water is scarce and demand for vegetables can be met through hydroponics.

10.5 Municipal Uses

In municipal practice the effluent can be reused for road washing, arboriculture (along roads), and watering of lawns and parks. Effluents to be used for watering golf courses, street flushing, lawns, etc., should be chlorinated to keep the coliform count below 1000/100 ml.

10.6 Water Hyacinth

Water hyacinth has been used successfully to upgrade effluents, particularly for the removal of algae. Water hyacinth is able to take up large amounts of nutrients (Stowell, 1981), i.e. nitrogen and phosphorus and heavy metals. At the same time its roots provide support for a gelatinous biomass which further stabilizes organic matter, producing CO_2, inorganic substances and other materials. Bacteria and other organisms adhere to the gelatin-covered paste. When the hyacinth is harvested, all these substances are removed from the water.

Hyacinth grows very rapidly in hot climates, doubling its mass in about 6 days (Hanser, 1981). One hectare of a hyacinth-covered pond can produce more than 4 tonnes wet weight of plants or 200 kg of dry solids per day; production of more than 290 kg/ha/day has been reported (Mara, 1981). Reductions of 80% nitrogen and

44% total phosphorus have been achieved by 0·55 ha of a hyacinth pond 0·6 m deep, with a detention time of 24–48 h and fed with 1000 m^3/day of facultative WSPs (Chughtai & Khurshid, 1981). Very low concentrations of ammonia nitrogen are present in water hyacinth ponds, which is important for fish rearing, and clear, low-BOD effluents are produced.

10.7 Upgrading of Liquid Effluents

Pond effluents will more often than not fail to meet stringent standards and would require upgrading before reuse. Moreover, many effluent quality improvement methods are not appropriate for use in developing countries and are too costly. Examples of such technologies are microstraining, rapid sand filtration, chlorination, flotation, and treatment with activated carbon.

10.8 Reclaimed Wastewater Reuse

Reusing reclaimed wastewater is becoming essential in most parts of the world, where waste-borne sewerage exists and reuse is desirable. Wastewater treatment process selection should consider pathogen removal as a primary objective.

Where space and climate allow, the use of waste stabilization ponds can achieve satisfactory pathogen removals for many reuse applications, without the need for chlorination.

Groundwater augmentation and recovery from percolation basins provide a means of producing a higher quality water suitable for most applications.

In general, conventional processes such as activated sludge and biological filtration do not produce effluents of an acceptable quality for reuse except under carefully controlled conditions. Tertiary treatment and heavy chlorination are required.

11 CHOICE OF A PROCESS FOR TREATMENT OF LIQUID EFFLUENTS

The overall progress in effluent treatment in the developing countries is rather insignificant. The major bottleneck seems to be the high costs of conventional methods which are prohibitive to the sustaining economy of the developing world.

The conventional treatment processes employing trickling filters and activated sludge are expensive, energy intensive, costly to operate and sophisticated, and require well-trained operators. These treatment processes are generally not relevant to the developing countries. In Pakistan, as in other developing countries, the experience of conventional treatment plants has been disillusioning as their performance was quite poor. For these reasons, the developing countries with limited resources need to use suitable low-cost technologies, which must be tailored to the local conditions.

High ambient temperatures in arid and semi-arid countries make it possible to use simple and economical treatment processes. For developing countries the most realistic and appropriate methods in order of performance are waste stabilization ponds, aerated lagoons, and oxidation ditches. Waste stabilization ponds are appropriate for both small and large communities, whereas aerated lagoons and oxidation ditches are generally used in large cities.

Waste stabilization ponds have found worldwide application and in some respect are more efficient than the conventional processes in the removal of pathogenic organisms. They can even eliminate protozoan cysts and helminth eggs and reduce the concentration of excreted bacteria and viruses to low levels. Hence, this type of treatment will be able to produce an effluent which meets the recommended quality guidelines for unrestricted irrigation, both at low cost and with minimal operational and maintenance requirements.

The recently developed new technologies of soil-aquifer treatment systems and the root zone method are also considered to be cheaper than conventional treatment plants. Research and development of these low-cost treatment methods should be undertaken in developing countries to assess their performance and suitability under local conditions.

REFERENCES

Amin, M. M. (1989). In *Proceedings of International Seminar on Wastewater Treatment Plants*. Pakistan Engineering Council, Multan, p. 244.

Chughtai, M. I. D. & Khurshid, A. (1980). *Pakistan J. Biochem.*, **13**, 69.

Chughtai, M. I. D. & Khurshid, A. (1981). *Pakistan J. Biochem.*, **14**, 39.

Chughtai, M. I. D. & Khurshid, A. (1982). Paper presented at *12th International Congress of Biochemistry*, Perth, Australia, 15–21 August.

Chughtai, M. I. D. & Khurshid, A. (1985). *Engg. News*, **23**, 106.

Chughtai, M. I. D. & Khurshid, A. (1988). *Pakistan J. Biochem.*, **21**, 1.

Downing, A. L. (1989). In *Proceedings of International Seminar on Wastewater Treatment Plants*. Pakistan Engineering Council, Multan, p. 34.

Ducan, M. (1985). Paper presented at *FAO Regional Seminar on the Treatment and Use of Sewage Effluent for Irrigation*, Cyprus, 7–9 Oct.

Farooqi, A. (1987). *Investigations on performance for the upgradation of activated sludge treatment plant at Islamabad.* MSc Thesis, University of Engineering and Technology, Lahore, Pakistan.

Feacham, R. G., Bradley, D. J., Garelick, H. & Mara, D. (1983). *Sanitation and Disease.* John Wiley & Sons, Chichester, UK, pp.129–37.

Gloyna, E. F. (1976). In *Proceedings of Water Resource Symposium 9*, eds E. F. Gloyna, J. F. Malina & E. M. Davis. The University of Texas at Austin, p. 144.

Hammer, M. J. (1977). *Water and Wastewater Technology*. John Wiley & Sons, New York, pp. 31–99.

Hanser, R. J. (1981). *J. WPCF*, **56**, 222.

Khurshid, A. (1979). *Pbl. Hlth. Engg*, **2**, 34.

Khurshid, A. (1984). *J. Sc. Tech. & Dev.*, **3**, 98.

Khurshid, A. (1985). *Asian Environment*, **6**, 124.

Khurshid, A. (1988). *J. Sci. Tech. & Dev.*, **7**, 249.

Khurshid, A. (1989*a*). *J. Arabian Sci. & Engg.* **14**, 21.

Khurshid, A. (1989*b*). *J. Sci. Tech. & Dev.*, **8**, 17.

Khurshid, A. & Chughtai, M. I. D. (1986). *Oxidation Ponds in Arid and Semi-arid Regions.* Pakistan Society of Biochemists, Lahore, pp. *xi*, 107.

La Riviere, J. W. M. (1981). In *Proceedings of the Third Conference on Global Impacts of Applied Microbiology*, eds Y. M. Freitas & F. Fernandes. University of Bombay, p. 148.
Mara, D. (1981). *Water, Wastes and Health in Hot Climates.* John Wiley & Sons, New York, p. 404–13.
McGarry, M. G. & Pescod, M. B. (1970). In *Proceedings of the 2nd International Symposium on Waste Treatment Lagoons*, Kansas City, MO, p. 188.
Overman, A. R. (1979). *J. ASCE, Environ. Engg. Div.*, **185**, 535.
Pescod, M. B. & Alka, U. (1985). Paper presented at *FAO Regional Seminar on the Treatment and Use of Sewage Effluent for Irrigation*, Cyprus, 7–9 Oct.
Saw, C. B. (1989). Paper presented at *National Workshop on Industrial Waste Management and Stabilization Pond Treatment*, Lahore, Pakistan, 25–30 March.
Schroeder, E. D. (1977). *Water and Wastewater Treatment.* McGraw-Hill, New York, pp. 709–16.
Shuval, H. I. (1977). *Health Considerations in Water Renovation and Reuse.* Academic Press, New York, pp. 236–42.
Stowell, R. (1981). *J. ASCE, Environ. Engg. Div.*, **107**, 919.
Tariq, M. N. & Khurshid, A. (1980). In *Proceedings of WHO Seminar: WSPs Design and Operation. WHO EMRO Tech. Publ. No. 3*, WHO Regional Office, Alexandria, p. 124.
World Health Organization (1987). *WHO/EMRO. Tech. Publ. No. 10*, WHO Regional Office, Alexandria, p. 19.

Chapter 4

BIOLOGICAL NUTRIENT REMOVAL

YERACHMIEL ARGAMAN

Department of Civil Engineering, Technion, Israel Institute of Technology, Haifa, Israel

CONTENTS

1 INTRODUCTION

This chapter reviews the fundamental principles and the available technologies for biological removal of nitrogen and phosphorus from municipal wastewater. Environmental nuisances associated with these pollutants include eutrophication in receiving water by all forms of nitrogen and phosphorus, oxygen depletion and aquatic toxicity by ammonia, and public health implications of nitrates in drinking water. Efforts to control the release of nutrients into the aquatic environments have led to the development of physical, chemical and biological wastewater treatment technologies. To date the most widely used and cost effective technology for nitrogen control is biological treatment. Phosphorus removal, on the other hand, has been effectively achieved by chemical precipitation too. Biological control of phosphorus has been known for some time and is gaining popularity as more data become available and the process is better understood.

1.1 Sources of Nitrogen and Phosphorus in Municipal Wastewater

Municipal wastewater of predominantly household origin contains nitrogen in the organic and ammonium forms. These are primarily waste products from protein metabolism in the human body. Approximately 60% of the nitrogen in fresh wastewater is in the organic form and 40% in the ammonium form. Microbial decomposition of proteinaceous matter and hydrolysis of urea transform organic nitrogen to ammonium. Normally a very small fraction of the nitrogen in fresh wastewater is in the oxidized forms of nitrite or nitrate.

The daily per capita discharge of nitrogen varies from 10 to 20 g. An estimate of the total amount of nitrogen discharged into sewerage systems in the United States in 1973 was 840 000 tons per year (USEPA, 1973). This is equivalent to approximately 15 g per capita daily (gpcd). The concentration of nitrogen in municipal wastewater depends on the per capita wastewater flow rate. Thus, for a

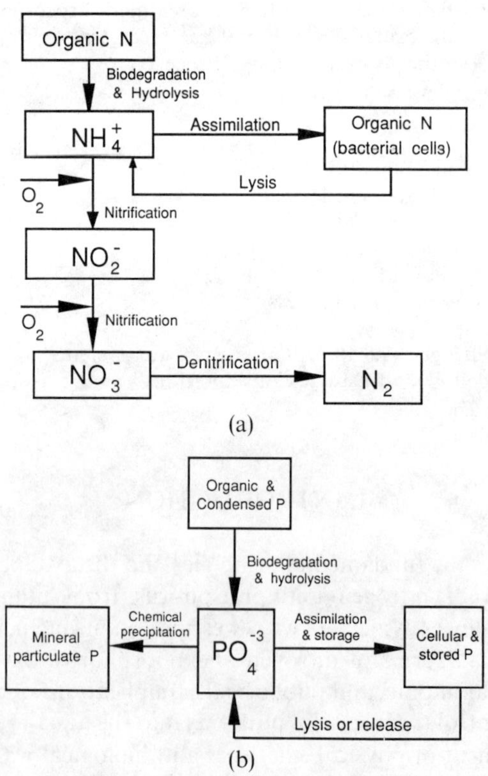

Fig. 1. (a) Nitrogen and (b) phosphorus transformations in biological treatment systems.

wastewater flow rate of 150–600 litres per capita daily (lpcd) and a nitrogen discharge of 15 gpcd, the nitrogen concentration would vary between 17 and 100 mg/litre. In the United States nitrogen concentration values in municipal wastewater range from 15 to 50 mg/litre (USEPA, 1975).

Phosphorus occurs in municipal wastewater as various forms of phosphate: orthophosphate (PO_4^{3-}), condensed phosphates (e.g. pyrophosphate, tripolyphosphate, trimetaphosphate) and organic phosphates (e.g. phospholipids, sugar phosphates, nucleotides). The main sources of phosphorus in municipal wastewater are human faecal and waste materials, and household detergents. The approximate contribution from these two sources in the United States are 1·6 gpcd from human wastes and 1·1 gpcd from detergents. The contribution of phosphorus from detergents is sharply reduced in areas where regulations are imposed that limit the phosphorus content of laundry detergents. The concentration of phosphorus in municipal wastewater varies from 3 to 15 mg/litre. Typical considerations in the United States, where no phosphorus limitations in detergents exist, are from 6 to 8 mg/litre.

1.2 Nitrogen Transformations in Biological Treatment Processes

Raw municipal wastewaters normally contain nitrogen in the organic and ammonium forms. Transformations of nitrogen forms that may occur in biological treatment systems are illustrated in Fig. 1(a). These transformations include:

- —hydrolysis and bacterial breakdown of organic nitrogen to ammonium
- —assimilation of ammonium nitrogen in bacterial cells
- —release of ammonium due to cell lysis
- —nitrification of ammonium to nitrites and nitrates
- —denitrification of nitrates to nitrogen gas

Treatment processes may be designed and operated to influence this transformation pattern in order to achieve the desired treatment goals.

1.3 Phosphorus Transformations in Biological Treatment Processes

Raw municipal wastewater contains phosphorus in the ortho, condensed and organic forms. Transformations of phosphorus forms that may occur in biological treatment plants are illustrated in Fig. 1(b). These transformations include:

- —hydrolysis and bacterial breakdown of organic and condensed phosphorus to orthophosphate
- —assimilation or storage of orthophosphate in bacterial cells
- —release of orthophosphate (due to lysis and other reasons)
- —chemical precipitation with naturally occurring or added cations

In contrast to nitrogen, which can be removed from wastewater as a gaseous product, phosphorus removal can only be achieved through its incorporation in particulate matter such as bacterial cells or mineral precipitates.

2 NITROGEN REMOVAL IN THE ACTIVATED SLUDGE PROCESS

Nitrogen entering an activated sludge system in the organic or ammonium forms may be removed through assimilation or oxidized to nitrate through nitrification. Nitrates can subsequently be denitrified to nitrogen gas. These processes are discussed subsequently.

2.1 Assimilation of Nitrogen

The nitrogen content of microbial cells is approximately 12·5% on a dry weight basis. Thus volatile solids produced and disposed of as excess sludge carry 12·5% nitrogen which is removed from the wastewater. The amount of nitrogen that can be removed by this mechanism is limited by the amount of net growth, which in turn depends on the organic content of the wastewater and the system's operating conditions.

The net amount of biological sludge in an activated sludge plant per unit of BOD removed is given by

$$\Delta X_v/S_r = a - b\chi_d X_v t/S_r \quad (1)$$

where ΔX_v = biological sludge produced (kg VSS/m^3)
S_r = BOD removal (ΔBOD) (kg/m^3)
X_v = mixed liquor volatile suspended solids (MLVSS) (kg/m^3)
t = hydraulic residence time (days)
a = yield coefficient (kg VSS produced/kg BOD removed)
b = autooxidation decay rate (kg VSS/kg VSS . day)
χ_d = degradable fraction of biomass

Since the BOD removal in most activated sludge plants is over 90%, i.e. $S_r \approx S_0$, where S_0 is the influent BOD, and using the 12·5% nitrogen content of excess biological sludge, eqn (1) can be rewritten as follows:

$$\Delta(NH_3 - N)\text{ assim.}/\Delta BOD = 0{\cdot}125a - 0{\cdot}125\chi_d/(F/M) \quad (2)$$

where F/M = organic loading rate (kg BOD/kg VSS . day).

For typical municipal wastewater with a yield coefficient of 0·6 kg VSS/kg BOD, a decay rate of 0·1/day and activated sludge systems operated at F/M of 0·1–0·3 kg BOD/kg VSS . day, the ammonia to BOD removal ratio will range from 0·015 to 0·050. Based on influent BOD and nitrogen concentrations of 150 and 30 mg/litre, respectively, the assimilative removal of nitrogen can range from 8 to 25%. This removal mechanism may become quite significant in some industrial wastewaters, or municipal wastewaters with a large industrial contributor.

To maximize nitrogen removal by assimilation one should maximize the net sludge production, i.e. increase the F/M loading rate. It should be noted that assimilated nitrogen may be released from the sludge in some sludge treatment processes (e.g. heat treatment, anaerobic digestion) and returned to the liquid wastewater stream. This may reduce considerably the assimilative removal of nitrogen.

2.2 Nitrification

Nitrification is the biological oxidation of ammonia to nitrate with nitrite formation as an intermediate. The microorganisms involved are the autotrophic genera *Nitrosomonas* and *Nitrobacter*, which carry out the reaction in two steps:

$$2NH_4 + 3O_2 \xrightarrow{Nitrosomonas} 2NO_2^- + 2H_2O + 4H^+ + \text{new cells}$$

$$2NO_2^- + O_2 \xrightarrow{Nitrobacter} 2NO_3^- + \text{new cells}$$

Since a buildup of nitrite is rarely observed, it can be concluded that the rate of conversion to nitrite controls the rate of the overall reaction.

The cell yield for *Nitrosomonas* has been reported as 0·05–0·29 kg NVSS/kg NH_3-N and for *Nitrobacter* 0·02–0·08 kg NVSS/kg NO_2-N. Here NVSS represents nitrifiers volatile suspended solids. A value of 0·15 g NVSS/g NH_3-N oxidized is usually used for design purposes (USEPA, 1975). The empirical overall (oxidation and synthesis) reaction is

$$NH_4^+ + 1{\cdot}83O_2 + 1{\cdot}98HCO_3^- \rightarrow$$
$$0{\cdot}98NO_3^- + 0{\cdot}021C_5H_7NO_2 + 1{\cdot}88H_2CO_3 + 1{\cdot}04H_2O$$

Thus the stoichiometric equation for nitrification indicates that for 1 kg of ammonia nitrogen removed approximately

—4·3 kg of O_2 are consumed
—7·1 kg of alkalinity are destroyed
—0·15 kg of new cells are formed
—0·08 kg of inorganic carbon are consumed

In all domestic wastewaters and most industrial wastes the concentration of carbonaceous organics greatly exceeds that of nitrogen. The heterotrophic organism yield also exceeds that of the autotrophs. Hence the autotrophic population normally constitutes a small fraction of the total biomass. Neglecting the endogenous decay process, the nitrifier (autotroph) fraction can be estimated by

$$F_N = a_N A_r/(aS_r + a_N A_r) \tag{3}$$

where F_N = nitrifiers' fraction
a_N = nitrifiers' yield coefficient (kg NVSS/kg NH_4^+-N)
A_r = ammonia nitrogen removal (kg/m^3)

In order to maintain a population of nitrifying organisms in a mixed culture of activated sludge, the system sludge age, or solids retention time (SRT), must exceed the reciprocal of the nitrifiers' net specific growth rate (Downing *et al.*, 1964):

$$SRT_c = 1/(\mu_N - b_N) \tag{4}$$

where SRT_c = critical solids retention time required for nitrification (days)
μ_N = nitrifiers' specific growth rate (day^{-1})
b_N = nitrifiers' decay rate (g NVSS/g NVSS . day)

Hence the minimum solids retention time or sludge age required for nitrification is

$$SRT_{c,min} = 1/\mu_{N,max} \tag{5}$$

where $SRT_{c,min}$ = minimum solids retention time required for nitrification (days)
$\mu_{N,max}$ = maximum specific growth rate of nitrifiers (g NVSS/g NVSS . day)

The actual growth rate of nitrifiers in an activated sludge plant depends on the concentrations of ammonium nitrogen and dissolved oxygen (DO), as well as temperature and pH. The effects of DO and ammonium can be expressed by Monod kinetics:

$$\mu_N = \mu_{N,max}\{NH_4^+\text{-}N/(K_N + NH_4^+\text{-}N)\}\,.\,\{DO/(K_O + DO)\} \tag{6}$$

where K_N, K_O = half saturation values for ammonium and DO, respectively (kg/m^3). The effect of temperature is given by

$$\mu_N(T) = \mu_N(15°C) \times \exp[0{\cdot}098(T - 15)] \tag{7}$$

where $\mu_N(T)$, $\mu_N(15°C)$ = nitrifiers' growth rates at temperatures T°C and 15°C, respectively (kg NVSS/kg NVSS . day).

Typical values of the various kinetic parameters for nitrification in domestic wastewater are (USEPA, 1975)

$$K_N = 0{\cdot}5 \times 10^{-3}\ \text{kg/m}^3$$

$$K_O = 1{\cdot}0 \times 10^{-3}\ \text{kg/m}^3$$

$$\mu_{N,max}(15°C) = 0{\cdot}45\ \text{kg NVSS/kg NVSS . day}$$

The theoretical SRT required for nitrification is obtained by substituting eqns (5) and (6) into eqn (3). The hydraulic retention time t is related to the SRT through

$$t = (aS_r + X_{vi})/X_v[0{\cdot}8b/(1 + 0{\cdot}2b\,SRT) + 1/SRT] \tag{8}$$

where X_{vi} = influent VSS (kg/m^3).

Equation (7) assumes the following empirical relationship between SRT and χ_d (Quirk & Eckenfelder, 1986):

$$\chi_d = 0{\cdot}8/(1 + 0{\cdot}2b\,SRT) \tag{9}$$

2.3 Denitrification

Denitrification is the biological conversion of nitrate-nitrogen to more reduced forms such as N_2, N_2O and NO. The process is brought about by a variety of facultative heterotrophs which can utilize nitrate instead of oxygen as the final electron acceptor.

It was shown that the breakdown of carbonaceous organics in the denitrification process is similar to that in the aerobic process, the only difference being in the final stages of the electron transfer. This would indicate the need for strict anoxic conditions in a denitrifying system. However, it has been shown that under acidic conditions denitrification can take place in the presence of oxygen. Moreover, fixed film reactors, as well as suspended growth systems, may consist of aerobic biomass layers and anoxic sublayers so that aerobic processes and denitrification may occur simultaneously.

The stoichiometric reaction describing denitrification depends on the carbonaceous matter involved. Thus, for methanol, which is the most extensively used and studied external carbon source, the empirical reaction including synthesis is

$$NO_3^- + 1{\cdot}08CH_3OH + 0{\cdot}24H_2CO_3 \rightarrow 0{\cdot}056C_5H_7NO_2 + 0{\cdot}47N_2 + 1{\cdot}68H_2O + HCO_3^-$$

This reaction indicates that for 1 g of nitrate-nitrogen that is denitrified

—2.47 g of methanol (or approximately 3·0 g of BOD) are consumed
—0·45 g of new cells are produced
—3·57 g of alkalinity are formed

Nitrate will also replace oxygen in the endogenous respiration reaction. The rate of denitrification depends primarily on the nature and concentration of the carbonaceous matter undergoing degradation. Most investigators agree that denitrification is a zero-order reaction with respect to nitrate down to very low nitrate concentration levels (Beccari *et al.*, 1983). Hence the nitrate removal in an anoxic basin when carbon is not limiting can be expressed by

$$\Delta NO_3^-\text{-N} = R_{DN} X_v t \qquad (10)$$

where ΔNO_3^--N = denitrified nitrate-nitrogen (kg/m^3)
R_{DN} = zero-order denitrification rate (kg NO_3^--N/kg VSS . day)

The effects of temperature and DO on the rate of denitrification are described by the empirical expression:

$$R_{DN}(T) = R_{DN}(20°C)\Theta^{T-20}(1 - DO) \qquad (11)$$

where Θ = temperature coefficient for denitrification.

Reported values of R_{DN} vary from 0·12 to 0·90 kg NO_3^--N/kg VSS . day when methanol is used as the carbon source (USEPA, 1975; Beccari *et al.*, 1983). For systems using municipal wastewater as a carbon source, the reported rate of denitrification varies from 0·03 to 0·11 kg NO_3^--N/kg VSS . day (USEPA, 1975). When the readily degradable portion of domestic wastewater is utilized as a carbon source, the rate of denitrification was reported to be as high as 0·72 kg NO_3^--N/kg VSS . day (Ekama *et al.*, 1984).

The denitrification rate under aerobic conditions will depend on the anoxic fraction of the biological floc and the availability of carbon substrate. The DO term in eqn (11) indicates that the denitrification rate decreases linearly to zero when the dissolved oxygen concentration reaches 1·0 mg/litre. Denitrification rates of 0·006 kg NO_3^--N/kg VSS . day have been reported under aerobic conditions (Christensen, 1975). For practical purposes denitrification can be ignored when dissolved oxygen concentrations are greater than 1·0 mg/litre. Recent experience in South Africa has shown significant denitrification occurring in an anaerobic plug flow process in which surface aerators are used and the surface dissolved oxygen maintained at 1·0–2·0 mg/litre. The hypothesis is that in the lower levels of the tank denitrification occurs at reduced oxygen levels.

2.4 Combined Nitrification and Denitrification

Nitrogen removal can be accomplished through denitrification of nitrified wastewater. In recent years the system of choice for this purpose is the single-sludge system with mixed liquor recycle. This configuration is preferred over the two-sludge system for economic and technical reasons. Flow schemes for the two configurations are shown in Fig. 2.

In the two-sludge system carbonaceous organic removal and nitrification take place in the first aerobic activated sludge unit. The clarified effluent of this stage is passed to the second stage, where anoxic conditions prevail and denitrification occurs. Since the organic carbon of the raw wastewater has been largely removed in the first stage, an external carbon source (e.g. methanol) is required to serve as electron acceptor in the denitrification basin.

In the single-sludge system the influent first enters an anoxic basin, where denitrification occurs with raw wastewater organics serving as carbon source. Nitrates are supplied to the anoxic basin by the mixed liquor that is recycled from the aerobic basin. Thus the influent ammonium passes through the anoxic basin to the aerobic basin, where it is nitrified to nitrate. A key feature of the single-sludge

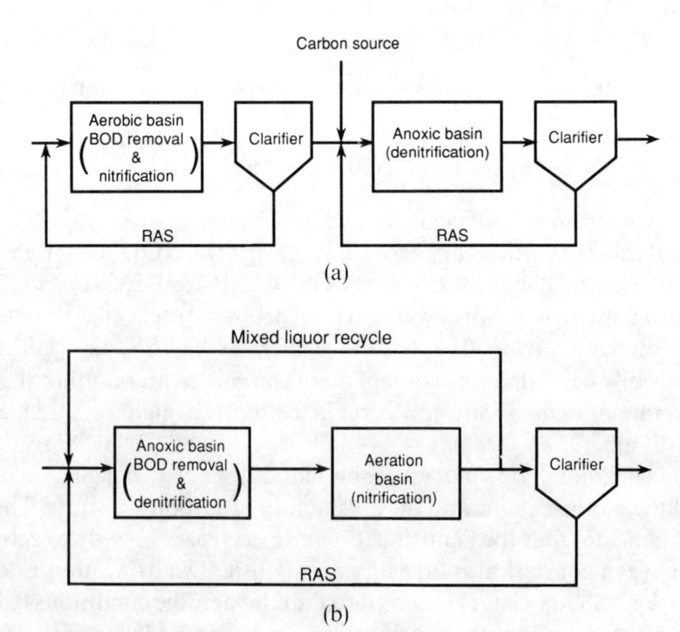

Fig. 2. Alternative systems for biological nitrogen removal: (a) two-sludge system; (b) single-sludge (recycle) system.

system is the high rate of mixed liquor recycle from the aerobic to the anoxic basin (200–500% in domestic wastewater applications).

The single-sludge recycle system offers several economic advantages over the two-sludge system in that it uses only one clarification step, no external carbon source, lower neutralization chemical requirements and lower oxygen requirements. Alternative calculation procedures for the single-sludge system are presented elsewhere (Barnard, 1974; Argaman, 1981; Grady *et al.*, 1986). A simplified trial and error method is described below.

Assuming complete denitrification of nitrates in the anoxic basin and neglecting assimilation, the overall recycle ration required for a specific performance is

$$r = (\Delta NH_4^+\text{-N}/NO_3^-\text{-N,e}) - 1 \quad (12)$$

where r = overall (mixed liquor + return sludge) recycle ratio
ΔNH_4^+-N = ammonium nitrogen removed in the process (kg/m^3)
NO_3^--N,e = effluent concentration of nitrate-nitrogen (kg/m^3)

The aerobic fraction of the overall reactor volume is assumed. Since the nitrifiers can only grow in the aerobic basin, the minimum SRT required for nitrification in a single-sludge system is given by

$$SRT_c' = SRT_c/\eta \quad (13)$$

where SRT_c', SRT_c = minimum SRT required for nitrification in a single-sludge system and in a conventional nitrification system, respectively (days)
η = aerobic volume fraction

The overall system hydraulic residence time (t) is calculated by substituting SRT_c' for SRT in eqn (8).

The required residence time in the anoxic basin is calculated by

$$t_{DN} = \Delta NO_3^-\text{-N}/X_v R_{DN} \quad (14)$$

where t_{DN} = required hydraulic residence time in the anoxic basin (days)
ΔNO_3^--N = concentration of nitrate-nitrogen to be denitrified (kg/m^3)

The value of t_{DN} is compared to $(1-\eta)t$. If the two are not equal, a new value of η is selected and the calculations are repeated.

3 NITROGEN REMOVAL IN FIXED-FILM PROCESSES

Fixed-film systems such as trickling filters and rotating biological contactors (RBC) can be used to nitrify secondary effluents. The biomass which accumulates on the medium surface consists of both heterotrophic and autotrophic microorganisms. The proportion of nitrifiers in this biomass reflects the relative removal of carbonaceous organics and ammonia. In a trickling filter this ratio may vary along the filter depth as the carbonaceous organics are gradually depleted.

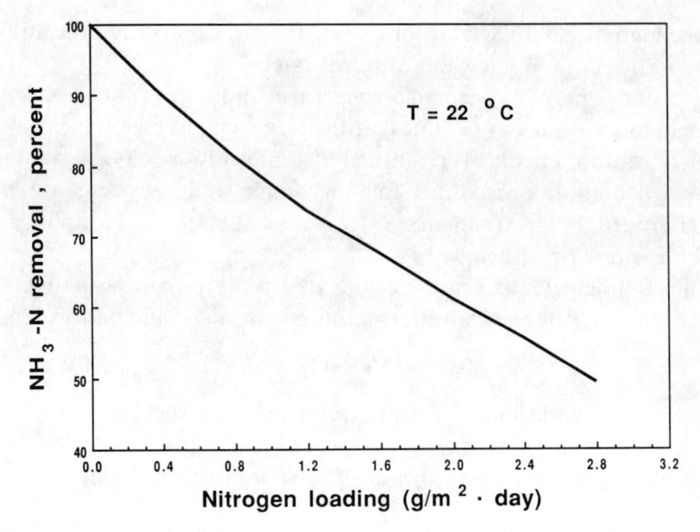

Fig. 3. Effect of trickling filter loading on nitrogen removal in tertiary filtration on plastic media (after Jiumm *et al.*, 1982).

Data on nitrification rates and efficiencies in trickling filters are scarce and somewhat confusing. Recently attempts were made to evaluate trickling filter performance in a systematic and consistent way to allow use of such data for design (Gullicks & Cleasby, 1986). Boller and Gujer (1986), based on pilot-plant studies of tertiary trickling filters, recommend a medium surface loading rate of 0·4 g NH_3-N/m^2. day for complete nitrification (effluent NH_3-N $<$ 2·0 mg/litre) at a water temperature of 10°C. Data compiled by Barnes and Bliss (1983) recommend a loading range of 0·5–1·0 g NH_3-N/m^2. day for plastic medium filters at temperatures ranging from 10 to 20°C. The effect of nitrogen loading on nitrification efficiency is shown in Fig. 3. This figure was made using pilot-plant data reported by Jiumm *et al.* (1982).

Nitrification in an RBC system normally takes place only when the bulk liquid BOD concentration is below 15 mg/litre. Data compiled by Heidman *et al.* (1984) show that for liquid ammonium concentrations exceeding 5 mg/litre nitrification is zero-order at a constant rate of $1{\cdot}5 \times 10^{-3}$ kg NH_4^+-N/m^2. day. This rate drops gradually as the ammonium concentration is decreased and is practically zero at an ammonium concentration of 0·5 mg/litre. It has been concluded that the maximum nitrification rate is limited by oxygen transfer and hence is unchanged with increasing temperature above 13°C.

Denitrification in fixed-film reactors can be accomplished in a variety of column configurations using various media to support the growth of denitrifiers. In all cases oxygen must be excluded from the column and an adequate carbon source must be present.

Submerged packed-bed reactors use granular media (e.g. gravel) or plastic media similar to that used in trickling filters. Fluidized-bed reactors typically use sand as support medium. Gas-filled columns use plastic media and nitrogen gas to fill the void space. The rate of denitrification in fixed-film reactors depends on the concentration of biomass which is related to the specific surface area of the support medium. It also depends on the nature of the carbon source, the temperature and other environmental factors. For submerged packed beds with plastic media the reported rates range from 0·06 to 0·4 kg NO_3^--N/m^3. day for temperatures between 5 and 20°C. Fluidized beds using fine media have denitrification rates as high as 20 kg NO_3^--N/m^3. day (USEPA, 1975).

4 BIOLOGICAL PHOSPHORUS REMOVAL

As indicated in the preceding sections, the only way soluble phosphorus can be removed from wastewater is through its incorporation in particulate matter which is subsequently separated from the liquid stream. In a biological phosphorus removal system the particulate matter is bacterial cells, while in chemical systems it is a mineral precipitate. Some phosphorus removal processes combine chemical and biological mechanisms.

In conventional activated sludge processes the typical phosphorus content of microbial cells is 1·5–2·0% on a dry weight basis. Since the net sludge yield in such systems is approximately 0·5 kg VSS/kg BOD, the anticipated ratio between phosphorus and BOD removal ranges from 0·0075 to 0·01. Thus in a wastewater containing 150 mg/litre BOD the anticipated biological removal of phosphorus ranges from 1·0 to 1·5 mg/litre. This is often insufficient to meet effluent quality requirements.

Experience in several full-scale plants has shown significant phosphorus removal capabilities, far in excess of what could be anticipated based on the normal phosphorus uptake for cellular growth. The term 'luxury uptake' of phosphorus was coined by Levin and Shapiro (1965) to describe this phenomenon. Although the exact mechanism of biological phosphorus removal was still unknown, some of the key requirements of such systems were identified. These included the need for plug flow conditions, inclusion of an initial anaerobic contact zone, elimination of nitrates from the anaerobic zone, and the presence of readily degradable organics in the anaerobic zone. These observations led subsequently to a better understanding of the phosphorus removal mechanism and the development of several treatment processes, including some proprietary schemes, for biological phosphorus removal.

4.1 Biological Phosphorus Removal Mechanism

All biological phosphorus removal processes involve the alternate exposure of the activated sludge to anaerobic and aerobic conditions. The simplest continuous flow system incorporating this feature is the one shown in Fig. 4(a). The anaerobic contact zone in this system functions as a biological selector for phosphorus-storing microorganisms. It provides a competitive advantage to those microorganisms since they can take up carbonaceous substrate in this zone before other microorganisms can. The energy required for this uptake is derived from the depolymerization of

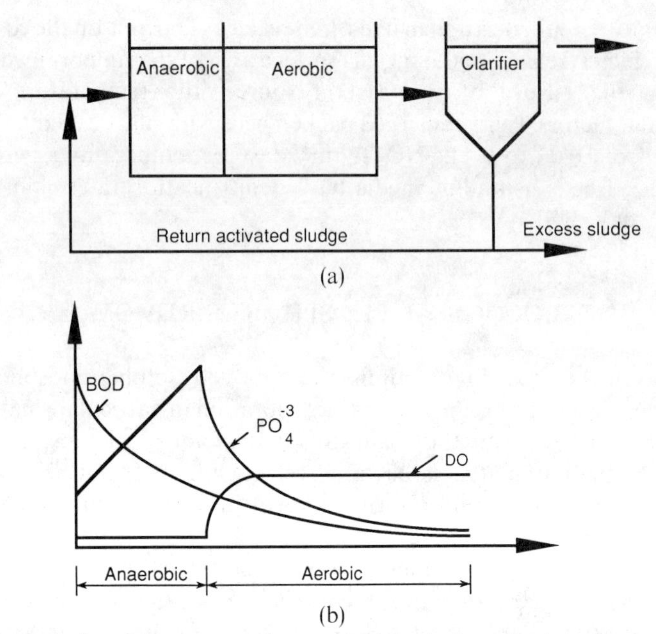

Fig. 4. Biological phosphorus removal system: (a) basic flow diagram (Phoredox or A/O® process); (b) profile of soluble constituents.

polyphosphates which were generated previously under aerobic conditions. A simplified illustration of the commonly accepted mechanism is depicted in Fig. 5. A more detailed biochemical pathway for biological phosphorus removal was proposed by Tracy and Flammino (1987). A profile of DO, BOD and phosphorus concentrations as observed in biological phosphorus removal plants is shown in Fig. 4(b). It is in agreement with the proposed phosphorus removal mechanism.

It is generally accepted that the carbonaceous substrate necessary for preferential uptake by phosphorus removing organisms should consist of simple, readily degradable compounds. In domestic wastewater such compounds are the volatile fatty acids (VFAs) which are the fermentation products of soluble organics. Thus, in the absence of VFAs in the raw wastewater, the anaerobic basin must also function as an anaerobic fermenter where soluble organics are converted to VFAs. The molar ratio of VFA utilization to phosphorus release varies from 0·6 (Rabinowitz, 1985) to 1·3 (Rensink *et al.*, 1981).

Phosphorus uptake in the aerobic basin is related to the biodegradation of carbonaceous substrates. These substrates consist of storage products as well as residual organics which were not taken up in the anaerobic basin. The phosphorus uptake rate was found to be zero-order with respect to phosphorus when the incoming BOD consists of simple organic compounds, and first-order when the influent BOD is made of more complex organics (Tracy & Flammino, 1985).

The presence of nitrates in the anaerobic basin may interfere with phosphorus removal, since denitrifying organisms can compete with phosphorus removing organisms for the available organics. It was shown, however, that phosphorus

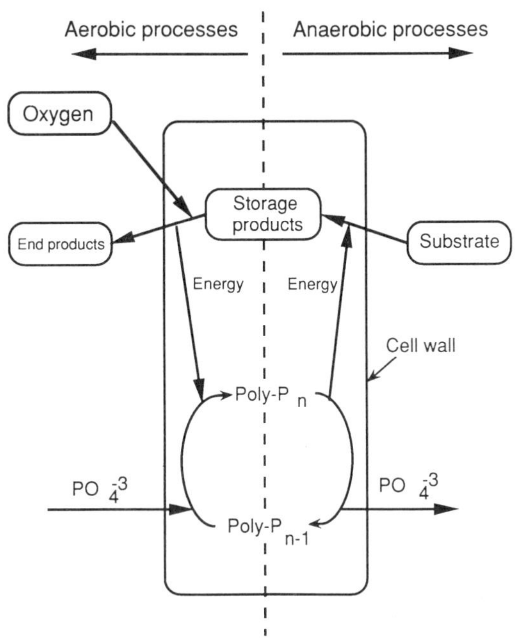

Fig. 5. Simplified illustration of biological phosphorus removal mechanism.

removing organisms such as *Acinetobacter* and others are capable of denitrification (Lotter, 1984), and uptake of phosphorus can take place in a denitrifying environment (Gerber *et al.*, 1986). Based on energetic considerations, Tracy and Flammino (1987) concluded that under a high denitrification rate sufficient ATP (adenosine triphosphate) is generated to drive the polyphosphate polymerization reaction so that phosphate uptake and denitrification take place simultaneously.

4.2 Phosphorus Removal Systems

Two activated sludge systems were developed for phosphorus removal. One is based on biological processes only while the other combines biological and chemical components.

4.2.1 Phoredox or A/O® Process[a]

The flow scheme for this process is shown in Fig. 4(a). This is a high rate system operated at short SRT in order to minimize nitrification. The anaerobic stage is added in order to enhance the selective growth of polyphosphate storing organisms. Since nitrification is minimized there are no significant quantities of nitrates in the return activated sludge (RAS). Hence the competition between denitrifiers and phosphorus removing organisms for the organic substrate in the anaerobic zone is minimized. Since the system is operated at a high organic loading rate the net sludge yield is relatively high, thus increasing the phosphorus/BOD removal ratio.

[a] ® Air Products and Chemicals, Inc., Allertown, PA, USA.

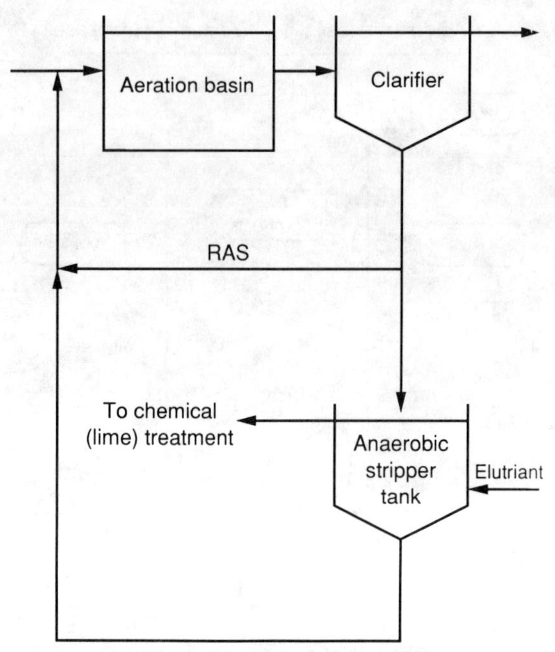

Fig. 6. PhoStrip® process.

4.2.2 PhoStrip® Process[a]

This process combines biological and chemical removal of phosphorus. It is also classified as a 'side-stream' process since phosphorus removal is actually accomplished by treating a side-stream flow of sludge. A flow diagram of the standard PhoStrip process is shown in Fig. 6. The main flow stream of the process is essentially an activated sludge process consisting of an aeration basin and a secondary clarifier. A portion of the RAS is diverted into a stripper tank where anaerobic conditions exist. Under these conditions soluble phosphorus is released from the RAS microorganisms. The soluble phosphorus is washed from the RAS by the addition of an elutriant. The overflow from the stripper tank is treated chemically by lime addition to precipitate the phosphorus. The underflow is returned to the aeration basin with the remaining RAS. The PhoStrip process offers significant saving in chemical usage compared to mainstream lime treatment for phosphorus removal.

4.3 Combined Nitrogen and Phosphorus Removal Systems

Most effluent discharge permits require reduced levels of both nitrogen and phosphorus. Removal of both contaminants can be achieved in a single activated sludge system which is properly designed and operated. A variety of processes have been developed for this purpose. All processes consist of the basic activated sludge process with additional anaerobic and anoxic zones and various flow and recycle

[a] ®Biospherics, Inc., Rockville, MD, USA.

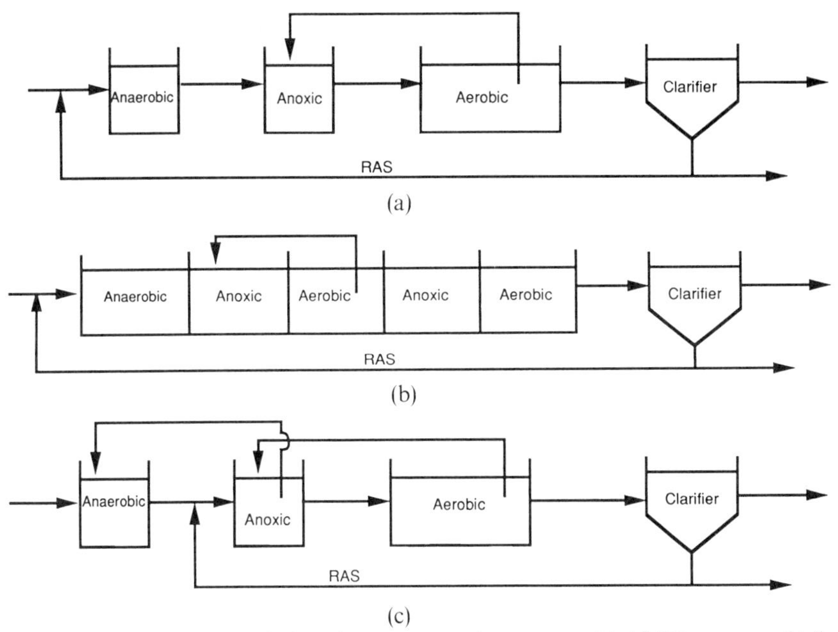

Fig. 7. Combined nitrogen and phosphorus removal processes: (a) A^2/O^R process; (b) five-stage Bardenpho^R process; (c) UCT (and VIP) process.

configurations. For phosphorus removal and nitrification an anoxic zone is not essential. The PhoStrip^R and A/O^R processes described above can achieve these treatment goals if the SRT is sufficiently long to assure nitrification. If complete phosphorus and nitrogen removals are required, the anoxic zone is essential.

As indicated above, several alternative process configurations have been developed. Three alternative processes are shown in Fig. 7. The A^2/O^R process, shown in Fig. 7(a), is similar to the A/O^R process with an additional anoxic stage between the aerobic and anaerobic basins. This process can also be applied in situations where only nitrification and phosphorus removal are required. The anoxic stage, in this case, is included to reduce the nitrate loading on the anaerobic stage through the RAS.

The five-stage Bardenpho^R process is used for improved nitrogen and phosphorus removals. In addition to the basic anaerobic/anoxic/aerobic combination, two additional basins are included. A second anoxic stage is added for enhanced denitrification, and a second aerobic stage is aimed at ensuring aerobic conditions in the final clarifier, thus avoiding phosphorus release.

The University of Cape Town (UCT) and the Virginia Initiative Plant (VIP) processes also include the three-basin series similar to the A^2/O^R process but with a different recycle system. To achieve nitrification and denitrification, the mixed liquor from the aerobic basin is recycled to the anoxic basin and from the anoxic to the anaerobic basin. The RAS is recycled to the anoxic basin. With this flow scheme the nitrate loading to the anaerobic basin is minimized, thus avoiding interference with the phosphorus removal process.

4.4 Process Modelling and Design Considerations

The mathematical analysis and process design procedures for nitrification and nitrogen removal were discussed in Section 2. These same procedures apply to the nitrogen removal elements of a combined nitrogen and phosphorus removal system. Thus the aerobic and anoxic basins, and the mixed liquor recycle rate of the A^2/O®, Bardenpho® or UCT processes, are designed basically in the same way as in a nitrogen removal system. The design for phosphorus removal is based largely on empirical findings rather than a rigorous mathematical approach.

The only add-on element to a nitrogen removal system that is required for phosphorus removal is the anaerobic basin. This basin is designed for a hydraulic residence time of 0·9–2·0 h, based on influent flow rate (USEPA, 1987). However, the exact role of this basin and the effect of anaerobic residence time on phosphorus removal have yet to be investigated.

Another factor which appears to have a major impact on phosphorus removal is the ratio of fermentation products, i.e. VFAs, to phosphorus in the anaerobic basin. Although the role of VFAs in the phosphorus removal mechanism is fairly well understood, the quantitative aspects of this subject are still vague. Because VFAs generated in an anaerobic basin are rapidly used, it is difficult to measure the quantity of VFAs available in such basins. Hence alternative parameters, such as readily degradable BOD (RDBOD), soluble BOD (SBOD) or total BOD (TBOD), were used by various investigators. Data compiled by the USEPA (1987) indicate that a minimum total BOD to total phosphorus (TBOD:TP) ratio of approximately 20 is required for significant phosphorus removal. It also shows that the effluent soluble phosphorus decreased from approximately 1·5 to 0·3 mg/litre as the TBOD:TP ratio increased from 20 to 40.

One approach to increasing the available VFAs in the anaerobic basin is to supplement the feed wastewater with the supernatant from a thickener where partial fermentation of primary sludge takes place. Alternatively, primary clarifiers may be operated as deep settler/thickener tanks and accomplish the same objectives. Such systems were termed activated primary sedimentation tanks (Barnard, 1984).

REFERENCES

Argaman, Y. (1981). *Water Research*, **15**, 841.

Barnard, J. L. (1974). *Water and Wastes Engineering*, **11**(7), 36; **11**(8), 41.

Barnard, J. L. (1984). *Water SA*, **10**(03).

Barnes, D. & Bliss, P. J. (1983). *Biological Control of Nitrogen in Wastewater Treatment*. E.& F. N. Spon, London.

Beccari, M., Passino, R., Ramadori, R. & Tandoi, V. (1983). *JWPCF*, **55**, 58.

Boller, M. & Gujer, W. (1986). *Water Research*, **10**, 1363.

Christensen, M. H. (1975). *Progress in Water Technology*, **7**(2), 339.

Downing, A. L., Painter, H. A. & Knowels, G. (1964). *Jour. Inst. Sew. Purif.*, **32**, 130.

Ekama, G. A., Marais, G. v. R. & Seibritz, I. P. (1984). In *Design and Operation of Nutrient Removal Activated Sludge Processes*. Water Research Commission, Pretoria 0001, South Africa.

Gerber, A., Mostert, E. S., Winter, C. T. & de Villiers, R. H. (1986). *Water Sci. Tech.*, **19**, 183.
Grady, C. P. L. Jr, Gujer, W., Henze, M., Marais, G. v. R. & Matsuo, T. (1986). *Water Sci. Tech.*, **18**, 47.
Gullicks, H. A. & Cleasby, J. L. (1986). *JWPCF*, **58**, 60.
Heidman, J. A., Brenner, R. C. & Gilbert, W. G. (1984). *Summary of design information on rotating biological contactors.* Office of Water, USEPA 430/9-84-008, US Environmental Protection Agency, Washington, DC.
Jiumm, M. H., C. Wu Yeun & Molof, A. (1982). In *Proceedings of First International Conference on Fixed-Film Biological Processes*, Kings Island, OH.
Levin, G. V. & Shapiro, J. (1965). *JWPCF*, **37**, 800.
Lotter, L. H. (1984). *Water Sci. Tech.*, **17**(11/12), 127.
Quirk, T. P. & Eckenfelder, W. W. Jr (1986). *JWPCF*, **58**, 932.
Rabinowitz, B. (1985). *The role of specific substrates in excess biological phosphorus removal.* PhD thesis, The University of British Columbia, Vancouver, British Columbia, Canada.
Rensink, J. H., Donker, H. J. G. W. & de Vries, H. P. (1981). Presented at *5th European Sewage and Refuse Symposium*, Munich, FRG, Proc. 487-502.
Tracy, K. D. & Flammino, A. (1985). Presented at *58th Annual Water Pollution Control Federation Conference*, Kansas City, MO.
Tracy, K. D. & Flammino, A. (1987). In *Biological Phosphorus Removal from Wastewaters*, ed. R. Ramadori. Pergamon Press, London, p. 15.
USEPA (1973). Nitrogen compounds in the environment. In *Hazardous Materials Advisory Committee*, EPA-ABS-73-001. US Environmental Protection Agency, Washington, DC.
USEPA (1975). *Process Design Manual for Nitrogen Control.* US Environmental Protection Agency, Washington, DC.
USEPA (1987). *Design Manual: Phosphorus Removal*, EPA/62-1/1-87/001. US Environmental Protection Agency, Washington, DC.

Chapter 5

BIOSENSORS IN BIODEGRADATION OF WASTES

ISAO KARUBE & HIDEAKI ENDO

Research Center for Advanced Science and Technology, University of Tokyo, Japan

CONTENTS

1 INTRODUCTION

The monitoring of the pollutants in waste water is very important for environmental control. As various compounds are contained in test samples, methods for the selective determination of these compounds are required. Most analysis of pollutants such as organic compounds can be performed by spectrophotometric methods. However, these methods required a long reaction time and complicated procedures.

On the other hand, electrochemical monitoring devices employing immobilized biocatalysts (biosensors) have been developed in recent years. Biosensors have distinct advantages. Firstly, test samples do not need to be optically clear and can be measured over a wide range of concentrations without pretreatment. Secondly, biosensors also offer the possibility of real-time analysis, which is particularly important for the rapid measurement of analytes in industry, as in process monitoring and control where there is a demand for in-situ determination of flow rates, level of contaminants, etc. Therefore biosensors have been utilized in environmental, clinical and industrial process analyses.

In this chapter the principles and the application of biosensors for environmental control, mainly developed in our group, are described.

2 PRINCIPLES OF BIOSENSORS

The chemical reactions in organisms proceed smoothly because of the action of catalysts such as enzymes. They have specificity for chemical substances and can form complexes with them to promote reactions. These enzyme reactions consume the chemical substances and produce various chemicals which can be measured by electrodes and other devices. The enzyme and these devices may be combined to produce a biosensor with extremely good selectivity. A biosensor can be prepared by immobilizing the enzyme in a water-insoluble polymer membrane, and fixing it on an electrode (Fig. 1). For example, the enzyme glucose oxidase oxidizes glucose, causing oxygen to be consumed and leading to the production of gluconolactone and hydrogen peroxide. By measuring the oxygen consumed with an oxygen electrode, or by measuring the hydrogen peroxide with a hydrogen peroxide electrode, the concentration of glucose can readily be determined. This action forms the basis of a glucose sensor consisting of a glucose oxidase membrane and an oxygen or hydrogen peroxide electrode, and can be used for the diagnosis of diabetes. There are many other types of biosensors that have been developed using the same principles and devices.

Many organisms are known for their selectivity towards certain molecules. It is

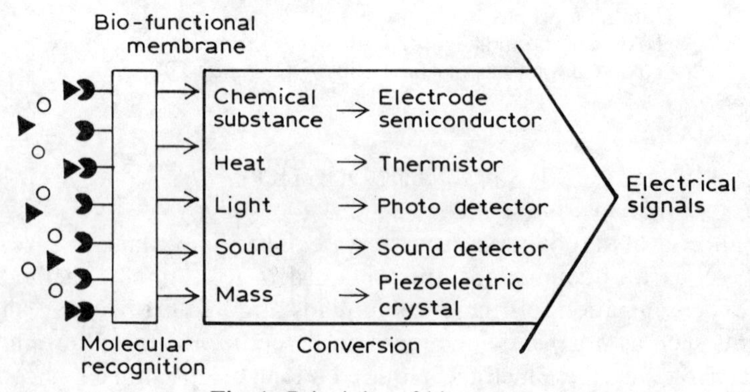

Fig. 1. Principles of biosensor.

also possible to use microorganisms as molecular recognition elements of sensors. Microbial sensors are currently being developed by combining a membrane on which microorganisms are immobilized with an electrode. Such sensors measure the assimilation capacity of the microorganisms and take as an index the respiration activity or metabolic activity. One such example is the combination of an alcohol-assimilating microorganism (yeast) immobilized on a membrane and an oxygen electrode, the oxygen consumption of which is measured at the oxygen electrode.

3 BIOSENSORS FOR ENVIRONMENTAL ANALYSES

3.1 BOD Sensor

Biochemical oxygen demand (BOD) is one of the most widely used and important indicators of organic pollution. The BOD test measures biodegradable organic compounds in waste waters. It requires a long incubation period of 5 days at 20°C, and considerable skill. Therefore rapid and reproducible methods are desirable for the BOD test. We have developed BOD sensors consisting of immobilized microorganisms and an oxygen electrode.

3.1.1 BOD Sensor Using Mesophilic Microorganisms

A microbial sensor for estimation of BOD was constructed from immobilized living whole cells of yeast and a Clark-type oxygen electrode (Karube *et al.*, 1977; Suzuki & Karube, 1978; Hikuma *et al.*, 1979, 1980). *Trichosporon cutaneum* was employed as the microbial sensor for BOD. The scheme of the BOD sensor is illustrated in Fig. 2. *T. cutaneum* was immobilized in porous acetylcellulose membrane by adsorption. A microorganism immobilizing membrane was placed on a gas permeable Teflon membrane of an oxygen electrode and covered with a dialysis membrane to construct a BOD sensor.

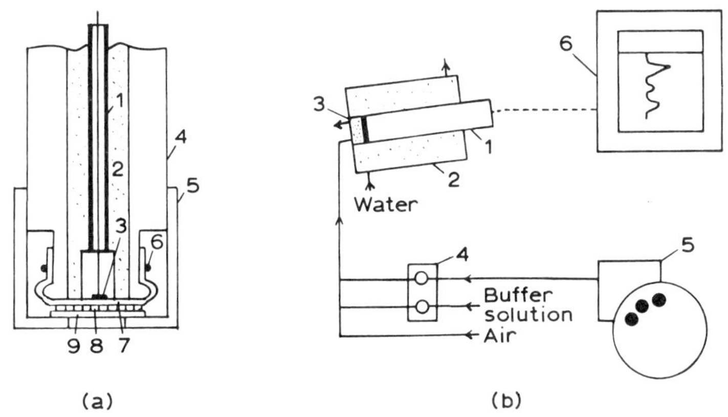

Fig. 2. Schematic diagram of the BOD sensor system. (a) Microbial electrode. **1**: Aluminum anode; **2**: electrolyte; **3**: platinum cathode; **4**: insulator; **5**: bored cap; **6**: O-ring; **7**: Teflon membrane; **8**: immobilized microorganisms; **9**: acetylcellulose. (b) Flow system. **1**: Microbial electrode; **2**: water jacket; **3**: flow cell; **4**: peristaltic pump; **5**: sampler; **6**: recorder.

When the sensor was immersed in the sample solution containing glucose and glutamic acid, organic compounds permeated through the porous membrane and were assimilated by the immobilized microorganisms. Consumption of oxygen by the immobilized microorganisms began and caused a decrease in the dissolved oxygen around the membrane. As a result the current of the sensor decreased markedly with time until a steady state was reached within 18 min. The steady state indicated that the consumption of oxygen by microorganisms and the diffusion from the sample solution to the membrane were in equilibrium.

The steady-state current depended on the BOD of the sample solution. The response time of the sensor (time required for the current to reach the steady state) was dependent on the kind of sample solution. For example, the response time was 8 min for acetic acid solution and 18 min for glucose solution. Thus an injection time of 20 min was employed for further work.

A linear relationship was obtained between the current difference and the 5-day BOD of the standard solution (150 mg liter^{-1} glucose + 150 mg liter^{-1} glutamic acid; as 220 mg liter^{-1} BOD (Japanese industrial standard)) below 60 mg liter^{-1}. The minimum detectable BOD was 3 mg liter^{-1}. The current was reproducible within $\pm 6\%$ of the relative error when the BOD standard deviation was 1·2 mg liter^{-1} over 10 experiments.

As the respiration activity of the microorganisms depended on pH and temperature, their effects on the microbial sensor were examined. BOD standard solution (30 mg liter^{-1}) was employed for experiments. The current difference of the microbial sensor was at a maximum between pH 6 and 8. Therefore a buffer solution of pH 7·0 was employed for further work. The influence of temperature on the output current of the sensor was examined. The amount of dissolved oxygen in the phosphate buffer decreased with an increase in temperature. Therefore the results were corrected on the basis of the amount of dissolved oxygen determined at various temperatures. The current difference of the sensor was almost constant from 25 to 35°C. However, the current difference decreased with an increase in temperature over 35°C because the yeasts in the membrane were inactivated by heat.

The influence of salts on the activity of microorganisms is well known; therefore the influence of sodium chloride, sodium sulfate, ammonium sulfate and potassium phosphate on the current of the sensor was examined. The current difference increased with an increase in the concentration of ammonium sulfate. The difference in current increase was found to be 0·13 μA (corresponding to BOD 17 mg liter^{-1}) per 1% of ammonium sulfate but 0·09 μA (corresponding to BOD 11 mg liter^{-1}) per 1% of sodium chloride, sodium sulfate and potassium phosphate monobasic chloride. Therefore the current of the sensor had to be corrected on the basis of the amount of salts in waste water. Furthermore, the difference in current was decreased when water was transferred to the system instead of phosphate buffer. Phosphate ion might activate the respiration activity of the microorganism. Therefore diluted phosphate buffer was used for the system.

The sensor system was applied to the estimation of 5-day BOD of waste water from a fermentation factory. The 5-day BOD of the waste water was determined by the JIS method (Japanese Industrial Standard Committee). Good correlation was

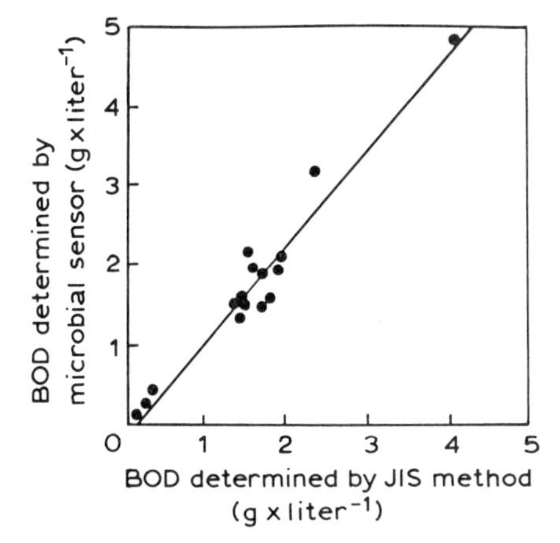

Fig. 3. Comparison of BOD values determined by microbial sensor and JIS 5-day method.

obtained between the values of BOD estimated by the sensor system and those determined by the JIS method (Fig. 3). The regression coefficient was 1·2 in 17 experiments and the ratios (BOD estimated by the microbial sensor/5-day BOD determined by the JIS method) were in the range from 0·85 to 1·36. This variation might have been caused by the change in composition of organic waste water compounds. Therefore the effect of various pure compounds on BOD estimated by the microbial sensor was examined. The results are shown in Table 1. The sensor showed low BOD value as compared with the 5-day BOD when lactose and soluble starch were employed for experiments. This can be due to the slow decomposition rate of these compounds by the immobilized yeasts. On the other hand, the sensor

Table 1
BOD Estimation by a Microbial Sensor and by Conventional 5-Day Method

Substrate	*BOD*	
	Microbial sensor (*g liter*$^{-1}$)	*5-day method* (*g liter*$^{-1}$)
Glucose	0·72	0·50–0·78
Fructose	0·54	0·71
Sucrose	0·36	0·49–0·76
Lactose	0·06	0·45–0·72
Soluble starch	0·07	0·22–0·71
Glycine	0·45	0·52–0·55
Glutamic acid	0·70	0·64
Histidine	0·35	0·55
Acetic acid	1·77	0·34–0·88
Lactic acid	0·72	0·63–0·88
Citric acid	0·17	0·40
Ethanol	2·90	0·93–1·67
Glycerol	0·51	0·62–0·83

showed high BOD values as compared with the 5-day BOD when acetic acid and ethyl alcohol were employed. These results suggested that the oxidation rates of acetate and ethyl alcohol are faster than those of some standard substrates such as glucose and glutamic acid. As shown in Table 1, no large difference was observed between the BOD as determined by both methods when other substrates were employed for experiments. Therefore waste waters containing substrates such as soluble starch showed low BOD values and those containing substrates such as acetic acid and alcohols showed high BOD values.

The long-term stability of the sensor was examined. Stable response to the standard solution (BOD 20 mg liter^{-1}) was observed for more than 17 days (400 tests). Fluctuations of the current and the base line (endogenous level) were within $\pm$20 and $\pm$15% respectively for 17 days. This sensor system was already commercialized.

3.1.2 A Novel BOD Sensor Using Thermophilic Microorganisms

The conventional BOD sensor system is ordinarily used at about 25–30°C since the microorganism used in the BOD sensor was a mesophilic microorganism. However, the temperature of the waste water is sometimes higher than the optimum temperature for mesophilic microorganisms. In such cases the BOD sensor using mesophilic microorganisms could not be applied.

Recently the use of thermophilic microorganisms (Karube *et al.*, 1989*a*) for industrial processes has received much attention. If thermophilic microorganisms are applied for the construction of the BOD sensor, the sensor can be used at a high temperature. In addition, employing high temperature may reduce the contamination of various microorganisms.

Considering this situation, we have recently developed a novel BOD sensor using thermophilic bacteria. The microorganism used in this biosensor was isolated from hot springs in Japan. These bacteria were cultured at 65°C with medium using glucose as a sole carbon source. The bacteria were immobilized on to a nitrocellulose membrane filter (0·45 μm pore size) by adsorption. This membrane was attached to an oxygen probe and was covered with a dialysis membrane to construct a BOD sensor.

When a sample solution containing glucose was injected into the sensor system at 50°C, the output current of the sensor gradually decreased, and within 7 min a steady-state current was observed.

The effect of temperature on this biosensor response was examined. The maximum current decrease was observed at 65°C, while almost no response was observed at 70°C. However, even after 2 h of incubation at 70°C, this sensor showed a good reproducible response to glucose at 50°C. Therefore this microbial sensor retained a high temperature stability. In further experiments this sensor was employed at 50°C.

The sensor response towards various organic compounds such as sugars, alcohol and amino acids was examined. The results are summarized in Table 2. The response values are expressed as the relative response (%) compared with glucose as the standard. The sensor system possessed broad specificity towards various samples. A

Table 2
Response of Microbial Sensor Using Thermophilic Bacteria

Substrate	*Relative response* (%)	*Substrate*	*Relative response* (%)
Glucose	100	Glycine	25
Fructose	29	Glutamic acid	120
Galactose	98	Glycerin	96
Maltose	96	Ethanol	0
Sucrose	95	Sodium acetate	0
Lactose	61		

good linear relationship was observed between current decrease and substrate concentration to each sample.

The measurement of BOD was carried out using this sensor system with the 5-day BOD standard solution. A good linear correlation was observed between sensor response and BOD value. Therefore this microbial sensor using thermophilic bacteria can be utilized as a BOD sensor.

The sensor stability was also examined. The sensor signals toward glucose were stable and reproducible for more than 40 days (Fig. 4). The high thermostability of this sensor promises wide application to BOD measurement at various temperatures.

We also developed a flow-type BOD sensor using immobilized thermophilic bacteria. In this sensor a polarographic-type oxygen electrode was used considering long-term usage. After 30 s of sample injection, the current started to decrease and within 4 min the minimum current was observed. Thus it takes 25 min to measure a sample using this flow-type BOD sensor. A good linear correlation was observed between the current decreases and BOD values. The detectable range of BOD was 5–15 g liter^{-1}. Measuring the BOD values of river and lake, further improvement of sensitivity is necessary. Figure 5 shows the reproducibility of the sensor response by varying the operational temperature. The measurement was attempted by injecting

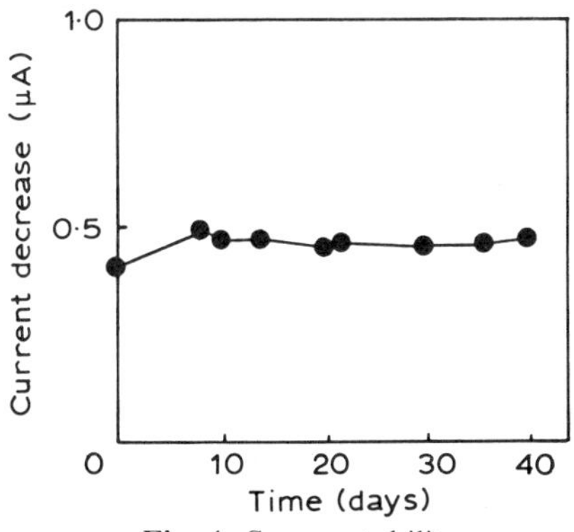

Fig. 4. Sensor stability.

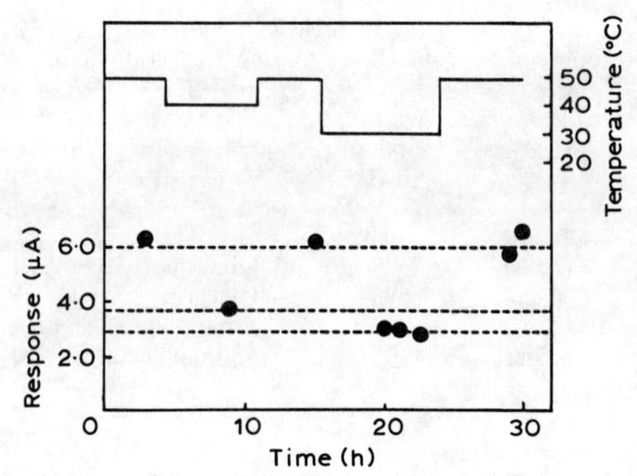

Fig. 5. Reproducibility of BOD sensor response by varying the operational temperature.

a sample containing 100 mM glucose. As can be seen from this figure, good reproducible signals were observed. Therefore, by monitoring the temperature of the sample simultaneously, the signal calibration is also made possible.

In conclusion, the BOD sensor using immobilized microorganisms appears quite promising and very attractive for the estimation of the 5-day BOD of industrial waste waters.

3.2 Dye Coupled Electrode System for Determination of Cell Number in Waste Water

For waste water processing it is necessary to detect cell numbers less than 10^6 cells ml^{-1}. A new electrode system using a fuel cell-type electrode, a redox dye and a porous acetylcellulose membrane for trapping microorganisms was developed (Nishikawa *et al.*, 1982). We have used this system to determine small numbers of cell populations in polluted waters.

The electrode was composed of a platinum anode (diameter 1·2 cm) and a silver peroxide cathode (0·6 × 4 cm). A phosphate buffer (0·1M, pH 7·0) was used as the catholyte. An anion-exchange membrane (Selemion type AMV; Asahi Glass Co.) was used as a separator. The electric current was measured with a millivolt ammeter, and the signal obtained was displayed on a reactor.

Cell population was determined as follows. The cultured cells were centrifuged at 4°C and 8000*g*. An appropriate quantity of cells was suspended in 50 ml of a phosphate buffer (0·05M, pH 7·0) and kept at 0°C. Sample water containing an appropriate number of cells was dropped on to the membrane filter with slight suction. The microorganisms were retained on the surface of the membrane filter, which was attached to the surface of the platinum anode of the electrode by a retaining cap. The electrode was inserted into 50 ml of a phosphate buffer solution (0·05M, pH 7·0) containing microorganism, and the current was measured at 30°C. Then 5 ml of a redox dye solution (4 mM) was introduced into the buffer solution, and the current was measured.

Table 3
Effect of Various Redox Dye Additions

Dye	*Current increase* (μA)
Methylene blue	0·02
Methyl viologen	0·02
Tetrazonium red	0·02
Phenazine methosulfate	0·02
DCIP	0·30

When the electrode was immersed in water containing $1{\cdot}4 \times 10^7$ cells ml^{-1} of *Escherichia coli*, scarcely any current was generated. To amplify the current a redox dye was added to the sample solution containing *E. coli*, and the steady-state current generated was measured. Table 3 shows the effect of various redox dyes on the current generated. The currents were low when methylene blue, methyl viologen, tetrazolium red and phenazine methosulfate were used. On the other hand, 2,6-dichlorophenolindophenol (DCIP) gave a high current.

Microorganisms in water (100 ml) were concentrated on a membrane filter, and the membrane was fixed on the anode of the electrode. When the electrode was immersed into the phosphate buffer solution, a small current was obtained. Then DCIP (40 μl) was added to the solution, and the current was continuously measured. The current of the electrode increased markedly with the increasing reaction time. The response time (the time required to reach a steady-state current) was 15 min. The current, however, decreased markedly when the microorganisms were inactivated by heat treatment (110°C, 10 min).

Figure 6 shows the calibration curve of the electrode. Various numbers of microorganisms were concentrated on the membrane and fixed on the anode of the

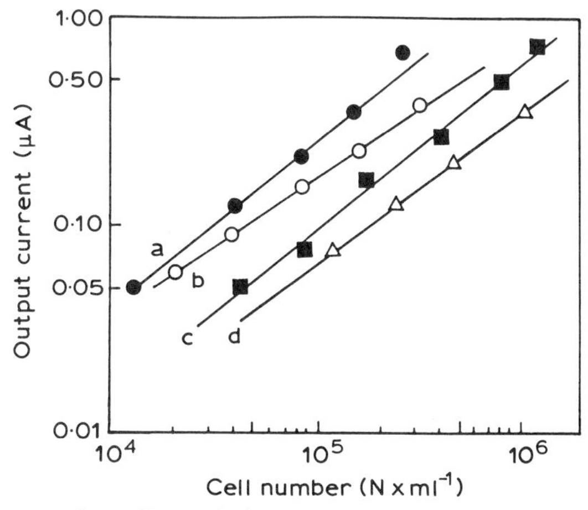

Fig. 6. Calibration curve for cell population. **a**: *Pseudomonas aeruginosa*; **b**: *Flavobacterium arborescens*; **c**: *Escherichia coli*; **d**: *Bacillus subtilis*.

electrode. Although the current generated varied from organism to organism when equivalent cell concentrations of different species were tested, a linear relationship was obtained between the current of the electrode and the number of cells measured by the colony count method above 10^4 cells ml^{-1}.

The electrode was then utilized for the determination of cell populations in industrial waters. Microorganisms in 2–100 ml (depending on the microorganism content) of industrial waters from a paper mill company and petrochemical company were concentrated on the membrane and attached to the anode of the electrode. The cell numbers were calculated from the calibration curve for *E. coli*. The cell population was also determined by the conventional colony count method. Although considerable deviation from a standard curve was observed, each test system showed a good correlation above cell numbers of 10^4 cells ml^{-1}.

3.3 Mutagen Sensor

Long-term carcinogenicity tests with laboratory mammals are not only time-consuming but also demanding on resources. In practice, it is inconceivable that resources could be made available (either men, monkeys or mice) on the necessary scale to screen all the tens of thousands of substances. Therefore testing with whole mammals is restricted to certain groups of suspected substances. If one is to screen for carcinogenic chemicals, one must use a short-term preliminary test with high predicted value.

The mutagenic activity of carcinogens has recently been confirmed in a great number of cases. The existence of a high correlation between the mutagenicity and carcinogenicity is now evident. The use of microbial systems is important for a survey of mutagenic chemicals. Recently a number of microbial methods for detecting the various types of mutagens have been developed. Microbial reversion assays using *Salmonella typhimurium* or *Escherichia coli* have been employed for screening tests of chemical carcinogens. A method named 'rec-assay' utilizing *Bacillus subtilis* has also been proposed for screening chemical mutagens and carcinogens. These methods are more rapid and simple than the carcinogen test using animals. However, the microbial reversion assays and the 'rec-assay' still require a lengthy incubation of bacteria and complicated procedures. Moreover, these methods are not suitable for continuous monitoring of environmental carcinogens and mutagens.

3.3.1 Mutagen Sensor Utilizing Rec-assay

In this sensor system *B. subtilis* recombination deficient strain M45 Rec^- and the wild strain H17 Rec^+ were utilized (Karube *et al.*, 1981*a*). The sensor system was composed of two microbial sensors immobilizing either Rec^- or Rec^+ *B. subtilis* (Fig. 7). The viability of immobilized microorganisms was determined by respiration using a Clark-type oxygen electrode.

When the Rec^- and Rec^+ electrodes were inserted into the glucose buffer solution (0·3 g $liter^{-1}$ glucose), steady-state currents were obtained. Then AF-2, a famous mutagen, was added to the solution. The time course of the electrode currents

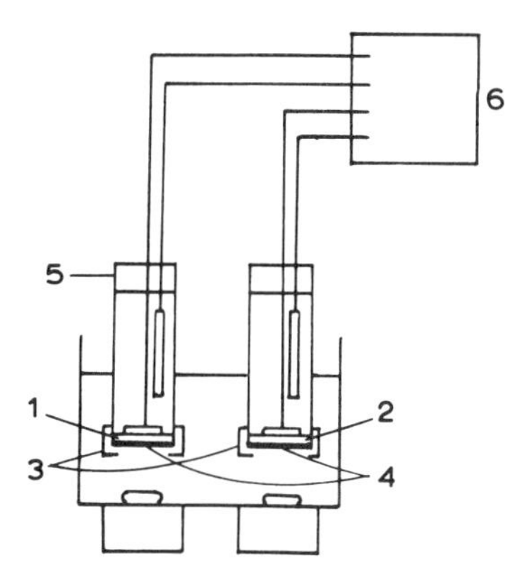

Fig. 7. Schematic diagram of a mutagen sensor based on rec-assay. **1**: *Bacillus subtilis* Rec$^+$; **2**: *Bacillus subtilis* Rec$^-$; **3**: Teflon membrane; **4**: membrane filter; **5**: oxygen electrode; **6**: recorder.

(response curves) is shown in Fig. 8. After 20–40 min, the current of the Rec$^-$ electrode began to increase, giving a sigmoidal curve. On the other hand, the current of the Rec$^+$ electrode did not increase. The rate of current increase is a measure of the mutagen concentration and is most easily measured as the linear slope at the midpoint of the sigmoidal curve. Table 4 summarizes the response of the electrode system to various typical chemical mutagens. When chemical mutagens such as AF-2, mitomycin, captan, 4NQO, *N*-methyl-*N'*-nitro-*N*-nitrosoguanidine (MNNG) and aflatoxin B_1 were added to the glucose buffer solution, the rates of current increase of the Rec$^-$ and Rec$^+$ electrodes were measured. The current of the Rec$^-$ electrode markedly increased when these reagents were added to the system. Therefore the mutagenicity of chemicals can be estimated with the electrochemical system.

Table 4 also shows the response of the Rec$^-$ and Rec$^+$ electrodes to chloramphenicol, streptomycin, penicillin, azide, cyanide and benzethonium chloride. Chloramphenicol and streptomycin are known as inhibitors of protein synthesis in bacteria. When 1·0 and 10 μg ml^{-1} of chloramphenicol were added to the system, the rates of current increase of both the Rec$^-$ and Rec$^+$ electrodes were 1 μA h^{-1}, respectively. The current of both electrodes did not increase when streptomycin (10 μg ml^{-1}) was employed. Therefore chloramphenicol and streptomycin are not mutagens. Penicillin is known as an inhibitor of the synthesis

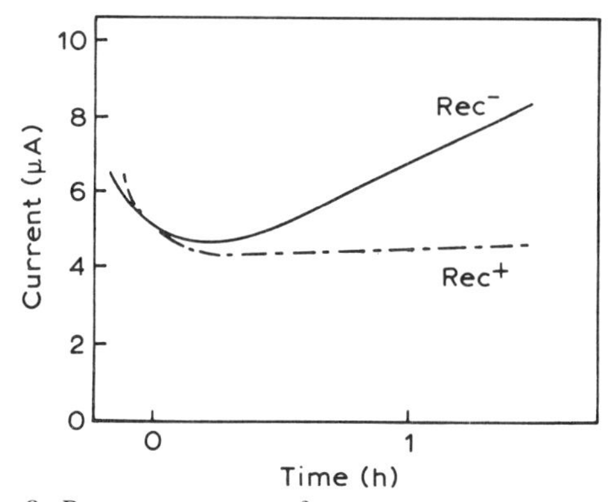

Fig. 8. Response curves of a mutagen sensor to AF-2.

Table 4
Response to Various Mutagens and Other Chemicals

Substrate	*Concentration* ($\mu g\ ml^{-1}$)	*Sensor response* ($\mu A\ h^{-1}$)	
		Rec$^-$	*Rec*$^+$
AF-2 (mutagen)	0·5	0	0
	1·6	2·7	0
	2·8	4·8	4·0
Mitomycin (mutagen)	0·9	0	0
	7·2	3·6	0·5
	14·4	12·0	11·0
Captan (mutagen)	0·5	0	0
	2·0	14·0	0
4-NQO	5·0	0	0
	16·0	11·0	0
N-methyl-*N'*-nitro-*N*-nitrosoguanidine	5·0	1·0	0·5
(mutagen)	20·0	22·0	1·0
Aflatoxin B1 (mutagen)	0·8	1·0	0
	12·0	12·0	1·0
NaN_3	50·0	8·0	9·0
Benzethonium chloride	20·0	16·0	16·0
Streptomycin	50·0	6·0	5·0
Cyanide	1·0	40·0	41·0
Chloramphenicol	10·0	0	0

of bacteria cell walls. Since current increase of the Rec$^-$ electrode was not observed, penicillin is also not a mutagen.

Azide and cyanide are inhibitors of cytochrome oxidase. They inhibit the respiration chain in the microbial cell. As shown in Table 4, the currents of both the Rec$^-$ and Rec$^+$ electrodes increased rapidly.

Benzethonium chloride is a known bactericide. When 20 and 50 mg ml^{-1} of benzethonium chloride were applied to the system, the currents of both the Rec$^-$ and Rec$^+$ electrodes increased at almost the same rates. Therefore, by utilizing this sensor system, mutagens and other toxic compounds can be simply distinguished.

3.3.2 Mutagen Sensor Utilizing Ames Test

A simple test using *Salmonella typhimurium* has been developed for testing chemical mutagens (Ames *et al.*, 1973*a,b*). The Ames test has shown that about 90% of the organic carcinogens tested thus far are mutagens. Now the Ames test is currently being used in over 2000 government, industrial and academic laboratories throughout the world.

S. typhimurium TA100 is a histidine-requiring bacterial mutant. This strain is reverted back to the wild type by chemical mutagens. Consequently it can grow in a medium even in the absence of histidine. The conventional Ames test is based on the colony method. Chemical mutagens are dispersed to the agar plate. Therefore 2 days of incubation are needed for the mutagen test. However, the growth of the

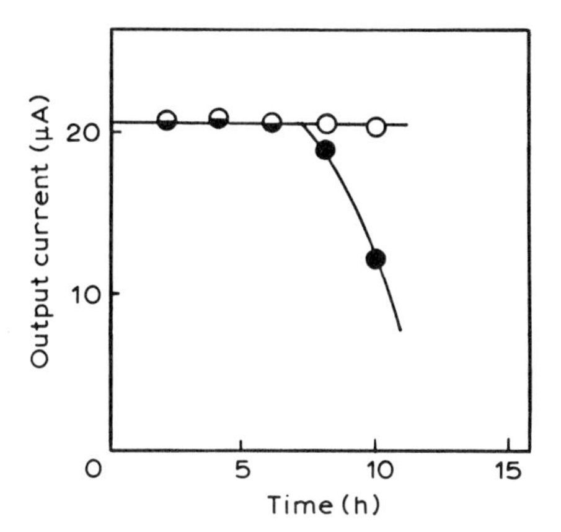

Fig. 9. Response to AF-2 for a mutagen sensor based on Ames test. ○: Control; ●: AF-2 (0·006 μg ml^{-1}).

revertants can be determined by the electrode system. Therefore we constructed a microbial sensor for mutagens utilizing *S. typhimurium.*

The following is the procedure for the determination of mutagens using this system (Karube *et al.*, 1982*a*). The histidine-free medium was employed for the rapid test of mutagen. The bacterial suspension (0·4 ml) was added to 9·5 ml of medium in a test tube. The mutagens, dissolved in 0·1 ml of water or dimethyl sulfoxide, were added to the mixture and incubated at 37°C for 10 h. After the incubation was completed, 1 ml of broth was dropped on to the membrane filter with slight suction, and the membrane filter retaining *S. typhimurium* was attached to the surface of the Teflon membrane of an oxygen electrode with holder. The electrode was inserted into the phosphate buffer solution (pH 7·0, 50 ml) containing 1 g liter^{-1} glucose, which was saturated with dissolved oxygen by stirring with a magnetic bar. The current was displayed continuously on a recorder and the steady-state current was measured.

S. typhimurium TA100 require histidine for their growth. However, the revertant of this strain can grow on the histidine-free medium. The electrode response depends on the numbers of viable bacteria retained on the membrane filter. Figure 9 shows the time course of the electrode response when various amounts of AF-2 were added to the medium. The current was measured at 2-h intervals. After 8 h of incubation, the current increased with increasing incubation time because the revertants grew above the minimum numbers of cells detectable by the electrode. A 10-h incubation gave the greatest sensitivity. On the other hand, there was no decrease in current from the medium in the absence of AF-2.

The relationship between the current decrease and AF-2 concentration was examined. The current decrease became large with increasing AF-2 concentration. The minimum measurable concentration of AF-2 was 0·001 μg ml^{-1}. The current decrease was reproducible within 5% of the relative error when a sample solution containing 0·006 μg ml^{-1} of AF-2 was employed. This sensor was tried out for the determinations of various mutagens. When *S. typhimurium* was incubated with chemical mutagens such as *N*-methyl-*N'*-nitro-*N*-nitrosoguanidine, nitrofurazone, methyl methanesulfonate and ethyl methanesulfonate, for 10 h the current decrease of the electrode was measured. The response of the electrode increased with increasing concentration of chemical mutagens. Therefore the mutagenicity of chemicals can be estimated with the microbial sensor.

3.3.3 Mutagen Sensor Utilizing a Phage Induction Test

A lysogenic strain of *Escherichia coli* was also utilized for the determination of mutagen. The Rec A protein was activated by the damage of DNA caused by some mutagen, and consequently caused phage induction of the lysogenic strain. Therefore, comparing the response toward chemicals of lysogenic and non-lysogenic strains, their mutagenicity could be estimated. Based on this theory a novel mutagen sensor was developed (Karube *et al.*, 1989*b*).

The sensor system was composed of two microorganism immobilizing oxygen electrodes. In each electrode either *E. coli* lysogenic strain or non-lysogenic strain were immobilized. The assay was employed as follows. AF-2, 4NQO and MNNG were dissolved in dimethylsulfoxide. Mitomycin C was dissolved in distilled water. The microbial sensor was immersed in a reaction vessel containing a cultivation medium. After the output current had reached the steady state, the mutagen prepared was injected into the reaction vessel. The electrode currents were displayed on a recorder. After 2 h of incubation, the output current from the electrode immobilized GY5027, lysogenic strain, gradually increased, while no current change was observed even after 4 h of incubation from the electrode immobilized GY5026, non-lysogenic strain. The current increase observed in GY5027 was reflected by the decrease of living cell numbers as the effect of AF-2 addition. However, cell numbers of GY5026 were not changed by the addition of AF-2. The difference in the effects of AF-2 addition between these two strains was due to the lysogenicity of GY5027. The addition of AF-2 caused the induction of phage lambda in GY5027, resulting in the lysis of GY5027. In other words, AF-2 caused the damage to GY5027 DNA, and activated Rec A protein. Therefore, by comparing the output current from two electrodes, the mutagenicity of AF-2 was estimated.

The effect of AF-2 concentration on the sensor response was then examined. Figure 10 shows the relationship between AF-2 concentration and the rate of current increase observed 2·5 h after the addition of AF-2. With high concentrations of AF-2, GY5026 also showed some deficiency in its respiration. This change was caused by mutation of some genes leading to deficiency in the respiration of both GY5027 and GY5026. However, the rate of mutation was less obvious than with phage induction. The response of GY5027 shows a good correlation between AF-2 concentration and the rate of current increase, in the range 0·01–0·2 μg ml^{-1}.

The estimation of mutagenicity of several mutagens by this sensor system was conducted. 4NQO, Mitomycin c and MNNG were tested by this sensor system. The output current from the electrode immobilized GY5027, lysogenic strain, gradually increased, while the output current from the electrode immobilized with non-lysogenic strain GY5026 did not. In other words, each substrate showed the typical response of the mutagen. The sensor responses to other chemicals, known to be non-

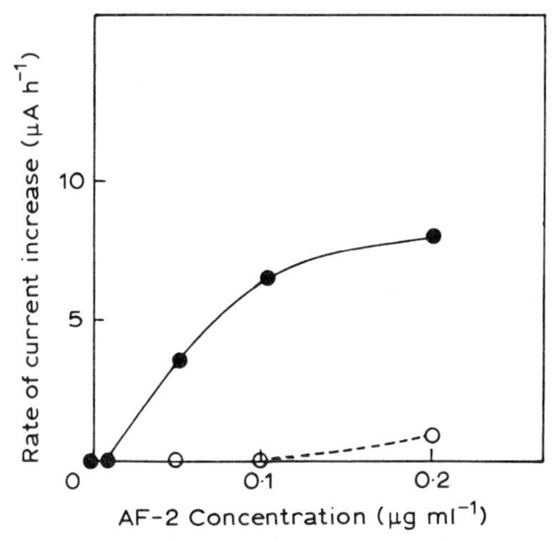

Fig. 10. Effect of AF-2 concentration on the response of a mutagen sensor based on induction test. ○: Response of GY5026 electrode; ●: response of GY5027 electrode.

mutagenic, were then investigated. Neomycin and kanamycin are the famous antibiotics inhibiting protein synthesis. Benzalkonium chloride is a bactericide. When these chemicals were added in the test solution, the output currents from both electrodes increased. There was no observed significant difference in the response between GY5026 and GY5027. The current increase was due to the decrease in the number of viable cells immobilized on to the electrode. These results suggested that using this sensor system it is easy to detect the difference between mutagens and other toxic compounds like antibiotics or bactericides.

3.4 Ammonia Sensor

The determination of ammonia is important in environmental, clinical and industrial process analysis. An ammonia gas electrode with a combined glass electrode and a gas permeable membrane is usually used for this purpose. In this case the determination must be performed under strong alkaline conditions (above pH 11). The ammonia electrode is based on potentiometric detection of ammonia. However, volatile compounds such as amines often interfere with the determination of ammonia. Therefore an ammonia sensor based on amperometry is desirable for the electrochemical determination of ammonia (Karube & Suzuki, 1980; Karube *et al.*, 1981*b*; Okada *et al.*, 1982).

Nitrifying bacteria comprise two genera: *Nitrosomonas* utilize ammonia as the sole source of energy and oxygen is consumed by the respiration; *Nitrobacter* oxidize nitrite to nitrate.

$$NH_3 + \tfrac{3}{2}O_2 \xrightarrow{\textit{Nitrosomonas}} NO_2^- + H_2O + H^+$$

$$NO_2^- + \tfrac{1}{2}O_2 \xrightarrow{\textit{Nitrobacter}} NO_3^-$$

The oxidation of both substrates (NH_3, NO_2^-) proceeds at a high rate, and oxygen

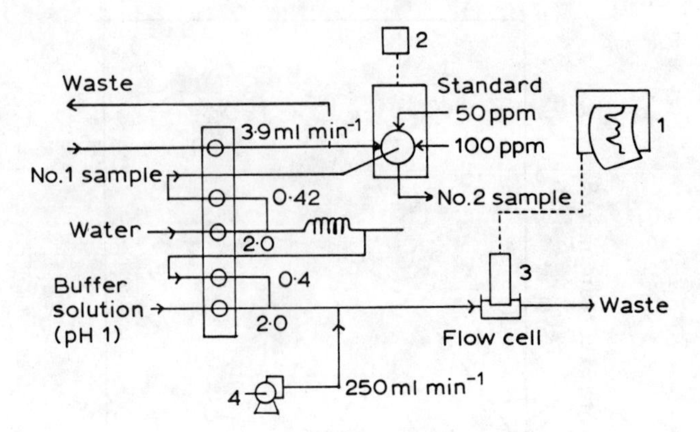

Fig. 11. Sensor system for continuous ammonia determination in waste water.

uptake by the bacteria can be directly determined by the oxygen electrode attached to the immobilized bacteria. Therefore ammonia is amperometrically determined by the microbial sensor using immobilized nitrifying bacteria and an oxygen electrode. The pH of a sample solution had to be kept sufficiently above the pK value for ammonia (9·1 at 30°C) because ammonium ions cannot pass through the gas permeable membrane.

The steady-state current was obtained depending on the concentration of ammonia. A good linear correlation was observed between the current decrease (the current difference between the initial and the steady state) and the ammonia concentration below 42 mg liter^{-1} (current decrease 4·7 μA). The minimum concentration for the determination of ammonia was 0·1 mg liter^{-1}. The reproducibility of the current decrease was within $\pm$4% of the relative error when a sample solution containing 21 mg liter^{-1} of ammonium hydroxide was used. The standard deviation was 0·7 mg liter^{-1} over 20 experiments.

Thus the amperometric determination of ammonia became possible by the microbial sensor. The sensitivity of the microbial sensor was almost at the same level as that of a glass electrode, and its minimum measurable concentration was 0·1 mg liter^{-1}. The selectivity of the microbial sensor for ammonia was examined. The sensor did not respond to volatile compounds such as acetic acid, ethyl alcohol and amines (diethylamine, propylamine and butylamine) or to involatile nutrients such as glucose, amino acids and metal ions (potassium ion, calcium ion and zinc ion).

Nitrifying bacteria, which utilize ammonia, did not assimilate acetic acid, ethyl alcohol, volatile amines, glucose, amino acids and metal ions. Therefore these substances did not affect the determination of ammonia by the sensor. Therefore selectivity of this microbial electrode became satisfactory. The long-term stability to the microbial sensor was examined with sample solution containing 33 mg liter^{-1} of ammonia. The microbial sensor was applied to the determination of ammonia in waste waters. The sample was diluted with glycine–NaCl–NaOH buffer (50 times) and employed for experiments. The concentration of ammonia was determined by electrochemical sensor and conventional methods.

Good comparative results were obtained between ammonia concentrations determined by both methods (correlation coefficient 0·9). Therefore the proposed sensor showed an economical and reliable method for the assay of ammonia. Figure 11 shows the sensor system for continuous ammonia determination in waste water. Ammonia in waste waters of a nitrification test plant could be determined by the system.

3.5 Nitrogen Dioxide Sensor

The principal gaseous oxides of nitrogen (NO, NO_2) are of great importance in air pollution sampling and analysis. During the combustion of all types of fossil fuels at flame temperature, less than 0·1% up to about 0·5% nitric oxide is formed, along with much smaller amounts of nitrogen dioxide. When discharged to the atmosphere, nitric oxide oxidizes at a measurable rate to nitrogen dioxide. Nitrogen dioxide is the most reactive of the gaseous oxides of nitrogen and is primarily an absorber of sunlight in atmospheric reactions that produce photochemical smog. Therefore the determination of nitrogen dioxide is important in environmental analyses.

We developed a microbial sensor for the determination of nitrite based on *Nitrobacter* sp. (Karube *et al.*, 1982*b*; Okada *et al.*, 1983). This microorganism utilizes nitrite as the source of energy, and oxygen is consumed by the respiration as follows:

$$2NO_2^- + O_2 \xrightarrow{\textit{Nitrobacter}\ \text{sp.}} 2NO_3^-$$

Oxygen uptake by the immobilized bacteria can be directly determined by an oxygen electrode, forming the basis of the measurement of NO_2 generated in the buffer (pH 2·0).

When the sample solution (sodium nitrate solution) is introduced into the flow cell, nitrous ions are changed to nitrogen dioxide gas at pH 2·0. The nitrogen dioxide passes through the gas permeable membrane, and is changed to nitrous ions in the bacterial layer, which are then utilized by *Nitrobacter* sp. as their sole source of energy. The consumption of oxygen around the membrane is determined by the oxygen electrode. As a result the concentration of sodium nitrite can be indirectly determined from the current decrease at the oxygen electrode.

The background current at time zero was obtained with the buffer solution saturated with dissolved oxygen and it showed the endogenous respiration level of the immobilized bacteria. When the sample solution containing sodium nitrite was injected into the system for 2 min, nitrogen dioxide was produced in the flow cell and permeated through the gas permeable membrane. Nitrite was formed in the bacterial layer and was assimilated by the immobilized bacteria. Consumption of oxygen by the bacteria began and caused a decrease in dissolved oxygen around the membrane. As a result the current of the electrode decreased markedly with time until a steady state was reached, which was usually within 10 min. At this time the consumption of oxygen by the bacteria and the diffusing species from the sample solution to the membrane were in equilibrium.

When a sufficient quantity of the bacteria is immobilized at the sensor, the current of the sensor for a sodium nitrite solution depends mainly on the rate of diffusion of nitrite from the sample solution to the immobilized bacteria. The O_2 contents in the sample solutions were checked using a differential sensor system with a sensor without an immobilized bacteria layer as a reference to negate the influence of oxygen in the sample solution, so that the steady-state current depended on the concentration of sodium nitrite alone. The differences between the initial and steady-state currents were directly proportional to the concentration of sodium nitrite. When only buffer solution was transferred to the flow cell, the current of the microbial sensor returned to its initial level within 12 min. The next sample could then be determined by continuously using the same sensor system.

A linear relationship was obtained between the current decrease and the sodium nitrite concentration below 0·59 mM. At greater than 0·65 mM sodium nitrite a linear relationship was not observed between the current and concentrations. The minimum concentration for the determination of sodium nitrite was 0·01 mM. The reproducibility of the current decrease was examined by using the same sample. The current decrease was reproducible within $\pm 4\%$ of the relative error and the standard deviation was 0·01 mM in 25 experiments when a sample solution containing 0·25 mM of sodium nitrite was employed.

Thus the amperometric determination of sodium nitrite became possible using the microbial sensor. The selectivity of the sensor for sodium nitrite was examined. The sensor did not respond to volatile compounds such as acetic acid, ethyl alcohol and amines (diethylamine and butylamine) or to involatile nutrients such as glucose, amino acids and ions (potassium and sodium ions). Therefore the selectivity of this microbial sensor was satisfactory in the presence of these different substances. The long-term stability of the microbial sensor was examined with a sample solution containing 0·25 mM sodium nitrite. The current output of the sensor was constant for more than 21 days and 400 assays. In the same experiments the concentration of sodium nitrite was determined by both the sensor proposed and the conventional method (dimethyl-naphthylamine method) as reference. The results obtained for the sodium nitrite concentrations determined by the two methods showed a good correlation (correlation coefficient 0·99).

3.6 Methane Gas Sensor

Methane is a clean fuel and a major component of city gas (88% methane), but it forms an explosive mixture with air (5–14% methane). A methane oxidizing bacterium, which grows well, has been isolated from a natural source, grown in pure culture, and identified as a new species, *Methylomonas flagellata*.

M. flagellata utilizes methane as its sole source of energy, and oxygen is consumed by respiration as follows:

$$CH_4 + NADH_2 + O_2 \rightarrow CH_3OH + NAD + H_2O$$

M. flagellata was immobilized in acetylcellulose filters with agar. The schematic diagram of the methane gas sensor system (Okada *et al.*, 1981) is shown in Fig. 12.

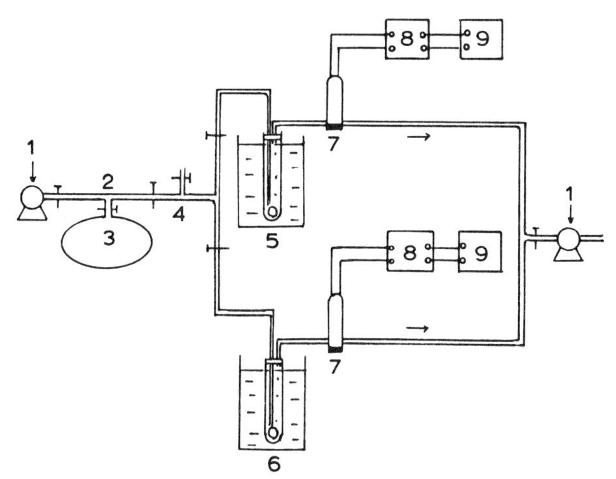

Fig. 12. Schematic diagram of methane gas sensor system. **1**: Pump; **2**: gas sampler; **3**: sample gas; **4**: cotton filter; **5**: reference reactor; **6**: methane oxidizing bacteria reactor; **7**: oxygen electrode; **8**: amplifier; **9**: recorder.

This system was composed of an immobilized microorganism reactor (300 mg immobilized cells per reactor), a control reactor and two oxygen electrodes. Methane gas was introduced into both reactors by a pump at a controlled flow rate. The difference between the output currents of the two electrodes was related to the amount of methane in the flow lines.

When the sample gas containing methane was transferred to the immobilized bacteria cells, methane was assimilated by the microorganisms. Oxygen was then consumed by the microorganisms so that the concentration of dissolved oxygen in the reactor decreased. The current decreased and reached a steady state, which indicated that the consumption of oxygen by the microorganisms and the diffusion of oxygen from the sample gas to the immobilized bacteria were in equilibrium. The steady-state current depended on the concentration of methane. When air was passed through the flow reactor, the current of the sensor returned to its initial level within 1 min. The response time required for the determination of methane gas was 1 min. The total time required for an assay of methane gas by this steady-state method was 2 min.

Figure 13 shows the calibration curve for this microbial sensor system. A linear relationship was observed between the current difference between the electrodes and the concentration of methane (below 6·6 mM). The minimum concentration for the determination was 13·1 μM. The current decrease was reproducible within $\pm 5\%$ in 25 experiments with sample gas containing 0·66 mM methane. The maximum current difference was observed at 30°C, which was constant below an air flow rate of 8·0 ml min^{-1} but above this flow rate the current difference decreased. The long-term stability of the microbial sensor was examined with a sample gas containing 0·66 mM methane. The current output was almost constant for more than 20 days and 500 assays. In the same experiment a good correlation was obtained between the methane concentrations determined by the electrochemical sensor and by

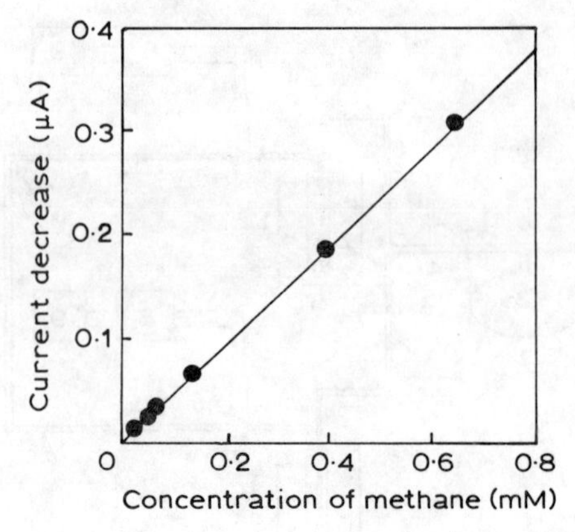

Fig. 13. Calibration curve for methane gas sensor. Operating conditions: 30°C, pH 7·2, gas flow velocity 80 ml min^{-1}, cell content 300 mg wet cells in 30 ml wet filter.

conventional gas chromatography methods (correlation coefficient 0·97). This sensor system can be used to determine the content of methane gas in the atmosphere. In conclusion, this microbial sensor system which used immobilized *M. flagellata* appears promising and is a very attractive system for the rapid and continuous determination of methane.

3.7 Phosphate Ion Sensor

The monitoring of phosphate ions is a very important factor in environmental and industrial processes. Inorganic phosphorus (Pi) has been determined by the colorimetric method; however, this method requires complicated and time-consuming procedures. Therefore the establishment of a rapid and simple method for the determination of phosphate ions was desired. An enzyme sensor for the determination of phosphate ions utilizing the following enzyme reaction was developed (Watanabe *et al.*, 1988):

$$\text{inosine} \xrightarrow[\text{nucleoside phosphorylase}]{\text{Pi}} \text{hypoxanthine} \xrightarrow[\text{xanthine oxidase}]{O_2} \text{uric acid}$$

The principle of measurement is based on the nucleoside phosphorylase catalyzed reaction, for which the presence of inorganic phosphorus is indispensable.

Xanthine oxidase and nucleoside phosphorylase were immobilized simultaneously on the triacetyl cellulose membrane containing 1,8-diamino-4-aminomethyloctane. The enzyme membrane was placed on the surface of an oxygen electrode and covered with a dialysis membrane to construct an enzyme electrode.

The measurement of phosphate ion was carried out with a flow system. The system consisted of an enzyme electrode, a flow cell, a peristaltic pump and a recorder. Tris-HCl buffer containing inosine was delivered continuously to the sensor system by a peristaltic pump at a constant flow rate. After the output current became steady, the standard solution of phosphoric acid (0·1–1·0 mM) or sample solution was injected directly into the flow line.

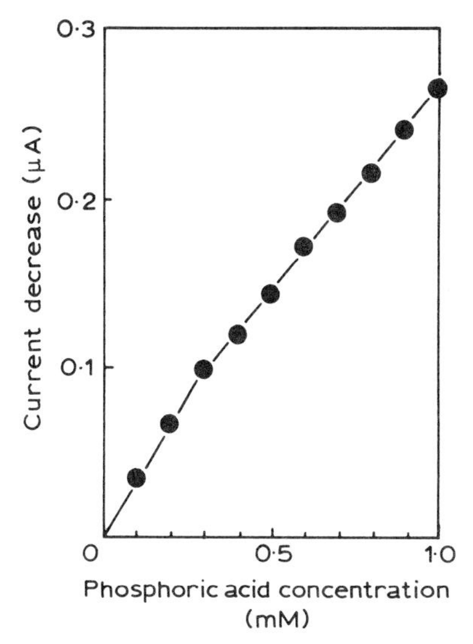

Fig. 14. Calibration curve for phosphate ion sensor. Operating conditions: 30°C, pH 7·0, flow rate 1·4 ml min^{-1}, inosine 200 mg liter^{-1}, sample volume 50 μl.

A 50-μl aliquot of 25-nM potassium dihydrogen orthophosphate (KH_2PO_4) solution was injected into the flow line with an inosine concentration of 0·2 mg ml^{-1}. After the injection of the sample, the output current began to decrease within 30 s, and then a minimum current was obtained within 60 s. One assay could be completed within 180 s.

The calibration curve for KH_2PO_4 is shown in Fig. 14. A good linear correlation was obtained in the range 0·3–1·0 mM KH_2PO_4.

The stability of the sensor was investigated using a standard solution (0·5 mM KH_2PO_4) under the optimum conditions. The current decrease was reproducible within 10% error. The standard deviation was ±0·04 mM in 70 assays. When the enzyme membrane was stored at 5°C, no appreciable decrease of the current was observed after 30 days.

3.8 Sulfite Ion Sensor

The determination of sulfite and sulfur dioxide in waste waters and atmosphere is important in any environmental analysis. There is an ever-increasing demand for new, simple and inexpensive methods for the measurement and control of sulfite and sulfur dioxide pollution.

Hepatic microsome is a subcellular organelle which contains many different oxidases and enzymatically oxidizes sulfite to sulfate with consumption of molecular oxygen. An amperometric organelle sensor for the determination of sulfite is composed of immobilized microsome particles, a gas permeable Teflon membrane and an oxygen electrode (Karube *et al.*, 1983).

Rat liver S9 fraction (100 μl) containing microsome was filtered through the porous acetylcellulose membrane. The quantity of organelle immobilized was equivalent to 2·7 mg protein. The microsome was retained on the acetylcellulose membrane. The organelle membrane was attached to the surface of the Teflon membrane of the oxygen electrode and covered with another Teflon membrane.

When the sample solution of sulfite was injected, sulfur dioxide permeated through the Teflon membrane and was oxidized by the microsome containing sulfite oxidase. Consumption of oxygen by the microsome began simultaneously and caused a decrease in the dissolved oxygen around the membrane. As a result the current of the sensor decreased markedly with time until a steady state was reached after 10 min. When a sufficient quantity of the microsome was employed in the sensor, the current of the sensor depended mainly on the rate of diffusion of sulfur dioxide from the sample solution to the immobilized organelle. The steady-state current thus depended on the concentration of sulfite ions.

The response of the organelle sensor increased with increasing amount of organelle, and became constant at an amount of organelle exceeding 2·7 mg protein. An organelle amount equivalent to 7·7 mg protein was therefore immobilized on the acetylcellulose membrane. In this case diffusion of substrate (sulfite) may also be rate determining.

A linear relationship was obtained between the steady-state current and the sulfite ion concentration below $3{\cdot}4 \times 10^{-4}$ M. The minimum concentration for determination was $0{\cdot}6 \times 10^{-4}$ M. The currents were reproducible with an average relative error of 7% when a sample solution containing $2{\cdot}8 \times 10^{-4}$ M of sulfite ion was used. The standard deviation was $0{\cdot}3 \times 10^{-4}$ M in 30 experiments.

The selectivity of the organelle sensor for sulfite is presented in Table 5. The sensor did not respond to non-volatile compounds such as sucrose, glucose, pyruvic acid, sulfate, phosphate ions and ammonium ions, since the organelle sensor was covered with the gas permeable Teflon membrane, and non-volatile nutrients could not penetrate through this membrane. Volatile compounds such as formic acid, acetic acid, propionic acid, *n*-butyric acid, methyl alcohol and ethyl alcohol can permeate through the Teflon membrane. However, no current was obtained from these compounds, because they are not oxidized by the enzyme system of the microsome. When the sensor was inserted into a sample solution containing $2{\cdot}8 \times 10^{-4}$ M nitrite, a current decrease was observed (2·0 μA). This shows that nitrite could interfere with

Table 5
Response of Sulfite Sensor to Various Compounds

Compound	*Current decrease (μA)*	*Compound*	*Current decrease (μA)*
SO_3^{2-}	8·5	CH_3OH	0
SO_4^{2-}	0	HCOOH	0
NO_2^-	2·0	NH_3	0
NO_3^-	0	CH_3COOH	0
C_2H_5OH	0	$C_6H_{12}O_6$	0

the determination of sulfite, but the response of the sensor to nitrite ion was only 23·5% of that to sulfite ion.

Although many enzyme electrodes have been developed and applied to clinical, food and environmental analyses, no such electrode has previously been developed for the determination of sulfite. This might be due to the difficulty of isolating sulfite oxidase, and for this reason the organelle was used as a source of the enzyme. The life of the organelles depends on storage conditions; they retain sulfite oxidase activity for 6 months when stored in a frozen state at −20°C. The electrode current is reproducible within an average relative error of 7% when the organelle membrane used for analysis is replaced with new membrane prepared from stored organelles. The stability of the organelle sensor proposed is still poor (life of 2 days and 20 assays) when used and stored at 30°C, because of the instability of the sulfite oxidase at this temperature. However, the preparation and exchange of an organelle membrane are very easy and replacement takes only 1 min.

3.9 Toxic Compounds Sensor

The quantification and identification of chemical toxicity are important in environmental control. Conventionally, the most reliable test for chemical toxicity is the test using the whole animal. However, animal testing requires a complicated and time-consuming procedure and is also very costly. Currently, ethical aspects of biological testing on whole animals are also an issue which has to be considered. Therefore the development of a simple test for the preliminary evaluation of chemical toxicity is required. For these purposes several tests utilizing cultured animal tissue cells have been proposed. These tests reduce the time and cost. However, there are some difficulties in the routine procedure and evaluation of such tests. Therefore we have developed biosensor systems for toxic compounds.

3.9.1 Toxic Compounds Sensor Utilizing Animal Cells

We have developed two types of biosensors utilizing Chang Liver cells for toxic compounds (Karube *et al.*, 1989*c*). The first sensor was based on the measurement of respiratory activity of tissue cells, and the second was based on measuring cell enzyme activity.

3.9.1.1 SENSOR SYSTEM BASED ON THE MEASUREMENT OF RESPIRATORY ACTIVITY

This sensor system was composed of immobilized Chang Liver cell membrane and oxygen electrode. The principle of the sensor is based on measuring the respiratory activity. Chang Liver cells were immobilized in porous membrane (pore size 8 μm) with a slight suction. The membrane retaining the cells was placed on the surface of the oxygen electrode and covered with a dialysis membrane to construct a sensor system.

In this study formalin was chosen as the typical toxic substance, because it especially inhibits respiration of the cells. When the sensor system was immersed in a sample solution containing formalin, the current of the sensor increased immediately. After 2 min, the output current reached a steady state. When 20 mM of

Table 6
Response of the Electrode to Various Compounds: Respiratory Activity Sensor

Compound	*Concentration* ($mg\ ml^{-1}$)	*Current increase* (μA)	*Concentration* ($mg\ ml^{-1}$) *at 3·0 μA output*	LD_{50} ($mg\ ml^{-1}$)
No compound		<0·9		
Formalin	0·015	1·0	0·05	800[a]
	0·47	12·5		
Sodium azide	0·02	1·2	0·22	18[b]
	0·49	5·6		
	1·0	8·5		
Benzethonium chloride	0·2	2·3	0·5	420[a]
	1·5	5·0		
Sodium benzoate	0·5	0·8	4·3	4 100[a]
	5·5	3·6		
Food red No. 3	1·0	−1·0	>5·0	300[c]
	1·9	0·3		
Food blue No. 1	2·5	−0·4	>5·0	2 000[a]
Food yellow No. 4	0·8	0·0	>10·0	12 750[d]
	10·0	−0·7		
Glucose	0·6	2·6	5·0	
	5·0	3·0		

[a] Oral administration in rats.
[b] Intraperitoneal in mice.
[c] Intraperitoneal in rats.
[d] Oral administration in mice.

sodium pyruvate was injected in the sensor system, the output current decreased with time. Autoclaved or sonicated cells were also immobilized on to the electrode and measurement applied. However, no current change was observed. In addition, the cell viability was also examined microscopically. After the measurement for formalin, the cells were resuspended in Dulbecco phosphate buffer (PBS) and stained by trypan blue. The cells which were treated with formalin had a deficiency in the selective permeation of the cell membrane. Therefore those cells were stained with trypan blue. The results showed that 80–90% of cells were stained after formalin treatment. The observed current increase of the sensor system thus represented the number of dead cells. In other words, this sensor system, which is based on the respiratory activity of tissue cells, can be utilized as the toxicity test. In further experiments the sensor response was normalized to the observed steady-state current after 30 min.

The sensor response towards various toxic substances was examined. Table 6 summarizes the response of the sensor to various chemical toxicities. Sodium azide and benzethonium chloride are known to inhibit specifically cytochrome a and a_3. The current increase obtained shows inhibition of respiration. Food dye showed lower current increase as compared with the other test compounds. Since these compounds must be metabolized by cells, the toxicity of these substances was

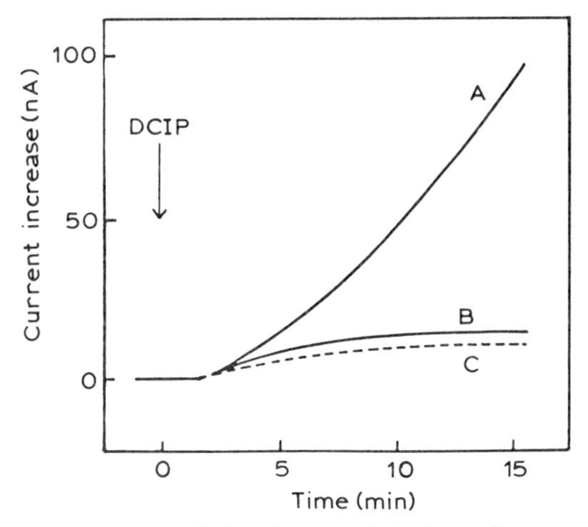

Fig. 15. Typical response curves of the dye coupled electrode system. **A**: Viable cells; **B**: autoclaved cells; **C**: without cells.

estimated to be low. However, the toxicity of glucose obtained from this sensor system was high. The effect of glucose on the respiration was widely accepted as the Crabtree effect. In this system this action was not distinguished from other inactivation by toxic compounds.

This sensor system can detect slight inactivation of cells caused by toxic compounds, and their toxicity.

3.9.1.2 SENSOR SYSTEM BASED ON THE MEASUREMENT OF CELLS' ENZYME ACTIVITY

We have developed another type of toxic compounds sensor utilizing Chang Liver cells. The sensor system was constructed from an immobilized Chang Liver cell, a platinum counter working electrode and a saturated calomel electrode. The principle of the sensor was based on sensing a redox dye, 2,6-dichlorophenolindophenol (DCIP), reduced by living cells with the electrode.

The typical response curves of the sensor system toward DCIP are shown in Fig. 15. Line A represents the response curve of the sensor using a membrane containing $1{\cdot}5 \times 10^7$ cells. In this case, after the addition of DCIP, the output current increased linearly, while the response of the sensor without cells (line B) and with autoclave treated cells (line C) showed only slight current increase. The current increase obtained was caused by the oxidation of the reduced form of DCIP, which occurred on the surface of living cells.

A toxicity test was carried out by the sensor system. Acetone (2 ml) was added to the sensor system (40 ml), and after 15 min of incubation the current increase was only 40% of that observed without acetone. This difference was caused by the inactivation of the cell viability by acetone. This equation was used to normalize the inactivation of cells' viability by toxic compounds:

$$\mathrm{AD}\,(\%) = \frac{A_1 - A_2}{A_1} \times 100$$

Table 7
Response of the Electrode to Various Compounds: Enzyme Activity Sensor

Compound	*Concentration (mg ml⁻¹)*	*Activity decrease (%)*	*AD_{50}[a] (mg ml⁻¹)*	*LD_{50} (mg kg⁻¹)*
No compound		20		
Acetone	7·8	18	43	5 300[b]
	103·0	90		
Benzethonium chloride	5·2	21	50	420[c]
	48·0	40		
Maneb[d]	0·07	21	0·083	
	0·2	92		
Sodium selenite	0·37	31	3·5	7[c]
	3·3	48		
Glucose	4·9	0		
	9·8	3		

[a] AD_{50} = concentration at 50% activity decrease.
[b] Oral administration in rabbits.
[c] Oral administration in rats.
[d] Dithiocarbamate pesticides.

where AD is activity decrease (%), A_1 is the observed current increase after 15 min incubation with DCIP, and A_2 is the observed current after 15 min incubation with DCIP and test substances.

A good correlation was obtained between acetone concentration and sensor response. Trypan blue test was also performed for the determination of cell viability after incubation with acetone. Cells incubated in solutions containing 7·1, 50 and 75 mg ml^{-1} of acetone were used. The ratios of dead cells were 22%, 42% and 75% respectively. Therefore the decrease in activity as measured by the sensor system reflected the number of dead cells.

Table 7 shows the results of toxicity for various compounds by the sensor system. AD_{50} was determined as the substance concentration giving 50% activity decrease. In each substance, with increase of concentration, activity also decreased. From this AD_{50} value the degree of toxicity of the chemicals was obviously distinguished. However, correlation with LD_{50} values obtained by conventional methods was not so good, because in this method tissue cells were used for the test, so the amount of toxicity could be overestimated compared with that obtained from animal tests. Since the impact of toxic compounds on tissue cells is more sensitive, this system is more useful for the preliminary screening of toxic compounds.

3.9.2 Toxic Compounds Sensor Utilizing Luminous Bacteria

A novel biosensor for the determination of toxic compounds has been developed using luminous bacteria (Sode *et al.*, 1991). The photomicrobial sensor system consists of luminous bacteria, *Photobacterium phosphoreum* MT10204, immobilized membrane and a photomultiplier. Measurement is based on the in-vivo intensity emitted by luminous bacteria affected by external conditions.

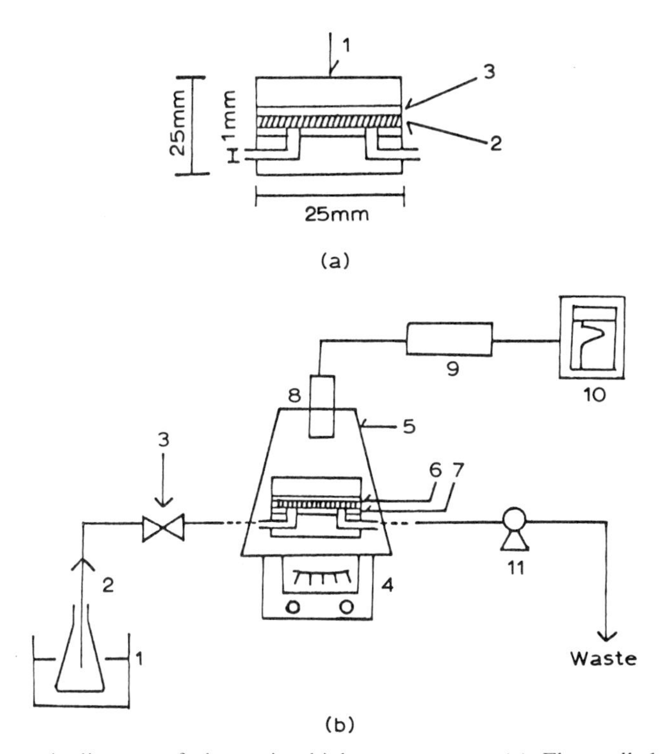

Fig. 16. Schematic diagram of photomicrobial sensor system. (a): Flow cell. **1**: Acryl-plate cover; **2**: microorganisms immobilizing membrane; **3**: silicon rubber spacer. (b): Flow system. **1**: Incubator; **2**: buffer solution reservoir; **3**: injection port; **4**: luminometer; **5**: dark box; **6**: immobilized bacteria; **7**: silicon rubber space; **8**: photomultiplier; **9**: electrometer; **10**: recorder; **11**: peristaltic pump.

P. phosphoreum MT10204 was immobilized in a cellulose nitrate membrane filter (pore size 0·45 μm) with slight suction. The immobilized membrane was fixed to a flow cell consisting of an acryl-plate cover and silicon rubber spacer (Fig. 16(a)). A schematic diagram of the sensor system is shown in Fig. 16(b). The system was constructed from a flow cell (cell volume 50 μl) containing the immobilized microorganism, a photometer equipped with photomultiplier, a peristaltic pump and a recorder. The flow cell surface faced the photomultiplier tube of a photometer.

The assay was employed as follows. The temperature of the flow cell was maintained at 30°C. The measurement was done by flow injection analysis (FIA). Using the peristaltic pump, phosphate buffer (pH 7·0) containing 3% NaCl was continuously transferred to the flow cell at a flow rate of 1·2 ml min^{-1}. When the base line of the luminescence intensity had reached a constant value, a sample was injected to the flow cell by microsyringe, and the observed luminescence intensity change was monitored by the recorder. The luminescence intensity change was expressed in arbitrary units.

The sensor response toward glucose was examined, since glucose is known to be an assimilable sugar of this microorganism. After the injection of 20 μl of 0·55 mM glucose solution, the emission intensity increased immediately, and within 1 min a maximum response was observed. The sensor response was recovered to the base line within 6 min at a flow rate of 1·2 ml min^{-1}.

The bioluminescence of the photobacteria was catalyzed by bacterial luciferase. This reaction required several co-factors, such as NADH, ATP and $FMNH_2$. Therefore the bioluminescence was strongly affected by the regeneration of the co-factors. It was assumed that the addition of glucose resulted in the co-factors' regeneration, subsequently increasing luminescence.

The time required for the recovery from the maximum sensor response to the base line was more rapid than with the other types of microbial sensors. For example, a microbial glucose sensor based on respiration change measurement required more than 10 min.

Since the photomultiplier was separated from the arrested flow-cell, the response properties of this sensor were determined only by the nature of the microorganism. In other words, the in-vivo reaction of photobacteria can be measured without the hindrance of other physico-chemical factors. In addition, the signal can be immediately detected with a high sensitivity by a photomultiplier.

Measurement of certain toxic compounds was carried out by this sensor system. Benzalkonium chloride (BC), sodium dodecyl sulfate (SDS) and chromium[VI] were chosen as test toxic compounds. When these compounds were injected into the system, the luminescence intensity decreased within several seconds and the base line shifted to a negative value. Figure 17 shows the relationship of the luminescence intensity and BC concentration. A good correlation was obtained between the decreased luminescence intensity and the amount of sample injected. BC is a famous

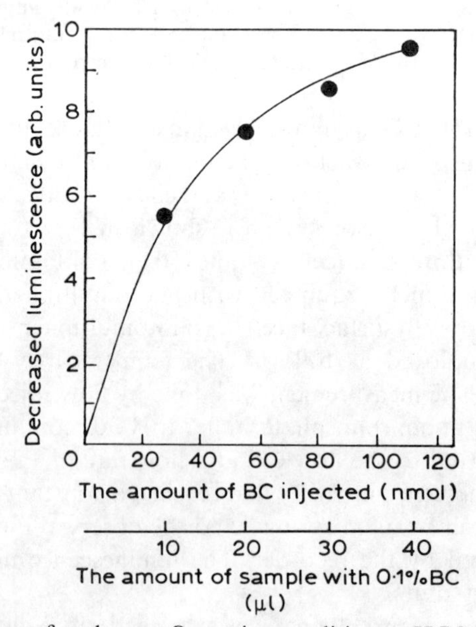

Fig. 17. Calibration curve for glucose. Operating conditions: pH 7·0, flow rate 1·2 ml min^{-1}, sample volume 20 μl, glucose concentration 0·55 mM.

bactericide. When BC was injected into the sensor system, it caused a decrease in the number of living cells or inhibited the enzyme related to the co-factor regeneration. In the same manner, measurement of SDS and chromium[VI] were also attempted. A good correlation was observed between the decreased light intensity and SDS concentration. The optimization in the measurement procedure (i.e. the amount of substrate injected) will improve the sensitivity and the detection spectra of environmental pollutants. The mechanism of luminescence inhibition by these environmental pollutants remains unknown; however, this system was found to be very useful for the estimation of substrate toxicity.

REFERENCES

Ames, B. N., Durston, W. E., Yamasaki, E. & Lee, F. D. (1973*a*). *Proc. Natl Acad. Sci. USA*, **70**, 2281.

Ames, B. N., Lee, F. D. & Durston, W. E. (1973*b*). *Proc. Natl Acad. Sci. USA*, **70**, 782.

Hikuma, M., Suzuki, H., Yasuda, T., Karube, I. & Suzuki, S. (1979). *European J. Appl. Microbiol. Biotechnol.*, **8**, 289.

Hikuma, M., Suzuki, H., Yasuda, T., Karube, I. & Suzuki, S. (1980). *European J. Appl. Microbiol. Biotechnol.*, **9**, 305.

Karube, I. & Suzuki, S. (1980). *Anal. Chem.*, **52**, 1020.

Karube, I., Matsunaga, T. & Suzuki, S. (1977). *J. Solid-Phase Biochem.*, **2**, 97.

Karube, I., Matsunaga, T., Nakahara, T. & Suzuki, S. (1981*a*). *Anal. Chem.*, **53**, 1024.

Karube, I., Okada, T. & Suzuki, S. (1981*b*). *Anal. Chem.*, **53**, 1852.

Karube, I., Nakahara, T., Matsunaga, T. & Suzuki, S. (1982*a*). *Anal. Chem.*, **54**, 1725.

Karube, I., Okada, T., Suzuki, S., Suzuki, H., Hikuma, M. & Yasuda, T. (1982*b*). *European J. Appl. Microbiol. Biotechnol.*, **15**, 127.

Karube, I., Sogabe, S., Matsunaga, T. & Suzuki, S. (1983). *European J. Appl. Microbiol. Biotechnol.*, **17**, 216.

Karube, I., Yokoyama, K., Sode, K. & Tamiya, E. (1989*a*). *Anal. Lett.*, **22**, 791.

Karube, I., Sode, K., Suzuki, M. & Nakahara, T. (1989*b*). *Anal. Chem.*, **61**, 2388.

Karube, I., Hiramoto, K., Kawarai, M. & Sode, K. (1989*c*). *Membrane*, **14**, 311.

Nishikawa, S., Sakai, S., Karube, I., Matsunaga, T. & Suzuki, S. (1982). *Appl. Environ. Microbiol.*, **43**, 814.

Okada, T., Karube, I. & Suzuki, S. (1981). *European J. Appl. Microbiol. Biotechnol.*, **12**, 102.

Okada, T., Karube, I. & Suzuki, S. (1982). *Anal. Chim. Acta*, **135**, 159.

Okada, T., Karube, I. & Suzuki, S. (1983). *Biotechnol. Bioeng.*, **25**, 1641.

Sode, K., Lee, S., Nakanishi, K., Marty, J. L., Tamiya, E. & Karube, I. (1991) (in preparation).

Suzuki, S. & Karube, I. (1978). *Enzyme Engineering*, **4**, 329.

Watanabe, E., Endo, H. & Toyama, K. (1988). *Biosensors*, **3**, 297.

Chapter 6

EXPLOITING COMPUTERS IN BIOLOGICAL WASTE TREATMENT

G. L. JONES

4 Grovelands Avenue, Hitchin, Hertfordshire, UK

CONTENTS

1 INTRODUCTION

The first attempts at the controlled treatment of sewage in purpose built reactors were made towards the end of the nineteenth century; previously formal treatment had been by distribution of the sewage to land on sewage farms. Since then the technology of wastewater treatment has undergone a continuous process of refinement, from the initial sole use of fixed film reactors, for example percolating filters, which are still employed today, to the activated sludge process in all its various guises (Ainsworth, 1987).

Of the biologically based industrial processes the treatment of wastewaters is unique for a number of reasons:

(1) The user/operator has no control over the feedstock in either quality or quantity, both of which will vary with time.
(2) Output from the process, whatever the quality of the input, is expected to be reasonably consistent and will increasingly be required to meet exacting quality standards.
(3) There is rarely an identifiable, or saleable, product.

To meet these constraints treatment plants, particularly the biological stages of them, have often been considerably overdesigned. As a result it has frequently been possible, with increasing experience of the performance of the plant, to absorb some of the future increases in load solely by improvements to the method of operation.

2 TREATMENT SYSTEMS

The complete treatment of a wastewater, such as sewage, comprises a number of individual treatment processes, both physico-chemical and biological (Fig. 1). The first stage involves removal of suspended material from the incoming wastewater by screening and primary sedimentation. The second is biological oxidation, usually aerobic, to remove the soluble constituents and most of the remainder of the suspended material. Solids produced in the biological processes, together with solids from the primary sedimentation stage, are then further treated by microbial digestion or by physico-chemical methods, before final dewatering and discharge to the environment. In some cases the liquid effluent from the aerobic biological stage may undergo tertiary treatment, for example sand filtration. Other processes may be included, for example ozonisation as a means of reducing smell nuisance, and chlorination for disinfection.

The biological stage of wastewater treatment is only one of several unit processes involved in full treatment. Performance is therefore constrained by the presence and effectiveness of these other processes. To assess properly the scope for exploiting the potential of computers in the operation of the biological stage of wastewater treatment, it is beneficial to consider in some detail both the objectives of treatment and the properties of the attendant biomass.

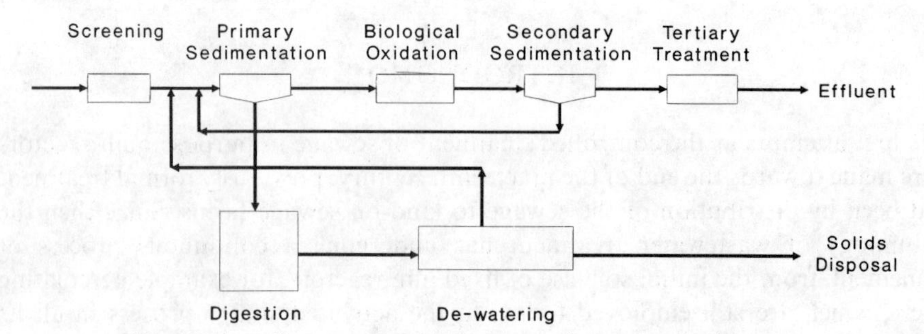

Fig. 1. Flow diagram of unit processes of wastewater treatment.

2.1 Treatment Objectives

Typically wastewaters contain material at relatively low concentration, by definition below the level at which economic recovery or exploitation of, say, biomass production is possible. For example, what would be considered to be a strong sewage, with a biological oxygen demand (BOD) of 380 mg BOD/litre, contained less than 200 mg carbon/litre (Painter *et al.*, 1961).

The objective of treatment is to achieve as far as is economically possible the maximum removal of the BOD of the wastewater, with the minimum production of solids from satisfying that BOD. Removing the BOD limits the polluting load on the environment. Minimising the production of solids reduces the problems associated with the disposal of solids. The final objective will also be constrained by whether the effluent is to be discharged to a sewer for possible further treatment, to an estuary or tidal region, or to a watercourse which may be used as a source of potable water downstream.

Equation (1) describes the removal of substrate by a growing culture of microorganisms, X, where K_s is the concentration of the rate limiting substrate, S, at which the specific growth rate is half the maximum, μ_m, and Y is the mass of cells produced per unit mass of substrate consumed:

$$ds/dt = -\mu_m XS/Y(K_s - S) \quad (1)$$

For a high quality effluent, with low potential biological activity, the concentration of substrate in that effluent will be significantly lower than the value of the half rate constant. From eqn (1) it is evident that the rate of reaction is governed mainly by the concentration of active biomass, and the extent of treatment by the length of time for which the biomass is in contact with the wastewater. Therefore the best treatment is achieved when the product of biomass and contact time is maximised.

2.2 The Plant

All successful biological wastewater treatment plants have as a common feature the retention of the biomass within the active zone, whether by provision of a support medium for a biological film, as in percolating filters and biological fluidised beds, or by separation and recycling of the solids as in activated sludge. The designer of a wastewater treatment plant has no control over either the rate of flow or the strength of the wastewater. In the case of sewage both flow and strength will vary diurnally, and the rate of change will depend on the size of population served and on how extensive is the sewage collection system.

The activated sludge system was introduced by Ardern and Lockett (1914) as an alternative to the percolating filter. As described, the process was in the form of an intermittently fed batch reactor, a fill and draw system, but it quickly evolved into the continuous process so well known today. There have been a number of variants on the initial concept, ranging from the original batch fed reactors, through fully enclosed oxygen enriched systems, as in the UNOX process, to the ICI Deep Shaft, and oxidation ditches. All are unmistakably activated sludge systems, dependent upon some form of solids separation and subsequent recycling to reduce the specific growth rate significantly below the dilution rate (D = flow rate of waste/volume of aeration tank, having units of reciprocal time). In addition various degrees of longitudinal mixing, from practically completely mixed to near plug-flow conditions, have been used. For the full treatment of domestic sewage, possibly with a proportion of added industrial wastes, restricted longitudinal mixing has been shown to be probably the most effective (Chambers & Tomlinson, 1982).

Some difficulties in achieving effective plant during the original development of the activated sludge process were possibly self inflicted. For example, it was not always recognised that the treatment process was biologically based. Indeed, the name 'filter' was given to the percolating filter since it was believed that the process of purification was in fact mechanical. However, the use of BOD as a measure of the effectiveness of treatment provided the means to develop the process on sound microbiological principles without the necessity to comprehend fully the microbiology involved. BOD determines the demand for oxygen, and is a measure of overall respiratory activity; it is not therefore confined to a single substance, or to substrate alone. The respiration of microorganisms in the effluent also contributes towards the BOD. Thus, by judging the effect of process modifications in terms of the reduction of both BOD and suspended solids in the final effluent, the cultural conditions required for the development of those bacteria which would flocculate, and therefore settle readily, in the case of activated sludge, or adhere to the medium in a precolating filter, and also protozoa capable of scavenging any free swimming bacteria present, were ensured.

Early work on the performance and behavior of laboratory scale activated sludge systems by Mohlman (1917) demonstrated well two basic principles governing the process:

(1) Sludge production is linked, inversely, to specific growth rate;
(2) Full nitrification requires both adequate aeration, and a suitably high concentration of suspended solids.

In spite of this and other early work on the biology of the process by the same group (Bartow & Mohlman, 1915; Russel & Bartow, 1916), it was still being argued as late as 1931 (Baly) that the major role of bacteria in the treatment of sewage by the activated sludge process was to poise the conditions for the most efficient precipitation of colloids.

2.3 The Biomass

The biomass present in a biological reactor consists of two major components: protozoa and bacteria.

2.3.1 Protozoa

The population of protozoa in a plant producing a well treated effluent consists in the main of active colonies of attached ciliate forms which feed on the free swimming microorganisms present (Curds & Vandyke, 1966). Protozoa occur in numbers of the order of 50 000/ml, comprising some 5% of the dry weight of the mixed liquor solids. Although there have been some claims that protozoa make significant reductions in the soluble BOD of a wastewater (Pillai & Subrahmanyan, 1944), it has been established that, even though protozoa will grow axenically on a soluble substrate, in wastewater treatment they have the primary function of producing a well clarified effluent (Curds & Cockburn, 1971).

2.3.2 Bacteria

Russel and Bartow (1916) showed that full treatment of sewage required both the heterotrophic bacteria to oxidise the carbonaceous material, and the, by that time, well known autotrophic microorganisms, *Nitrosomonas* and *Nitrobacter*, to oxidise ammonia. Nitrification was shown to be the limiting stage in obtaining full treatment, although it was not absolutely essential to achieve a stable effluent.

The bacterial biomass, far from being active, has the properties of a culture well into the stationary phase of growth. Analysis of activated sludge gives values for protein, DNA, RNA and fats consistent with an outgrown culture. Wooldridge and Standfast were the first to recognise that activated sludge was effectively an old culture of microorganisms, with a large proportion of the population being moribund (Wooldridge & Standfast, 1932, 1933; Wooldridge, 1933). They demonstrated that actively growing microorganisms were not essential for the removal of BOD from sewage, and concluded that non-viable bacteria were present in the system in such a high proportion that they could be responsible for a significant part of the total biochemical activity of activated sludge.

Direct counting of bacteria in activated sludge has repeatedly shown the low level of viable bacteria in activated sludge. Heukelekian (1934) repeated the observation that production of solids declined as the specific growth rate decreased, and also showed that this reduction in yield was accompanied by a decrease in the proportion of viable cells present. Additionally, Heukelekian concluded that the proportion of viable cells present in activated sludge solids was less than that found in solids contained in the incoming sewage. More recent, and more quantitative, studies have confirmed these observations (DOE, 1972; Pike & Carrington, 1972). Work at the Water Pollution Research Laboratory showed that typically less than 10% of the bacterial population in activated sludges producing good quality effluents are viable.

3 COMPUTER APPLICATIONS

The many different stages in the treatment process offer considerable scope, each in their different ways, for the application of automatic control systems. Using the term 'computer' in its generic sense, the wastewater treatment industry has applied computers to virtually all the unit processes involved. Use had been made of mini-computers, micro-computers, and sophisticated PLCs. However, most of the reports have been concerned with the theory, rather than the practice, of automation and control. Hamilton (1983) carried out a review which covered actual and planned applications of such devices within the industry. Excluded from the bibliography were general papers on how computers should be used, pilot scale studies, simulations and off-line applications. In all there were citations for significant applications of computers to sewage treatments works at only 72 locations in a total of 10 countries. Since this review there has been no great increase in the number of implementations. The success rate for computer applications, and the proportion of treatment plants which take advantage of real-time interaction, is still low (Alleman *et al.*, 1989).

The control may be, and most frequently is (Alleman *et al.*, 1989), a relatively simple affair based on a timer for, say, the desludging of a sedimentation tank. The more complex systems attempt to exert tight control over part of the operation, particularly the biological stage, for example aeration in the activated sludge plant. Most control systems have been confined to the activated sludge process, either as an individual unit (Korelin, 1983), or as part of a more ambitious project to monitor and control the complete works (Burns & Fielden, 1989). By their nature percolating filters offer little opportunity for short term operational control. The biomass once established remains relatively constant, and the distribution of the wastewater over the surface of the filter is readily achieved, usually by provision of a suitable hydrostatic head when the plant is built. Aeration is achieved by diffusion of atmospheric oxygen through the surface of the biological film.

More recent developments in fixed film technology, for example the fluidised bed reactor, do permit some degree of control (Cooper & Wheeldon, 1981). Rates of recycle, and the upward flow velocity through the bed, can be modified, and some degree of control over the amount of biomass present should be possible (Atkinson, 1974). However, although most of the process problems of using fluidised bed reactors have been overcome (Cooper, 1985), such systems are largely confined to the treatment of industrial wastewaters and are not widely used at present for sewage treatment.

In contrast to the percolating filter the activated sludge process has a high energy requirement, mainly for aeration which consumes between 60 and 80% of the total. In addition there are two flow streams, the sludge recycle and the sludge wastage, which, if the facility has been provided at the design stage, are capable of being adjusted at the will of the operator. However, a characteristic of the activated sludge process is its considerable inertia which makes control difficult. Any change to the operating conditions, even a small one, can have a prolonged effect on the system. Firstly the specific growth rate is low. With cell doubling times normally in excess of 3·5 days, the time required for the system to reach a new equilibrium in response to an operational adjustment is much greater than the interval between changes in conditions brought about the diurnal fluctuation in quantity and/or quality of the wastewater. Secondly the effect of the final sedimentation tank is to provide a degree of feedback to the aeration tank which can cause instability.

Typically the modern process takes the form shown diagrammatically in Fig. 2. Biological treatment is achieved by bringing the wastewater into contact with a suitable population of microorganisms for a sufficient period of time for those microorganisms to metabolise the polluting constituents to the required extent (eqn (1)). The incoming wastewater is mixed with a flocculent suspension of

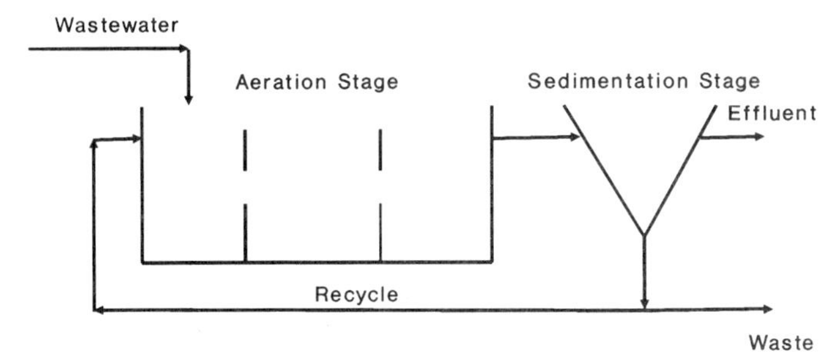

Fig. 2. Diagram of conventional activated sludge plant.

microorganisms, the activated sludge, which is separated by sedimentation after reaction. A small proportion of the separated sludge, determined by the rate of growth within the system, is wasted and the remainder recycled to the inlet of the reactor.

The final sedimentation tank forms a critical stage of the process. Full nitrification requires specific growth rates usually less than 0·008/h, while dilution rates of more than 0·05/h are usual to avoid undesirably large aeration stages. Thus recycling of the solids, which reduces the specific growth rate below that of the dilution rate, D, is essential.

3.1 Aeration

The major expense in the treatment of wastewaters by the activated sludge process is the provision of oxygen. Early in its development it was recognised that the supply of oxygen was a critical stage in the process. Of all the control systems so far applied to activated sludge this is the one that has received the most attention.

One of the earliest examples of dissolved oxygen control was described before the ready availability of computers. Briggs *et al.* (1967) employed a method of controlling the dissolved oxygen in a pilot plant by switching between two alternate rates of oxygen transfer. Application of this technique was tried on the full scale, but, although it showed promise, difficulties were experienced in evaluating the system fully because of shortcomings in the equipment employed in others parts of the plant.

Simple techniques such as this have fallen into disuse as more sophisticated computer based control systems have become available. Typical of these are methods based either on the direct measurement of dissolved oxygen (Chambers & Thomas, 1985), or on the estimation of the respiration rate, and the profile of dissolved oxygen along the aeration tank (Olsson & Andrews, 1978).

Holmberg *et al.* (1989) described a method of estimating the respiration rate while at the same time controlling the level of dissolved oxygen and presented some results of two experiments, each lasting three days, on a full-scale plant. This technique (Holmberg & Olsson, 1985) has attracted the attention of a number of workers with a view to further development (Bocken *et al.*, 1989) but has been confined to simulation studies.

Charpentier *et al.* (1989) reported the use of oxidation–reduction potential as a means of controlling aeration. They claimed benefits over normal dissolved oxygen control because of the nature of the signal obtained, which also provided information on the transition between full and partial nitrification.

3.2 Plant Loading

As shown by the work of Wooldridge and Standfast mentioned earlier, much of the biochemical activity of activated sludge is unlikely to be due to actively growing organisms. Consequently the response of the process to a change in loading will differ depending upon whether the change is due to an increase in flow or to an increase in concentration. Increasing the flow reduces the contact time between the biomass and the wastewater. To compensate for the increased load there must therefore be an increase in biomass (eqn (1)). With typical doubling times of 3·5–8 days, an increase in the concentration of biomass present cannot be dramatic. Increasing load in the form of concentration of substrate, on the other hand, will provoke an immediate increase in activity merely by increasing the saturation of enzyme with substrate. Figure 3 shows the response for both BOD and ammonia in the final effluent for a 50% increase in loading on the plant predicted by the WRc mathematical model of the activated sludge process. For the case where load was changed by increasing the rate of flow, there was an immediate increase in the effluent BOD due to the increased loss of solids from the system as a result of changing the upward flow velocity in the final sedimentation tank. The peak in the concentration of ammonia was delayed by the hydraulic retention of the system.

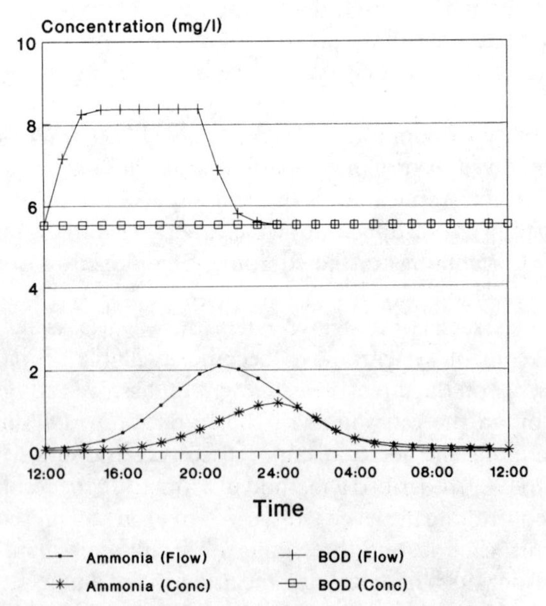

Fig. 3. Predicted effect of a 50% increase in flow or concentration on effluent BOD and ammonia.

With the increase in load due to concentration there is no detectable response for BOD, while that for ammonia is similar to, but slightly less than, that for flow.

The difference between the behaviour of BOD and ammonia in response to the concentration increase can be accounted for mainly by the influence of the final sedimentation tank. Changes in the concentration of soluble BOD compared with ammonia are very small because of the much higher maximum specific growth rates of the heterotrophic organisms (of the order of 0·03/h compared with 0·015/h). Being very small, these changes are masked by the BOD exerted by suspended microorganisms in the effluent, and the effluent BOD is therefore influenced very strongly by clarification in the sedimentation tank.

Naghdy and Helliwell (1989) investigated the possibilities of controlling treatment plant by smoothing flow and load of the wastewater applied to the aeration stage. Using an off-line balancing tank they showed the technique had potential for both energy saving and process improvement. A disadvantage of this technique is the additional cost of providing a separate balancing tank. Forecasting flow and load with an auto-regressive moving average model allowed a reduction in the volume of the balancing capacity required.

3.3 Biological control

While aeration control is well established and proven, the prospects for control of the biological stage by manipulation of the process streams is more difficult, if not impossible. Bernard and Eymard (1978) observed that in the absence of large storage capacity, either for the wastewater or for the activated sludge, for an urban wastewater 'an optimisation of the biological treatment, utilising elaborate computing and memorising means, remains illusive'. They stressed that, as treatment standards became more stringent, the provision of sufficient capacity becomes an important factor.

Simulations, using a version (Bidstrup & Grady, 1987) of the IAWPR model of the activated sludge process (Henze *et al.*, 1987), illustrate this problem. Figure 4 compares the predicted output of ammoniacal nitrogen in response to a diurnal variation in flow and strength as it is affected by an increase in retention time at a fixed sludge age (10 days); as retention time is increased, from 5 to 10, 15 and 20 hours, so the peak concentration of ammonia decreases.

To increase retention time requires new works to provide the extra capacity. However the distribution of concentration through the 24-hour period is not necessarily improved by increasing the retention time alone. Concentrations in excess of the mean value, 0·94 mg/litre, achieved at 5 hours retention, are obtained more frequently at the higher retention times (see Table 1). Therefore, in addition to providing additional tank capacity, provision would also need to be made to increase the sludge age involving both additional sedimentation capacity and aeration.

3.3.1 Suspended Solids

The concentration of mixed liquor suspended solids (MLSS) has always been one

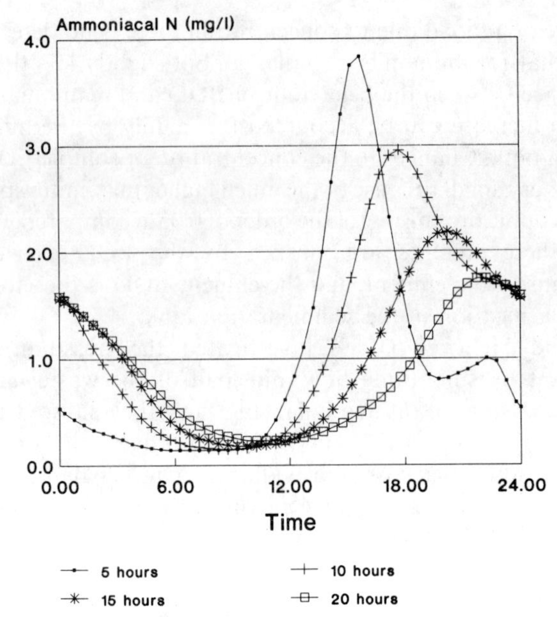

Fig. 4. Effect of retention time on effluent ammonia predicted by the IAWPR activated sludge model.

of the most popular of control subjects. Intuitively the concentration of the biomass would seem to be a rewarding area of control. However, this is one of the more difficult parameters in activated sludge to control. Firstly it depends not only on the load onto the plant, both strength and flow, but also on the composition of the wastewater, and on the performance of the sedimentation stage. Recycling of solids may well be intermittent, and the concentration of solids in the underflow somewhat variable.

3.3.1.1 WASTAGE RATE

Automatic control of suspended solids, albeit by a passive but what is still arguably

Table 1
Predictions of the Clemson Implementation of IAWPR Activated Sludge Model on the Effect of Retention Time on the Concentration of Ammoniacal N

	Retention time (h)			
	5	*10*	*15*	*20*
Mean (mg/litre)	0·94	1·20	1·01	0·86
Standard deviation	1·02	0·90	0·68	0·54
Maximum (mg/litre)	3·83	2·95	2·18	1·74
Minimum (mg/litre)	0·14	0·15	0·18	0·20
Proportion > 0·94 mg/litre	0·30	0·54	0·50	0·50

Table 2
Comparison of Constant Wastage Rate and Suspended Solids Control on Concentration of Ammoniacal N in Final Effluent

	Concentration (mg/litre)	
	Constant wastage rate	*Suspended solids control*
Mean	3·2	3·5
Standard deviation	2·1	2·3
Maximum	6·1	6·8
Minimum	0·1	0·0
MLSS	2 976·0	3 006·0

the most effective method, was implemented by Garrett (1958). Garrett showed that MLSS could be most easily controlled by wasting solids from mixed liquor. Reasoning on the basis of Monod kinetics, which were just becoming widely known, he devised a method of automatic control, using a weir at the outlet to the aeration stage, which ensured that the specific growth rate of the sludge in the system was properly matched to the requirements of the process. This technique enables the specific growth rate at which the plant is to operate to be easily defined (Jones, 1973).

Simulation using the WRc activated sludge model indicates that controlling the level of MLSS by active adjustment of the wastage rate provides no better contol over the variability of effluent quality than maintaining a fixed wastage rate known to maintain the suspended solids in the required range (see Table 2).

Vaccari and Christodoulatous (1989) considered a number of algorithms for controlling the wastage rate of activated sludge. They examined three types of control:

(1) Static control, either constant flow, similar to that described by Garrett, or constant proportion, which takes account of waste sludge concentration;
(2) Feedback control;
(3) Feedforward control.

For the feedback control they tested as process controllers both the dynamic sludge age (Vaccari *et al.*, 1985) and mass. In simulation studies they concluded that, while control over wastage rate was feasible, lowest variability was obtained using static control.

As part of an extensive study into the application of instrumentation, control and automation (ICA), based on a DEC PDP 11/44, to sewage treatment, Butwell *et al.* (1989) investigated sludge wastage control for maintaining the required level of suspended solids. Using an algorithm which computed the rate of sludge wastage required from the concentration of reactor and return solids, they were able to control to within 10% of the required concentration for 90% of the time. This was significantly better than recorded for a manual control period, although effluent quality was similar during both periods. Increased maintenance of the instrumentation was required when computer control was employed.

3.3.1.2 RECYCLE RATE

Of the flow streams built in to the activated sludge plant, recycle of sludge from the sedimentation tank to the aeration tank is the other possible candidate for control action. In existing plants there may be no provision for controlling the flow other than on/off control of the pumps used for recycling the sludge. Assuming such adjustment is possible, the major effect is on both the contact time between biomass and wastewater in the aeration stage, and the loading on the final sedimentation tank. Increasing the rate of return in response to a change in load will have the effect not only of decreasing the contact time between biomass and wastewater but also of increasing the dilution of the wastewater, which would in fact decrease the overall rate of reaction (eqn (1)). Reduction in recycle, while improving the possibilities for biological activity, could have an adverse effect on the loading of the final sedimentation tank. Proposals for overcoming these problems have included applying the wastewater simultaneously to different parts of the plant, thereby improving the match between wastewater loading and biomass (Andrews, 1976).

Fielden (1988) reported the experience of recycle control at the Witney EDF which showed it was possible to maintain the required proportional flow for 96% of the time. This study was concerned entirely with the feasibility of asserting control in response to diurnal flows. No assessment was made of its impact on either effluent quality or sedimentation tank performance. Simulation has suggested the possibility that recycle control could influence distribution of effluent quality (Jones, 1982), although Stephenson (1985) reported that on the whole manipulation of recycle rate was likely to result in overall diminution of effluent quality.

3.4 Power Saving

Actual examples of energy saving as a result of using a control computer are not very frequent. Often considerable savings can be made by more accurate matching of power input to power requirements. The simplest method of control is to provide, in a plug flow system, tapering of the supply such that the major input is provided at the inlet to the works where the demand is greatest, and providing sufficient power in each region of the aeration tank so that the concentration of dissolved oxygen never falls below the desired minimum. Inevitably, because of diurnal variations in flow and strength, for much of the time there will be an excess of power over the immediate requirements and higher levels of dissolved oxygen than necessary will be obtained. Control of the level of dissolved oxygen, so that at times of low demand the power input is reduced, while still producing an effluent of the required quality will achieve further saving in energy as shown. Figure 5, obtained from a simulation using the WRc activated sludge model, illustrates the advantage that can accrue from controlling dissolved oxygen; in this case the extra saving is 15%.

Using this technique, Chambers and Thomas (1985) achieved a 50% improvement in energy transfer on a full-scale plant. The aeration system installed was a fine-bubble diffused air system, optimised for the plant, coupled with an efficient system of dissolved oxygen control employing a sophisticated programmable logic controller incorporating a 3-term PID algorithm.

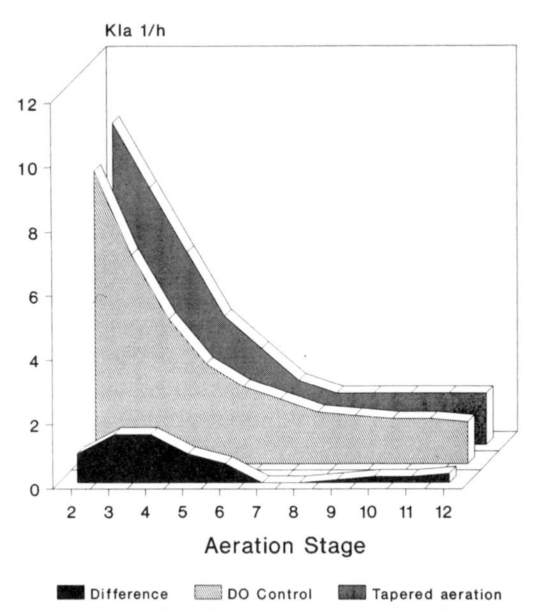

Fig. 5. Comparison of rate of transfer of oxygen (Kla) required for tapered aeration system with and without dissolved oxygen control.

Butwell *et al.* (1989), as part of the ICA program for the Witney Evaluation and Demonstration Facility (EDF), used submerged turbine aerators. Although inefficient compared with fine bubble diffused air systems, the equipment was amenable to on/off control. Using triplicated dissolved oxygen probes to control the switching (Burns, 1986), some 19–25% saving in aeration costs, compared with constant aeration, was achieved, whilst maintaining similar effluent quality. Similar improvements in electricity consumption were obtained by Charpentier *et al.* (1989) on full-scale works when controlling aeration by reference to oxidation–reduction potential.

In pilot-scale studies Naghdy and Helliwell (1989) reported the possibility of considerable energy savings using the technique of flow and load balancing. Overall process improvement, compared with the control, was achieved even with a 25% reduction in the air supply to the experimental plant.

4 FUTURE DEVELOPMENTS

For the wastewater treatment industry, notwithstanding more than 20 years of effort, the full and effective use of computers has yet to be achieved. This is not unexpected considering the economics of the process and the fact that it has no readily identifiable commercial end product. Wastewater treatment is primarily a low technology operation and it is inevitable that there should be problems in making effective use of the high technology available to it. Many of the treatment works in operation today were designed, and built, before the ready availability of computers, and may therefore make no provision for any form of on-line control. It should not be surprising then that the application of computers in the industry has been headed by data management and report generation, with logic-based control of the process at the bottom of the list (Alleman *et al.*, 1989).

To exploit fully the benefits of computerisation requires full consideration not only of the process being controlled, but also of how it is affected by other parts of the system. The need for a properly integrated approach is demonstrated by some of the implementations of dissolved oxygen control. Certainly improvements can be obtained merely by exercising control, as was shown independently by Butwell *et al.* (1989) and Charpentier *et al.* (1989), both of whom claimed power savings in the range 19–25% with existing aeration systems. Chambers and Thomas (1985), on the other hand, achieved a much greater saving, 50%, when the control system was installed as an integral part of an optimised aeration system.

Available evidence suggests that, other than for aeration control, process optimisation using on-line control techniques, at least of the biological stage, is not readily achieved; oxygen is the only substrate for the biomass over which the operator can have complete control. One possible exception to this arises from the work of Charpentier *et al.* (1989). Their use of oxidation–reduction potential as a control parameter, rather than dissolved oxygen itself, indicated the potential for both monitoring and influencing nitrogen and phosphorus metabolism. With increasing concern about the discharge of these elements to the environment, the ability to control them, whilst at the same time optimising oxygen consumption, is a worthwhile objective for a computer controlled system.

As was pointed out by Bernard and Eymard (1978), improvements in the degree of treatment are constrained by the size of the plant, and the ability to provide storage capacity for biomass and wastewater. In effect the quality of the final effluent to be obtained is fixed at the design stage. Sludge wastage and sludge recycle are the two process streams available for continuous adjustment. Both impinge on the performance of the final sedimentation tank. Decreasing the proportion of the sludge wasted from the system increases the concentration of biomass within the aeration stage. While this will allow increased removal of pollutants, assuming sufficient aeration capacity, it will also increase the load on the final sedimentation tank. The maximum possible load that can be placed on the final sedimentation tank may be increased by increasing the rate of recycle, the underflow rate (White, 1975).

Simulations using the steady state version of the WRc model of the activated sludge process (Jones & Paskins, 1979), shown in Table 3, demonstrate this effect. As recycle ratio was increased, sludge age, mixed liquor suspended solids, the oxygen demand and maximum loading on the sedimentation tank also increased, accompanied by a reduction in the concentration of ammonia in the effluent. Some tuning of the performance of the biological stage, in response to diurnal variations, could perhaps be achieved by close monitoring of the final sedimentation tank, and optimising its performance, consistent with an adequate concentration of dissolved oxygen in the aeration stages.

Green *et al.* (1989) have described an information system, incorporating individual process control modules at remote telemetry outstations, being

Table 3
Effect of Recycle Ratio on Sludge Age, Biomass Concentration and Effluent Quality

	Recycle ratio			
	0·75	*1·0*	*1·25*	*1·5*
MLSS (g/litre)	3·9	4·1	4·2	4·3
Sludge age (days)	8·8	9·6	10·1	10·4
Effluent ammonia (mg N/litre)	1·0	0·9	0·8	0·7
Oxygen demand (kg/h)	115·8	117·6	118·6	119·1
Sedimentation tank loading (kg/m^2/h)	9·7	11·7	13·6	15·4

developed in the United Kingdom by Anglian Water. These outstations are to be installed not only at treatment plants but also at pumping stations which are part of the sewer network. Among the objectives of this project is the development of pump scheduling software to reduce power costs by exploiting the possibility of storage in sewer systems. Successful implementation of such a scheme would take a significant step towards the requirements for the additional storage capacity identified as necessary by Bernard and Eymard (1978) to allow optimisation of the treatment process. Also included in the Anglian Water project is the monitoring of rainfall which, combined with the possibilities of planned storage in the sewer, would fulfil the requirements of the forecasting and load smoothing techniques described by Naghdy and Helliwell (1989).

Data collection, instrumentation and instrument reliability are a critical area for the control of wastewater treatment systems. Briggs (1985) discussed the availability of a wide range of sensors; and investigations of sensor reliability, together with cost of ownership, were among the objectives of the work carried out at the Witney EDF (Butwell *et al.*, 1989). It was established that a fully integrated computer control and monitoring system could produce benefits, in both cost savings and reliability. However, there were warnings concerning equipment maintenance. Butwell *et al.* reported that the use of a triple validation procedure for dissolved oxygen resulted in additional maintenance requirements, which were already greater than had been expected. Similarly, when using automatic control for sludge wastage, additional maintenance costs were incurred for the suspended solids sensors, without any additional improvement in effluent quality over that obtained with manual control.

Difficulties associated with the maintenance requirements for more complex sensors suggest a critical role for computers in data and fault logging. Simpler, therefore more robust, and fewer, therefore less total maintenance, sensors connected to a computer could be used to provide much of the information required. The records obtained from strategically placed sensors, whether measuring flow, suspended solids, dissolved oxygen or any other useful determinand, stored in the computer will provide an increasingly representative body of data describing the performance of the plant. Fault analysis (Jefferis, 1982) and pattern recognition techniques, particularly in conjunction with computer simulation of plant performance, could then be used to identify necessary control or maintenance actions.

The application of computers to wastewater treatment is now emerging from its infancy. Successful exploitation of computers in this field depends upon a proper understanding not only of the biological process itself, but also of the interaction between all the other unit processes upon which effective treatment depends. To optimise the performance of a wastewater treatment plant requires more than the simple provision of instrumentation and computer control. It is essential that there should be an integrated systems approach involving designers, operators and control engineers (Beck, 1977). It must also be recognised that the prime requirement is treatment of the wastewater, and the plant must not be so tightly coupled to the computer and control system that it cannot function without them.

REFERENCES

Ainsworth, G. (1987). *Wat. Pollut. Control*, 220.

Alleman, J. E., Sweeney, M. W. & Kamber, D. M. (1989). *Wat. Sci. Tech.*, **21**, 1271.

Andrews, J. F. (1976). *Prog. Water Technol.*, **8**(6), 451.

Ardern, E. & Lockett, W. T. (1914). *J. Soc. Chem. Ind.*, **33**, 523, 1122.

Atkinson, B. (1974). *Biochemical Reactors*, chapter 7. Pion Press, London.

Baly, E. C. C. (1931). *Chemy Ind.*, **50**, 22T.

Bartow, E. & Mohlman, F. W. (1915). *J. Ind. Eng. Chem.*, **7**, 318.

Beck, M. B. (1977). *Prog. Water Technol.*, **9**(5/6), 13.

Bernard, J. & Eymard, A. (1978). *Prog. Water Technol.*, **9**(5/6), 519.

Bidstrup, S. M. & Grady, C. P. L., Jr (1987). *A Users Manual for SSSP. Simulation of Single-sludge Processes for Carbon Oxidation, Nitrification and Denitrification.* Clemson University, Environmental Systems Engineering, Clemson, SC.

Bocken, S. M., Braee, M. & Dold, P. L. (1989). *Wat. Sci. Techn.*, **21**, 1197.

Briggs, R. (1985). In *Comprehensive Biotechnology*, **4**. *The Practice of Biotechnology: Speciality Products and Service Activities*, eds C. W. Robinson & J. A. Howell. Pergamon Press, Oxford, chapter 64.

Briggs, R., Jones, K. & Oaten, A. B. (1967). *Effluent and Water Treatment Convention*, London.

Burns, J. M. (1986). *An assessment of the dissolved oxygen control system at Witney STW.* Water Research Centre, Report 542-S.

Burns, J. M. & Fielden, R. S. (1989). *International Biodeterioration*, **25**(1/3), 79.

Butwell, A. J., Burns, J. M., Fielden, R. S. & Berry, M. J. (1989). *Wat. Sci. Tech.*, **21**, 1239.

Chambers, B. & Thomas, V. K. (1985). Presented at *8th Symposium on Wastewater Treatment*, Montreal.

Chambers, B. & Tomlinson, E. J. (1982). In *Bulking of Activated-Sludge—Preventative and Remedial Methods*, ed. B. Chambers & E. J. Tomlinson. Ellis Horwood, Chichester, UK.

Charpentier, J., Godart, H., Martin, G. & Mogno, Y. (1989). *Wat. Sci. Tech.*, **21**, 1209.

Cooper, P. (1985). In *Comprehensive Biotechnology*, **4**. *The Practice of Biotechnology: Speciality Products and Service Activities*, eds C. W. Robinson & J. A. Howell. Pergamon Press, Oxford, chapter 57.

Cooper, P. F. & Wheeldon, D. H. V. (1981). In *Biological Fluidised Bed Treatment of Water and Wastewater*, eds P. F. Cooper & B. Atkinson. Ellis Horwood, Chichester, UK, p. 121.

Curds, C. R. & Cockburn, A. C. (1971). *J. Gen. Microbiol.*, **66**, 95.

Curds, C. R. & Vandyke, J. M. (1966). *J. Appl. Ecol.*, **3**, 127

Fielden, R. S. (1988). *An assessment of the return sludge control system at Witney STW*. Water Research Centre, Report 692-S.
Garrett, M. T. (1958). *Sewage Ind. Wastes*, **30**, 253.
Green, J. W., Page, C., Eastman, G. M. & Howes, D. (1989). *Wat. Sci. Technol.*, **21**, 1283.
Hamilton, I. M. (1983). *The application of computers at sewage treatment works, water treatment works and pumping stations. An annotated bibliography covering the period 1971 to 1982*. WRc Technical Report TR194.
Henze, M., Grady, C. P. L., Jr, Gujer, W. J., Marais, G. V. R. & Matsuo, T. (1987). *Wat. Res.*, **21**(5), 505.
Heukelekian, H. (1934). *Sewage Wks J.*, **6**(4), 676.
DOE (1972). In *Water Pollution Research, 1971*, Department of the Environment, HMSO, London, p. 62.
Holmberg, U. & Olsson, G. (1985). in *1st IFAC Symposium on Modelling and Control of Biotechnological Processes*. Noordwijkerhout, The Netherlands, 11–13 Dec., p. 985.
Holmberg, U., Olsson, G. & Andersson, B. (1989). *Wat. Sci. Tech.*, **21**, 1185.
Jefferis, R. P. (1982). In *Computer Applications in Fermentation Technology*, Society of Chemical Industry, London, p. 199
Jones, G. L. (1973). *Wat. Res.*, **7**, 1475.
Jones, G. L. (1982). In *Computer Applications in Fermentation Technology*, Society of Chemical Industry, London, p. 31.
Jones, G. L. & Paskins, A. R. (1979). *Mathematical model of activated sludge—a comparison of predictions and experimental results*. WRc Laboratory Report 1002.
Korelin, A. (1983). *Vesitalous*, **24**(3), 21.
Mohlman, F. W. (1917). *Ill. State Wat. Surv.*, **14**, 75.
Naghdy, G. & Helliwell, P. (1989). *Wat. Sci. Tech.*, **21**, 1225.
Olsson, G. & Andrews, J. F. (1978). *Water Res.*, **12**, 985.
Painter, H. A., Viney, M. & Bywater, A. J. (1961). *J. Inst. Sew. Purif.*, 302.
Pike, E. B. & Carrington, E. G. (1972). *Wat. Pollut. Cont.*, **71**, 583.
Pillai, S. C. & Subrahmanyan, N. V. (1944). *Nature, London*, **154**, 179.
Russel, R. & Bartow, E. (1916). *Ill. State Wat. Surv.*, **13**, 348.
Stephenson, J. P. (1985). In *Comprehensive Biotechnology*, **4**. *The Practice of Biotechnology: Speciality Products and Service Activities*, eds C. W. Robinson & J. A. Howell. Pergamon Press, Oxford, chapter 67.
Vaccari, D. A. & Christodoulatous, C. (1989). *Wat. Sci. Tech.*, **21**, 1249.
Vaccari, D. A., Fagedes, T. & Longtin, J. (1985). *Biotechnology and Bioengineering*, **27**, 695.
White, M. J. D. (1975). *Settling of activated sludge*. WRc Technical Report TR11.
Wooldridge, W. R. (1933). *Biochem. J.*, **27**(1), 193.
Wooldridge, W. R. & Standfast, A. F. B. (1932). *Nature, London*, **130**, 664.
Wooldridge, W. R. & Standfast, A. F. B. (1933). *Biochem. J.*, **27**(1), 183.

Chapter 7

WATER QUALITY MODELLING OF BOD AND DO IN RIVERS USING A HYDROLOGICAL MODEL

DENIS COUILLARD & GUY MORIN

Institut National de la Recherche Scientifique (INRS-Eau), Université du Québec, Sainte-Foy, Québec, Canada

CONTENTS

1 INTRODUCTION

Among the various water quality parameters, dissolved oxygen is frequently used and is generally considered to be a good indicator of the health of aquatic ecosystems (Couillard, 1979, 1980, 1983; Couillard & Tyagi, 1988; Hassan, 1989). It is not only easy to monitor on a continuous basis, but it is also dependent on a whole series of chemical and biological processes such as the respiration of aquatic organisms and the bio-degradation of organic wastes. There are thousands of studies dealing with this parameter in the scientific literature.

Almost all of the models that have been developed for rivers are derived from the work of Streeter and Phelps (1925), who defined a two-equation system of interactions between dissolved oxygen (DO) and the biochemical oxygen demand (BOD). The scientists who followed added additional terms to the basic equations, specified the methods to evaluate the parameters and extended the field of application of the original model.

The proposed DO and BOD models are part of a global water quality model (Morin *et al.*, 1986*a*,*b*, 1987, 1988; Couillard *et al.*, 1988), which is based on a conceptual hydrological model. The hydrological model allows daily simulation of streamflow at any point within a watershed. For this purpose, the watershed is divided into square elements on which are evaluated the physiographic characteristics and the meteorological data required to calculate the water balance. The DO and BOD models developed are applied to the Sainte-Anne River (Québec, Canada).

2 LITERATURE REVIEW OF RIVER DISSOLVED OXYGEN MODELS

2.1 Major Dissolved Oxygen Models

Given x, the horizontal axis of a river reach; t, the time; Q, the flow; U, the mean current speed; and T, the water temperature; and assuming a permanent, uniform regime for Q and T, Streeter and Phelps (1925) defined two equations describing the evolution of dissolved oxygen (DO) and of biochemical oxygen demand (BOD):

$$\text{DO:} \quad \frac{\partial C}{\partial t} = U\frac{\partial C}{\partial x} = k_2(C_s - C) - k_1 L \tag{1}$$

$$\text{BOD:} \quad \frac{\partial L}{\partial t} = U\frac{\partial L}{\partial x} = -k_1 L \tag{2}$$

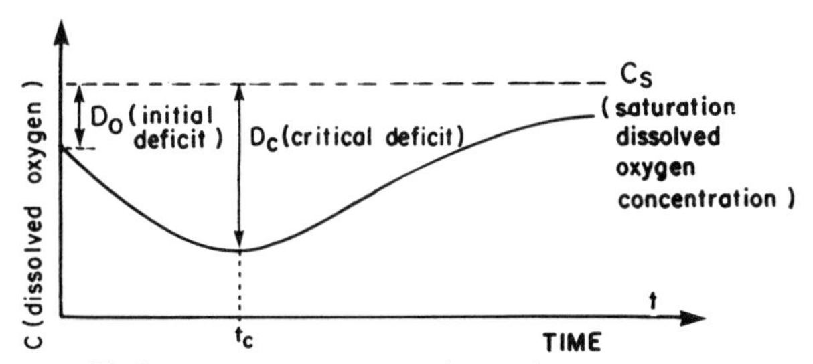

Fig. 1. Streeter–Phelps sag curve representing a river response to an organic load.

where C = dissolved oxygen concentration (mg/litre)
C_s = dissolved oxygen concentration at saturation (mg/litre)
L = biochemical oxygen demand (mg/litre)
k_1 = BOD reduction coefficient, or DO consumption coefficient (day^{-1})
k_2 = re-oxygenation coefficient (day^{-1})

Assuming that k_1, k_2 and C_s are independent of x and t, the solutions are:

$$D = \frac{k_1 L_0}{k_2 - k_1} - (e^{-k_1 t} - e^{-k_2 t}) + D_0 e^{-k_2 t} \tag{3}$$

$$L = L_0 e^{-k_1 t} \tag{4}$$

where D = dissolved oxygen deficit $= C_s - C$ (mg/litre)
D_0 = initial dissolved oxygen deficit (mg/litre)
L_0 = initial BOD concentration (mg/litre)

Equation (3) can be represented by the well-known sag curve illustrating the interaction between the de-oxygenation and re-oxygenation forces in a river (Fig. 1). The new terms introduced in Fig. 1 are D_c = critical deficit, and t_c = time at which the critical deficit is reached. These two terms are easily derived in mathematical form.

Equation (4) assumes an exponential decrease of the organic load at a rate dependent on the coefficient k_1 (Fig. 2). In parallel, this implies a geometric increase in the population of consuming bacteria and thus the need for adequate mixing of the river water.

By considering only bacterial oxygen consumption and natural re-oxygenation in a river, the Streeter–Phelps model constitutes a gross oversimplification of reality.

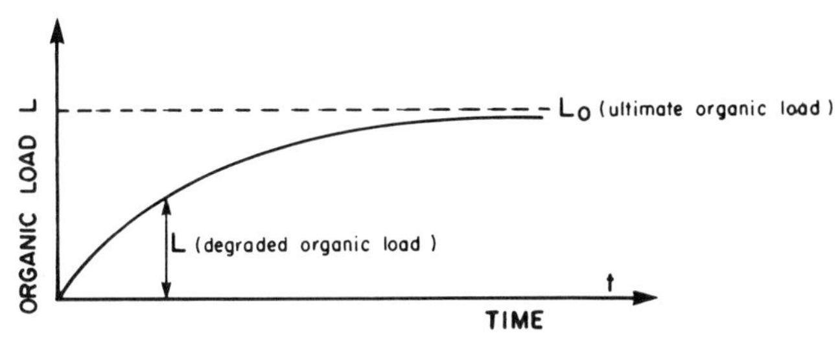

Fig. 2. Exponential decrease of the organic load *L*.

However, with the addition of an error term, it has been widely used. A number of natural processes were neglected, and eventually introduced by other workers. They are:

—the decrease in BOD resulting from sedimentation and adsorption;
—the addition of BOD by the diffusion of degraded organic matter from the river bottom into the water column;
—the addition of BOD resulting from local run-off;
—the decrease in DO resulting from sludge respiration;
—the increase in DO resulting from photosynthesis;
—the decrease in DO resulting from the respiration of plants, algae and phytoplankton.

Streeter and Phelps (1925) presented the problem in the form of first-order differential equations. The contributions of Dobbins (1964) and O'Connor (1961, 1967) are representative of the next stage in the evolution of DO–BOD models:

$$\frac{\partial M}{\partial t} = D_{\mathrm{L}} \frac{\partial^2 M}{\partial x^2} - U \frac{\partial M}{\partial x} \pm \sum S \tag{5}$$

where M = concentration of the simulated parameter, dissolved oxygen (C) or BOD (L)
S = dissolved oxygen (C) or BOD (L) sources and sinks
D_{L} = longitudinal dispersion coefficient
U = mean current speed
x = longitudinal distance
t = time

and for DO

$$\sum S = k_2(C_s - C) - k_1 L + D_{\mathrm{B}} \tag{6}$$

while for BOD

$$\sum S = -k_1 L - k_3 L L_{\mathrm{A}} \tag{7}$$

where k_1 = coefficient of BOD reduction (day^{-1})
k_2 = re-oxygenation coefficient (day^{-1})
k_3 = constant for BOD reduction by sedimentation or adsorption (day^{-1})
C = dissolved oxygen concentration (mg/litre)
C_s = dissolved oxygen concentration at saturation (mg/litre)
L = biochemical oxygen demand (mg/litre)
L_{A} = rate of BOD increase by local run-off (mg/litre . day)
D_{B} = net rate of dissolved oxygen increase resulting from the combined effects of bottom sediment degradation (g/m^2 . day) and plant photosynthesis and respiration (mg/litre . day)
S = dissolved oxygen and BOD sources and sinks

Dobbins (1964) gives the solutions for the case of a permanent regime. Later, with the help of his co-worker (Dresnack & Dobbins, 1968), he used the finite differences approach to deal with the dynamic conditions found in rivers. Within the D_{B} term (eqn (6)), he distinguishes the rate of dissolved oxygen decrease resulting from sediment degradation and plant respiration, and the rate of DO increase resulting from photosynthesis.

This second generation of models was refined at the end of the 1960s with a more complete description of the diurnal variations in dissolved oxygen and of the coefficients found in the basic equations. O'Connor and DiToro (1970) give for DO

$$\sum S = k_2(C_s - C) - k_c L_c - k_n L_n - D'_B - R + P \tag{8}$$

and for BOD

$$\sum S = -k_c L_c - k_n L_n \tag{9}$$

where k_2 = re-oxygenation coefficient (day^{-1})
k_c = coefficient of de-oxygenation resulting from carbon-BOD sources (day^{-1})
k_n = coefficient of de-oxygenation resulting from nitrogen-BOD sources (day^{-1})
C = dissolved oxygen concentration (mg/litre)
C_s = dissolved oxygen concentration at saturation (mg/litre)
D'_B = bottom sediment dissolved oxygen demand (mg/litre . day)
P = DO source resulting from algal photosynthesis (mg/litre . day)
R = DO sinks resulting from algal respiration (mg/litre . day)
S = sources and sinks (mg/litre . day)
L_c = BOD concentration from carbon sources (mg/litre)
L_n = BOD concentration from nitrogen sources (mg/litre)

O'Connor (1961, 1967, 1971) and most users of his model substituted eqns (8) and (9) into first-order continuity equations, in which temporal variations in the concentration of the parameter being modelled (C or L) are assumed equal to the balance of sources and sinks. In eqn (8), the periodicity of the term P can be simulated by identification to a Fourier series. Furthermore, the distinction between carbon-BOD and nitrogen-BOD better corresponds to typical experimental BOD curves (Fig. 3). In practice, however, in comparison to carbon-BOD the degradation of nitrogen-BOD in rivers is considered to occur without a lag.

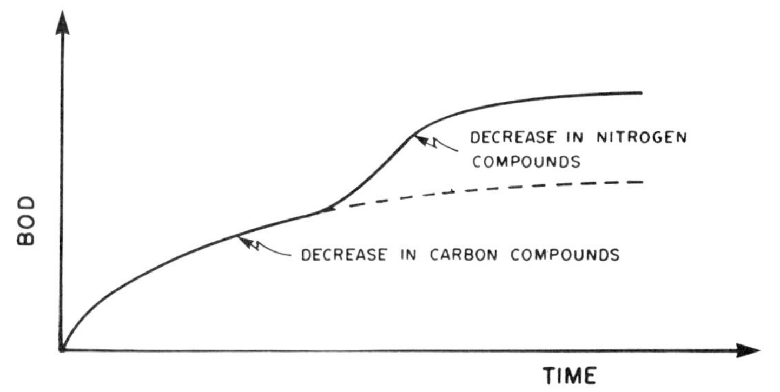

Fig. 3. Typical BOD response curve to organic carbon and nitrogen wastes.

2.2 Estimation of Parameters

2.2.1 Estimation of Oxygen in Water

The DO concentration at saturation is mainly a function of temperature. In order of increasing accuracy, a few of the proposed equations are as follows:

$$C_s = 14{\cdot}48 - 0{\cdot}36T + 0{\cdot}0043T^2 \tag{10a}$$

$$C_s = 14{\cdot}652 - 0{\cdot}410\,222T + 0{\cdot}007\,99T^2 - 0{\cdot}000\,077\,77T^3 \tag{10b}$$

$$C_s = 14{\cdot}619\,96 - 0{\cdot}404\,20T + 0{\cdot}008\,42T^2 - 0{\cdot}000\,09T^3 \tag{10c}$$

where C_s = dissolved oxygen concentration at saturation (mg/litre)
T = water temperature (°C)

Equation (10a) was proposed by Markofsky and Harleman (1971), eqn (10b) by Rich (1973), and eqn (10c) by Lawrence *et al.* (1978).

Equations are also available for adjusting the term C_s as a function of atmospheric pressure and the water chloride ion concentration.

2.2.2 Re-oxygenation and Oxygen Consumption Coefficients

There are numerous formulae and methods for estimating the re-oxygenation (k_2) and oxygen consumption (k_1) coefficients. They are essentially empirical and valid for the precise conditions in which they were established. It is particularly important

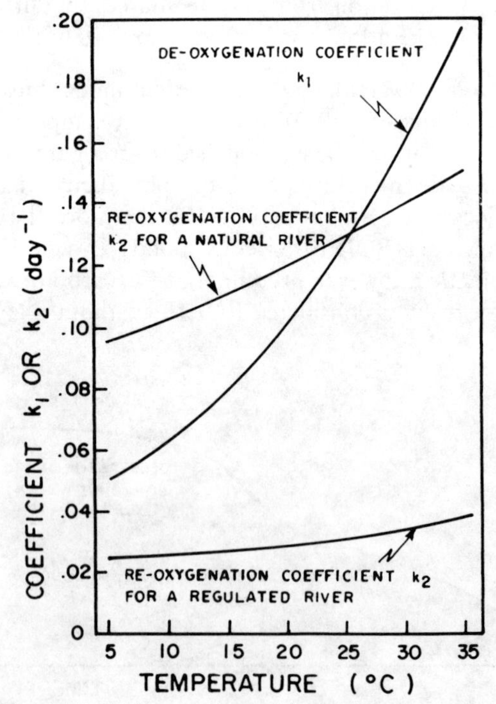

Fig. 4. Variations of k_1 and k_2 with temperature.

to estimate the oxygen consumption coefficient (k_1) from in-situ observations and data.

These coefficients may vary widely depending on the river's characteristics and particularly, as illustrated in Fig. 4, on its temperature (Cluis, 1973). It can be noted that k_1, which depends on the biological activity of microorganisms, is much more sensitive to temperature variations than is the river's re-oxygenation coefficient (k_2).

The determination of k_1, the dissolved oxygen consumption coefficient, is usually based on the interpretation of bacterial oxygen consumption curves obtained with representative samples of the rivers studied. The most common laboratory method is the BOD_5(20°C) test which measures the dissolved oxygen consumption at the end of a 5-day incubation period at 20°C. Ratios of known values for various types of pollution can be used to relate the BOD_5 value to the final BOD. The inclusion of an additional coefficient to account for nitrogen-BOD requires extending the BOD tests past 5 days, thereby reaching the time when nitrogen compounds begin to be reduced (c. Fig. 3), 7–8 days after the test's initiation. In cases for which in-situ BOD values cannot be used, typical values for various types of pollution are available in the literature. Examples for the agricultural, municipal and industrial sectors are available in Cluis *et al.* (1974) and in Couillard (1974). The chemical oxygen demand (COD) test is sometimes used instead of the BOD test (Couillard *et al.*, 1989; Couillard & Gariépy, 1990). Although the COD test can be carried out more quickly since it consists in an artificial oxidation of the water being analysed, greater care must be taken in interpreting the results.

Equation (11) is usually used to correct the estimated k_1 coefficient for temperature. The value of the constant (1·045) varies somewhat with the authors:

$$k_1(T) = k_1(20°C) \times 1{\cdot}045^{T-20} \qquad (11)$$

where $k_1(T)$ = BOD reduction coefficient at temperature T
$k_1(20°C)$ = BOD reduction coefficient at 20°C
T = water temperature (°C)

Most of the equations used to estimate k_2, the re-oxygenation coefficient, are as follows:

$$k_2 = C\frac{U^m}{H^n} \qquad (12)$$

where k_2 = re-oxygenation coefficient, usually for a 20°C temperature
C = empirical constant which, according to the authors, is a function of the diffusion coefficient or of the Froude number
U = mean current speed
H = depth
m = empirical constant, usually varying between 0·5 and 1·0
n = empirical constant, usually varying between 1·5 and 2·0

Formulae proposed by various workers are provided in Table 1 (Bansal, 1973). Estimated values for a number of these appear in Table 2.

Table 1
Empirical Equations for the k_2 Re-oxygenation Coefficient at 20°C (Day^{-1})

Ref.	*Equation*
Streeter & Phelps (1925)	cU^nH^{-2}
Miyamoto (1932)	$r\exp(-E_a/RT_0)$
Higbee (1935)	$2/H(D_m/\pi t)^{0\cdot5}$
Kalinske & Levich (1944)	$f(D_eH^2)$
Gameson & Truesdale (1955)	$9{\cdot}41U^{0\cdot67}H^{-1\cdot85}$
Dobbins (1956)	$10{\cdot}09U^{0\cdot73}H^{-1\cdot75}$
O'Connor & Dobbins (1958)	$127\,D_m^{0\cdot5}U^{0\cdot5}H^{-1\cdot5}$ isotopic turbulence $480\,D_m^{0\cdot5}s^{0\cdot25}H^{-1\cdot25}$ non-isotopic turbulence
Krenkel & Orlob (1962)	$1{\cdot}318\times10^{-5}D_1^{1\cdot321}H^{-2\cdot32}$
Churchill *et al.* (1962)	$5{\cdot}026U^{0\cdot969}H^{-1\cdot673}$
Thackston (1966)	$0{\cdot}000\,125(1+F^{0\cdot5})U/H$
Langbein & Durum (1967)	$3{\cdot}3UH^{-1\cdot33}$

Symbols:

c = empirical constant
D_e = turbulent diffusion coefficient
D_1 = longitudinal dispersion coefficient
D_m = molecular diffusion coefficient
E_a = activation energy of oxygen molecules
f = function
F = Froude number
H = depth of the water body in feet (1 ft = 0·3048 m)
n = empirical constant
r = turnover rate in surface area
R = universal gas constant
s = slope
t = penetration period of gas molecules in a liquid film
T_0 = temperature in Kelvin
U = mean current speed in ft/s (1 ft/s = 0·3048 m/s)
π = 3·1416

Table 2
Calculated Values of the k_2 Re-oxygenation Coefficients at 20°C According to Various Equations[a] (Day^{-1})

Mean current speed U (m/s)	*Mean depth H (m)*														
	H = 1 Equation					*H = 2 Equation*					*H = 3 Equation*				
	A	*B*	*C*	*D*	*E*	*A*	*B*	*C*	*D*	*E*	*A*	*B*	*C*	*D*	*E*
0·5	1·11	1·81	1·46	1·12	2·80	0·35	0·54	0·40	0·44	0·99	0·18	0·26	0·19	0·26	0·54
1·0	2·18	3·00	2·32	2·23	3·96	0·68	0·89	0·64	0·89	1·40	0·35	0·44	0·30	0·52	0·76
1·5	3·23	4·04	3·04	3·35	4·85	1·01	1·20	0·84	1·33	1·72	0·51	0·59	0·40	0·78	0·93
2·0	4·26	4·98	3·68	4·46	5·60	1·34	1·48	1·02	1·77	1·98	0·68	0·73	0·48	1·03	1·08

[a] Equation A: $2{\cdot}178U^{0\cdot969}\ H^{-1\cdot673}$ (Churchill *et al.*, 1962)
B: $3{\cdot}003U^{0\cdot73}\ H^{-1\cdot75}$ (Dobbins, 1956)
C: $2{\cdot}316U^{0\cdot67}\ H^{-1\cdot85}$ (Gameson & Truesdale, 1955)
D: $2{\cdot}230U\ H^{-1\cdot33}$ (Langbein & Durum, 1967)
E: $3{\cdot}962U^{0\cdot5}\ H^{-1\cdot5}$ (O'Connor, 1971)

Table 3
Experimental Values of the k_2 Correction Constant (σ) as a Function of Temperature Obtained from the Literature

Ref.	*Constant σ*	*Site*
Streeter *et al.* (1936)	1·047	laboratory
Truesdale & Van Dyke (1948)	1·018	laboratory
Downing & Truesdale (1955)	1·024	canal with mechanical aerator
Elmore & West (1961)	1·016	canal with mechanical aerator
Streeter (1926)	1·016	natural water body

As is the case for the k_1 coefficient, the re-oxygenation coefficient (k_2) is usually corrected for temperature according to eqn (13):

$$k_2(T) = k_2(20°\text{C})\sigma^{T-20} \tag{13}$$

where $k_2(T)$ = re-oxygenation coefficient at temperature T
$k_2(20°\text{C})$ = re-oxygenation coefficient at 20°C
σ = empirical constant
T = water temperature (°C)

The constant σ is usually set at 1·025. A few of the typical values obtained experimentally are presented in Table 3 (Bansal, 1973).

2.2.3 *Other Parameters*

In certain rivers where bottom organic sediments of natural or of anthropogenic origin are of importance, the BOD associated with the sediments may account for up to 50% of the reduction in dissolved oxygen. Therefore, in such cases it is necessary to adequately evaluate the role of the river bed.

The benthic oxygen demand (D'_B in eqn (8)) may depend on the following factors:

(1) oxygen consumption resulting from the diffusion into the upper water layer of the soluble product of benthic degradation;
(2) oxygen consumption by aerobic organisms (bacteria and higher forms) at the sediment–water interface;
(3) a reduction of the re-aeration surface by anaerobic gas bubbles originating from these sediments (particularly CH_4 and H_2S).

The D'_B term is influenced by other factors such as water temperature, dissolved oxygen concentration, the nature of the biological community, etc.

Fair and Geyer (1954) give an empirical equation for the benthic oxygen demand in which appear an estimate of the BOD_5(20°C) of the volatile substances from the sediments and an index of the rate of sedimentation. Their formula can be used to obtain a preliminary evaluation of the benthic demand but it is always preferable to use values measured *in situ* with samples from the river bed. Laboratory simulations with undisturbed sediment can also provide estimates with an accuracy varying around 30% (Ogunrombi & Dobbins, 1970).

The effects of photosynthesis and respiration on dissolved oxygen are more difficult to simulate and are usually only considered when a representation of diurnal DO variations is needed. Photosynthesis acts as a source of oxygen when sunlight is available whereas respiration is a sink, particularly at night. According to the limiting factor theory, the nutrient loads (C, N, P), on which the aquatic flora depends, can have a direct influence on dissolved oxygen.

The approach usually chosen to simulate the combined evolution of the respiration (R) and photosynthesis (P) terms (eqn (8)) is to use a periodic function (e.g. sinusoidal), its extension expressed as a Fourier series (O'Connor & DiToro, 1970). The calibration procedure is iterative and requires a few manipulations before a good fit between the predicted and observed values can be obtained.

3 CEQUEAU HYDROLOGICAL MODEL[a]

The objective of simulating DO and BOD in rivers at all points on a watershed, under either natural or regulated conditions, requires the coupling of a water temperature model and DO and BOD models with a distributed parameter model. for this we chose the CEQUEAU model.

The CEQUEAU model (Girard *et al.*, 1972*a*,*b*; Charbonneau *et al.*, 1975; Morin *et al.*, 1981; WMO, 1986) is a flexible model which takes into account the physical characteristics of the watershed through subdivisions in elements. It is a water-balance type conceptual model with distributed parameters. The spatial subdivision procedure makes it possible to follow the time–space evolution of the phenomena to predict the effect of any physical modification of the basin and to take into account actual or future reservoirs (Couillard, 1988). Finally, subdividing basins into elements facilitates the use of remote sensing for the determination of physiographic characteristics and the estimation of variables such as precipitation and snowcover.

3.1 Schematic Representation of a Basin

The geographical subdivisions of a basin may vary significantly in shape and density. The CEQUEAU model uses a set of square areas with identical dimensions (Fig. 5). In practice, this is accomplished by the superposition of a square grid on a map of the basin so as to define surface elements called 'whole squares' (Fig. 6) (Cluis *et al.*, 1979; Couillard & Cluis, 1980*a*,*b*). For each 'whole square', the physiographic characteristics (altitude of the southwest corner of the square, forested area, lake area, marsh area) required by the model to calculate the water balance are estimated.

The dimensions of the whole squares depend on the watershed's surface and on the number of meteorological stations. The model has previously been used with square dimensions varying from 0·25 to 30 km. Each whole square resulting from the first subdivision is further divided into 'partial squares' according to sub-basin divides. This allows the determination of downstream routeing from one partial square to the next. The only physical characteristics attached to this partial square are its area, given in percentage of the whole square, and the direction of flow.

[a] From *Encyclopedia of Fluid Mechanics*, Volume 10: Surface Groundwater Phenomena, by N. P. Cheremisinoff.

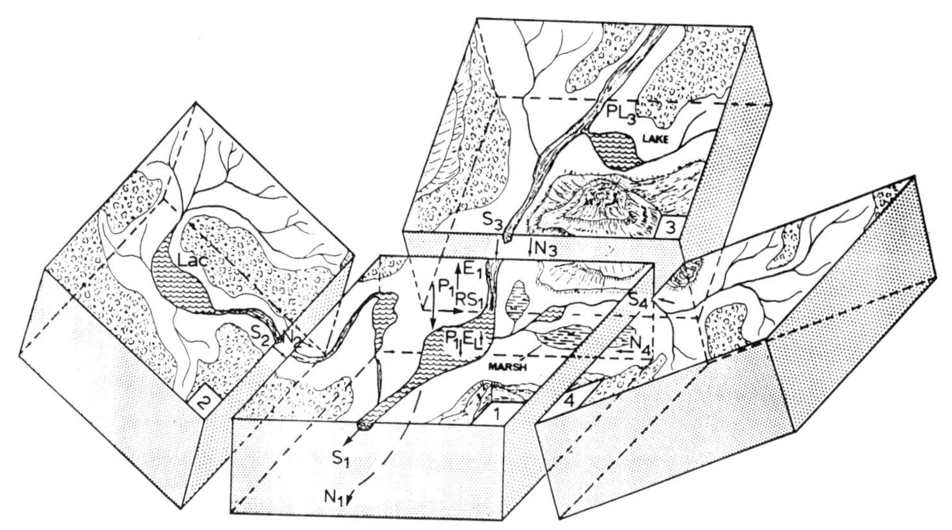

Fig. 5. Schematic representation of a watershed's subdivision into squares.

This second subdivision allows one to:

—follow the formation and the evolution of streamflow in time and space;
—introduce any artificial modification of streamflow in rivers;
—calculate the discharge at any point of the drainage network.

3.2 Meteorological Data

Besides physiographic data from whole squares, the CEQUEAU model requires for each of the squares the solid and liquid precipitation as well as the maximum and

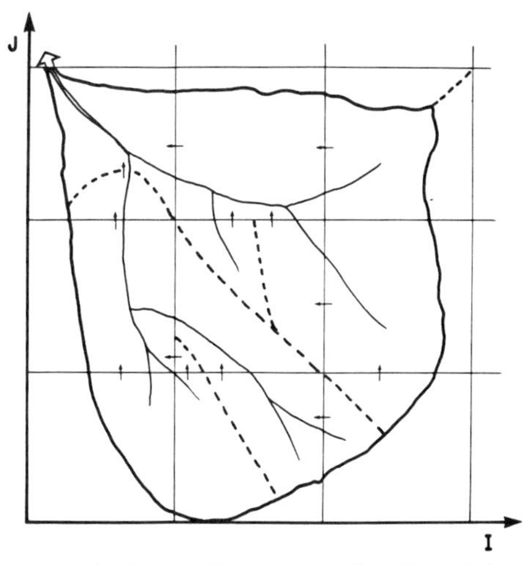

Fig. 6. Subdivision of the watershed into 'whole squares' and 'partial squares' as a function of sub-basins. The dashed lines show water divides between sub-basins.

minimum air temperatures. Those data being available only at a limited number of stations, their estimation is obtained by interpolation for each square.

3.3 Production Function

The production function represents in a simple and realistic manner what happens to liquid or solid precipitation from the moment it falls on the ground to the moment it reaches the river. Water from rainfall is in principle available immediately for infiltration; but for snowfall it is necessary to define a snowmelt

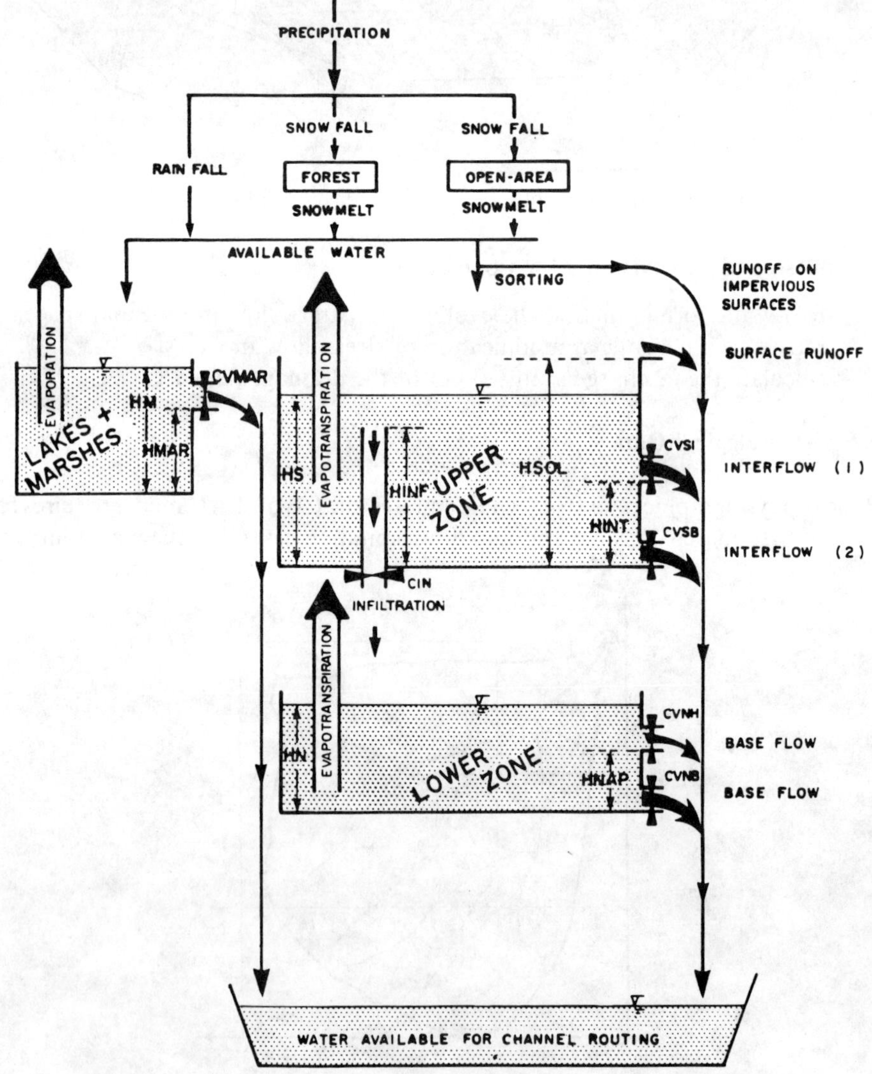

Fig. 7. Diagram of the production function of the CEQUEAU model.

model. Whatever may be the origin of atmospheric water entering a square before it is available for downstream routeing, it will be submitted to various processes that will have a direct influence on the synthesis of streamflow.

The model represents the ground as a series of interconnected reservoirs. Mathematical relations are used to calculate, on a daily basis, the different mass transfers (Fig. 7). The objective of those mathematical relations is to simulate the different components of the hydrological water-balance, which are:

—formation and melt of the snowpack;
—evaporation and evapotranspiration;
—water in the unsaturated zone;
—water in the saturated zone;
—water in lakes and marshes.

The hydrological balance is calculated on a daily basis on each whole square. For the day J, we have:

$$Q_J = P_J - \mathrm{ET}_J + (\mathrm{HS}_J - \mathrm{HS}_{J-1}) + (\mathrm{HN}_J - \mathrm{HN}_{J-1}) \qquad (14)$$

where Q = flow from upper and lower zones (mm)
P = liquid precipitation or snowmelt (mm)
ET = evapotranspiration (mm)
HS = water accumulated in the upper reservoir (mm)
HN = water accumulated in the lower reservoir (mm)

Equation (14) does not explicitly consider infiltration towards the lower zone; however, this part of the hydrological balance is taken into account for the determination of the water level in both reservoirs. After the diagram of the production function has been established, the values of the parameters of the mathematical relations simulating the various mass transfers between reservoirs have to be determined in order to obtain calculated flows as close as possible to observed flow. This is done by calibration of the model.

3.4 Routeing Function

The production function described in Fig. 7 yields water volumes ready for channel routeing. This is realized through the routeing of water from one partial square to the other. The volume of water available on a partial square is obtained from the product of the volume of water produced on the whole square by the percentage of area occupied by the partial square. This volume is added to the inflow to this square from the partial squares immediately upstream. The resulting volume becomes the volume available for routing to the next square downstream. Figure 8 shows the diagram of the routeing function. If the volume V_i is the volume accumulated in the partial square i, it will yield to the next partial square downstream a volume v_i proportional to its own value V_i and to the specific routeing coefficient of the partial square i.

The routeing coefficient of each partial square is related to the hydraulic characteristics of the flow within its boundaries and, more specifically, to the storage

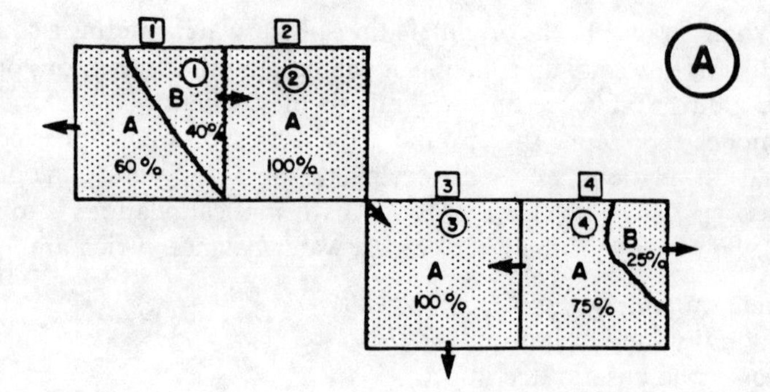

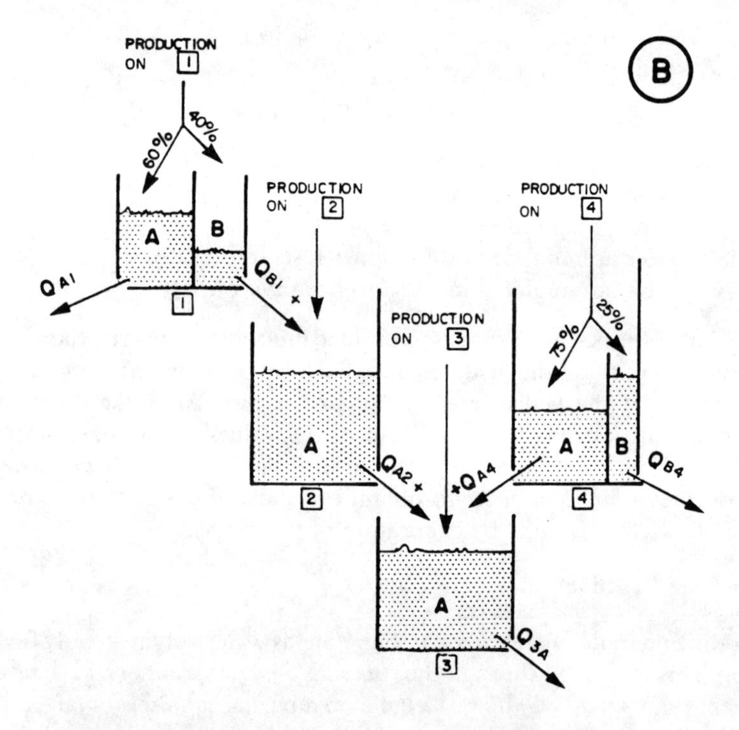

Fig. 8. Diagram of the routeing function.

capacity of the drainage network. A good index of this storage capacity is the areal extent of surficial waters on that square. For instance, a large lake has a potentially large storage capacity; however, its relative influence for flood control depends also on the surface area of the upstream basin. For example, a 10 km^2 lake does not have much influence on the flow of a 5000 km^2 basin, but will have a much greater storage effect on a 100 km^2 basin. In the CEQUEAU hydrological model, the

Table 4
Parameters and Constants of the CEQUEAU Model

Parameters for the upper and lower zone reservoirs over land and the lakes and marshed reservoirs:	
CIN[a]	percolation coefficient from the upper zone to the lower zone
CVMAR[a]	lakes and marshes drainage coefficient
CVNB[a]	lower-zone lower drainage coefficient
CVNH[a]	lower-zone upper drainage coefficient
CVSB[a]	upper-zone lower drainage coefficient
CVSI[a]	upper-zone intermediate drainage coefficient
HINF[a]	percolation threshold from the upper to the lower zone
HINT[a]	upper-zone intermediate drainage threshold
HMAR[a]	lakes and marshes threshold
HNAP[a]	lower-zone upper threshold
HRIMP[a]	precipitation threshold for direct run-off on impervious surface
HSOL[a]	upper-zone capacity
Snowmelt parameters:	
STRNE[b]	rain–snow threshold
TFC[b]	melting rate in forested areas
TFD[b]	melting rate in open areas
TSC[b]	melting threshold in forested areas
TSD[b]	melting threshold in open areas
TTD[b]	cold content coefficient for the snowpack
TTS[b]	priming threshold of the snowpack
Evapotranspiration parameters:	
EVNAP[a]	percentage of total daily evapotranspiration taken from lower-zone
HPOT[a]	threshold over which water is evapotranspirated at the potential rate
XAA[b]	exponent of the Thornthwaite formula
XIT[b]	thermal index of the Thornthwaite formula
Routeing:	
EXXKT[a]	fitting parameter for the routeing parameter XKT
ZN[c]	concentration time for the basin
Other parameters and constants:	
COET[b]	temperature lapse rate
COEP[b]	precipitation lapse rate
FACT[a]	coefficient allowing the modification of the mean areal precipitation for a group of whole squares
JOEVA[a] JONEI[a]	phasing parameters for sooner or later occurrence of the maximum of the annual isolation curve: JOEVA for evapotranspiration and JONEI for snowmelt
XINFMA[b]	daily maximum infiltration
XLA[c]	mean latitude of the basin
TRI[c]	percentage of impervious area in the basin

[a] Parameter determined by trial and error.
[b] Parameter determined for the physical characteristics of the phenomenon.
[c] Constant determined from hydrologic and physiographic characteristics of the basin.

routeing coefficient of each partial square is determined by the following equation:

$$XKT_i = 1 - \exp\left(\frac{EXXKT \times SA_i}{SL_i} \times \frac{100}{CEKM2}\right) \tag{15}$$

where XKT_i = daily routeing coefficient of the partial square i
EXXKT = fitting parameter whose value is determined by trial and error
SA_i = area of the basin upstream of the partial square i (km^2)
SL = area of surficial water on the partial square i (km^2)
CEKM2 = area of the whole squares (km^2)

The determination of routeing coefficients by eqn (15) can be in error for a few particular partial squares (a large lake or an unusual cross-section); a more rigorous solution may then be used, such as the utilization of stage–storage and stage–discharge curves. The model may also take into account actual or future dams by introducing the main characteristics of the reservoir storage and the operating policy.

3.5 Model Calibration

Before a model can be used to simulate streamflows from a watershed, the values given to its parameters and constants have to be optimized so that it reproduces the observed flows as closely as possible; this operation is called the 'calibration of the model'. The definitions of the terms 'parameter' and 'constant' must be clear. A parameter may be:

(a) determined by trial and error, or
(b) derived from the physical characteristics of a phenomenon. It may be chosen by studies independent from the model (e.g. snowmelt parameters).

A constant is determined from the hydrologic and physiographic characteristics of the drainage basin (e.g. the time of concentration).

Parameters and constants used by the CEQUEAU model are listed in Table 4. The calibration of these parameters is normally done by trial and error until simulated flows are sufficiently close to observed flows. However, in order to facilitate the calibration procedure, graphs of observed and calibrated streamflows as a function of time are printed out, as well as a number of other graphs and criteria describing and quantifying the differences between observed and calculated streamflows. The parameters may also be calibrated by automatic optimization.

4 INTEGRATION OF DO AND BOD SUBROUTINES TO THE CEQUEAU HYDROLOGICAL MODEL

The dissolved oxygen (DO) and biochemical oxygen demand (BOD) subroutines adapted to the CEQUEAU hydrological model are characterized by two major concepts. The first concept is related to the production of daily BOD and DO loads on each portion of the watershed being studied. In the case of the BOD, its production from non-point sources is related to land use and the accumulated loads will be entirely or partly transported to the river only if there is sufficient surface water run-off. Loads from point sources will be added to the loads transported from non-point sources. The production of DO is related to local water sources.

The second concept is related to the degradation and transfer of loads within the river. Degradation is calculated using equations proposed in the literature. Downstream load transfer is simulated with the procedure used for flow.

4.1 DO and BOD Simulation in Rivers

4.1.1 Choice of the DO and BOD Models

The model used is derived from the approach of O'Connor and DiToro (1970) and of O'Connor (1971). As defined in eqns (8) and (9), temporal variations in BOD and DO concentrations are equal to the sum of the sources and sinks for these two parameters. The combined term for photosynthesis and respiration, which in eqn (8) represents the diurnal variations in DO, will not be considered here. The river water temperature model (Morin *et al.*, 1987; Morin & Couillard, 1990) is needed to define the dissolved oxygen concentration at saturation and to apply a temperature correction factor to the DO and BOD consumption coefficients and to the benthic oxygen demand. According to O'Connor (1971), for a homogeneous river reach of constant area and flow solutions to the DO–BOD model can be expressed as:

$$
\begin{aligned}
D = {} & D_0\,\mathrm{e}^{-k_2 t} && \text{(re-oxygenation)} \\
& + \frac{k_D L_0}{k_2 - k_R}\left(\mathrm{e}^{-k_R t} - \mathrm{e}^{-k_2 t}\right) && \text{(carbon BOD)} \\
& + \frac{k_n N_0}{k_2 - k_n}\left(\mathrm{e}^{-k_n t} - \mathrm{e}^{-k_2 t}\right) && \text{(nitrogen BOD*)} \\
& + \frac{B}{k_2}\left(1 - \mathrm{e}^{-k_2 t}\right) && \text{(benthic demand*)} \qquad (16)
\end{aligned}
$$

$$
\mathrm{BOD_T} = L_0\,\mathrm{e}^{-k_R t} + \left(N_0\,\mathrm{e}^{-k_n t}\right)^* \qquad (17)
$$

where an asterisk indicates a part of the equation not introduced in the present version of the model, and

k_2 = re-oxygenation coefficient (day^{-1})
k_n = nitrogen-BOD reduction coefficient (day^{-1})
k_D = DO consumption coefficient (day^{-1})
k_R = carbon-BOD reduction coefficient (day^{-1})
t = time (days)
B = benthic dissolved oxygen demand (mg/litre)
$\mathrm{BOD_T}$ = total BOD concentration (mg/litre)
D = dissolved oxygen deficit at time t (mg/litre)
D_0 = initial dissolved oxygen deficit at time 0 (mg/litre)
L_0 = initial carbon-BOD concentration (mg/litre)
N_0 = initial nitrogen-BOD concentration (mg/litre)

O'Connor (1971) mentions that the rate of oxygen consumption is not necessarily equal to the rate of BOD decrease which can also be reduced by sedimentation, volatilization or through other processes. In the case of carbon-BOD, this accounts for the distinction between the k_R and k_D coefficients.

Due to a lack of data these two coefficients are initially set equal. They are distinguished, however, to provide greater freedom in adjusting the model. At present, the terms associated with nitrogen-BOD and benthic oxygen demand have not been introduced. They can be added, however, when considered important for a particular river or when data are available.

4.1.2 Determination of the DO–BOD Model Coefficients

4.1.2.1 DO AND BOD CONSUMPTION COEFFICIENTS

Whenever possible, the values of the coefficients of DO consumption (k_D) and carbon-BOD decrease (k_R) should be based on the interpretation of typical BOD_5 test results for the rivers being studied. First set equal, slightly different values may be given to these two coefficients during the calibration of the DO–BOD model.

4.1.2.2 RE-OXYGENATION COEFFICIENT

The re-oxygenation coefficient, k_2, is determined with the equation of Dobbins (1956) (c. Tables 1 and 2) modified with two adjustment coefficients:

$$k_2(20°\text{C}) = C_a C_h \frac{3{\cdot}0 U^{0{\cdot}67}}{H^{1{\cdot}75}} \tag{18}$$

where $k_2(20°\text{C})$ = re-oxygenation coefficient at 20°C
C_a = annual adjustment coefficient
C_h = adjustment coefficient under ice cover
U = mean current speed (m/s)
H = mean depth (m)

The C_a adjustment coefficient allows an increase or a reduction in the calculated value of the re-oxygenation coefficient. During winter, the effect of the ice cover can be taken into account by the C_h coefficient which allows a reduction in the calculated value of the re-oxygenation coefficient. The freeze-up and ice break-up dates are obtained with the water temperature model (Morin *et al.*, 1987; Morin & Couillard, 1990). During the ice-free period, the value of C_h is set at 1·0.

The mean current speed is calculated as the average number of partial squares travelled each day multiplied by the mean length of a river reach on a partial square. The average number of partial squares travelled is provided by the CEQUEAU hydrological model, whereas the mean length of a reach is estimated by calculating the square root of the surface area of a whole square. The mean depth is the value defined by the water temperature model (Morin & Couillard, 1990).

4.1.2.3 COEFFICIENT CORRECTION AS A FUNCTION OF TEMPERATURE

Equation (11) is used to correct the DO and BOD consumption coefficients (k_D and k_R) as a function of temperature. The k_2 re-oxygenation coefficient is adjusted with eqn (13) with the constant σ set at 1·025;

$$k_D(T) = k_D(20°C) \times 1{\cdot}045^{T-20} \quad (19)$$

$$k_R(T) = k_R(20°C) \times 1{\cdot}045^{T-20} \quad (20)$$

$$k_2(T) = k_2(20°C) \times 1{\cdot}025^{T-20} \quad (21)$$

where $k_D(T)$ = DO consumption coefficient at temperature T (day^{-1})
$k_R(T)$ = BOD consumption coefficient at temperature T (day^{-1})
$k_2(T)$ = re-oxygenation coefficient at temperature T (day^{-1})
T = water temperature (°C)

4.1.3 Advective BOD Transport to the River

The user of the DO–BOD model must define, for the watershed being studied, the mean daily BOD loads of human, animal, industrial, agricultural and natural origin. It is important to distinguish between point and non-point loads (Couillard & Cluis, 1980*a,b*). Point loads are those arriving directly into a river; examples are loads related to the outlet of a sewage network or to an industrial effluent discharge. Non-point loads, such as animal manure, are more spatially spread out and can travel a certain time on or in the soil before reaching a major river reach (Cluis *et al.*, 1979).

In accordance with the grid defined for the CEQUEAU hydrological model (Morin *et al.*, 1981), the partial square to which point loads belong must be identified. In the case of non-point sources, the whole square must be identified. Once identified, the loads become entry data for the DO–BOD model.

The total BOD load is the sum of the point and non-point loads that will reach the river.

4.1.3.1 ESTIMATION OF THE EFFECTIVE NON-POINT LOAD ON A WHOLE SQUARE

On each whole square, only a portion of a non-point load will reach the river. Therefore, a procedure was devised using the hydrological variables provided by the CEQUEAU model to estimate the effective non-point load in the river.

We consider that only surface and delayed run-off water can contain a BOD load. It is assumed that no load is carried by groundwater or by overflow water from lakes and marshes.

Each day, on each whole square, the mean daily non-point load is added to the load accumulated the previous day. Daily, the accumulated load undergoes a partial degradation that varies according to rain or snowmelt events.

Degradation of the accumulated load: When no sheath of rain or meltwater is present on the ground, the accumulated load undergoes a linear degradation:

$$CH_i = (1 - P_1)CH_{i-1} \quad (22)$$

where CH_i = accumulated non-point load at the end of the time interval i (kg)
CH_{i-1} = accumulated non-point load at the beginning of the time interval i (kg)
P_1 = degradation coefficient, varying between 0·0 and 1·0

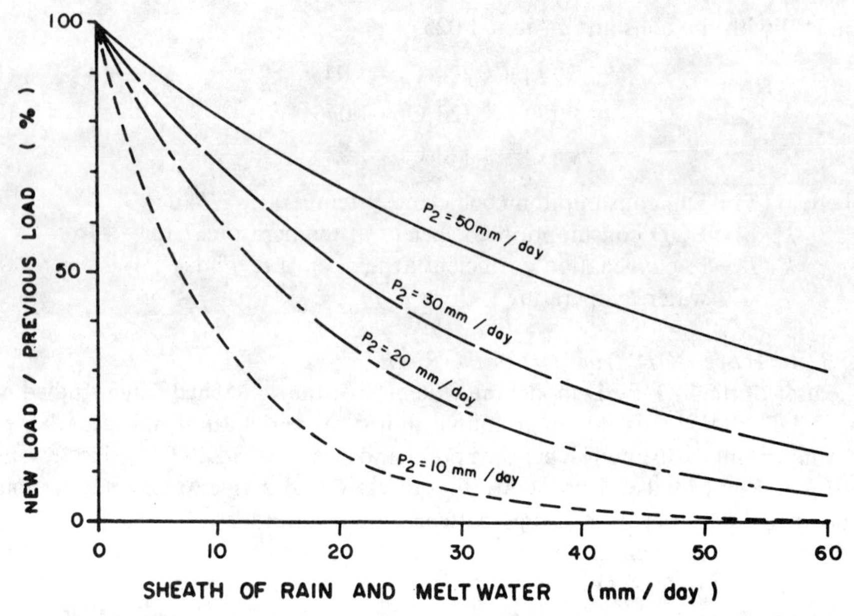

Fig. 9. Effect of the P_2 degradation parameter on the accumulated load.

When rain or meltwater is observed at ground level, the load remaining on the ground is estimated with the following equation:

$$CH_i = CH_{i-1} e^{-L/P_2} \tag{23}$$

where CH_i = accumulated load at the end of the time interval i (kg)
CH_{i-1} = accumulated load at the beginning of the time interval i (kg)
L = sheath of rain or meltwater (mm)
P_2 = adjustment parameter

Figure 9 illustrates the influence of the P_2 parameter.

Transport of the non-point load to the river: If, on a given day, there is surface and/or delayed run-off, a portion of the load is carried towards the river and is undergoing a certain amount of degradation.

The load carried towards the river is a function of the load accumulated on the ground, of the water run-off as calculated by the hydrological portion of the CEQUEAU model and of an adjustment parameter:

$$CH_e = CH(1 - e^{-R/P_3}) \tag{24}$$

where CH_e = load transported towards the river (kg)
CH = load accumulated on the whole square (kg)
R = water run-off calculated by the hydrological model (mm)
P_3 = adjustment parameter (mm)

Figure 10 illustrates the influence of the P_3 parameter.

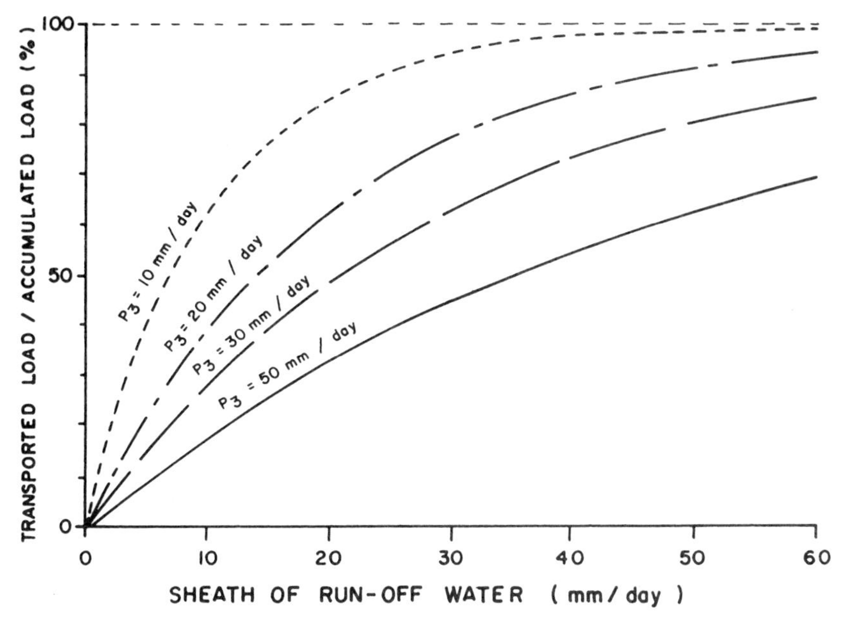

Fig. 10. Effect of the P_3 transport parameter on the accumulated load.

The load carried by run-off water is being degraded and only a portion reaches the river. This portion, called the 'effective non-point load', is calculated with the following equation:

$$CH_d = (1 - P_4)CH_e \tag{25}$$

where CH_d = effective non-point load on a whole square reaching the river (kg)
CH_e = load carried towards the river (kg)
P_4 = degradation coefficient, varying between 0·0 and 1·0.

4.1.3.2 TOTAL BOD LOAD ON A PARTIAL SQUARE

The effective non-point load on a whole square is shared between its partial squares in relation to their relative surface area. The total BOD load daily reaching a given river reach is the sum of the effective non-point loads thus defined and of the point loads on the partial square being considered.

4.1.4 Advective DO Contributions to the River

The advective DO loads on each partial square are those contained in run-off and delayed run-off water, in groundwater and in the overflow water of lakes and marshes. They are all considered to be saturated. The quantities associated with these various sources are provided by the hydrological model.

4.1.5 Degradation and Transfer in Rivers

The aim of the last section of the dissolved oxygen and biochemical oxygen demand model is to reproduce the downstream transfer of the load. Two phenomena are considered: the degradation of the load accumulated on each reach and its displacement downstream.

The DO and BOD concentrations of each reach at the beginning of the day will be degraded in accordance with the O'Connor (1971) model (eqns (16) and (17)). These concentrations are obtained by dividing the total DO and BOD loads of the given reach by the corresponding dilution volume.

The total DO and BOD load at the beginning of a time interval is the sum of the reach's load, of the upstream load and of the load produced locally on the partial square. As mentioned previously, in the case of the BOD the point and non-point loads constitute the total load, whereas in the case of the DO it is constituted of the loads associated with the locally produced water.

The volume of dilution water on a partial square is the sum of the set and variable volumes. The set volume is the product of the reach's length multiplied by its width, both estimated from physiographic data, and multiplied by the depth, determined during calibration. The variable water volume in a reach is defined as the variable volume stored during the previous time interval, plus the upstream and local contributions. These last two quantities are provided by the CEQUEAU hydrological model.

For the first simulated day, the mean BOD concentration on the watershed is set by the user. In the case of the DO concentration, it is set at saturation for the temperature initially determined by the user.

For a given partial square, the BOD or DO load transferred to the downstream partial square is the product of the water volume at the outlet of the partial square being considered, multiplied by the corresponding concentration having been degraded in the river. These transfers are carried out from the most upstream partial square to the one farthest downstream.

5 MODEL ADJUSTMENTS AND RESULTS

For the present study, the physiographic and meteorological data needed to operate the CEQUEAU hydrological model were prepared for the Sainte-Anne watershed. The location of the watershed is presented in Fig. 11 and Table 5 summarizes the state of the data banks constituted for this application.

5.1 Physiographic Data

In addition to the physiographic data normally used in the CEQUEAU hydrological model (see preceding section), for each river reach, other physical characteristics were measured on topographical maps. The characteristics retained for each partial square are:

—the length of the main river
—the width of the river at the partial square's exit
—the altitude of the river at the partial square's exit

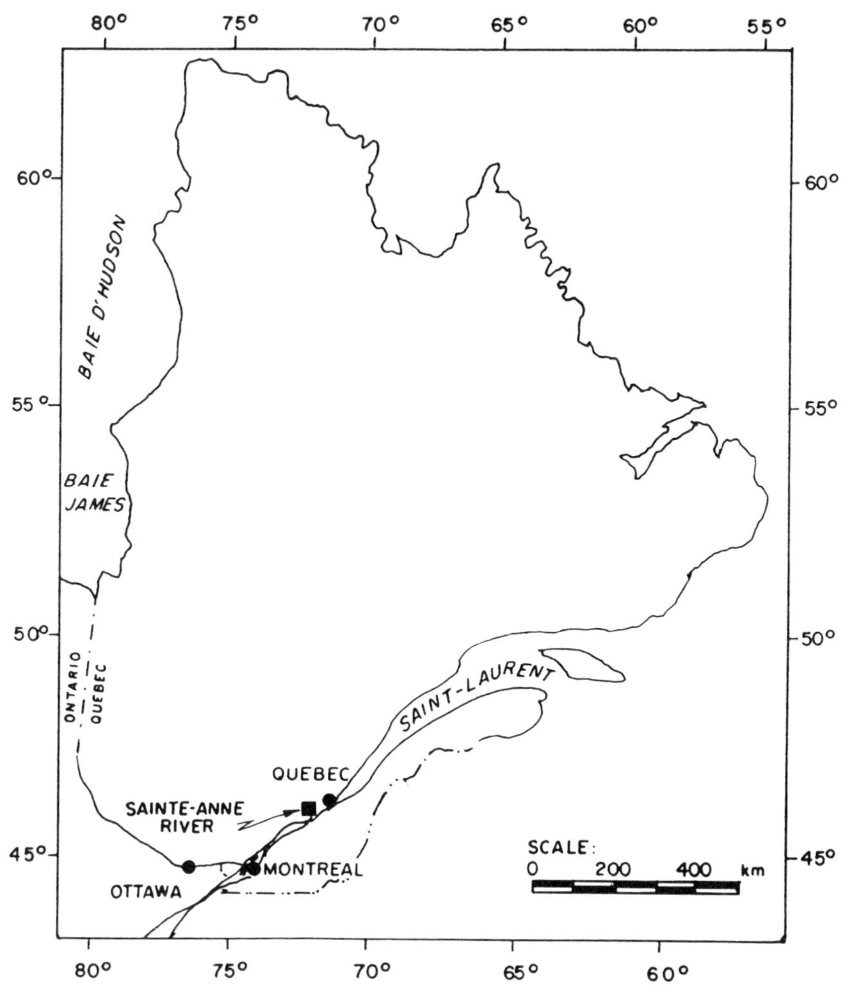

Fig. 11. Geographic location of the watershed.

Table 5
Characteristics of the Data Banks Constituted for the Operation of the Model for the Drainage Basin of the Sainte-Anne River

Drainage basin (km^2)	*Grid size (km × km)*	*Number of squares*		*Period*	*Number of stations*	
		Whole	*Partial*		*Meteorological*	*Hydrographic*
2 700	10 × 10	43	69	1968–1980	10	4

Note that if there is more than one river at a partial square's exit, the estimated width takes into account all the rivers that can be located at a square's exit.

These data will be used to calculate the surface of free-running water and the slope of the river on each partial square. Width and length data will eventually be estimated with the help of relations linking these values to the physiographic data.

5.2 Meteorological Data

In addition to daily maximum and minimum air temperature and liquid and solid precipitation data already used in the CEQUEAU hydrological model, the operation of the subroutine to calculate water temperature requires the following data: solar radiation, vapour pressure, cloud cover, and wind velocity (Morin *et al.*, 1987; Morin & Couillard, 1990). Since these data are not available on a daily basis in many of Québec's regions, the river water temperature calculation subroutine uses monthly mean values. Table 6 presents the data used for the Sainte-Anne river basin.

5.3 Model Adjustment

Prior to adjusting the dissolved oxygen (DO) and biochemical oxygen demand (BOD) model, it is necessary in order to adequately reproduce the observed flows to calibrate the hydrological model (Morin *et al.*, 1981).

Afterwards, the DO–BOD model is adjusted by modifying, as needed, the empirical coefficients associated with the terms of production and degradation.

In the case of the Sainte-Anne river watershed, we used the dissolved oxygen data

Table 6
Mean Monthly Meteorological Data Used for Water Temperature Simulations in the Sainte-Anne River

Parameter	*Station*	*Jan.*	*Feb.*	*Mar.*	*Apr.*	*May*	*June*
Solar radiation[a] (MJ/m²)	Montréal–Normandin mean	5·22	8·95	13·47	16·90	19·01	20·38
Cloud cover[b]	Estimated on a map	0·62	0·63	0·61	0·56	0·52	0·52
Wind (km/h)[c]	Valcartier	5·1	5·5	6·1	6·0	6·1	5·6
Vapour pressure[d]	Québec	1·73	1·88	2·63	3·90	5·55	9·00
		July	*Aug.*	*Sept.*	*Oct.*	*Nov.*	*Dec.*
Solar radiation[a] (MJ/m²)	Montréal–Normandin mean	20·19	17·07	12·56	7·35	4·48	3·92
Cloud cover[b]	Estimated on a map	0·47	0·48	0·58	0·63	0·75	0·71
Wind (km/h)[c]	Valcartier	5·3	4·8	4·7	4·9	4·9	5·0
Vapour pressure[d]	Québec	11·10	10·58	8·25	5·70	3·75	2·10

[a] Environment Canada (1982*a*).
[b] Gariépy *et al.* (1981)
[c] Environment Canada (1982*b*).
[d] Environment Canada (1976).

measured at La Pérade during 1979 and 1980 (Ministère du Loisir, de la Chasse et de la Pêche, 1981; Environnement Québec, 1983).

5.4 Model Parameters

As described in the preceding section, in the major equations we have used factors that allow an adjustment of the various components involved in the calculations for each whole and partial square.

The parameters needed to operate the dissolved oxygen and biochemical oxygen demand model are described below.

5.4.1 DO and BOD Consumption Coefficients

The values of these coefficients should be based on the interpretation of BOD_5 tests for the rivers being studied. Initially, these coefficients are set equal. It is possible, however, to give different values when calibrating the model. These coefficients are modified according to temperature with eqns (19) and (20). Trials were carried out with the values of each coefficient varying between 0·15 and 0·55.

5.4.2 Adjustment Parameters for the Re-oxygenation Coefficient

As calculated in the model, the re-oxygenation coefficient takes two adjustment factors into account. The first is applicable year-round, while the second is only applicable when there is an ice-cover on the river.

The re-oxygenation coefficient is calculated daily with eqns (18) and (21). Trials were carried out with values of the annual coefficient varying between 0·5 and 1·5, and with values for the ice-cover coefficient varying between 0·0 and 1·0.

5.4.3 Degradation Parameters for the Accumulated Non-point Load

With time, the accumulated loads at each whole square undergo degradation as a function of the sum of the sheaths of precipitation and meltwater.

There is a linear degradation with time (eqn (22)) that depends on the adjustment parameter. Trials were carried out with values varying between 0·1 and 0·5.

Degradation by the sheath of water available at ground level follows an exponential function estimated with the help of the sheath of precipitation and meltwater and of an adjustment parameter (eqn (23)). Trials were carried out with values varying between 5 and 50 mm.

5.4.4 Transport Parameter of the Non-point Load to the River

In the proposed model, the non-point load accumulated on the whole square can only be carried towards the river if there is surface or delayed run-off. The load carried varies exponentially as a function of run-off which is provided by the hydrological model, and as a function of an adjustment parameter (eqn (24)). Trials were carried out with values varying between 5 and 50 mm.

5.4.5 Degradation Parameter of the Transported Load

The load carried by run-off water undergoes a linear degradation and consequently

Table 7
BOD_5 Production by Human Populations per Partial Square (CP) or per Whole Square (CE) on the Sainte-Anne River Basin

References			Site	Population		Unit production (kg/day/person)	Total production (kg/day)	
CE	CP	IJ		With sewer	Without sewer		Point	Non-Point
1	1	12 11 A	La Pérade—village	1 039	—			
			Ste-Anne-de-la-Pérade—parish	1 218	233			
			TOTAL	2 257	233	0·076	172	18
2	2	11 11 A	St-Prosper-de-Champlain—parish	454	209	0·076	35	16
3	3	12 12 A	St-Casimir—village	1 042	91			
			St-Casimir—parish	—	444			
			St-Casimir est—village	362	—			
			St-Thuribe—parish	226	184			
			TOTAL	1 630	719	0·076	124	55
6	6	13 12 A	St-Alban—village	673	—			
			St-Alban—parish	—	583			
			TOTAL	673	583	0·076	51	44
7	7	11 13 A	St-Ubalde sd	281	1 324	0·076	21	101
13	16	14 13 A	Ste-Christine—parish	94	255	0·076	7	19
14	18	14 14 B	St-Léonard-de-Portneuf sd	191	868	0·076	15	66
18	27	15 14 A	St-Raymond—town	3 551	54			
			St-Raymond—parish	942	2 983			
			TOTAL	4 493	3 037	0·076	341	231

Sources: Statistique Canada, 1982*a*; OPDQ, 1979; MAS, 1978, 1981; Mascolo *et al.*, 1973.

only a portion reaches the river. This degradation is adjusted with the help of a parameter (eqn (25)). Trials were carried out with values varying between 0·1 and 0·5.

5.4.6 Other Parameters

For all simulations, the ratio of the ultimate BOD to the BOD_5 was set at 1·47. The river BOD_5 concentration on the first day of calculations was set at 1 mg/litre.

5.5 Point and Non-point Loads

Operating the DO–BOD model requires entry data consisting of the mean daily BOD loads of human, animal, industrial, agricultural or natural origin whenever

Table 8

BOD_5 Production on Whole Squares (CE) by Animal Populations of the Sainte-Anne River Basin

References		*Estimated population*				*Production*[a] *(kg/day)*
CE	*IJ*	*Cattle*	*Hogs*	*Horses*	*Chickens*	
1	12 11	1 518	—	20	200	1 017
2	11 11	501	—	24	31	346
3	12 12	3 600	50	21	334	2 401
4	11 12	1 635	717	48	190	1 233
5	12 13	2 054	26	12	181	1 370
6	13 12	1 366	—	6	148	907
7	11 13	1 518	597	17	169	1 116
8	12 14	951	—	9	90	635
9	13 13	1 457	516	9	177	1 057
10	11 14	140	—	2	11	93
11	12 15	—	—	—	—	—
12	13 14	217	1 153	6	21	343
13	14 13	122	577	4	18	182
14	14 14	394	791	9	1 351	414
15	15 13	—	—	—	—	—
16	13 15	648	—	10	20	434
17	14 15	309	10	5	6 697	276
18	15 14	155	8	2	5 357	158
19	13 16	92	5	1	3 348	96
20	15 15	257	14	4	9 375	269
21	14 16	147	8	2	5 357	153
22	15 16	202	11	3	7 366	211
23	16 15	73	4	1	2 679	77
24	16 16	55	3	1	2 009	58
25	15 17	—	—	—	—	—
26	16 17	—	—	—	—	—
27	17 16	—	—	—	—	—
28	14 17	110	6	2	4 018	115
29–43		—	—	—	—	—

Sources: Statistique Canada, 1982*b*; Cluis *et al.*, 1974.

[a] Unit BOD_5 production used: 0·66 kg/day/cattle; 0·17 kg/day/hog; 0·63 kg/day/horse; 0·01 kg/day/chicken.

Table 9
BOD_5 Production per Partial Square (CP) by Industries Located on the Sainte-Anne River Basin

References CP	References IJ	Site	Industry	Employees	Water consumption (litre/day/employee)	Effluent BOD_5[a] (mg/litre)	BOD_5 production[a] (kg/day)
1	12 11 A	Ste-Anne-de-la-Pérade	dairy	83	3 593	3 700	1 103
			candy	2	1 109	250	1
			bakery	4	1 001	260	1
			margarine	28	3 593	62 000	6 237
			clothing	45	—	~0	~0
			animal feed	2	—	~0	~0
			metal products	10	—	~0	~0
							TOTAL 7 342
2	12 12 A	St-Casimir	sawmill equipment	20	—	~0	~0
			concrete piping	4	—	~0	~0
			lumber planing	3	—	~0	~0
			plywood	67	—	~0	~0
			animal feed	13	—	~0	~0
							TOTAL ~0
6	13 12 A	St-Alban	butter	7	3 593	62 000	1 559
			raw lumber	15	—	~0	~0
			sawmill	6	—	~0	~0
			hydraulic equipment	7	—	~0	~0
							TOTAL 1 559
7	11 13 A	St-Ubalde	bakery	13	1 001	260	3
			clothing	60	—	~0	~0
			lumber	12	—	~0	~0
			sawmill	15	—	~0	~0
							TOTAL 3

16	14 13 A	Ste-Christine	charcoal	11	—	~0	~0
18	14 14 B	St-Léonard-de-Portneuf	doors, windows	2	—	~0	~0
			starters and generators	4	—	~0	~0
			timber	3	—	~0	~0
			lumber	12	—	~0	~0
						TOTAL	~0
27	15 14 A	St-Raymond	charcoal	10	—	~0	~0
			charcoal	12	—	~0	~0
			cheese-dairy	4	3 593	12 000	172
			printing	6	—	~0	~0
			printing	1	—	~0	~0
			lumber planing	30	—	~0	~0
			newsprint	150	11 050	200	332
			sawmill	40	—	~0	~0
			dairy	50	3 593	3 700	665
			gloves	20	—	~0	~0
			sawmill, soap	3	—	~0	~0
			wood cupboard	6	—	~0	~0
			wood furniture	200	—	~0	~0
			sawmill	50	—	~0	~0
			concrete, gravel	100	—	~0	~0
			wood treatment	12	—	~0	~0
						TOTAL	1 169

Sources: Scott's Industrial Directories, 1980; Couillard, 1974; INRS-Eau Groupe Système Urbain, 1973.
[a] Estimated values.

applicable. It is necessary to distinguish between non-point loads distributed on the whole square which will eventually be carried to the river, and point loads arriving directly to the river.

The non-point loads are of natural, agricultural, human or animal origin. On the watershed of the Sainte-Anne river, the loads of natural origin or originating from cultures are considered negligible and were not taken into account.

Human population density was estimated on each partial square of the Sainte-Anne river watershed with the help of Statistique Canada (1982*a*) publications. For municipalities equipped with a sewer system (MAS, 1978, 1981; OPDQ, 1979), the point load is associated with the corresponding partial square. For municipalities without sewer systems, the non-point load is applied to the whole square.

Table 7 summarizes the spatial distribution of the point and non-point loads of human origin on the watershed. The unit production of 0·076 kg/day/person is taken from Mascolo *et al.* (1973).

Based on Statistique Canada (1982*b*) data, Table 8 shows the spatial distribution of non-point loads of animal origin on the Sainte-Anne river watershed.

The unit production values of 0·66 kg/day/cattle, 0·17 kg/day/swine, 0·63 kg/day/horse and 0·01 kg/day/chicken are taken from Cluis *et al.* (1974).

Point loads originate from municipalities with sewer systems and from industries. Table 9 summarizes the spatial distribution and the type of industries operating on the Sainte-Anne watershed. The data were obtained from the industry year-book (Scott's Industrial Directories, 1980).

Water consumption per employee was obtained from a report by the INRS-Eau

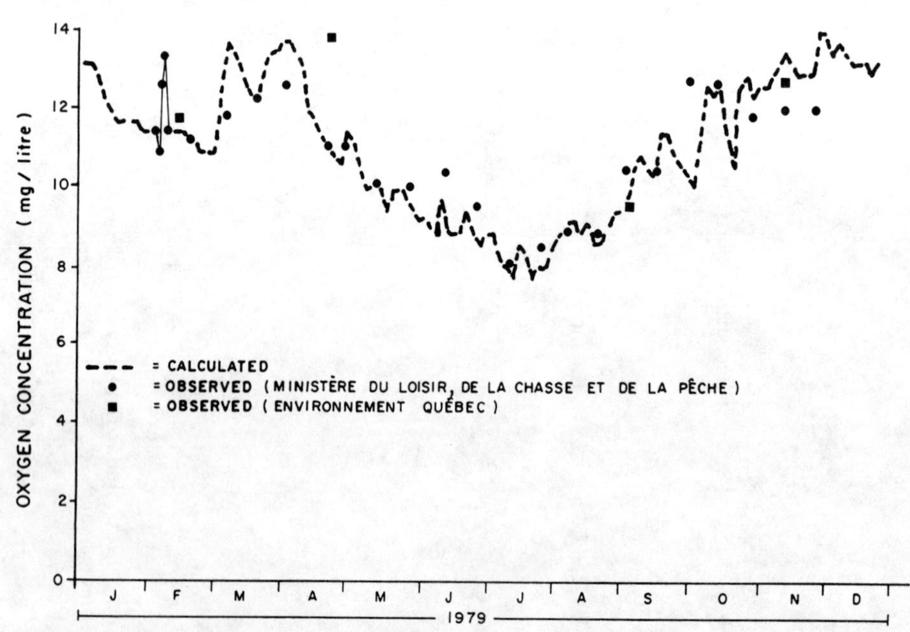

Fig. 12. Calculated (3-day mean) and observed dissolved oxygen values for the year 1979, at La Pérade on the Sainte-Anne river.

Groupe Système Urbain (1973), while the typical effluent concentrations in mg/litre were obtained from Couillard (1974).

5.6 Calibration and Analysis of Results

The model was calibrated for the Sainte-Anne river by using the few dissolved oxygen measurements obtained at La Pérade during 1979 and 1980.

The trial and error method of calibration was used; after a first simulation a few parameters are modified for the next simulation, and the results are then compared. This process is repeated until satisfactory results are obtained. Knowledge of the model and of the interrelationships between the various parameters allows the user to determine which parameter should be modified.

The values of the parameters used for the simulations presented are:

k_D, DO consumption coefficient (eqn 19)	0·35
k_R, BOD consumption coefficient (eqn 20)	0·35
C_a, Annual re-oxygenation coefficient (eqn 18)	0·50
C_h, Winter re-oxygenation coefficient (eqn 18)	0·15
P_1, Temporal degradation coefficient (eqn 22)	0·10
P_2, Coefficient of degradation by precipitation and meltwater (eqn 23)	20 mm
P_3, Load transport coefficient (eqn 24)	5 mm
P_4, Coefficient of degradation of the transported load (eqn 25)	0·10
Ratio of ultimate BOD to BOD_5	1·47
Initial BOD concentration	1 mg/litre

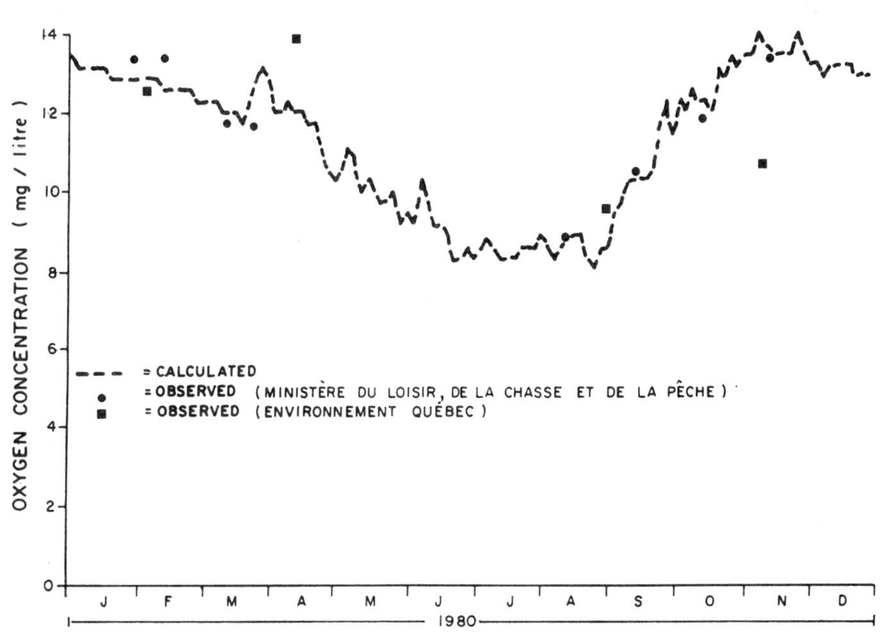

Fig. 13. Calculated (3-day mean) and observed dissolved oxygen values for the year 1980, at La Pérade on the Sainte-Anne river.

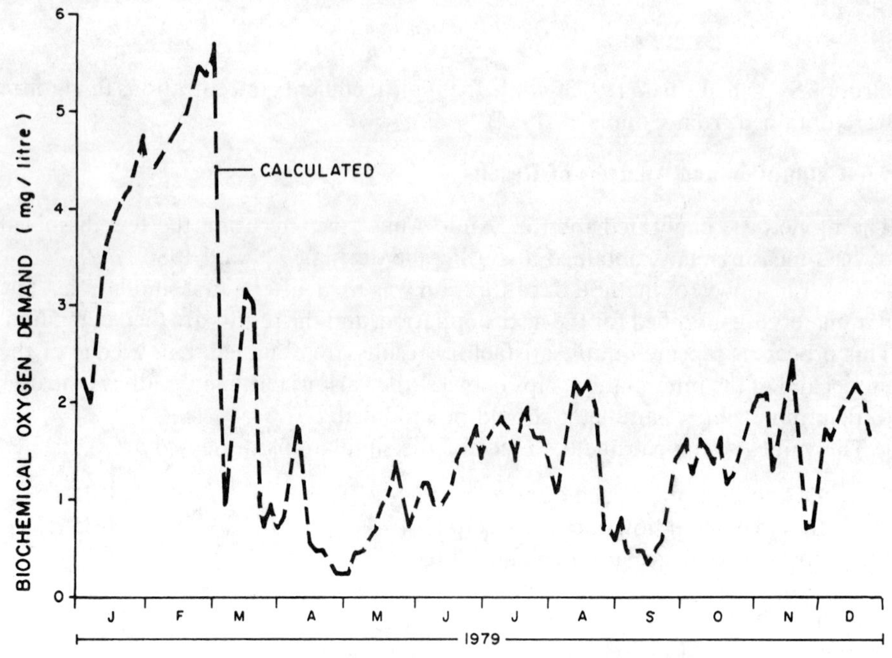

Fig. 14. Biochemical oxygen demand (3-day mean) calculated for the year 1979 at La Pérade on the Sainte-Anne river.

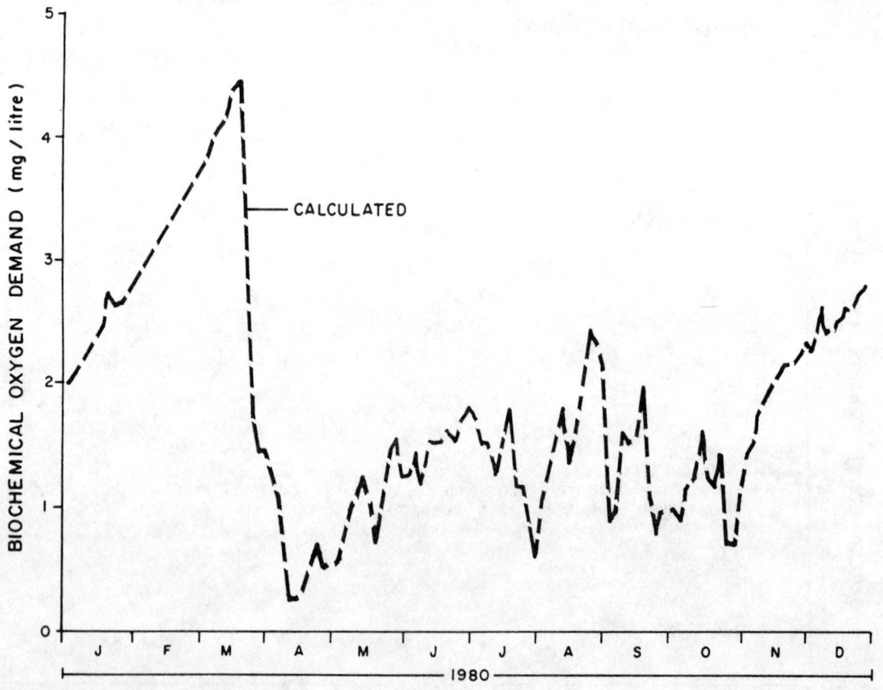

Fig. 15. Biochemical oxygen demand (3-day mean) calculated for the year 1980 at La Pérade on the Sainte-Anne river.

Figures 12 and 13 show the dissolved oxygen concentrations (3-day mean) calculated at La Pérade on the Sainte-Anne river for the years 1979 and 1980. The few measured values available are also shown on these figures. Generally, there is a good fit between the calculated and measured values. Result analysis is hazardous due to the lack of observations and to the fact that the values are near saturation. Nevertheless, the model is considered adequate for the simulation of this parameter.

The biochemical oxygen demand values (3-day mean) calculated at La Pérade on the Sainte-Anne river for the years 1979 and 1980 are presented in Figs 14 and 15. These figures are only indicative since no measured value is available with which to verify the adjustment of the model. The variations in BOD observed in these figures are related to the intensity of river flow. The reduction observed toward the end of March is due to the large flow increase following snowmelt. Similarly, increases in BOD observed during the months of November to March were due to a reduction in river flow.

6 CONCLUSION

The model appears to adequately simulate dissolved oxygen at La Pérade on the Sainte-Anne river. It is, however, obvious that a good adjustment of the model will require other simulations for rivers where ample data are available and particularly for rivers where dissolved oxygen concentrations are well below saturation.

The biochemical oxygen demand model also needs to be verified in rivers where data are available.

ACKNOWLEDGEMENTS

The authors wish to express their appreciation for the financial support of Hydro-Québec and the Natural Sciences and Engineering Research Council of Canada. Special thanks are due to D. Cluis and J. M. Gauthier for their assistance.

REFERENCES

Bansal, M. K. (1973). *Wat. Res.*, **7**, 769.

Charbonneau, R., Fortin, J. P. & Morin, G. (1975). *Hydrol. Sci. Bull.*, **22**(1), 193.

Churchill, M. A., Elmore, H. L. & Buckingham, R. A. (1962). *J. Sanit. Engng Div., Am. Soc. Civil Engrs*, **88**(SA4), 1.

Cluis, D. (1973). *Analyse des réactions en rivière: modèles mathématiques de qualité—revue de littérature.* Scientific report no. 23, INRS-Eau, Université du Québec, Sainte-Foy, Québec, Canada. 127 pp.

Cluis, D., Couillard, D. & Potvin, L. (1974). *Planification de l'acquisition des données de qualité de l'eau au Québec. Tome 4: Utilisation du territoire d'un bassin et modèle d'apports.* Publ. no. Q.E.-9, Ministère des Richesses naturelles, Service de qualité des eaux, Gouvernement du Québec, Québec, Canada. 135 pp.

Cluis, D. A., Couillard, D. & Potvin, L. (1979). *Wat. Resource. Res.*, **15**(3), 630.

Couillard, D. (1974). *Compilation de certains rejets industriels: bilan des polluants*. Scientific report no. 47, INRS-Eau, Université du Québec, Sainte-Foy, Québec, Canada. 246 pp.
Couillard, D. (1979). *Sci. Total Envir.*, **12**(2), 169.
Couillard, D. (1980). *Sci. Total Envir.*, **14**(2), 167.
Couillard, D. (1983). *J. Envir. Syst.*, **13**(1), 43.
Couillard, D. (1988). *J. Envir. Mgnt.*, **26**(2), 95.
Couillard, D. & Cluis, D. (1980*a*). *Wat. Supply Mgmt.*, **4**(4), 263.
Couillard, D. & Cluis, D. (1980*b*). *Wat. Res.*, **14**(11), 1621.
Couillard, D. & Gariépy, S. (1990). *Can. J. Chem. Engng*, **68**(8), 300.
Couillard, D. & Tyagi, R. D. (1988). *Envir. Technol. Lett.*, **9**(12), 1327.
Couillard, D., Cluis, D. & Morin, G. (1988). *Wat. Res.*, **22**(8), 991.
Couillard, D., Gariépy, S. & Tran, F. (1989). *Wat. Res.*, **23**(5), 573.
Dobbins, W. E. (1956). In *Biological Treatment of Sewage and Industrial Wastes*, eds J. A. McCabe & W. W. Eckenfelder, Jr. Reinhold, New York.
Dobbins, W. E. (1964). *J. Sanit. Engng Div., Am. Soc. Civil Engrs*, **90**(SA3), 53.
Downing, A. L. & Truesdale, G. A. (1955). *J. Appl. Chem.*, **5**, 570.
Dresnack, R. & Dobbins, W. E. (1968). *J. Sanit. Engng Div., Am. Soc. Civil Engrs*, **94**(SA5), 789.
Elmore, H. L. & West, W. F. (1961). *J. Sanit. Engng Div., Am. Soc. Civil Engrs*, **87**(SA6), 59.
Environment Canada (1976). *Station, sea level and vapour pressure normals based on the period 1953–1972.* Government of Canada: Atmospheric Environment Service, Ottawa, Canada. 17 pp.
Environment Canada (1982*a*). *Normales climatiques au Canada—Volume 1: Rayonnement solaire 1951–1980.* Gouvernement du Canada: Service de l'environnement atmosphérique, Ottawa, Canada. 57 pp.
Environment Canada (1982*b*). *Normales climatiques au Canada—Volume 5: Vent 1951–1980.* Gouvernement du Canada: Service de l'environnement atmosphérique, Ottawa, Canada. 283 pp.
Environnement Québec (1983). *Banque de données de la qualité du milieu aquatique.* Environnement Québec: Service de la qualité des eaux, Gouvernement du Québec, Québec, Canada.
Fair, G. M. & Geyer, J. C. (1954). *Water Supply and Waste Water Disposal.* John Wiley & Sons, New York.
Gameson, A. & Truesdale, G. A. (1955). *J. Inst. Wat. Engrs*, **9**(7), 123.
Gariépy, J., Clavet, C. & Leduc, R. (1981). *L'ensoleillement au Québec.* Ministère de l'Environnement du Québec: Service de la météorologie. Publ. MP-60, Québec, Canada. 32 pp.
Girard, G., Morin, G. & Charbonneau, R. (1972*a*). *Cahier ORSTOM, série hydrologie*, Paris, **IX**(4), 35.
Girard, G., Charbonneau, R. & Morin, G. (1972*b*). In *Proceedings of the International Symposium on Modelling Techniques in Water Resources System*, ed. A. K. Biswas. Environment Canada, Ottawa, Canada, p. 190.
Hassan, M. F. (1989). In *Encyclopedia of Environmental Control Technology, Vol. 3: Wastewater Treatment Technology*, ed. N. P. Cheremisinoff. Gulf Publ. Co., Houston, TX, p. 77.
Higbie, R. R. (1935). *Trans. Am. Inst. Chem. Engrs*, **31**, 365.
INRS-Eau Groupe Système Urbain (1973). *Systèmes urbains de distribution d'eau—Étude du système et de la demande.* Scientific report no. 12, INRS-Eau, Université du Québec, Sainte-Foy, Québec, Canada. 49 pp.
Kalinske, A. A. & Levich, C. L. (1944). *Ind. & Engng Chem.*, **36**, 220.
Krenkel, P. A. & Orlob, G. T. (1962). *J. Sanit. Engng Div., Am. Soc. Civil Engrs*, **88**(SA2), 53.
Langbein, W. B. & Durum, W. H. (1967). *The reaeration capacity of streams.* US Geol. Surv. Circular 542. US Govt Printing Office, Washington, DC.
Lawrence, K. W., Vielkind, D. & Wang, M. H. (1978). *Ecological Modelling*, **5**, 115.

Markofsky, M. & Harleman, D. R. F. (1971). *A predictive model for thermal stratification and water quality in reservoirs.* Technical report no. 134, MIT Hydrodynamics Laboratory, Department of Civil Engineering, Massachusetts Institute of Technology, Boston, USA, 283 pp.

MAS (1978). *Inventaire national des équipements en eau des municipalités du Canada 1977.* Publ. Cat. no. EM44-10/1977, Ministère des Approvisionnements et Services, Gouvernement du Canada, Ottawa, Canada. 339 pp.

MAS (1981). *Inventaire national des équipements en eau des municipalités du Canada 1981.* Publ. Cat. no. EN44-10/81, Ministère des Approvisionnements et Services, Gouvernement du Canada, Ottawa, Canada. 389 pp.

Mascolo, D., Meybeck, M., Cluis, D. & Couillard, D. (1973). In *L'eau et l'environnement,* Conf. conjointe AQTE–FACE, Montréal, 30 April–3 May 1972, p. 541.

Ministère du Loisir, de la Chasse et de la Pêche (1981). *Caractéristiques physico-chimiques de l'eau de la rivière Sainte-Anne à La Pérade, Québec.* Comité d'étude sur le poulamon atlantique. Technical report no. 3, Gouvernement du Québec, Québec, Canada. 49 pp.

Miyamoto, S. (1932). *Bull. Chem. Soc. of Japan,* **7**(1), 8.

Morin, G. & Couillard, D. (1990). In *Encyclopedia of Fluid Mechanics,* ed. N. P. Cheremisinoff. Gulf Publ. Co., Houston, TX, p. 171.

Morin, G., Fortin, J. P., Lardeau, J. P., Sochanska, W. & Paquette, S. (1981). *Modèle CEQUEAU: manuel d'utilisation.* Scientific report no. 93, INRS-Eau, Université du Québec, Sainte-Foy, Québec, Canada. 449 pp.

Morin, G., Cluis, D., Couillard, D., Jones, H. G. & Gauthier, J. M. (1986*a*). In *La Recherche en Hydrologie au Québec,* eds V. T. V. Nguyen & Y. Faucher. Les Presses de l'Université du Québec, Québec, Canada, p. 134.

Morin, G., Couillard, D., Cluis, D., Jones, G. H. & Gauthier, J. M. (1986*b*). *Can. J. Civil Engng,* **13**(2), 196.

Morin, G., Couillard, D., Cluis, D., Jones, G. H. & Gauthier, J. M. (1987). *Hydrol. Sci. J.,* **32**(1), 31.

Morin, G., Cluis, D., Couillard, D., Jones, G. H. & Gauthier, J. M. (1988). *Can. J. Civil Engng,* **15**(3), 315.

O'Connor, D. J. (1961). *Trans. Am. Soc. Civil Engrs,* **126**, 3.

O'Connor, D. J. (1967). *Wat. Resourc. Res.,* **3**(1), 65.

O'Connor, D. J. (1971). *Stream and Estuarine Analysis.* Manhattan College, New York. 299 pp.

O'Connor, D. J. & DiToro, D. M. (1970). *J. Sanit. Engng Div., Am. Soc. Civil Engrs,* **96**(SA2), 547.

O'Connor, D. J. & Dobbins, W. E. (1958). *Trans. Am. Soc. Civil Engrs,* **123**, 641.

Ogunrombi, J. A. & Dobbins, W. E. (1970). *J. Wat. Pollut. Control Fed.,* **42**(4), 538.

OPDQ (1979). *Banque de données statistiques par bassin géographique.* Office de planification et de développement du Québec. Éditeur officiel du Québec, Gouvernement du Québec, Québec, Canada. 252 pp.

Rich, L. G. (1973). *Environmental Systems Engineering.* McGraw-Hill, New York, p. 138.

Scott's Industrial Directories (1980). *Répertoire industriel du Québec,* 10th edition. Penstock Publications, Oatville, Canada.

Statistique Canada (1982*a*). *Recensement du Canada de 1981—Population—Répartition géographique—Québec.* Ministère des approvisionnements et services Canada. Publ. Cat. no. 93–905, vol. 2, série provinciale, Gouvernement du Canada, Ottawa, Canada.

Statistique Canada (1982*b*). *Recensement du Canada de 1981—Agriculture—Québec.* Ministère des approvisionnements et services Canada, Service d'aide aux utilisateurs, Montréal. Données municipales de la banque CANSIM, Gouvernement du Canada, Ottawa, Canada.

Streeter, H. W. (1926). *Trans. Am. Soc. Civil Engrs,* **89**, 1351.

Streeter, H. W. & Phelps, E. B. (1925). *A study of the pollution and natural purification of the Ohio river.* Bull. no. 146, US Public Health Service, Washington, DC. 96 pp.

Streeter, H. S., Wright, C. T. & Kehr, R. W. (1936). *J. Sew. Works*, **8**(2), 282.
Thackston, E. L. (1966). *Longitudinal mixing and reaeration in natural streams.* PhD Thesis, Vanderbilt University, Nashville, TN.
Truesdale, G. A. & Van Dyke, K. G. (1948). *Wat. Waste Treat. J.*, **7**(9), 133.
WMO (1986). *Intercomparison of models of snowmelt runoff.* Operational hydrology report no. 23, WMO-No. 646, Secretariat of the World Meteorological Organization, Geneva, Switzerland.

Chapter 8

DEHALOGENATION REACTIONS CATALYZED BY BACTERIA

LAWRENCE P. WACKETT

Department of Biochemistry and Gray Freshwater Biological Institute, College of Biological Sciences, University of Minnesota, Navarre, Minnesota, USA

CONTENTS

1 INTRODUCTION

Halogenated organic compounds comprise the largest group of Environmental Protection Agency priority pollutants (Leisinger, 1983). They contain fluorine, chlorine, bromine and/or iodine substituents but chloro-substitution is most common. These compounds have attracted widespread attention due to their toxicity and their general persistence in the environment. For example, chlorinated pesticides were expected to last for prolonged periods in soils and waters. Also polychlorinated biphenyls (PCBs) were used as transformer fluids because of their great stability (Hutzinger *et al.*, 1974). A flurry of patents were issued for newly synthesized chlorinated organic compounds during the period from 1950 to 1970 (Hutzinger & Veerkamp, 1981). However, the often toxic and/or carcinogenic potential of halogenated compounds has more recently become apparent. In this context, the environmental longevity of halogenated organic molecules is now perceived to be a curse rather than a blessing. New applications for chlorinated compounds have waned. Furthermore, many older compositions have been banned. Despite this, chlorinated compounds are still prevalent in the environment as a result of the persistence of old applications and contamination from continued used of less

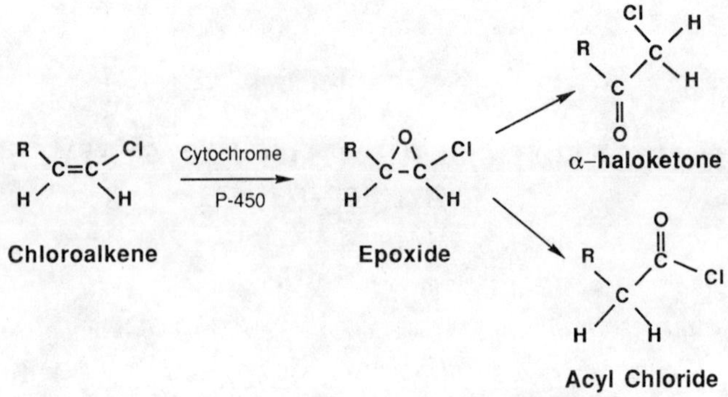

Fig. 1. Activation of chloroalkenes by cytochrome P-450 monooxygenase to form reactive intermediates.

toxic materials. For example, trichloroethylene, an industrial solvent currently used at a rate of 178 million pounds per year in the USA (Storck, 1987), is being found increasingly in drinking water supplies (Roberts *et al.*, 1982).

In general, chlorinated organic compounds exert both toxic and carcinogenic effects in mammals as a result of metabolic activation which can produce reactive intermediates (Henschler, 1985). The first metabolic step is often catalyzed by cytochrome P-450 monooxygenase. Although this enzyme sometimes serves to detoxify xenobiotic compounds, the monooxygenation of chlorinated compounds often produces very unstable, highly reactive products such as epoxides, acyl chlorides and α-haloketones (Fig. 1). The reaction of these metabolites with critical cellular molecules, most significantly DNA, can mediate cellular toxicity and potentially carcinogenicity.

The metabolism of chlorinated molecules by bacteria can contribute to environmental detoxification but many halogenated compounds are biodegraded very poorly. This metabolic failure is clearly due to the presence of halogen substituents, as shown in Fig. 2. Hydrocarbons, which are natural products formed by diagenesis (Blumer, 1975), are now known to be biodegraded by numerous environmental microorganisms (Gibson, 1982). Many halogenated natural products have been described (Neidleman & Geigert, 1986) and most of these molecules are likely to be susceptible to biodegradation. However, the most highly halogenated molecules appear to be metabolized exceedingly slowly. In the list of perchlorinated compounds shown in Fig. 2, carbon tetrachloride is biotransformed most rapidly (Egli *et al.*, 1988; Fathepure *et al.*, 1988). It is of interest that halogenated methanes are natural products formed by fungi and marine algae (Gschwend *et al.*, 1985; Wuosmaa & Hager, 1990). Tetrachloroethylene and hexachlorobenzene, which are used commercially as a solvent and a fungicide, respectively, are in the environment as the result of industrial contamination or agricultural application. These compounds undergo very slow biotransformation and are thought to be biodegraded only in anaerobic environments (Tiedje *et al.*, 1987; Vogel *et al.*, 1987).

Perhaps chloroalkanes and chloroalkanoic acids are more readily biodegraded

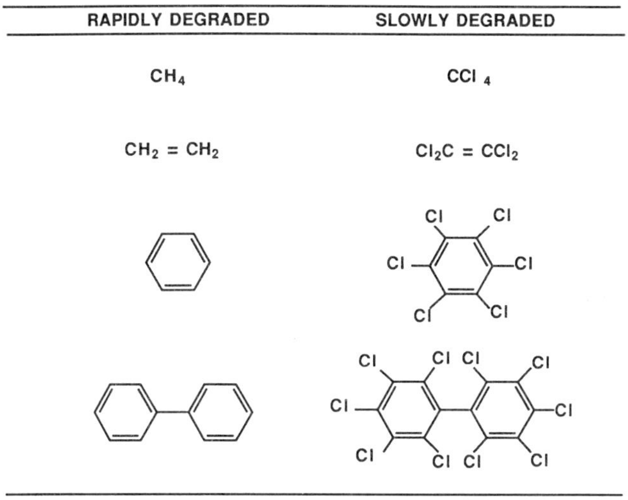

RAPIDLY DEGRADED	SLOWLY DEGRADED
CH_4	CCl_4
$CH_2 = CH_2$	$Cl_2C = CCl_2$

Fig. 2. Relative biodegradability of hydrocarbons and their perchlorinated analogs.

than chloroalkenes and chlorobenzenes due to the relatively facile nucleophilic displacement reactions the former undergo (Reineke, 1984). Haloalkanes and haloalkanoic acids are sufficiently reactive that biological defluorination reactions are known. Hydrolysis rates of haloalkanes decrease in the order iodo > bromo > chloro > fluoro. The carbon-to-fluorine bond is remarkably strong with a typical bond dissociation energy ranging from 106 to 115 kcal/mol (Reineke, 1984). Despite this, many bacteria are known to metabolize fluoroacetate via enzymatic hydrolysis to yield glycolate (Tonomura *et al.*, 1965; Kawasaki *et al.*, 1981). The stereochemical outcome of fluoroacetate processing by an enzyme from *Pseudomonas* was investigated by Au and Walsh (1984). They observed the reaction to occur with inversion of configuration, consistent with a direct S_N2 displacement of fluoride ion by hydroxide anion. The hydrolytic cleavage of the carbon-to-chlorine bond in chloroacetate and other chloroalkanoic acids has also been investigated (Goldman *et al.*, 1968).

2 CLASSES OF BIOLOGICAL DEHALOGENATION REACTIONS

In general, bacteria catalyze carbon–halogen bond cleavage via four general mechanisms (Fig. 3). Hydrolytic mechanisms occur with the net replacement of a halogen substituent with a hydroxyl group derived from water. Many of the known dehalogenating enzymes purified from bacteria utilize this type of mechanism.

A second class of enzymes that carry out dehalogenation chemistry are oxygenases and the mechanism is sometimes denoted as oxygenolytic or oxidative. Many of the known dehalogenation reactions catalyzed by oxygenases are non-specific or gratuitous reactions. That is, biosynthesis of the bacterial oxygenases is often not induced by the halogenated compounds that are ultimately oxidized. In

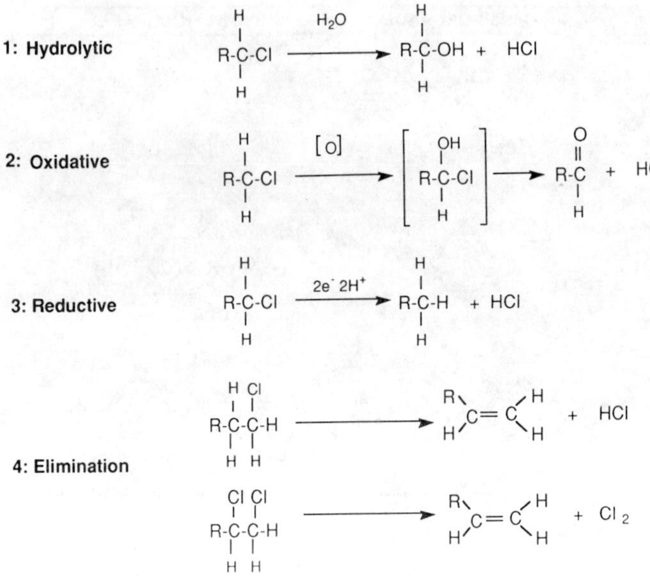

1: Hydrolytic

$$\mathrm{R\text{-}CH_2\text{-}Cl} \xrightarrow{H_2O} \mathrm{R\text{-}CH_2\text{-}OH} + \mathrm{HCl}$$

2: Oxidative

$$\mathrm{R\text{-}CH_2\text{-}Cl} \xrightarrow{[O]} \left[\mathrm{R\text{-}CH(OH)\text{-}Cl}\right] \longrightarrow \mathrm{R\text{-}CH{=}O} + \mathrm{HCl}$$

3: Reductive

$$\mathrm{R\text{-}CH_2\text{-}Cl} \xrightarrow{2e^-\ 2H^+} \mathrm{R\text{-}CH_2\text{-}H} + \mathrm{HCl}$$

4: Elimination

$$\mathrm{R\text{-}CHCl\text{-}CH_2\text{-}H} \longrightarrow \mathrm{R(H)C{=}CH_2} + \mathrm{HCl}$$

$$\mathrm{R\text{-}CHCl\text{-}CHCl\text{-}H} \longrightarrow \mathrm{R(H)C{=}CH_2} + \mathrm{Cl_2}$$

Fig. 3. Four general mechanistic classes of biological dehalogenation reactions.

many examples the bacteria producing the oxygenase cannot utilize the halogenated compounds as a source of carbon and energy. This type of metabolism, which has sometimes been called cooxidation or cometabolism, has important ramifications for the biodegradation of haloorganic compounds (Wackett *et al.*, 1989*b*; Fox *et al.*, 1990) and will be further discussed in a subsequent section.

The reductive dehalogenation of a carbon center bearing a chlorine atom occurs with the net input of two electrons and two protons yielding a carbon-to-hydrogen bond and hydrochloric acid. This type of dechlorination mechanism has only more recently been observed to be catalyzed by biological systems (Chacko *et al.*, 1966; Suflita *et al.*, 1982). The importance of reductive dechlorination to detoxify environmental toxins is underscored by observations that many highly chlorinated compounds are only biodegraded reductively.

A fourth mechanism of dehalogenation is via elimination reactions. Two types of elimination reactions are shown in Fig. 3. The first class occur with the elimination of a proton and a halide atom from adjacent carbon atoms to yield an olefin and a hydrogen halide. The second class are characterized by the loss of vicinal halide atoms to yield an olefin. The former type of elimination reaction has been observed in the metabolism of chlorinated pesticides such as lindane and DDT (Lal & Saxena, 1982).

3 DICHLOROMETHANE DEHALOGENASE—A HYDROLYTIC DEHALOGENASE

Hydrolytic dehalogenases figure most prominently in dehalogenation reactions catalyzed by bacteria. Thus it is crucial to characterize the chemistry underlying

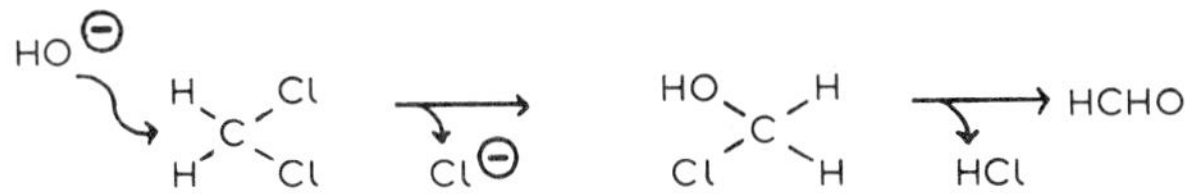

$$HO^{\ominus} + H_2CCl_2 \xrightarrow{-Cl^{\ominus}} HO(Cl)CH_2 \xrightarrow{-HCl} HCHO$$

Fig. 4. Hydrolytic reaction pathway for dichloromethane dehalogenase.

hydrolytic dehalogenation to develop a broad understanding of the biodegradation of halogenated organic compounds. Most hydrolytic dehalogenases produce alcohols from haloorganic compounds. In those instances where the carbon atom undergoing hydrolysis is bound to a second halogen atom, a ketone can be formed (Berry *et al.*, 1979; Stucki *et al.*, 1981; Anders & Pohl, 1985). The first halogen atom is displaced enzymatically by hydroxide anion and the second one eliminates spontaneously as hydrogen halide (Fig. 4). Certain methylotrophic bacteria are now known to utilize the reaction sequence of Fig. 4 to enable them to grow on dihalomethanes as the sole source of carbon and energy (Fig. 5). Methylotrophs utilize C_1 compounds such as methanol, methylamine and dimethylsulfide as growth substrates (Anthony, 1982). Each substrate is oxidized to formaldehyde as an intermediate in metabolism. Formaldehyde can be assimilated into more complex biomolecules via the serine or ribulose monophosphate pathways. Oxidation of formaldehyde to formate and carbon dioxide provides electrons for energy transducing reactions. A subset of methylotrophic bacteria contain the hydrolytic enzyme dichloromethane dehalogenase. The product of the dehalogenase reaction is formaldehyde (Stucki *et al.*, 1981). In this way dichloromethane or dibromomethane can serve as a source of carbon and energy for certain methylotrophs (Scholtz *et al.*, 1988).

The growth rates of methylotrophic strains have been examined with different carbon sources. Most dichloromethane-utilizing strains grow more slowly with dihalomethanes than with methanol. The type A dehalogenase purified from these strains was shown to have a turnover number (K_{cat}) of 30 min^{-1} (Kohler-Staub & Leisinger, 1985). The dichloromethane-utilizing strain DM4 compensates for this slow catalytic rate by biosynthesizing dichloromethane dehalogenase in extraordinary amounts, approximately 15–20% of the total soluble protein (Scholtz *et al.*, 1988). Despite this natural enzyme overproduction, the conversion of dichloromethane to formaldehyde is still thought to be rate determining in growth. Thus, it

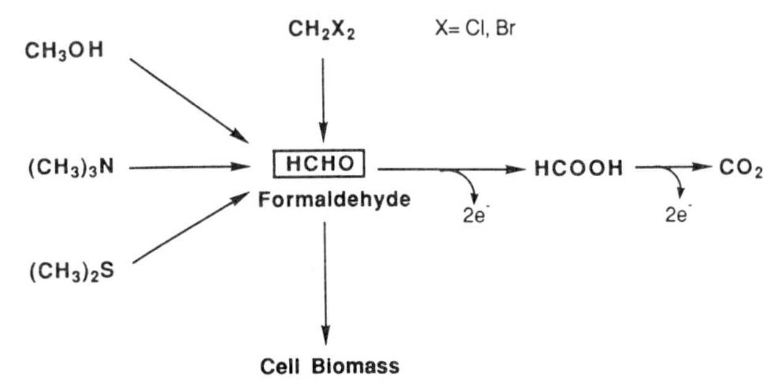

Fig. 5. Metabolism of C_1 compounds, including dihalomethanes, by methylotrophic bacteria.

Table 1
A Comparison of Dichloromethane Dehalogenases Purified from Two Different Dichloromethane-degrading Bacteria

Property	*Dehalogenase from two strains*	
	DM4 (*Type A*)	*DM11* (*Type B*)
Similarities		
Substrate	CH_2X_2 Glutathione	CH_2X_2 Glutathione
Subunit molecular weight	35 000	34 000
Differences		
N-terminal sequence	Ser–Pro–Asn–Pro	Thr–Lys–Leu–Arg
ELISA titer to DM4 antibody	$>10^5$	$4{\cdot}5 \times 10^2$
K_{cat} (min^{-1})	33	197

was of interest that another dichloromethane-degrading strain, called DM11, was isolated which grew more rapidly on dichloromethane (Scholtz *et al.*, 1988). The dehalogenase was purified from strain DM11, denoted as a type B enzyme and shown to have a higher K_{cat} (197 min^{-1}) than similar enzymes purified from other strains. The comparative properties of type A and type B dichloromethane dehalogenases are summarized in Table 1. Both enzymes are composed of a single type of subunit with similar molecular weights of approximately 34 000. Both dehalogenase types require glutathione, or γ-glutamylcysteinylglycine, for catalytic activity. Despite these similarities, the two dehalogenases are structurally and catalytically distinct (Table 1). Further insights into the relationship between these two types of dichloromethane dehalogenases will be obtained from further studies on the cloning and sequencing of the dehalogenase genes (LaRoche & Leisinger, 1990).

It is instructive to examine the reaction rates of the dichloromethane dehalogenases, which at 30–197 min^{-1} are slow compared to some other enzyme-catalyzed reactions which occur as fast as $2{\cdot}4 \times 10^9$ min^{-1} (Ferscht, 1985). However, the enzyme-catalyzed rate must be calibrated against an uncatalyzed reaction rate to reflect the true rate-enhancing ability of the enzyme. It is most appropriate to compare the second-order rate constant for the uncatalyzed reaction of dichloromethane and potassium hydroxide to the enzyme-catalyzed second-order rate constant (V_{max}/K_m) for the hydrolytic reaction carried out by bacterial dichloromethane dehalogenase. Dichloromethane hydrolyzes in potassium hydroxide at 79°C with a rate constant of $2{\cdot}1 \times 10^{-5}$ liter/mol/s (Salomaa, 1966), while the enzyme-catalyzed V_{max}/K_m rate is $4{\cdot}0 \times 10^5$ liter/mol/s. Thus the bacterial dehalogenase effects an overall 10^{10}-fold rate enhancement over the uncatalyzed reaction. By this yardstick dichloromethane dehalogenase is a highly efficient biocatalyst.

Efforts to elucidate the reaction mechanism of dichloromethane dehalogenase

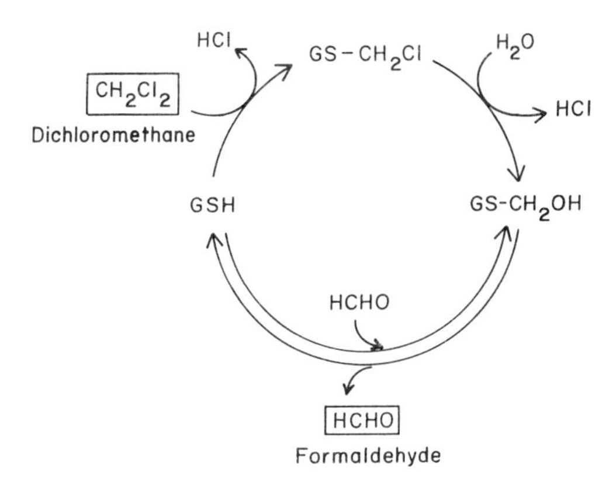

Fig. 6. Possible role of glutathione in the reaction catalyzed by dichloromethane dehalogenase.

initially focused on the role of glutathione. Glutathione is obligatory in the reaction; the enzyme shows saturation effects with it, but glutathione is not consumed as dichloromethane is processed to formaldehyde. A possible role for glutathione is that of a diffusible small molecule containing a thiolate capable of acting as a nucleophile to displace the first chloride atom (Fig. 6). The resultant chloromethylthioether is susceptible to hydrolysis non-enzymatically at rate constants greater than that of the overall enzyme-catalyzed reaction (Bohme *et al.*, 1949). The hydroxymethylthioether thus produced is an equilibrium form with free glutathione and formaldehyde. A plausible reaction scheme, as yet unproven, is for enzyme enhancement of glutathione attack on dichloromethane followed by spontaneous reactions. This possibility is rendered more intriguing with the recent sequencing of a dichloromethane dehalogenase gene and demonstrating its homology with some mammalian and plant glutathione S-transferases (LaRoche & Leisinger, 1990). Glutathione S-transferases have been shown to enhance the nucleophilicity of the glutathione thiolate for reaction with various electrophilic substrates (Keen *et al.*, 1976).

Experiments have been conducted with bacterial dichloromethane dehalogenases to demarcate whether the reaction occurs via nucleophilic displacement or, alternatively, by proceeding through a carbene intermediate. As shown in Fig. 7, enzyme processing of dideuterio-dichloromethane (CD_2Cl_2) to formaldehyde occurs with retention of both deuterium atoms if a nucleophilic mechanism is operative. A carbene intermediate would arise from *gem*-elimination of DCl from CD_2Cl_2, necessitating the loss of at least one deuterium in the reaction product formaldehyde. The two mechanistic pathways can thus be clearly delineated by trapping formaldehyde generated enzymatically from CD_2Cl_2 and determining whether the deuterium atoms are retained or lost (Fig. 7(C)). Formaldehyde was trapped by the Nash reaction (Nash, 1953) and the adduct, 3,5-diacetyl-1,4-dihydrolutidine, was analyzed by mass spectrometry. The lutidine derivative prepared with dihydridoformaldehyde has a molecular weight of 193 and shows a significant parent ion by mass spectrometry (Fig. 8). In contrast, the adduct obtained

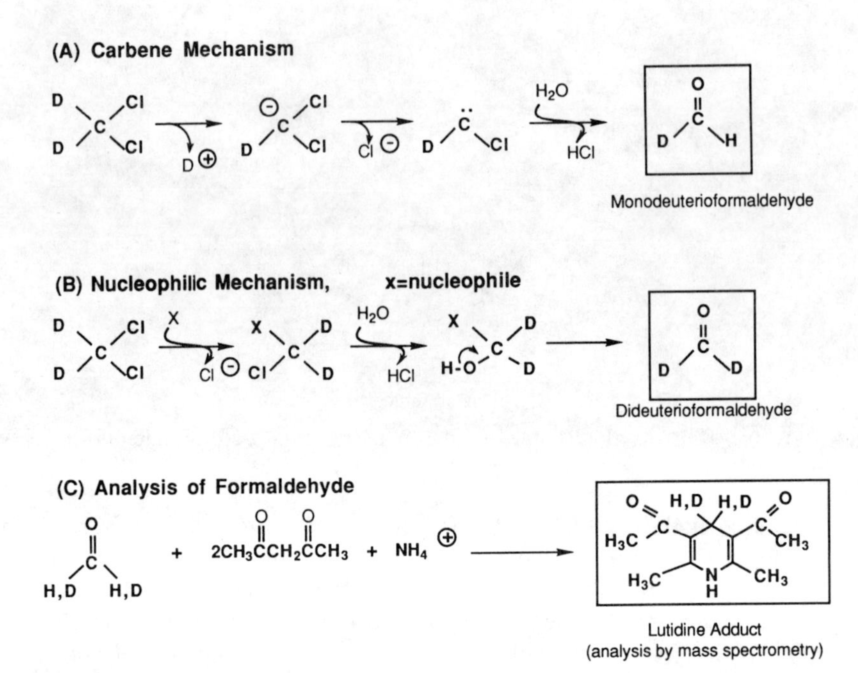

Fig. 7. Distinction between (A) carbene and (B) nucleophilic reaction mechanisms for dichloromethane dehalogenase using CD_2Cl_2, and (C) analysis of the reaction product.

from formaldehyde derived via enzymatic dechlorination of CD_2Cl_2 showed a parent ion of 195. These data indicated both deuterium atoms were retained. Similar results were obtained with both type A (Gälli *et al.*, 1982) and type B dehalogenases (Bao-li and Wackett, unpublished data). The data are consistent with a nucleophilic displacement of the first halogen atom from dihaloalkane substrates. Further studies are under way to determine if the glutathione thiolate is the initial nucleophile serving in halide displacement. Alternatively, hydroxide anion or a nucleophilic amino acid side chain group in the enzyme active site could act as the attacking nucleophile.

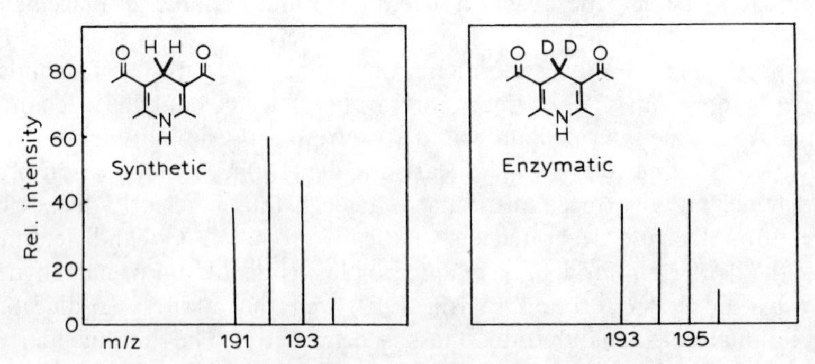

Fig. 8. Mass spectral analysis of lutidine adduct from synthetic formaldehyde and enzyme-generated formaldehyde derived from CD_2Cl_2.

4 TRICHLOROETHYLENE OXIDATION BY OXYGENASES

Oxidative mechanisms of dehalogenation are often catalyzed by oxygenases and typically occur in bacteria which derive no metabolic dividend from the oxidation reaction. Oxygenases are enzymes that incorporate oxygen from the atmosphere into organic compounds. This important enzyme class was independently codiscovered in 1955 by Hayaishi *et al.* and by Mason *et al.* Bacteria utilize oxygenases in the metabolism of many natural products and man-made compounds to carbon dioxide and this contributes to the cycling of carbon in the biosphere. For example, approximately 2×10^{12} pounds of biologically produced natural gas (methane) are recycled in the world by the action of methane monooxygenase found in methanotrophic bacteria. Recently, oxygenases have been implicated in the bacterial degradation of trichloroethylene (TCE), a widespread industrial solvent that has become an important pollutant of drinking water supplies (Nelson *et al.*, 1988). The bacteria that oxidize TCE are not able to grow on TCE. This contrasts with the utilization of dichloromethane by methylotrophic bacteria which was discussed previously. Non-specific oxidation of organic molecules by bacteria has been termed cooxidation (Perry, 1979). The phenomenon is largely due to the non-specific attack of bacterial oxygenases on diverse substrates to yield products which cannot be metabolized further. Cooxidative attack on widespread pollutants such as TCE may have important implications for bioremediation.

Two types of oxygenases, monooxygenases and dioxygenases, are known and both kinds are involved in bacterial TCE oxidation. Monooxygenases incorporate one atom from molecular oxygen into an organic substrate with the second oxygen atom being reduced to form water. Dioxygenases incorporate both atoms from the same molecule of atmospheric oxygen into an organic molecule. The initial products expected to arise from TCE oxidation by a monooxygenase and a dioxygenase are shown in Fig. 9. Both initial oxidation products shown are very unstable and they would decompose spontaneously with the liberation of chloride ion. Moreover, reactive intermediates can arise from oxygenase-catalyzed reactions, as discussed previously. Thus, it is imperative to fully account for all reaction products arising from oxidative dechlorination before a TCE biotreatment process could be

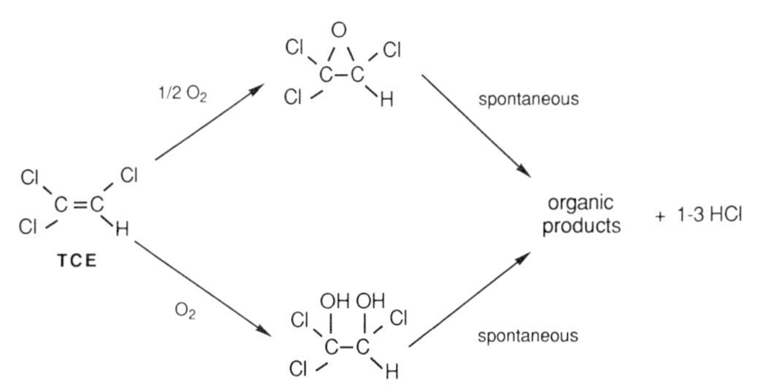

Fig. 9. Proposed pathways of trichloroethylene oxidation by dioxygenases and monooxygenases.

Table 2
Relative Rates of Oxidation of Trichloroethylene by Different Bacteria Containing Different Oxygenases

Organism	*Enzyme*	*Rate (nmol/min/mg)*	*Reference*
Methylosinus trichosporium OB3b	Soluble methane monooxygenase	20–150	Oldenhuis *et al.*, 1989 Tsien *et al.*, 1989
Pseudomonas cepacia G4	Toluene 2-monooxygenase	8	Folsom *et al.*, 1990
P. mendocina	Toluene 4-monooxygenase	2	Winter *et al.*, 1989
P. putida	Toluene dioxygenase	2	Wackett and Gibson, 1988
Nitrosomonas europaea	Ammonia monooxygenase	1	Arciero *et al.*, 1989
Mycobacterium sp.	Propane monooxygenase	0·5	Wackett *et al.*, 1989*b*
Alcaligenes eutrophus JMP 134	Phenol hydroxylase	0·2	Harker and Kim, 1990

considered safe. A significant number of bacterial oxygenases are able to oxidize TCE. As indicated in Table 2, monooxygenases are most often involved. Toluene dioxygenase has also been demonstrated to catalyze TCE oxidation both *in vivo* and *in vitro*. The greatest rates of TCE oxidation are reported for methanotrophic, or methane-oxidizing, bacteria (Fox *et al.*, 1990).

Methanotrophs biosynthesize two distinct monooxygenases with the ability to oxidize methane. One type of methane monooxygenase is membrane bound and the other is freely soluble. Studies by Oldenhuis *et al.* (1989) and Tsien *et al.* (1989) established that the soluble methane monooxygenase (sMMO) is the enzyme responsible for high rates of TCE oxidation. This conclusion was confirmed by studies with purified sMMO protein components (Fox *et al.*, 1990). Methanotrophs preferentially express sMMO rather than particulate methane monooxygenase (pMMO) only when cultures contain limited quantities of copper (Scott *et al.*, 1981; Stanley *et al.*, 1983). The molecular basis of this metal-dependent enzyme regulation is currently unknown. However, it is a key issue that is crucial to the effective use of methanotrophs in TCE bioremediation.

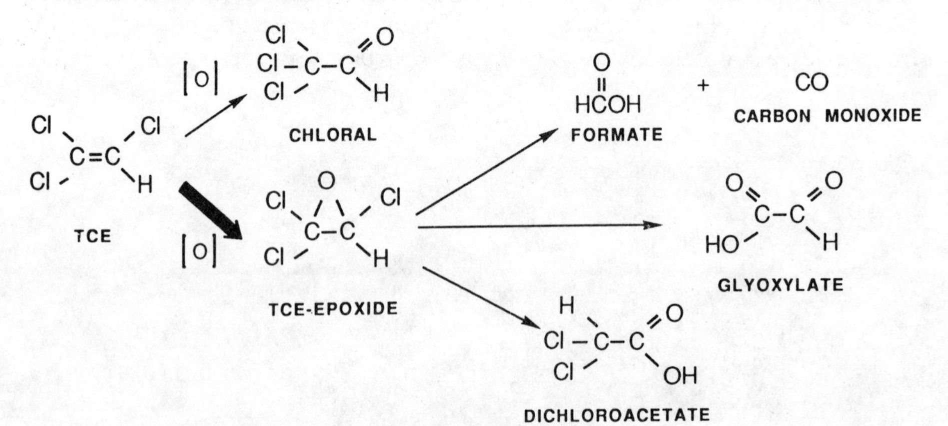

Fig. 10. Experimentally derived products of trichloroethylene oxidation by soluble methane monooxygenase.

The oxidation of TCE and other chlorinated alkenes by sMMO was investigated in detail as a model for understanding monooxygenase-dependent chloroolefin dechlorination (Fox *et al.*, 1990). The use of purified enzyme components, obtained from *Methylosinus trichosporium* OB3b, allowed an unambiguous determination of the reaction products. As predicted, sMMO oxidized haloalkenes largely to epoxide intermediates. This was demonstrated by the formation of diagnostic adducts upon reaction with 4-(*p*-nitrobenzyl)pyridine. Furthermore, expected stable decomposition products of the epoxides were identified in enzyme reaction mixtures. The total intermediates and stable products detected during the oxidation of TCE by sMMO are shown in Fig. 10. The predominant fate of the epoxide appears to be hydration followed by carbon–carbon bond scission. These spontaneous reactions led to the formation of the major products carbon monoxide and formic acid. It should be noted that methanotrophs oxidize carbon monoxide and formic acid to

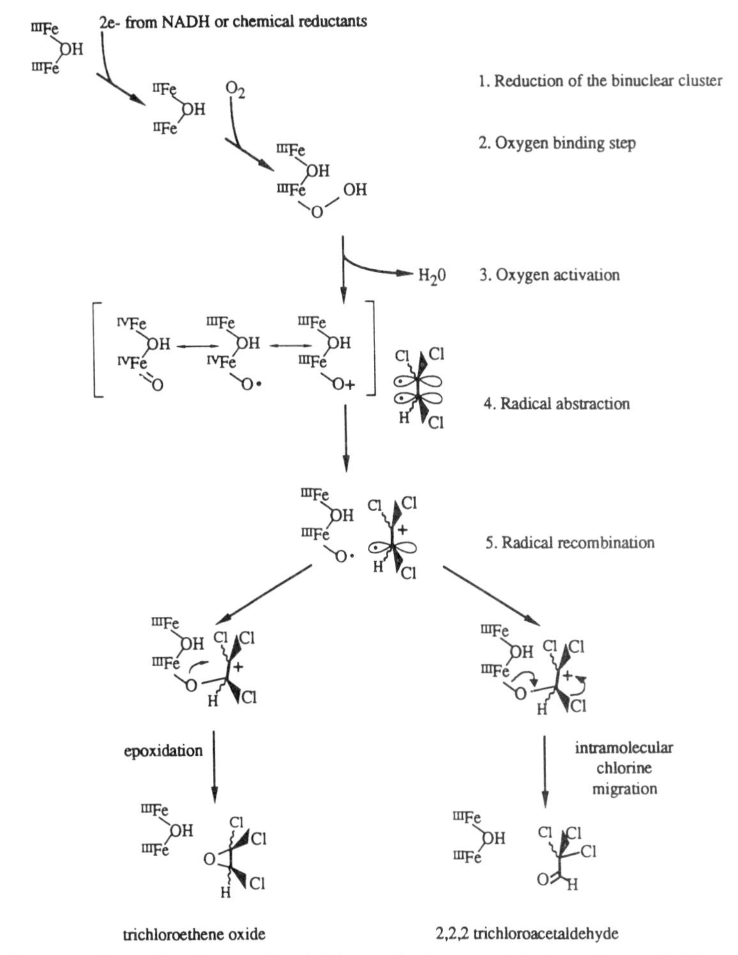

Fig. 11. Proposed mechanism of trichloroethylene oxidation by soluble methane monooxygenase.

carbon dioxide (Colby *et al.*, 1977). This explains the reported stoichiometries of $^{14}C-CO_2$ formation from ^{14}C-TCE in previous in-vivo studies (Little *et al.*, 1988). In addition to TCE-epoxide, chloral was observed as a minor (6%) product. Chloral is derived from TCE via an intramolecular chloride migration reaction. Product analysis of reaction mixtures containing synthetic TCE-epoxide showed no detectable chloral formation, indicating that chloral was likely generated at the sMMO active site. A proposed mechanism for the oxidation of TCE by methane monooxygenase is shown in Fig. 11. The key features of the mechanism are (1) the generation of a reactive monoatomic oxygen species resonance stabilized by the binuclear iron center of sMMO, (2) electron abstraction and rebound to form a cationic intermediate, and (3) partitioning of the intermediate to yield either TCE-epoxide or chloral.

A turnover- and time-dependent inactivation of the sMMO protein components occurred during the oxidation of TCE *in vitro* (Fox *et al.*, 1990). The inactivation was shown to be due to a diffusible reaction product by demonstrating covalent modification of all sMMO components with [1,2-^{14}C]-TCE. In-vivo experiments with a toluene dioxygenase-containing strain of *Pseudomonas putida* indicated that toxic products can profoundly disrupt metabolism and diminish observed rates of TCE biodegradation (Wackett & Householder, 1989). Further in-vivo experiments need to be conducted with methanotrophs containing sMMO to evaluate the effects of toxic intermediates on TCE biodegradation.

5 REDUCTIVE DEHALOGENATION CATALYZED BY TRANSITION METAL COENZYMES

The reductive dehalogenation of hazardous compounds represents a crucial environmental detoxification mechanism. Recent studies have indicated that heavily chlorinated compounds are generally resistant to aerobic biodegradation. As an example, the oxygenases described to oxidize TCE (Table 2) all fail to metabolize tetrachloroethylene. Similarly, highly chlorinated polychlorinated biphenyls (PCBs) are not degraded by aerobic bacteria capable of metabolizing PCB congeners bearing fewer chlorine atoms (Bedard *et al.*, 1986). PCB congeners with six or more chlorines, which resist aerobic attack, are also the most toxic of the PCBs. Thus, it is highly significant that tetrachloroethylene and polychlorinated PCBs undergo biologically-mediated reductive dechlorination in anaerobic environments (Vogel & McCarty, 1985; Brown *et al.*, 1987). These observations have led to the suggestion that anaerobic and aerobic biodetoxification mechanisms could be combined to treat polyhalogenated wastes via a sequential anaerobic/aerobic treatment process (Tiedje *et al.*, 1987).

The significance of microbial-catalyzed reductive dehalogenation processes to the environment and to human health has fostered greater efforts to understand the mechanisms underlying these biotransformations. In general, biochemical studies have been rendered difficult by the inability to obtain pure cultures of bacteria capable of reductively dehalogenating pollutants. There is one notable exception, however. DCB-1, a unique anaerobic bacterium, is now in pure culture and it reductively dechlorinates tetrachloroethylene and halobenzoates (Fathepure *et al.*, 1987). Furthermore, DCB-1 biosynthesizes ATP (adenosine triphosphate) in the coupling of formate and acetate oxidation to the reduction of 3-chlorobenzoate to benzoate (Dolfing, 1990). A number of other halogenated benzenes are not processed by DCB-1. There is evidence for inducibility of the dechlorinating activity. These data suggest the involvement of specific enzymes in the reductive dechlorination of 3-chlorobenzoate by DCB-1.

Other studies with pure and mixed cultures of bacteria indicate that some reductive dechlorination reactions occur by non-specific mechanisms. For example, the reduction of carbon tetrachloride is mediated by live methanogens or by heat-killed cells supplied with titanium(III) citrate as an exogenous reductant (Krone *et al.*, 1989*a*). Purified cobalamins and coenzyme F_{430} also catalyze the reduction of carbon tetrachloride to methane (Krone *et al.*, 1989*a*,*b*). This reaction and other reductive dechlorination processes may be non-specifically mediated *in vivo* by transition metal coenzymes biosynthesized by anaerobic bacteria. The non-specific dehalogenation of polyhalogenated compounds by metallocofactors could be of supreme importance for anaerobic bioremediation practices. Currently anaerobic biodegradation is limited by the slow observed rates. A further understanding of reductive dehalogenation mechanisms will be imperative for developing rational approaches to enhancing the rates of these processes.

There are two general classes of transition metal cofactors produced by bacteria growing under anaerobic conditions (Fig. 12). In one type the metal is coordinated by a stable macrocyclic ligand system. The metallomacrocycle can be bound by specific enzyme systems. Examples of this type of cofactor are the heme-iron complex, cobalt-containing vitamin B_{12} (cobalamin), and the nickel coenzyme F_{430} found in methanogenic bacteria. These cofactors have been shown to mediate electron transfer as well as catalytic reactions (Wackett *et al.*, 1989*a*). All three metallomacrocycles shown in Fig. 12 catalyze the reductive dechlorination of chloromethanes *in vitro*. In the second type of cofactor the metal atom(s) is directly coordinated to protein ligands. Iron–sulfur clusters are often coordinated by protein cysteines but Reiske-type clusters are coordinated by two histidine residues and two cysteines (Gurbiel *et al.*, 1989).

Studies have been conducted to determine the potential of transition metal coenzymes for catalyzing the reductive dechlorination of polychlorinated alkanes, alkenes and benzenes. To test their reactivity the cofactors were reduced using titanium(III) citrate. The midpoint potential of titanium(III) citrate is sufficiently low at −480 mV to reduce significant amounts of all of the cofactors tested (Krone *et al.*, 1987*a*). Anaerobic reaction mixtures containing reduced cofactors and a chlorocarbon were analyzed by gas chromatography (GC) and/or high pressure liquid chromatography (HPLC) for dechlorinated products. The rates of dechlorination were determined by monitoring the disappearance of the starting material and/or the formation of products. Control experiments were conducted without coenzymes to preclude the occurrence of titanium-mediated reductive dechlorination. In some reactions different isomeric products could be formed. For

METAL COORDINATED TO NON-PROTEIN LIGANDS

Coenzyme F_{430}

Vitamin B_{12}

Hematin

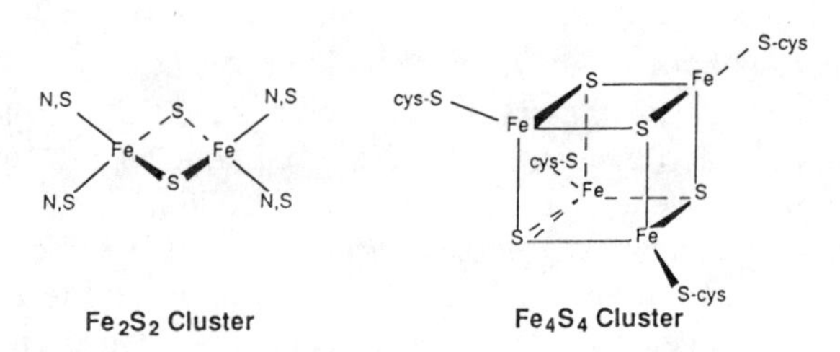

Fig. 12. Transition metal coenzymes examined as potential catalysts in reductive dechlorination.

example, reductive removal of one chlorine atom from trichloroethylene could yield *cis*-1,2-dichloroethylene, *trans*-1,2-dichloroethylene, 1,1-dichloroethylene or some mixture of these isomers. GC was conducted with authentic standards to insure that all three dichloroethylene isomers could be resolved and quantitatively determined (Fig. 13). The in-vitro isomeric product distribution is of particular interest in the light of published results on the biological formation of products. Comparisons of this type could aid in discerning which cofactors participate in reductive dechlorination reactions *in vivo*.

The macrocyclic coenzymes shown in Fig. 12 were active in the reductive dechlorination of chloroalkanes, chloroalkenes and chlorobenzenes whereas iron–sulfur proteins and azurin were inactive. The rates of dechlorination differed with the three coenzymes surveyed (Table 3). However, several trends emerged. First, vitamin B_{12} and coenzyme F_{430} catalyzed dechlorination of perchlorinated compounds with relative rates in the order carbon tetrachloride > tetrachloroethylene > hexachlorobenzene. For hematin the rate order was carbon tetra-

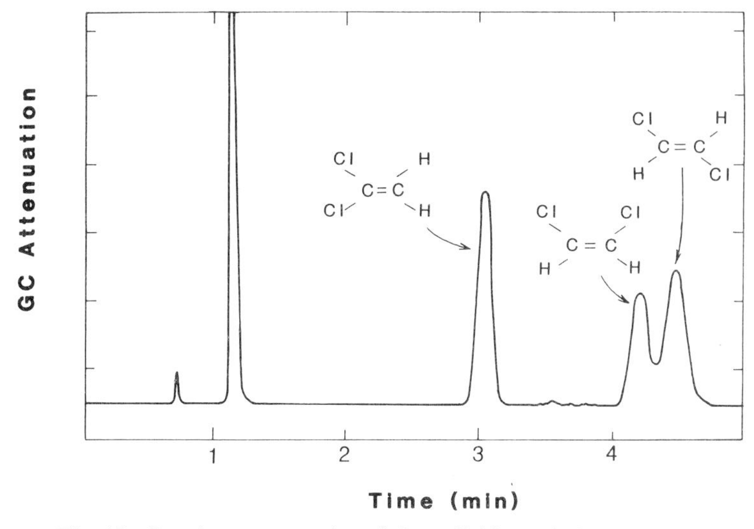

Fig. 13. Gas chromatography of three dichloroethylene isomers.

chloride > hexachlorobenzene > tetrachloroethylene. In biological systems carbon tetrachloride is dechlorinated more rapidly than chlorinated ethylenes or benzenes. Thus, the in-vitro data obtained here are generally consistent with reports on biological reductive dechlorination. Another trend in the in-vitro data was the decrease in dechlorination rates with decreasing chlorine substitution of the chloroethylenes or chlorobenzenes. This has been observed previously with the dechlorination of chloromethanes by Krone *et al.* (1989*a*,*b*). Biological systems also show rate diminution with decreasing chlorination of the chlorocarbon (Parsons *et al.*, 1984; Freedman & Gossett, 1989).

Tetrachloroethylene (perchloroethylene, PCE) and trichloroethylene (TCE) are significant environmental pollutants so their dechlorination was investigated in greater detail. PCE is an important dry cleaning solvent and TCE is widely used in degreasing applications (Storck, 1987). Anaerobic environments polluted with these compounds sometimes become contaminated with vinyl chloride. Laboratory studies with anaerobic consortia have shown that PCE and TCE are reductively

Table 3
Relative Rates for the Dechlorination of Perchlorinated Compounds with Titanium(III) Citrate Using Three Different Coenzyme Catalysts

Compound	*First-order rate constants (h^{-1}) with three coenzymes*		
	Vitamin B_{12}	*Coenzyme F_{430}*	*Hematin*
Carbon tetrachloride	74 ± 4	100 ± 4	2·4 ± 0·1
Tetrachloroethylene	7·3 ± 1·5	3·7 ± 0·5	0·10 ± 0·03
Hexachlorobenzene	1·53 ± 0·13	<0·02	0·48 ± 0·20

Table 4
Isomeric Distribution of Dichloroethylenes Formed in the Reductive Dechlorination of Trichloroethylene by Coenzymes

Coenzyme	*% Dichloroethylene isomer*		
	cis-	trans-	*1,1*
Vitamin B_{12}	90	5	5
Coenzyme F_{430}	97	3	<1
Hematin	100	<1	<1

dechlorinated to form vinyl chloride (Parsons *et al.*, 1984; Freedman & Gossett, 1989). This is of great concern to human health. Vinyl chloride is a potent carcinogen as demonstrated by epidemiological and animal studies (Maltoni & Lefemine, 1974). In one recent study a bioreactor containing mixed anaerobic cultures dechlorinated PCE completely to ethylene (Freedman & Gossett, 1989). It was of interest to examine the extent of PCE dechlorination by transition metal coenzymes *in vitro*. Vitamin B_{12} and coenzyme F_{430} catalyzed dechlorination of all chloroethylenes at demonstrable rates. Thus vinyl chloride was dechlorinated to ethylene, albeit very slowly. The rate of dechlorination of vinyl chloride by coenzyme F_{430} was 33 times slower than the rate of its formation from *cis*-1,2-dichloroethylene. This would explain why vinyl chloride accumulates to a significant extent in anaerobic ecosystems.

The stereochemical course of TCE dechlorination was investigated to allow a comparison of the products obtained with each of the three cofactors to the products obtained *in vivo*. As shown in Table 4, *cis*-1,2-dichloroethylene was the predominant product formed by all three cofactors. Most studies with anaerobic consortia have identified *cis*-1,2-dichloroethylene as the major product (Vogel *et al.*, 1987). Freedman and Gossett (1989) reported *trans*-1,2-dichloroethylene as the major intermediate in their bioreactor. 1,1-Dichloroethylene appears to be formed only in trace amounts by anaerobic consortia and by our in-vitro systems using coenzymes.

6 CONCLUSIONS

Microorganisms are key agents for recycling organic matter in the biosphere. Many chlorinated organic compounds are relatively resistant to microbial attack and thus accumulate in the environment. Increasingly bacterial systems are being identified which have the ability to biodegrade chloroorganic molecules. The mechanisms bacteria use to detoxify chlorinated compounds are being studied extensively. Understanding dechlorination mechanisms will pay important dividends in efforts to develop more efficient biological processes for treating hazardous wastes.

Four fundamental mechanisms of biodehalogenation have been identified. These are (1) hydrolytic, (2) oxidative, (3) reductive and (4) eliminative reactions. Most dehalogenating enzymes studied to date catalyze net hydrolytic carbon–halogen bond cleavage. Dichloromethane dehalogenase catalyzes the hydrolysis of dihaloalkanes to yield formaldehyde. Oxidative dechlorination is often mediated by monooxygenases or dioxygenases. These reactions are usually cooxidative. That is, the chlorinated compound undergoing oxidation is not the primary substrate of the enzyme and it does not support growth of the organism producing the oxygenase.

Reductive dechlorination may be very important environmentally. Anaerobic dechlorination is typically very slow, but highly chlorinated molecules undergo dechlorination via this mechanism almost exclusively. Some biologically relevant reductive dechlorination reactions may be mediated by transition metal coenzymes found in anaerobic bacteria. The reduced cofactors shown to be reactive *in vitro* are hematin, vitamin B_{12} and coenzyme F_{430}. An understanding of these and other mechanisms of biodehalogenation will allow the use of innovative biotechnological approaches for enhancing bioremediation.

ACKNOWLEDGEMENTS

I wish to thank the talented coworkers in my laboratory who contributed to our further understanding of dehalogenation reaction mechanisms. This work was supported by National Institutes of Health Grant GM 41235 and by Air Force Office of Scientific Research Grant 89-0457.

REFERENCES

Anders, M. W. & Pohl, L. R. (1985). In *Bioactivation of Foreign Compounds*, ed. M. W. Anders. Academic Press, New York, p. 283.

Anthony, C. (1982). *The Biochemistry of Methylotrophs*. Academic Press, London.

Arciero, D., Vannelli, T., Logan, M. & Hooper, A. B. (1989). *Biochem. Biophys. Res. Commun.*, **159**, 640.

Au, K. & Walsh, C. T. (1984). *Bioorg. Chem.*, **12**, 197.

Bedard, D. L., Untermanx, R., Bopp, L. H., Brennan, M. J., Haberl, M. L. & Johnson, C. (1986). *Appl. Environ. Microbiol.*, **51**, 761.

Berry, E. K. M., Allison, N., Skinner, A. J. & Cooper, R. A. (1979). *J. Gen. Microbiol.*, **110**, 39.

Blumer, M. (1975). *Angewandte Chemie, Internat. Edit.*, **14**, 507.

Bohme, H., Fischer, H. & Frank, R. (1949). *Justus Liebig Annln. Chem.*, **563**, 54.

Brown, J. F., Bedard, D. L., Brennan, M. J., Carnahan, J. C., Feng, H. & Wagner, R. E. (1987). *Science*, **236**, 709.

Chacko, C. I., Lockwood, J. L. & Zabik, M. J. (1966). *Science*, **154**, 893.

Colby, J., Stirling, D. I. & Dalton, H. (1977). *Biochem. J.*, **165**, 395.

Dolfing, J. (1990). *Arch. Microbiol.*, **153**, 264.

Egli, C., Tschan, T., Scholtz, R., Cook, A. M. & Leisinger, T. (1988). *Appl. Environ. Microbiol.*, **54**, 2819.

Fathepure, B. Z., Nengu, J. P. & Boyd, S. A. (1987). *Appl. Environ. Microbiol.*, **53**, 2671.

Fathepure, B., Tiedje, J. M. & Boyd, S. A. (1988). *Appl. Environ. Microbiol.*, **54**, 327.

Ferscht, A. (1985). *Enzyme Structure and Mechanism*, 2nd edn. W. H. Freeman, New York.

Folsom, B. R., Chapman, P. J. & Pritchard, P. H. (1990). *Appl. Environ. Microbiol.*, **56**, 1279.

Fox, B. G., Borneman, J. G., Wackett, L. P. & Lipscomb, J. D. (1990). *Biochemistry*, **29**, 6419.

Freedman, D. L. & Gossett, J. M. (1989). *Appl. Environ. Microbiol.*, **55**, 2144.
Gälli, R., Stucki, G. & Leisinger, T. (1982). *Experientia*, **38**, 1378.
Gibson, D. T. (1982). *Toxic. Environ. Chem.*, **5**, 237.
Goldman, P., Milne, G. W. A. & Keister, D. B. (1968). *J. Biol. Chem.*, **243**, 428.
Gschwend, P. M., MacFarlane, J. K. & Newman, K. A. (1985). *Science*, **227**, 1033.
Gurbiel, R. J., Batie, C. J., Sivaraja, M., True, A. E., Fee, J. A., Hoffman, B. M. & Ballou, D. P. (1989). *Biochemistry*, **28**, 4861.
Harker, A. R. & Kim, Y. (1990). *Appl. Environ. Microbiol.*, **56**, 1179.
Hayaishi, O., Katagiri, M. & Rothberg, S. (1955). *J. Am. Chem. Soc.*, **77**, 5450.
Henschler, D. (1985). In *Bioactivation of Foreign Compounds*, ed. M. W. Anders. Academic Press, New York, p. 317.
Hutzinger, O. & Veerkamp, W. (1981). In *Microbial Degradation of Xenobiotic and Recalcitrant Compounds*, ed. T. Leisinger, A. Cook, R. Hutter & J. Nuesch. Academic Press, London, p. 3.
Hutzinger, O., Safe, S. & Zitko, V. (1974). *The Chemistry of PCBs*. CRC Press, Cleveland, OH.
Kawasaki, H., Miyoshi, K. & Tonomura, K. (1981). *Agric. Biol. Chem.*, **45**, 543.
Keen, J. H., Habig, W. H. & Jakoby, W. B. (1976). *J. Biol. Chem.*, **251**, 6183.
Kohler-Staub, D. & Leisinger, T. (1985). *J. Bacteriol.*, **162**, 676.
Krone, U. E., Laufer, K., Thauer, R. K. & Hogenkamp, H. P. C. (1989*a*). *Biochemistry*, **28**, 10061.
Krone, U. E., Thauer, R. K. & Hogenkamp, H. P. C. (1989*b*). *Biochemistry*, **28**, 4908.
Lal, R. & Saxena, D. M. (1982). *Microbiol. Rev.*, **46**, 95.
LaRoche, S. D. & Leisinger, T. (1990). *J. Bacteriol.*, **172**, 164.
Leisinger, T. (1983). *Experientia*, **39**, 1183.
Little, C. D., Palumbo, A. V., Herbes, S. E., Lidstrom, M. E., Tyndall, R. L. & Gilmer, P. J. (1988). *Appl. Environ. Microbiol.*, **54**, 951.
Maltoni, C. & Lefemine, G. (1974). *Environ. Res.*, **7**, 387.
Mason, H. S., Fowlkes, W. L. & Peterson, E. (1955). *J. Am. Chem. Soc.*, **77**, 2914.
Nash, T. (1953). *Biochem. J.*, **55**, 416.
Neidleman, S. L. & Geigert, J. (1986). *Biohalogenation: Principles, Basic Roles and Applications*. John Wiley, New York.
Nelson, M. J., Montgomery, S. O. & Pritchard, P. H. (1988). *Appl. Environ. Microbiol.*, **54**, 604.
Oldenhuis, R., Vink, R. L., Vink, J. M., Janssen, D. B. & Witholt, B. (1989). *Appl. Environ. Microbiol.*, **55**, 2819.
Parsons, F., Wood, P. R. & DeMarco, J. (1984). *J. Am. Water Works Assoc.*, **76**, 56.
Perry, J. J. (1979). *Microbiol. Rev.*, **43**, 59.
Reineke, W. (1984). In *Microbial Degradation of Organic Compounds*, ed. D. T. Gibson. Marcel Dekker, New York, p. 319.
Roberts, P. V., Schreinger, J. E. & Hopkins, G. C. (1982). *Water Res.*, **16**, 1025.
Salomaa, P. (1966). In *The Chemistry of the Carbonyl Group*, ed. S. Patai. Wiley Interscience, New York, p. 177.
Scholtz, R., Wackett, L. P., Egli, C., Cook, A. M. & Leisinger, T. (1988). *J. Bacteriol.*, **170**, 5698.
Scott, D., Brannan, J. & Higgins, I. J. (1981). *J. Gen. Microbiol.*, **125**, 63.
Stanley, S. H., Prior, S. D., Leak, D. J. & Dalton, H. (1983). *Biotechnol. Lett.*, **5**, 487.
Storck, W. (1987). *Chem. Eng. News*, **65**, 11.
Stucki, G., Gälli, R., Ebersold, H.-R. & Leisinger, T. (1981). *Arch. Microbiol.*, **130**, 366.
Suflita, J. M., Horowitz, A., Shelton, D. R. & Tiedje, J. M. (1982). *Science*, **218**, 1115.
Tiedje, J. M., Boyd, S. A. & Fathepure, B. Z. (1987). *Dev. Ind. Microbiol.*, **27**, 117.
Tonomura, K., Futai, F., Tanabe, O. & Yamaoka, T. (1965). *Agric. Biol. Chem.*, **29**, 124.
Tsien, H.-C., Brusseau, G. A., Hanson, R. S. & Wackett, L. P. (1989). *Appl. Environ. Microbiol.*, **55**, 3155.
Vogel, T. M. & McCarty, P. L. (1985). *Appl. Environ. Microbiol.*, **49**, 1080.
Vogel, T. M., Criddle, C. S. & McCarty, P. L. (1987). *Environ. Sci. Tech.*, **21**, 722.
Wackett, L. P. & Gibson, D. T. (1988). *Appl. Environ. Microbiol.*, **54**, 1703.

Wackett, L. P. & Householder, S. R. (1989). *Appl. Environ. Microbiol.*, **55**, 2723.
Wackett, L. P., Orme-Johnson, W. H. & Walsh, C. T. (1989*a*). In *Metal Ions and Bacteria*, ed. T. J. Beveridge & R. J. Doyle. John Wiley, New York, p. 165.
Wackett, L. P., Brusseau, G. A. & Hanson, R. S. (1989*b*). *Appl. Environ. Microbiol.*, **55**, 2960.
Winter, R. B., Yen, K.-M. & Ensley, B. D. (1989). *Biotechnology*, **7**, 282.
Wuosmaa, A. M. & Hager, L. P. (1990). *Science*, **249**, 160.

Chapter 9

BIODEGRADATION OF SANITARY LANDFILL LEACHATE

D. THIRUMURTHI

Department of Civil Engineering, Technical University of Nova Scotia, Halifax, Nova Scotia, Canada

CONTENTS

1 INTRODUCTION

Ground water, rainwater and snow melt, in and around a landfill, as they percolate through the solid wastes dissolve and extract organic and inorganic matter. The resulting liquid is called leachate because it has leached out the chemicals from the solid wastes. It has a potential to be a water pollutant of very significant environmental concern.

This chapter deals with leachate generated at landfills where municipal solid wastes (sometimes called 'domestic wastes' in the UK) are disposed of in a 'sanitary way'. Liquid sludges, industrial wastes and hazardous wastes (such as radioactive, highly ignitable, toxic and corrosive wastes) are not normally permitted in municipal landfills. A vast majority of the solids in the leachate is usually in the soluble form. Most suspended solids are filtered out, or trapped by the soil around the compacted refuse.

Table 1 summarizes the concentration ranges of pollutants reported in leachates generated by 15 different landfills located in Canada (Atwater, 1980; Atwater & Mavinic, 1986; Thirumurthi *et al.*, 1986*a,b*), France (Millot *et al.*, 1987), the UK (Maris *et al.*, 1985; Robinson, 1985; Robinson & Lucas, 1985) and the USA (Chian & DeWalle, 1977*a*; SMC-Martin, Inc., 1981). The French study of three landfills showed that significant differences can occur between various sampling locations in the same landfill.

1.1 Biodegradation Modes

Biodegradation modes could be either aerobic (in the presence of free oxygen) or anaerobic (in the absence of free oxygen). The aerobic path is accomplished by activated sludge processes, aerated lagoons, stabilization ponds, rotating biological contactors (RBC) or trickling filters. The anaerobic biodegradation path is achieved by anaerobic lagoons, upflow anaerobic sludge blanket units (UASB), anaerobic fixed film reactors (AFFRs) or anaerobic submerged biological filters (AnSBFs) and 'hybrid' reactors (where concepts of AFFR and UASB are combined).

The chosen treatment scheme is site-specific, depending on the quantity of leachate, available land space, age of the landfill, the required treatment efficiency (which depends on the ultimate receiving water or soil), the cost of available energy source and available capital as well as operating budget.

Aerobic biological decomposition is recommended for low-strength (COD < about 500 mg/liter) leachates. For medium-strength (COD range of approximately 500–5000 mg/liter) leachates, on the other hand, the choice could be either aerobic or anaerobic. Often, however, potentially toxic metals may have to be removed by pretreatment (such as chemical precipitation). The techniques for stabilization of biodegradable organic carbon have been well streamlined. However, removal of organic nitrogen (N) and ammonia nitrogen ($NH_4.N$) continues to be a challenge because it requires long hydraulic retention time (HRT), and mean cell residence time (MCRT), nitrification/denitrification and/or stripping of free ammonia (NH_3) gas.

Anaerobic treatment (AFFR or UASB units or 'hybrid' reactors) is strongly recommended for high-strength (COD above approximately 5000 mg/liter) leachate, provided toxic metals are removed by chemical precipitation.

While the aerobic biological treatment system is an energy consumer (to run the

Table 1
Characteristics of Raw Leachate

Constituents	*Ranges of concentrations*[a]		*Constituents*	*Ranges of concentrations*[a]	
	Minimum	*Maximum*		*Minimum*	*Maximum*
1. General			3. Non-metals		
pH	5·2	8·2	Ammonia N	1	1 700
Alkalinity as $CaCO_3$	37	14 000	Chloride	30	2 900
Conductivity (μmhos/cm)	400	50 000	Nitrate N	0·1	10
TDS	2 000	15 300	Orthophosphate	0·5	39
SS	100	700	Total phosphate	0·6	75
TS	500	15 800			
2. Metals			4. Organics		
Al	1·5	2·7	BOD (5-day)	11	38 000
As	0·006	0·2	COD	20	70 000
Ba	0·1	0·3	TOC	196	23 000
Ca	29	4 300	TVA	186	15 000
Cd	0·0005	0·007	Kjeldahl N	4	762
Cr	0·002	1·0	Organic N	3	770
Co	0·01	1·8			
Cu	0·01	0·3			
Fe	0·3	2 050			
Pb	0·002	12·3			
Ni	0·01	6·1			
Na	235	2 400			
Zn	0·01	130			

[a] With the exceptions of pH and conductivity, all parameters are in mg/liter.

compressor/blower, to pump the leachate, to turn the RBC shaft and other needs), the anaerobic system is a potential energy producer (by virtue of its ability to produce methane). The higher the strength of leachate, the higher is the energy requirement for the aerobic treatment. On the other hand, as the COD of the leachate increases, the amount of methane that can be generated increases. This is the reason for choosing the aerobic biodegradation path for low-COD wastes and anaerobic stabilization for high-COD wastes. For wastes with intermediate COD (about 500–5000 mg/liter) the choice is site-specific.

Chian and DeWalle (1977*a*) concluded, after monitoring leachate generated from landfills of different ages, that the proportion of biodegradable organics decreased but that of non-biodegradable 'fulvic-like' organics increased with landfill age. It is recommended, therefore, that the age of a landfill be taken into account while designing a plant. Forgie (1988) reviewed the literature on leachate treatment and suggested that biological processes are most appropriate when the molecular weight of the majority of the organics is less than 500 g/mol.

1.2 Pre-treatment

Most leachates need to be pre-treated, prior to biodegradation, to 'detoxify' them. Such a step renders the liquid amenable to biodegradation. The most common potentially toxic chemicals are heavy metals, such as zinc. A preferred method of handling the metals is precipitation by adjusting pH (to about 7·5–8·5), coagulation, flocculation and settling (Thirumurthi *et al.*, 1986*a,b*).

2 AEROBIC BIODEGRADATION

The leachate should be nutritionally balanced to support microbial life. Studies in Europe as well as North America have shown that leachate is usually deficient in the phosphate nutrient but not in nitrogen. For efficient aerobic treatment it is generally accepted that a BOD_5:N:P ratio of about 100:5:1 must be maintained in the influent stream. In several municipal landfill leachates the value of 5-day BOD/TP is more than about 300 (Table 1). Therefore it is essential to supplement the phosphate for successful aerobic biological treatment.

Total excess suspended solids (SS) produced as a result of aerobic biodegradation of leachate could be in the vicinity of 1 kg SS/kg of BOD_5 removed, which is about double that generated from municipal wastewater (Henry, 1985). Aeration of leachate precipitates Fe, Ca and other inorganic solids which account for the large rate of solids generation (Stegmann, 1980).

2.1 Activated Sludge Treatment

2.1.1 Introduction

Microbes responsible for biodegradation of organics in leachate grow in suspension in the aeration tanks of activated sludge treatment systems. The biota are called suspended growth or dispersed growth cultures. Three activated sludge plants treating leachate were operating in West Germany (Stegmann, 1980) and one in the USA (Pennsylvania) during the early 1980s (Keenan *et al.*, 1984). Very limited data, however, are available on biodegradation of organics in such full-scale plants.

2.1.2 The Pennsylvania Experience

A landfill at Falls Township, Pennsylvania, has a full-scale activated sludge plant designed to provide maximum operational flexibility to facilitate testing of a variety of degradation options (Keenan *et al.*, 1983, 1984). It has two aeration and two settling tanks, and the two systems can be operated in parallel or in series. They can be operated in the conventional mode (when the aeration tanks are in parallel) or extended aeration mode (when the aeration tanks are in series). The food: microorganism ratio was 0·12–0·32/day during the periods of satisfactory operation. The capacity of each aeration tank is 75·7 m^3, resulting in a hydraulic retention time (HRT) of 6·6 h at the design maximum flow of 380 liter/min (22·8 m^3/h). The system was operated most successfully at a mixed liquor volatile suspended solids (MLVSS) concentration between about 6000 and 12 000 mg/liter. When the two aeration tanks were operated in parallel, at a flow rate of 80 m^3/day and volumetric organic load of 1·87 kg BOD_5/m^3 . day, about 97% of BOD_5 was removed. On the other hand, when the tanks were in series mode (extended aeration), the average influent flow was about 38 m^3/day. The BOD_5 removal efficiency was 92% at a volumetric organic load of 0·3 kg BOD_5/m^3 . day.

A design engineer who is contemplating to recommend an activated sludge plant for high-strength leachate, such as in Falls Township (COD = 6000–21 000 mg/liter, BOD_5 = 3000–13 000 mg/liter and $NH_4 . N$ = 200–2000 mg/liter), should carefully review the numerous operational and start-up problems encountered at that plant. The solids tended to float in the secondary clarifier and the biodegradation of substrate was inhibited by excessive ammonia concentration. The cost of aeration and additions of phosphoric acid, lime and other chemicals to aeration tanks was about US$1/$m^3$ (US$3·49/1000 US gal) of leachate. The comparable cost of anaerobic treatment could have been much lower.

2.1.3 The West German Experience

Stegmann (1980) suggested the following guidelines based on his experiences with full-scale and pilot-scale activated sludge plants: (1) for complete degradation of organics and to produce an effluent of soluble $BOD_5 < 25$ mg/liter the recommended volumetric organic load is less than 0·15 kg BOD_5/m^3. day; (2) for partial biodegradation of organics the load is less than 0·5 kg BOD_5/m^3. day.

In some West German plants the partial treatment is used to 'deodorize' the liquid before transporting it to a municipal treatment plant for further biodegradation. In the absence of such pretreatment, unpleasant odors were common in the municipal plants.

The operational problems encountered in the West German aeration units were high foam production, precipitation of iron and carbonates that may eventually clog the air diffusers, excessive sludge production rates (0·6 kg dry sludge/kg of BOD_5 removed) and drop in efficiency during cold winter months. It is necessary, therefore, to cover the aeration tanks and conserve heat. Ammonia levels should be taken into account, in addition to organic matter, for estimating the oxygen requirements. Of course, phosphate nutrient supplementation is often essential.

2.2 Trickling Filters

2.2.1 Introduction

Microorganisms grow in fixed films (attached cultures), and are attached to solid surfaces provided for that purpose. The solid media could be the modern synthetic materials or rocks/gravel which were used in the older designs. The former are characterized by large void spaces, and therefore aerobic conditions can be maintained more easily. High values of porosities of synthetic media enable engineers to design deeper filters which can handle more volumes of leachate per unit area of land space. The conventional trickling filters with rock/gravel are limited in depth to about 2 m because in a deeper filter anaerobic environment could create odor problems. On the other hand, the modern filters with synthetic media can be designed to be fully aerobic even when the depth is as much as 8 m.

2.2.2 The UK Experience

An outdoor package trickling filter plant (in parallel with a similar activated sludge unit) was monitored by Knox (1985) for two years at a co-disposal landfill at Pitsea in the UK. It was a 60-year-old landfill at the time of the study, during which approximately 350 000 tonnes of municipal refuse and 70 000 m^3 of industrial input were handled per year. The 'stabilized leachate' generated at this 'old landfill' had typically a very low $BOD_5/NH_3 . N$ value of 1/3. Leachate $NH_3 . N$ concentrations ranged from 150 to 550 mg/liter while TOC levels were from 200 to 500 mg/liter. Very little of the TOC was biodegradable. Ammonia was identified as the primary constituent of concern because it was potentially toxic to rainbow trout.

The filter medium was Flocor R (randomly-packed 3-cm corrugated plastic cylinders). The specific surface area of the medium was 230 m^2/m^3 and its total volume was 16·5 m^3. The trickling filter was 2·7 m in depth, and the HRT varied between 15 h and 4·5 days.

The daily $NH_3 . N$ removal rate varied from 8·2 g N per *unit volume* (m^3) of plastic medium (at 1·7°C) to 71 g N/m^3 at 16°C. The daily $NH_3 . N$ removal rate varied from 36 g N per *unit surface area* (m^2) of the plastic medium at 1·7°C to 309 g N/m^2 at 16°C.

Unlike in the activated sludge plant (in which clarification was poor due, partly, to the low $BOD_5:NH_3$ ratio of 1:3) fewer operational problems occurred in the biological filter. The attached-growth systems are generally simpler to operate for treatment of leachates with low $BOD_5:NH_3$ ratios.

2.3 Aerated Lagoon Treatment

2.3.1 Introduction

In areas where inexpensive land is available, aerated lagoons would constitute the least expensive method of aerobic biological degradation. The path of biological stabilization is identical to that of all other aerobic processes; however, in aerated lagoon treatment the system is engineered with a lower degree of sophistication. An aerated lagoon is an excavation in the ground with earthen embankments around to hold the waste liquid. Impermeable liners are strongly recommended to eliminate exfiltration and pollution of ground water. The full-scale, pilot-scale and laboratory-model studies in the US, Canada, the UK, Australia and West Germany indicate that aerated lagoons can effectively treat leachate, provided, however, reductions in efficiencies of COD, BOD_5, TOC and N removal can be tolerated during the colder winter months. The recommended design parameters vary widely depending on the qualities of raw leachate, the required degree of stabilization and temperature. The rate of biodegradation is much slower in an aerated lagoon in comparison with activated sludge process or trickling filters. Aeration is accomplished by means of various types of commercially available devices, some of which are known as surface aerators and others as diffused aerators.

2.3.1.1 RATE OF DEGRADATION

Proper understanding and estimation of this rate, which is also expressed as BOD or COD removal coefficient (K), would enable a rational design of an aerated lagoon. The flow pattern in a lagoon can be either (1) plug flow (unaerated pond), (2) completely mixed aerobic aerated lagoon, or (3) 'incompletely mixed' aerobic/anaerobic aerated lagoon (Thirumurthi, 1979). The normally recommended

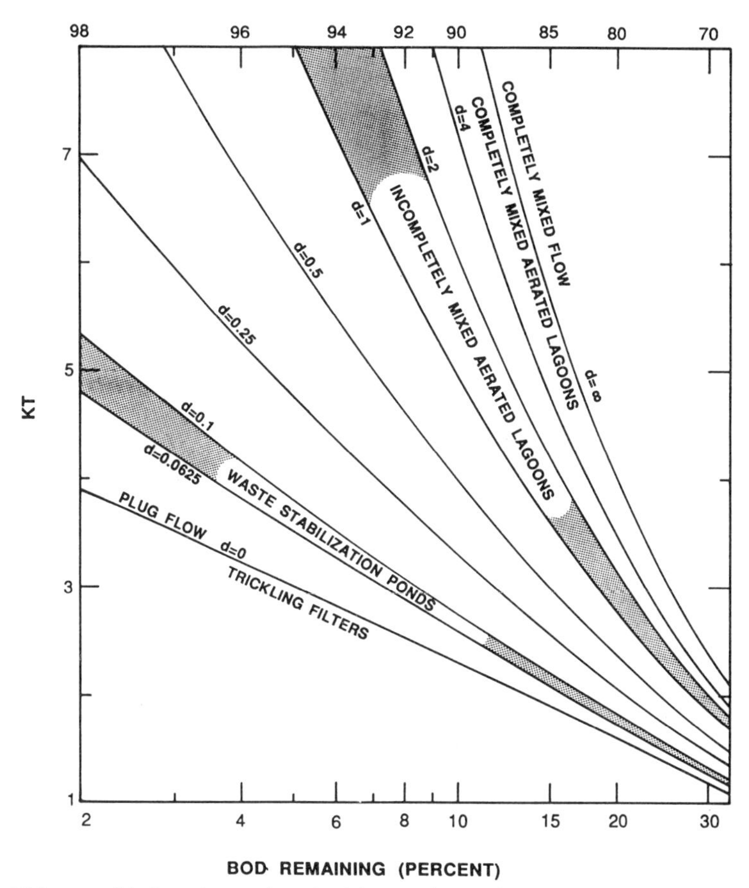

Fig. 1. Thirumurthi chart (reproduced with permission from John Wiley and Sons, Inc.).

equation for type (1) is given in Chapter 10. The type (2) lagoon can be designed by making use of either a design chart (Fig. 1) or eqn (1). An assumption is made that the type (2) aerated lagoon is a completely-mixed stirred tank reactor or a continuous-flow stirred tank reactor (CSTR):

$$\text{influent COD/effluent COD} = C_i/C_e = (1 + Kt) \tag{1}$$

where K = first-order COD removal coefficient (per day) and t = mean hydraulic retention time (HRT, days). The type (3) lagoon is designed using either Fig. 1 or eqn (2):

$$\frac{C_e}{C_i} = \frac{4a \exp(1/2d)}{(1+a)^2 \exp(a/2d) - (1-a)^2 \exp(-a/2d)} \tag{2}$$

where d = dispersion index which is estimated from a tracer test (Thirumurthi, 1969*a*) and $a^2 = (1 + 4Ktd)$. Equation (2) was originally derived by Wehner and

Wilhelm (1956) for chemical reactor design, and Thirumurthi (1969*b*, 1974) showed that it could also be employed for lagoons. Wehner and Wilhelm derived eqn (2) for chemical reactors with first-order reaction kinetics, and Thirumurthi suggested that lagoons could be designed as biochemical reactors with first-order BOD removal kinetics (Levenspiel, 1965) by using that equation. Additional details are given in Chapter 10.

2.3.2 The Experiences in the UK

Two aerated lagoons were extensively monitored in the UK, and the results constitute significant contributions to the better understanding of biodegradation of organics in leachate (Maris *et al.*, 1985; Robinson, 1985).

2.3.2.1 BRYN POSTEG LANDFILL

Leachate from Bryn Posteg Landfill (located in Wales) is treated by an aerated lagoon which was monitored by Robinson (1985) for 18 months. The landfill site handles primarily domestic solid wastes (about 50 t/day). The annual average rainfall in the vicinity of the landfill site is about 1200 mm and the site is located in an old lead mine. Various shafts in the old lead mine required sealing and the mine spoil in the area was moved out. The landfill was started in June 1982 and the full-scale treatment plant began operating in May 1983. The aerated lagoon has a capacity of 1000 m^3 and is lined with high-density polyethylene. The aeration is accomplished by two floating surface aerators. The HRT in the aerated lagoon depends on the amount of leachate generated, which varies from day to day. However, the minimum retention time in the aerated lagoon is about 10 days. Phosphoric acid is added as a nutrient. The sludge produced in the aerated lagoon is removed regularly and disposed on the landfill itself in a relatively unthickened form.

The effluent from the aerated lagoon is discharged to a nearby municipal treatment plant, where the quality is further improved. The COD values in the lagoon effluents never exceeded 350 mg/liter, although the raw leachate COD was as high as 24 000 mg/liter. Similarly, the effluent BOD_5 concentrations had always been less than 100 mg/liter compared to raw leachate BOD_5 values reaching a maximum of about 18 000 mg/liter. The author reports that liquid temperatures as low as 2–3°C during the winter months had no noticeable effect on effluent quality. The average overall biodegradation was high, resulting in significant removals of COD (97%), BOD_5 (99%) and $NH_3 . N$ (91%). The proportion of the volatile solids in the Bryn Posteg aerated lagoon was about 40–60% of the total solids, with a mean value of 45%. The food to microorganisms ratio (F/M ratio) was typically between 0·1 and 0·3 g COD/g MLVSS . day during the winter months and generally less than 0·05 g COD/g MLVSS . day during the summer months. These values compared closely with the results reported from a full-scale leachate treatment plant in Philadelphia (Keenan *et al.*, 1984), which operated satisfactorily when the F/M ratio was between 0·12 and 0·32 per day.

Sludge solids (excess biomass) in the lagoon exhibited excellent settling properties, and these were observed both in the laboratory cylinders and in the actual lagoon. Sludge settled to a concentration of about 3% solids (range 2·7–3·5%), and was pumped out by using a tanker and disposed of onto recently landfilled wastes at the site.

2.3.2.2 BIODEGRADATION OF AMMONIA

As described elsewhere in this chapter, degradation and removal of organic N and ammonia can be attained in different ways. The aerated lagoon at Bryn Posteg achieved ammonia removal, according to the author, not by nitrification but by bacterial uptake of N, incorporation in biomass and removal as sludge (Robinson, 1985). The pH of the raw leachate was typically about 5·8, and the pH of the aerated lagoon effluent was typically around 8. On this basis Robinson claims that relatively small quantities of nitrogen were lost as ammonia gas by volatilization. It is a well-known fact that at the pH value of about 8 less than 2% of the total ammonia (cationic ammonium plus molecular ammonia) would be in the molecular form at 10°C. Previous work (Robinson & Maris, 1983) on leachate indicated that removal of ammonia can be achieved by incorporation in sludge. Experience has shown that if this sludge is not removed at suitable times the ammonia can be re-released to appear in the effluent. When sufficient desludging takes place, however, this is not a problem and ammonia levels in effluent can be consistently low. An aerated lagoon at Falls Township Landfill, Pennsylvania (Steiner *et al.*, 1977), removed ammonia by the air-stripping method at a high pH of 10–11. Additional information is given in Section 3.4.3. The basic difference between the two observations is the pH of the liquid. As the pH increases a higher proportion of total ammonia is in the form of molecular ammonia, which can be removed from the liquid phase by volatilization or air-stripping.

2.3.2.3 NEW PARK LANDFILL

Aerated lagoon studies of treatment of leachate produced from this landfill (located in south-east England) were conducted by Maris and co-workers (1985). Based on previous laboratory-model studies, the authors concluded that the simpler aerated lagoon system was adequate for their application, and the more complex activated sludge process was considered to be unnecessary because the laboratory tests had indicated that the leachate could readily produce sufficient biomass without the need for sludge recirculation.

The lagoon model was 2 m deep and about 29 m^3 in volume. The oxygen supply and mixing were provided by coarse bubble aeration. Subsequent to aeration, the liquid was transferred to a clarifier of volume 1·62 m^3. The New Park Landfill leachate being deficient in phosphate, an external source was added, in the form of phosphoric acid, at the ratio of BOD to P of 100:1. Subsequent to completion of the initial phase, the model aerated lagoon was operated at different solids retention times (SRT). Six operating periods (each with an SRT of 20·6, 14, 10·7, 5 and 3 days) were monitored during the two-year study. The data are summarized in Tables 2 and 3. This pilot-scale study represents one of the most detailed research studies published on aerated lagoon treatment of leachate. In view of the fact that there was no sludge recycle, the HRT and the SRT were equal. The data of Table 3 indicate that the pilot aerated lagoon system, followed by sedimentation, performed well

Table 2
Aerated Lagoon Pilot-plant Operating Data[a]

	Solids retention (days) (SRT = HRT)					
	20·6	*14·0*	*10·7*	*7·0*	*5·0*	*3·0*
Organic load (applied)						
kg COD/m^3.day	0·64	0·85	1·66	1·68	1·25	1·41
kg COD/kg MLSS.day	0·13	0·15	0·12	0·18	0·11	0·12
kg COD/kg MLVSS.day	—	0·28	0·28	0·32	0·30	0·29
MLSS (mg/liter)	4 979	5 611	13 406	9 290	8 247	11 698
MLVSS (mg/liter)	—	3 007	5 899	5 267	4 222	4 808
Sludge (production)						
Vol. sludge/vol. leachate applied.day (liter/liter.day)	0·11	0·20	0·20	0·32	0·26	0·32
Sludge solids (clarifier) (kg/m^3)	40	26	62	27	38	43
Hydraulic retention time in clarifier (h)	10·6	7·2	5·5	3·6	2·6	1·5

Source: Maris *et al.*, 1985; adapted with permission from the publisher (University of British Columbia).
[a] Aeration tank volume = 28·9 m^3.

with impressive removals of COD, BOD_5, SS and Amm. N varying from about 81% to 99%. The only exception was a low removal of ammonia for the test with a low SRT of 5 days at 15·4°C. During that test 55% of ammonia was removed. The biomass (solids) settled readily in the clarifier to produce sludges consisting of 26–62 kg of dried solids per m^3. However, the volume of sludge requiring disposal was high, ranging between 10% and 30% of the total flow.

2.3.3 Laboratory-model Lagoons

Small-scale tanks in the laboratory enable researchers and design engineers to better understand the mechanisms of biodegradation of organics. Leachate from each landfill being different from the others, proper understanding of the microbial decomposition and economical design of the full-scale lagoon cannot be accomplished without extensive laboratory-model study, preferably followed by pilot-scale field investigation. In the following three sub-sections knowledge gained by investigators in the US, Canada and Australia from laboratory-model aeration tanks is described. Aerated lagoon treatment has a very suitable application in the overall scheme of leachate biodegradation, namely polishing or further treatment of anaerobically degraded effluents. Studies in Canada (Thirumurthi *et al.*, 1986*a*) and Australia (Bull *et al.*, 1983), summarized in sub-sections 2.3.3.2 and 2.3.3.3, indicate that aerated lagoons are appropriate in treating anaerobic effluents.

2.3.3.1 THE US STUDY

Chian and DeWalle (1977*a*) conducted research on biodegradation of organics in

Table 3
Removal Efficiencies of Pilot-scale Aerated Lagoons[a]
(concentrations are expressed in mg/liter)

SRT (days)	*Lagoon temp. (°C)*	*COD*			*BOD₅*			*Amm-N*		
		Inf.	*Eff.*	*% Removal*	*Inf.*	*Eff.*	*% Removal*	*Inf.*	*Eff.*	*% Removal*
20·6	9·8	13 265	224	98·3	10 721	29	99·7	348	7	98·0
14·0	15·9	11 928	448	96·2	9 674	17	99·8	353	41	88·4
10·7	25·3	17 717	529	97·0	11 713	30	99·7	590	10	98·3
7·0	9·8	11 931	443	96·3	7 650	38	99·5	374	13	96·5
5·0	15·4	6 240	562	91·0	2 944	38	98·7	388	175	54·9
3·0	21·4	4 141	365	91·2	1 663	12	99·3	87	3	96·5

Source: Maris *et al.*, 1985; adapted with permission from the publisher (University of British Columbia).
[a] Pilot-aerated lagoon followed by clarification (no sludge recycle).

leachate under various loading conditions and levels of nutrient addition. The leachate was obtained from a simulated landfill lysimeter located at the University of Illinois. Six completely mixed vessels were used as aerated lagoon models and the first three of these units were 3-liter plastic tanks, in which HRTs of 30, 60 and 86 days were maintained. The second three units were 30-, 15- and 7-liter tanks with HRTs of 30, 15 and 7 days, respectively. Each day these units were fed with predetermined amounts of leachate after the equivalent volume of the contents of the model aerated tanks was withdrawn. Nutrient additions were made to maintain a COD to P ratio of 164:1 and a COD to N ratio of 19:1 for the first three units, which required additions of 347 mg/liter of P and 1737 mg/liter of N. Lower levels of nutrients were added to units 4, 5 and 6, which were fed with 211 mg/liter of P and 1057 mg/liter of N. Nutrients were added, in the beginning, in the forms of dibasic potassium phosphate and ammonium nitrate. However, the concentration of K ions in the effluents was increasing rapidly, therefore the nutrient solutions were replaced by ammonium phosphate and ammonium nitrate. Aeration for each unit was accomplished through porous glass diffusers. Tanks 1, 2 and 3 were operated for 150 days and tanks 4, 5 and 6 for 70 days.

The aerated lagoon treatment of a strong leachate with a COD of 57 900 mg/liter could remove between 93% and 97% of the organic matter, without any pretreatment, at detention times ranging from 7 to 86 days. For the tanks with longer HRT the P requirement was lower, whereas for the units with shorter retention time the P requirement was higher. For instance, for the unit with a 30-day HRT the minimum requirement was a COD/P value of 300. On the other hand, for the tanks with HRTs of 86 and 60 days the requirement was lower, at a COD/P value of 1540. However, for the tanks with HRTs less than 30 days, when the value of COD/P exceeded 165, an immediate increase in effluent organic matter, a decrease in mixed liquor volatile suspended solids and a reduction in the sludge settling rate were observed.

2.3.3.2 THE CANADIAN STUDY

A laboratory-model aerated lagoon biodegradation study was conducted at

Table 4
Performance of a Laboratory-model Aeration Tank

Parameter	*Concentrations (mg/liter except pH)*		*Removal (%)*
	Partly treated feed leachate	*Aerated tank effluent*	
Alkalinity as $CaCO_3$	2 563	812	68
Al	<0·5	<0·5	—
Amm. N	382	1·3	99
As	<0·05	<0·05	—
Ba	<0·05	<0·05	—
B	6·1	4·3	30
BOD_5	254	4	98
Cd	<0·02	<0·02	—
Ca	108	86	20
Cl	1 015	790	22
Cr	<0·1	<0·1	—
COD	693	142	80
Cu	<0·1	<0·1	—
Hardness as $CaCO_3$	909	707	22
Humic acid	368	230	38
Fe	11·8	0·36	97
TKN	395	6·5	98
Pb	<0·02	<0·02	—
Mg	154	128	17
Mn	0·6	0·10	83
$(NO_3 + NO_2)N$	<0·5	138	—
PO_4 (*ortho*)	0·82	2·1	(156)
pH	7·43	8·56	—
K	388	299	23
Na	955	755	19
SO_4	28	210	650
SS	117	4	97
TDS	4 220	3 497	17
TOC	224	59	74
PO_4 (total)	5·2	3	42
TVA	55	20	64
TVS	714	414	—
VSS	86	3	97
Zn	0·52	0·13	75

Source: Thirumurthi *et al.*, 1986*a*; reproduced with permission from *Water Pollution Research Journal of Canada.*

Halifax, Canada, for the purpose of further biodegrading the organics in leachate which had been pretreated by physical–chemical treatment and anaerobic fixed film reactors (Thirumurthi *et al.*, 1986*a*). The results are summarized in Table 4. The HRT was about 70 days and the volumetric organic load was in the range 0·01–0·013 kg COD/m^3. day. A glass aquarium tank of about 21-liter capacity and 20-cm liquid depth was aerated using two diffuser stones supplied with compressed air. It was operated as a well-mixed aerated system. The aerated tank was kept at room temperature (18–22°C). A long HRT of 70 days had to be used because the feed to the aerated tank had a high COD of 694 mg/liter. (The studies done in West Germany have also shown that HRTs of 20–100 days are very common in that country.) The results indicate that removals of BOD and TKN were slightly above 98%; however, the removals of COD (79%) and TOC (73%) were lower, indicating the presence of non-biodegradable organics in the aerated lagoon effluent. About 36% of NH_4 . N was converted into $(NO_3 + NO_2)N$ (according to the data in Table 4, 138 divided by 382 multiplied by 100). However, the aerated aquarium tank removed more than 99% of NH_4 . N. Therefore it can be assumed that the biological assimilation and ammonia stripping resulted in the remaining 63% removal of Amm . N. The pH of the contents increased from a value of about 7·4 for the feed of the aerated lagoon to about 8·6 for the aerated lagoon effluent. Such increases in pH were commonly observed, as shown in the data published from the aerated lagoons in the UK and West Germany. The probable reason for such increase in the pH is due to the stabilization of volatile acids in the leachate.

2.3.3.3 THE AUSTRALIAN STUDY

A laboratory-model investigation of biodegradation of leachate from Lucas Heights Regional Landfill, Sydney, Australia, was completed by Bull and his co-workers (1983). The leachate was first anaerobically treated by a back-mix, suspended growth reactor. The effluent from the anaerobic unit was not suitable for discharge to a surface receiving body. Therefore aerobic treatment followed it, whereby 86% of BOD_5, 51% of TOC and 65% of TKN were removed at an HRT of 4 days and a temperature of 24°C. The average volumetric organic load was 50 g of BOD_5/m^3 . day, which was about four times the load maintained in the tank studied in Halifax, Canada (Thirumurthi *et al.*, 1986*a*).

3 ANAEROBIC BIODEGRADATION

Advantages and disadvantages of anaerobic biological treatment of leachate in comparison with aerobic treatment are summarized in the following paragraphs.

3.1 Advantages of Anaerobic Treatment

1. The anaerobic process has a potential to be a net energy producer. While the aerobic processes have significant energy requirements for aeration, mixing and/or pumping, the anaerobic systems produce methane gas, a useful energy source.

2. The fraction of organics in wastewater that is converted into biomass (waste sludge) is much smaller (about 0·1) for anaerobic stabilization than for aerobic stabilization (about 0·5). Therefore costs and problems associated with disposal of excess biological sludges from treatment plants are greatly reduced (McCarty, 1985).

3. The inorganic nutrient requirement (N and P) for anaerobic treatment is much less than that for aerobic treatment because the anaerobic system is characterized by

lower net growth of biomass. For example, the required COD/P value for anaerobic leachate treatment is in the range 15 000–30 000 (Thirumurthi & Groskopf, 1988). At BOD_5/P values of 6200–12 545 no deficiency of phosphorus was detected by Henry *et al.* (1987). The BOD_5:N:P requirement for aerobic treatment is about 100:5:1. Therefore, for leachate which is usually low in PO_4, anaerobic treatment is more attractive than aerobic treatment.

4. According to McCarty (1985), 'recent evidence suggests that many hazardous organic chemicals such as halogenated solvents can be biotransformed under anaerobic conditions'. Anaerobic followed by aerobic stabilization can combine the anticipated advantages of anaerobic and aerobic treatment. It is also likely that some xenobiotic chemicals that were previously considered to be refractory could be degraded.

5. For leachates of stronger concentrations the anaerobic rather than the aerobic decomposition is more practical. The studies completed at the University of Wisconsin (Ho *et al.*, 1974) indicated that of all the biological treatment methods investigated anaerobic treatment seemed to be the best for handling leachates of COD above 10 000 mg/liter.

3.2 Disadvantages of Anaerobic Treatment

1. Most of the leachates generated in various parts of the world tend to be acidic in nature, with pH ranging from 5·5 to 7·0. At such low pH values the slow-growing methanogenic bacteria are sensitive. The acid-forming bacteria decompose the complex organic matter into volatile acids and other simpler organic compounds. The methanogenic bacteria break down the acids and other simpler organic compounds into methane and carbon dioxide. While the acid formers are less sensitive to pH, and are faster-growing in nature, the second group of bacteria are pH-sensitive and they grow at a slower rate. As a result the methanogenic bacteria are usually the rate-limiting microorganisms and, unless the pH of the anaerobic systems is kept at 7 or above, the volatile acids tend to accumulate, which results in a drop in pH. Consequently the methanogenic bacteria start to die off or their growth rate is inhibited. Therefore a properly designed anaerobic biological treatment system should be able to sustain high pH (above 7) and retain the biomass, thereby increasing the MCRT or minimizing the wash-out of the slow-growing bacteria.

2. Anaerobic metabolism converts the organic N to ammonia N, which exists mostly in the form of NH_4^+ rather than free ammonia gas. At pH levels above 7 ammonia gas is formed and can be removed by air-stripping. However, anaerobic processes operate in the pH range between 7 and 7·5, at which range most of the ammonia N is present as NH_4^+. Unless the anaerobic effluent can be further treated, either by means of an aerobic process or by lime addition to increase the pH and to strip the ammonia gas, the effluent from an anaerobic system will be high in the concentration of ammonium, which may not be acceptable in several receiving waters because it exerts an oxygen demand and is toxic to fish.

3. The optimum temperature for anaerobic decomposition is usually around 35°C. As the temperature drops below 35°C the efficiency of treatment decreases

rapidly. Therefore most anaerobic treatment systems are accomplished in enclosed, heated digesters of very high capital cost. It is widely known that anaerobic systems are very sensitive, especially in the early stages when digesters are initiated.

4. High concentrations of some of the heavy metals could be toxic to anaerobic microorganisms. In such cases the leachate has to be pretreated by a physical–chemical process in order to render the leachate more amenable for anaerobic biological treatment (Thirumurthi *et al.*, 1986*a*). The inhibition of anaerobic digestion processes occurs at soluble (membrane-filtered) concentrations of approximately 3 mg/liter for Cr, 2 mg/liter for Ni, 1 mg/liter for Zn and 0·5 mg/liter for Cu (DeWalle *et al.*, 1979).

3.3 Anaerobic Fixed Film Treatment

Anaerobic fixed film reactors (AFFRs) are more suitable than anaerobic suspended growth reactors (the types used as sludge digesters in municipal plants) for treatment of high-strength leachate because AFFRs can retain biomass for a long time, resulting in better performance. For any wastewater with low suspended solids and high biodegradable organics (such as strong leachates, COD > 5000 mg/liter) an AFFR is suitable. The ability of an AFFR to maintain a high ratio of MCRT:HRT results in high-rate treatment. For example, in laboratory-model AFFRs Thirumurthi and Groskopf (1988) could achieve MCRT values as high as 921 and 984 days with a short HRT of 13·9 days, resulting in about 97% COD removal at 35°C and an organic load of 1·24 kg COD/m^3. day.

Chian and DeWalle (1977*b*), widely known for their pioneering research contributions on treatment of leachate, conducted numerous studies using anaerobic filters. (An AFFR is sometimes called an anaerobic filter or anaerobic submerged biological filter, AnSBF.) When plug flow anaerobic filters, without provisions for recycling the effluent or mixing the contents, are used to treat low-pH leachate, costly additions of chemicals to increase pH and alkalinity are often required. Chian and DeWalle (1977*c*) quote examples of studies of plug flow reactors to which sodium bicarbonate or other chemicals were added to increase the pH and alkalinity of the feed because they decreased initially as a result of acid fermentation. The acidic pH in the lower section of the AFFR can potentially inhibit the methane-fermenting bacteria. The authors hypothesized that a well-mixed AFFR would not experience the pH decrease observed in once-through plug flow units because the mixing and recycle of effluent maintain a fairly uniform pH throughout the depth of the filter, thus minimizing the need for additions of buffer solutions. To verify the hypothesis they conducted laboratory-model research on treatment of leachate using a completely mixed anaerobic filter to establish the advantages of such a system in comparison to once-through plug flow reactors. The medium used in the filter was a modular plastic 'Surpac' manufactured by Dow Chemical Co. The specific surface area of the medium was 206 m^2/m^3 of column volume. The total surface area in the column was 11·3 m^2, and only 6% of the column volume was taken up by the plastic material, resulting in a porosity of 94%. The leachate employed in this study was generated from a lysimeter filled with solid waste, to which simulated rainwater was added. It had a COD of 54 000 mg/liter and a pH of 5·4. The fatty acids constituted 49% of the total COD. The observed concentrations of metals were 2200 mg/liter Fe, 104 mg/liter Zn, 18 mg/liter Cr, 13 mg/liter Ni and 0·5 mg/liter Cu.

In Phase I of the study different start-up procedures and pH stabilities of the unit were tested, whereas in Phase II various operational difficulties of the completely mixed unit were evaluated. Phase III was used to study the effluent characteristics as affected by organic loading.

The research data showed that the authors' assumptions were indeed correct. During the start-up of the project the test was conducted without pH adjustment and without any seeding, to establish the fact that the filter could not perform well under such conditions. Later on the pH of the feed was increased to 7 and there was no seeding. During the subsequent period the system was seeded with municipal sludge. A recirculation ratio of 1:20 was used for this project.

It was observed that a high-strength leachate with a low pH could be successfully treated using a well-mixed anaerobic reactor, thus eliminating the need for costly additions of buffer solutions as was required by plug flow reactors. Although some metal toxicity was observed, this was eliminated by the addition of sulfides to the unit, which were effective in precipitating remaining soluble metals. The solids yield was estimated to be 0·012 g VSS/g of COD removed. During the first 60 days of Phase II 97% of the COD was removed at an organic loading rate of 0·62 kg COD/m^3.day. The biogas produced during that period contained about 78% methane. This corresponds to 89% of the theoretical gas production as calculated from total COD removals.

Two-stage anaerobic biodegradation of organics in leachate is a practical option because it has the advantage of higher efficiency and flexibility in operation. In the event that one reactor needs to be repaired, the second unit could provide at least partial treatment. The experiences of Wu and his co-workers (1982) in Pittsburgh are described in Section 3.4.1.

3.3.1 System Start-up

It is widely known that starting an AFFR is perhaps the most difficult period of operation. The start-up time in the laboratory model, as well as in full-scale units, could range from 10 to 60 days. The shorter times are normally associated with the use of large quantities of active seed materials, usually obtained from nearby anaerobic sludge digesters. The longer start-up times are usually associated with the use of low level additions of seeding materials, and low temperatures. Subsequent to the start-up time, these reactors usually take a very long time to reach a state of biological maturity, maximum performance potential and near-steady-state performance. For successful start-up of a reactor, and to enable it to reach its potential to stabilize the organic matter, several other factors are also important. Temperature should be around 35°C; pH of the feed, as well as the contents of the reactor, should be above 6·8, preferably around 7; and there should be adequate alkalinity to provide buffer capacity to the contents of the reactor. Of course, in addition to these, adequate nutrient addition is also essential. Another important factor relates to the physical features such as the reactor design, size and shape, and selection of the filter medium, such that it can retain the biomass for as long a time as possible. In general, near-ideal conditions of seeding, pH, temperature, alkalinity and nutritional balance should be maintained for a healthy start-up of the reactor, otherwise the researchers and operators will face substantial difficulties in achieving the full potentials of the AFFR.

3.4 Case Studies

3.4.1 Pittsburgh Case Study

A laboratory-model, two-stage, anaerobic submerged biological filter (AnSBF) to study the feasibility of using such units for the treatment of acidic leachate was completed in Pittsburgh. The effects of organic load, HRT and metal ions on the removals of BOD_5, COD and volatile acids, and the generation of methane gas were monitored. The two AFFRs (operated in series) consisted of Munters medium for the first stage and Goodrich medium for the second stage. The two AFFRs were operated in the upflow mode and the effluent from the second filter reached the settling basin, where both substrate and solids concentrations were monitored.

The leachate was obtained from the 16-year-old Elizabeth Township Municipal Landfill located south of Pittsburgh, Pennsylvania. The nutrients were added to the raw leachate in the forms of ammonium chloride and monobasic potassium phosphate at a COD:N:P ratio between 100:1:0·1 and 100:2:0·2. It should be noted here that the authors added a rather high concentration of the phosphate nutrient because, in 1982, information was not available on the minimum concentration of phosphate required to sustain anaerobiosis. However, the studies conducted at the University of Toronto and Technical University of Nova Scotia, both in Canada, showed later on (1983/85) that much lower phosphorus concentrations (COD/P = 15 000–30 000) are sufficient to sustain anaerobic metabolism and organic stabilization in AFFRs. In addition to the nutrients, the authors added sodium bicarbonate to increase the alkalinity in the leachate to 2200–2300 mg/liter. The Pittsburgh study consisted of three phases, and the organic load varied from 0·08 to 2·3 kg COD/m^3. day. The authors (Wu *et al.*, 1982) concluded that, with two-stage AFFR treatment, more than 96% of COD and BOD_5 can be removed from the leachate if the volumetric organic load is kept at about 1·2 kg COD/m^3. day (75 lb COD/1000 ft^3. day) in each filter, or 0·9 kg BOD_5/m^3. day (56 lb BOD_5/1000 ft^3. day), and when the HRT is about 17 days in each AFFR. During Phase III of the study the soluble COD of the raw leachate decreased to about 9600 mg/liter and therefore, with a lower organic load of 1·0 kg COD/m^3. day (62 lb COD/1000 ft^3. day) and an HRT of 9·5 days, 90% of the COD could be removed. The AFFRs were also efficient in removing more than 92% of Zn and Fe during the three phases. The volume of methane produced was 0·31–0·36 m^3/kg (5–5·8 ft^3/lb) of COD removed. It was also concluded that the reactors, which were kept between 36°C and 38°C, did not require any effluent recycle or sludge return. Biogas production

Table 5
Concentrations of Various Parameters in a Laboratory-model Study

Parameter	*Concentrations (mg/liter except pH)*				
	Raw leachate	*Feed leachate*	*AFFR-1*	*Aerated tank*	*Settling tank*
Alkalinity as $CaCO_3$	3 630	4 200	2 563	812	1 800
Al	1·9	0·2	<0·5	<0·5	<0·5
Amm. N	375	380	382	1·3	37·5
Sb	0·14	0·07	<0·5	<0·5	<0·5
As	0·03	0·005	<0·05	<0·05	<0·05
Ba	0·18	0·16	<0·05	<0·05	<0·05
BOD_5	17 500	17 400	254	4	15
Cd	0·004	<0·002	<0·02	<0·02	<0·02
Ca	1 776	2 025	108	86	27
Cl	1 060	—	1 015	790	1 080
COD	22 900	22 300	693	142	325
Co	0·39	0·25	<0·1	<0·1	<0·1
Cu	0·02	0·02	<0·1	<0·1	<0·1
Hardness as $CaCO_3$	5 534	6 044	909	707	857
Humic acid	445	405	368	230	381
Fe	1 002	155	11·8	0·36	1·39
TKN	490	470	395	6·5	49
Mg	257	—	154	128	187
Mn	63·2	—	0·6	0·10	0·07
$(NO_3)N$	<0·2	<0·2	<0·5	138	<0·5
PO_4 (*ortho*)	0·2	1·0	0·82	2·1	5·6
pH	5·6	7·5	7·43	8·56	8·82
K	423	—	388	299	412
Na	910	—	955	755	975
SO_4	930	—	28	210	54
SS	320	277	117	4	24
TDS	14 940	19 200	4 220	3 497	4 215
TOC	8 100	8 900	224	59	120
PO_4 (total)	17	60	5·2	3	7·0
TVA	10 100	10 200	55	20	28
TVS	—	—	714	414	611
VSS	—	—	86	3	20
Zn	80	4	0·52	0·13	0·11

Source: Thirumurthi *et al.*, 1986*a*; reproduced with permission from *Water Pollution Research Journal of Canada.*

fluctuated between 1·3 and 16 liter/day, with 71–83% methane content. The methane generation rate was 0·31–0·36 m^3/kg of COD removed, which was between 74% and 87% of the theoretical maximum.

3.4.2 Halifax Case Study

Thirumurthi *et al.* (1986*a*) completed a laboratory-model study of AFFRs treating leachate produced at a Halifax, Canada, landfill. The raw leachate, which had a low pH (5·5) and a potentially toxic Zn concentration (60–80 mg/liter), was pretreated by

Table 6
Performance Levels (% Removal) of the Treatment Units Used in Laboratory-model Studies at Technical University of Nova Scotia

Parameter	*System I(a)*	*System I(b)*	*AFFR-1*	*Aerated tank*	*Settling tank*
Alkalinity as $CaCO_3$	77·6	50·4	39·0	68·3	30·0
Amm. N	99·7	90·0	(0·5)[a]	99·7	90·2
BOD_5	99·9	99·9	98·5	98·4	94·1
Ca	96·2	98·5	94·7	20·4	75·0
Cl	25·5	2·8	—[b]	22·2	(1·5)
COD	99·4	98·6	96·9	79·5	53·1
Hardness as $CaCO_3$	87·2	84·5	85·0	22·2	5·7
Humic acid	48·3	14·4	9·1	37·5	(3·5)
Fe	99·6	99·8	92·4	96·9	88·2
TKN	98·7	90·0	16·0	98·4	87·6
Mg	50·2	27·2	—	16·9	(21·4)
Mn	99·8	99·9	—	83·3	88·3
$(NO_2 + NO_3)$ as N	n.d.[c]	n.d.	n.d.	n.d.	n.d.
PO_4 (*ortho*)	n.c.[d]	n.c.	18·0	(156·1)	(583)
K	29·3	2·6	—	22·9	(6·2)
Na	14·8	(7·1)	—	18·8	(2·1)
SS	98·8	92·5	57·8	96·6	79·5
SO_4	77·4	94·2	—	(650·0)	(92·9)
TDS	76·6	71·8	78·0	17·1	0·1
TOC	99·3	98·5	97·5	73·7	46·4
PO_4 (total)	n.c.	n.c.	91·3	42·3	(34·6)
TVA	99·8	99·7	99·5	63·6	49·1
TVS	—	—	—	42·0	14·4
VSS	—	—	—	96·5	76·7
Zn	99·8	99·8	87·0	75·0	78·8

Source: Thirumurthi *et al.*, 1986*a*; reproduced with permission from *Water Pollution Research Journal of Canada.*
[a] Values in parentheses indicate the percent increases.
[b] — Not monitored.
[c] n.d.: The concentrations being below detectable limits removals could not be calculated for these parameters, and for Al, Sb, As, Ba, Be, Cd, Cr, Co, Cu, Pb, Ni, Se, Sn and V.
[d] n.c.: These values were not calculated because phosphate was added as a pretreatment step.

(1) addition of lime or NaOH to increase pH to a range of 7·5–8·0, (2) flocculation, and (3) settling. The settled leachate was treated by two upflow AFFRs consisting of randomly packed loops or biorings of filamentous toroidal helical shape. Although these were manufactured originally for use in air pollution control devices (scrubbers and stripping columns), this is the first known investigation in which this bioring was used as the medium in an AFFR. Its light weight, high surface area (180 m^2/m^3) and 87% void volume render it an ideal medium for AFFRs. As summarized in Tables 5 and 6, system I(a) consisted of pretreatment, AFFR-1 and an aerated tank in series. System I(b) was similar to I(a) except that the aerated tank was replaced by a settling tank. Excellent removals of pollutants were achieved, as shown in Tables 4–6.

Thirumurthi and Groskopf (1988) studied the effects of low concentrations of PO_4 on AFFR performances. When three AFFRs (A, B and C) were fed, respectively, with *ortho*, organic and condensed PO_4 at COD/P values varying from 6100 to 64 300, it was observed that when the COD/P value exceeded 34 300 a decrease in COD removal efficiency occurred. During Phase I, when the COD/P ratio was 6100 in all three reactors, it was concluded that they performed at near-identical efficiencies. Anaerobes did not seem to prefer one form of PO_4 over the other. During Phase II the COD/P values were changed to 7700 in AFFR-A (*ortho*) and AFFR-B (organic), and to 64 300 in AFFR-C (condensed PO_4). While A and B continued to perform well, the effluent COD, SS and VSS increased in AFFR-C, indicating a deficiency in PO_4. During Phase III reactors A and B, with COD/P values of 10 200 and 15 200, respectively, continued to perform well. However, AFFR-C, with a COD/P ratio of 34 300:1, performed with a lower percent of COD removal. It is concluded that the critical COD/P value for anaerobic fixed film treatment of leachate is between 15 000 and 34 300. The major contribution of this study is the conclusion that several researchers have been adding excessive amounts of PO_4 (COD/P as high as 100) while the minimum requirement could be as low as 15 000.

3.4.2.1 LEACHATE BIODEGRADATION PLANT IN HALIFAX

The full-scale leachate treatment plant at Halifax is the first of its kind in North America (Wright *et al.*, 1985). This plant was perhaps the first facility where complete treatment consisting of pH increase, chemical precipitation, 'hybrid' anaerobic reactors, aerated lagoons, settling ponds and sludge thickening was accomplished in 1987. The landfill site was started in 1976. Refuse received is primarily municipal and commercial in nature, and hazardous or toxic wastes are not permitted. Leachate biodegradation was achieved initially (1976) by means of two aerated lagoons in series. The inability of the lagoon treatment system to handle the increasing leachate strength and quantities, in addition to concerns of impact on the receiving water quality, led to a program to upgrade treatment capability. Leachate from this landfill is characterized by high strengths (COD = 24 000 mg/liter) and low flows (average flows of 1·5 liter/s).

The treatment system consists of four stages, including a pretreatment stage, high-rate anaerobic reactors to reduce organic strength, sludge handling facilities and polishing lagoons. The pretreatment stage utilizes physical–chemical treatment to adjust pH of the leachate and remove inhibitory metals. Caustic soda is added as the leachate flows through the processes of rapid mix, flocculation and settling. The pretreated leachate, free of hindrances to biological treatment, is then pumped to the anaerobic reactors.

The anaerobic reactors were constructed as combination upflow sludge blanket/anaerobic filter systems (Fig. 2). There are two identical units constructed in concrete tanks of 5-m diameter by 8-m side wall depth. Flow is directed from the bottom upwards through the sludge blanket and anaerobic filter sections before exiting via an overflow box to the aerated lagoons. A large mass of microorganisms is active in the sludge blanket zone and provides the high rate degradation of organic

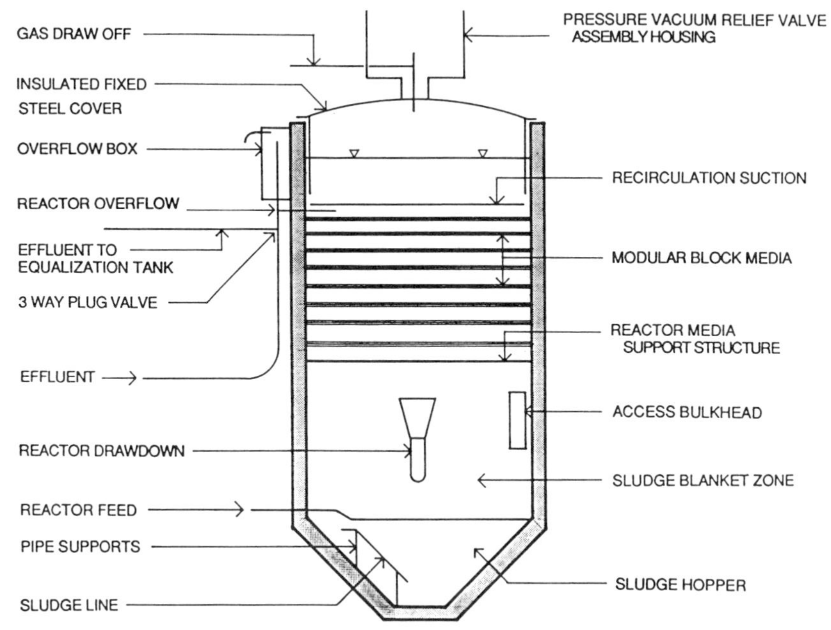

Fig. 2. Anaerobic reactor used in Halifax landfill (reproduced with permission from Porter-Dillon Ltd).

matter. The filter media section above the sludge blanket provides an additional degree of treatment while at the same time allowing for clarification of liquid and degasification. By-product gas collects in the atmospheric zone above the media, from where it is drawn off for recycling of the methane through the hot water heating boiler. Sufficient methane is generated to maintain the reactors at 35°C and provide all space-heating needs.

Sludge from both the pretreatment stage and the anaerobic reactors is removed to a sludge thickener. The thickened sludge is then dried by means of a precoat vacuum filter and returned for disposal to the landfill. Effluent from these processes is recycled to the head of the treatment facility.

The existing aerated lagoons were modified to improve their liquid retention capability and to modernize the air distribution systems. The lagoons serve as the final formal treatment step before effluent is discharged through the site's settling ponds to the receiving environment. The leachate treatment facility underwent start-up in the spring of 1987. The anaerobic reactor portion of the plant was designed to remove 90% of the COD of the incoming leachate. The start-up procedure for development of the biomass utilized conventional municipal anaerobic digester sludge as a 'seed' for the high rate anaerobic biomass.

3.4.3 Falls Township Case Study

The US Environmental Protection Agency (EPA) awarded a demonstration grant to investigate the effectiveness of various treatment sequences for leachate produced at the Geological Reclamation Operations and Waste Systems (GROWS) Landfill, located at Falls Township, Pennsylvania (Steiner *et al.*, 1977). A full-scale treatment system was built to handle a leachate flow rate of 380 liter/min. This facility had the flexibility to test alternative treatment sequences. The study included four different systems of treatment. System 1 consisted of physical–chemical treatment followed by activated sludge treatment. System 2 consisted of physical–chemical treatment only. System 3 included biological treatment followed by physical–chemical treatment. System 4 comprised biological treatment only, namely activated sludge treatment. The last system was a bench-scale operation including activated carbon treatment. The physical–chemical treatment systems studied included lime precipitation.

About 85% of the solid wastes were of municipal origin. The remainder was industrial and commercial. This landfill permits disposal of sewage sludge and some industrial liquid wastes. The discharge of the treated effluent directly to the Delaware River occurs during the months of December through April. During the remainder of the year the effluent is returned to the landfill, which has ample storage capacity in the pore space, so that storage does not create any difficulties. The effluent was spread on the landfill using aeration nozzles. As the age of the landfill increased, the quality also significantly varied.

The full-scale treatment plant consisted of a flow equalization lagoon, chemical precipitation using lime in an upflow solids contact reactor–clarifier, and an ammonia stripping lagoon (at an elevated pH of 10–11), nutrient (H_3PO_4) addition and acid (H_2SO_4) addition (to decrease pH), activated sludge treatment and chlorine disinfection. The flow equalization (aerated) lagoon, of 950 m^3 volume, was built to dampen the qualitative and quantitative fluctuations. The detention time during the design flow of 380 liter/min was 1·74 days. The lagoon was lined with a chlorinated polyethylene liner.

During the next stage (physical–chemical treatment) lime slurry was added into the upflow reactor–clarifier to raise pH to about 10 to precipitate metals such as Cd, Cr, Cu, Fe, Pb, Hg, Ni, K and Zn. The concentration of Amm . N in the clarified effluent was too high (average 890 mg/liter as N) to be removed by the subsequent activated sludge treatment mode. Therefore an ammonia stripping lagoon was included.

The pH of the lime treatment effluent was around 8·5–10, which was too low to strip off the ammonium. The higher the pH of the liquid, the higher the proportion of ammonia gas, rendering the air-stripping process more efficient. Therefore NaOH is occasionally added to the stripping lagoon (of capacity 950 m^3), especially during cold weather, to elevate the pH to about 11 and force off the NH_3 gas. Before the construction of the air-stripping lagoon, on average, the lime treatment step (system A) decreased Amm . N by about 24% (from 1167 to 890 mg/liter). However, after the completion of the lagoon, the Amm . N reduction by the combined lime treatment and ammonia stripping lagoon (system B) was increased to about 48% (from 785 to 412 mg/liter). The other benefits of the lagoon are summarized by the authors. The principle behind ammonia gas removal from the stripping lagoon is described elsewhere in this chapter.

The activated sludge treatment consists of two aeration tanks and two secondary clarifiers, which can be operated either in parallel or in series. Early attempts to develop an activated sludge culture were unsuccessful due to phosphorus limitation and ammonia toxicity (the feed to the aeration tank had an NH_3-N concentration between 149 and 423 mg/liter for 95% of the time). The final criterion for H_3PO_4 addition was to provide just enough to satisfy the biological demand and maintain an effluent concentration of 1 mg/liter (as PO_4). The operational and start-up problems as well as the solutions and cost data are given by the authors. During the periods of satisfactory operation the food to microorganisms ratio was between 0·12 and 0·32/day. The effluent BOD_5 concentration was low (<76 mg/liter) except during cold weather operation (temperature <10°C). The effluent quality criteria were NH_3-N < 35 mg/liter, BOD_5 < 100 mg/liter, Cd < 0·02 mg/liter and Zn < 0·6 mg/liter.

Overall the complete treatment sequence (physical–chemical followed by activated sludge) produced an effluent with the following characteristics. Organic matter was reduced to 153 mg BOD_5/liter, a 99% removal. The corresponding COD removal was 95%. The effluent BOD to COD ratio was 0·16. The effluent ammonia concentration was 75 mg/liter, which represents 90% removal. During warm weather operation active nitrification reduced the ammonia to less than 10 mg/liter.

4 CONCLUDING REMARKS

Sound scientific knowledge, engineering principles, intensive laboratory-model studies of all combinations of available biodegradation technologies and pilot-scale investigations are essential for successful and cost-effective design and maintenance of plants for biodegradation of organics in landfill leachates. Reluctance of some authorities in financing a well-equipped, long-term laboratory-model study could result in poor design and operational problems.

REFERENCES

Atwater, J. W. (1980). Impact on landfills. In *Fraser River Estuary Study, Water Quality* series. Prepared for the Fraser River Estuary Study Steering Committee, Environmental Protection Service, Environment Canada, Ottawa, Canada.

Atwater, J. W. & Mavinic, D. S. (1986). *Characterization and treatment of leachate from a west coast landfill.* Report prepared for Waste Management Branch (Ottawa) and Wastewater Technology Centre (Burlington), Environmental Protection Service, Environment Canada, Ottawa, Canada.

Bull, P. S., Evans, J. V., Wechsler, R. M. & Cleland, K. J. (1983). *Wat. Res.*, **17**, 1473.

Chian, E. S. K. & DeWalle, F. B. (1977*a*). *Evaluation of leachate treatment. Volume I: Characterization of leachate.* Report No. EPA-600/2-77-186a, prepared for Municipal Environmental Research Laboratory, Office of Research and Development, USEPA, Cincinnati, OH.

Chian, E. S. K. & DeWalle, F. B. (1977*b*). *Wat. Res.*, **11**, 295.

Chian, E. S. K. & DeWalle, F. B. (1977*c*). *Evaluation of leachate treatment. Volume II: Biological and physical–chemical processes.* Report No. EPA-600/2-77-186b, prepared for Municipal Environmental Research Laboratory, Office of Research and Development, USEPA, Cincinnati, OH.

DeWalle, F. B., Chian, E. S. K. & Brush, J. (1979). *J. Wat. Poll. Control Fed.*, **51**, 22.
Forgie, D. J. L. (1988). *Wat. Poll. Res. J. of Canada*, **23**, 308.
Henry, J. G. (1985). In *Proceedings of International Conference on New Directions and Research in Waste Treatment and Residuals Management*, University of British Columbia, Vancouver, BC, Canada, **1**, 139.
Henry, J. G., Prasad, D. & Young, H. (1987). *Wat. Res.*, **21**, 1395.
Ho, S., Boyle, W. C. & Ham, K. (1974). *J. Wat. Poll. Control Fed.*, **46**, 1776.
Keenan, J. D., Steiner, R. L. & Fungaroli, A. A. (1983). *J. Env. Eng. Div., Proceedings of the Amer. Soc. of Civil Engineers*, **109**, 1371.
Keenan, J. D., Steiner, R. L. & Fungaroli, A. A. (1984). *J. Wat. Poll. Control*, **56**, 27.
Knox, K. (1985). *Wat. Res.*, **19**, 895.
Levenspiel, O. (1965). *Chemical Reaction Engineering*. John Wiley, New York, p. 274.
Maris, P. J., Harrington, D. W. & Mosey, F. E. (1985). In *Proceedings of International Conference on New Directions and Research in Waste Treatment and Residuals Management*, University of British Columbia, Vancouver, BC, Canada, **1**, 280.
McCarty, P. L. (1985). In *Proceedings of the Seminar/Workshop on Anaerobic Treatment of Sewage*, University of Massachusetts, Amherst, MA, p. 3.
Millot, N., Garnet, C., Wicker, A., Faup, G. M. & Navarro, A. (1987). *Wat. Res.*, **21**, 709.
Robinson, H. D. (1985). In *Proceedings of International Conference on New Directions and Research in Waste Treatment and Residuals Management*, University of British Columbia, Vancouver, BC, Canada, **1**, 166.
Robinson, H. D. & Lucas, J. L. (1985). In *Proceedings of International Conference on New Directions and Research in Waste Treatment and Residuals Management*, University of British Columbia, Vancouver, BC, Canada, **1**, 31.
Robinson, H. D. & Maris, P. J. (1983). *Wat. Res.*, **17**, 1537.
Robinson, H. D. & Maris, P. J. (1985). *J. Wat. Poll. Control.*, **57**, 30.
SMC-Martin, Inc. (1981). *Evaluation of a leachate collection and treatment facility in Enfield, Connecticut*. Contract No. 68-01-4438, report submitted to USEPA, Office of Water and Hazardous Materials, Washington, DC.
Stegmann, R. (1980). In *Proceedings of Leachate Management Seminar*, University of Toronto, Ontario, November, p. 63.
Steiner, R. L., Keenan, J. E. & Fungaroli, A. A. (1977). *Demonstration of a leachate treatment plant—Interim report*. Report prepared for USEPA, Office of Solid Waste, Washington, DC.
Thirumurthi, D. (1969*a*). *Amer. Soc. Civil Eng. J. of the San. Eng. Division*, **95**(SA2), 311.
Thirumurthi, D. (1969*b*). *J. Wat. Poll. Control Federation*, **41**(2), R405.
Thirumurthi, D. (1974). *J. Wat. Poll. Control Federation*, **46**, 2094.
Thirumurthi, D. (1979). *Amer. Soc. Civil Eng. J. of the Env. Eng. Div.*, **105**, 135.
Thirumurthi, D. & Groskopf, G. R. (1988). *Can. J. Civ. Eng.*, **15**, 334.
Thirumurthi, D., Austin, T. P., Ramalingaiah, R. & Khakhria, S. (1986*a*). *Wat. Poll. Res. J. Canada*, **21**, 8.
Thirumurthi, D., Rana, S. M. & Austin, T. P. (1986*b*). In *Proceedings of the 1986 International Conference on Innovative Biological Treatment of Toxic Wastewaters*, Crystal City, Arlington, VA, June, p. 158.
Wehner, J. F. & Wilhelm, R. H. (1956). *Chemical Engineering Science*, **6**, 89.
Wright, P. J., Austin, T. P., Kennedy, K. & Robson, D. R. (1985). In *Proceedings of International Conference on New Directions and Research in Waste Treatment and Residuals Management*, University of British Columbia, Vancouver, BC, Canada, **1**, 262.
Wu, Y. C., Kennedy, J. C. & Smith, E. D. (1982). Paper presented at *First International Conference on Fixed Film Biological Processes*, Kings Island, OH, p. 1495.

Chapter 10

BIODEGRADATION IN WASTE STABILIZATION PONDS (FACULTATIVE LAGOONS)

D. THIRUMURTHI

Department of Civil Engineering, Technical University of Nova Scotia, Halifax, Nova Scotia, Canada

CONTENTS

1 INTRODUCTION

Waste stabilization ponds, also known as sewage lagoons or facultative ponds, constitute the least expensive method of biodegradation of organics in wastewater, provided, however, inexpensive land is available in the vicinity where the wastewater is generated. Although these lagoons are simple to operate, the biology and biochemistry involved are the most complex of all the engineered biodegradation systems known to man. This is the only method of wastewater biodegradation in which aerobic as well as anaerobic metabolisms occur, in addition to photosynthesis and sedimentation. All these four major processes occur simultaneously, of course, in different zones of a pond.

Of all the treatment options utilized by engineers, this is one of the options closest to nature's way of assimilation of organic wastes and is one of the most energy-efficient. This chapter highlights the scientific principles of biology/biochemistry of waste stabilization ponds and the engineering design aspects as well as the interrelationships between the two (science and engineering).

Waste stabilization lagoons are the most common systems of water pollution control in rural areas, for isolated industries and Third World countries because they are simple to design, operate and maintain.

The first recorded construction of a stabilization pond in the United States was at San Antonio, Texas, in 1901 (USEPA, 1983). As of 1975 there were approximately 7000 such ponds in operation in the United States. In Canada, there were about 868 ponds in use in 1981 (Government of Canada, 1981). Their popularity is due to simplicity of operation, low cost, and energy efficiency.

1.1 Nomenclature

The terms 'pond' and 'lagoon' are normally used synonymously in the research literature as well as in engineering usage. In this chapter the two words are used interchangeably. Smith and Finch (1983), however, do differentiate between the terms 'lagoon' and 'pond' in that the former uses a submerged outlet and frequently a submerged inlet, and the latter uses an outlet from the surface as in a lake.

1.1.1 General Classifications

Waste treatment lagoons are classified based on the mode of biodegradation (aerobic or anaerobic), presence or absence of aeration equipment, and other design features. Dissolved oxygen (DO) in upper layers supports aerobic metabolism, and its absence in lower layers encourages the anaerobic metabolic path. In the middle layers, if the DO level fluctuates from zero to supersaturation concentrations, facultative microbes respond according to the DO level, and choose between the aerobic and anaerobic metabolic paths. In most sewage lagoons DO levels vary from day to day, month to month, and location to location. As a result, a complex microbial system is continuously developing and changing. Therefore, when an aerobic top zone predominates in a lagoon, it is called an aerobic lagoon. In a deep pond, the anaerobic bottom zone would constitute the dominant portion, and very few centimeters of top layers may occasionally have DO and sustain aerobiosis. When a system is designed to have DO in most layers during most of the year it is called an 'aerobic lagoon'. A lagoon in which very little or no DO is expected to be present during most days and at most locations is named an 'anaerobic lagoon'. The following paragraphs review the various classifications commonly employed.

1.1.1.1 AEROBIC PONDS

Shallow systems (depth less than about 50 cm) can be operated with predominant aerobic layers in warmer climates. These are, however, not feasible in cold climates where the liquid can freeze throughout the full depth of the lagoon during the winter months. Settled sludge should be removed frequently (once in about 2–4 years) to minimize the anaerobic lower layers. *These ponds are not common* because they

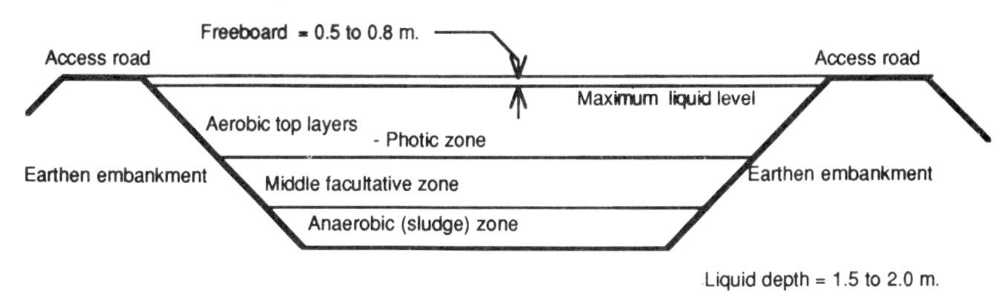

Fig. 1. A typical waste stabilization pond (not to scale).

require more space than the other types, and it is not practically feasible to maintain DO at all levels throughout the year.

1.1.1.2 FACULTATIVE LAGOONS

These units, also known as waste stabilization ponds (WSP), are deeper than the aerobic ponds. The liquid depth could vary from about 1·5 to 2 m, depending on the geographic location, climate and depth required for settled sludge storage. A 'freeboard' of about 0·5–0·8 m is recommended as shown in Fig. 1. Top layers (about 30 cm) are almost always aerobic, the lower sludge layers are consistently anaerobic and the dominant middle zone (from approximately 30 to 150 cm) is characterized by the facultative metabolism. Facultative microbes biodegrade the organics in wastewater aerobically whenever DO is present and anaerobically otherwise. Algae, of course, exist up to a depth where sunlight can penetrate. Facultative lagoons are more common than the aerobic or anaerobic ponds. *This chapter deals with such facultative lagoons.* The primary source of oxygen is algal photosynthesis. The secondary source is atmospheric air from which oxygen is transferred by the action of wind. The design is based on the first source.

1.1.1.3 ANAEROBIC LAGOONS

A deep holding unit (depth as much as 6 m), in which the bottom anaerobic layers are the dominant feature, is normally called an anaerobic lagoon. Top aerobic layers (less than about 50 cm) may have DO depending on wind, temperature and rate of organic load. In general, the top aerobic zone is either absent or insignificant in influencing the microbial dynamics of the aquatic environment.

1.1.1.4 HIGH-RATE ALGAL PONDS (HRAP)

Continuous mixing is required in these ponds where algal photosynthesis is predominant. These are similar to the shallow aerobic lagoons except that the degree of maintenance is higher. Properly engineered mixing systems should be included in the design of an HRAP. In some cases, the liquid is recycled to enhance efficiency. The first installation, with paddle wheel mixing, was in Manila, the Philippines (Oswald, 1988). Developed at the University of California (Oswald *et al.*, 1957*a*,*b*), these are not feasible in cold climates such as northern Canada and Europe because of the winter freezing problems.

1.1.1.5 AERATED LAGOONS

Artificial aeration, achieved by either mechanical aerators or diffused aeration devices, is used in these units. The rate of biodegradation is much higher than in the unaerated lagoons described in subsections 1.1.1.1–1.1.1.3. Unlike in HRAPs described in the preceding paragraph, algae are either absent or insignificant in these systems. Insignificant amounts of oxygen produced by algae are ignored in the design.

1.2 Theory of Waste Biodegradation

Waste organic matter, when discharged into rivers and lakes, can deplete the DO which is an essential resource to support the life cycles of fish and other desirable species of the natural environment. Biodegradable organic matter in municipal waste water and other waste liquids can be more efficiently decomposed by bacteria and other microbes within an engineered system than in the natural environment. The unstable organics are converted to relatively less offensive but more stable inorganics. While the biochemical reactions are almost similar, the natural environment decomposes the organic matter in an uncontrolled way. In a properly engineered system such as a waste stabilization pond, however, the rate of biodegradation can be controlled in an efficient manner.

A typical biological reaction which takes place in an aerobic environment may be simplified in the form of the following equation:

$$\text{organics} + O_2 + \text{microbes} \rightarrow CO_2 + H_2O + \text{energy} + \text{more microbes} \qquad (1)$$

The aerobic microbes shown in the left-hand side of eqn (1) are normally present in the waste liquid as well as in the receiving water. Under optimum or near-optimum environmental conditions (adequate DO, pH, temperature, nutritional balance and absence of toxic chemicals), the microorganisms grow and multiply, and produce the relatively less harmful inorganic end products.

In a typical anaerobic environment, on the other hand, two distinct reactions occur. Hydrolysis of organic compounds and conversion to intermediate organic acids are achieved by acid-forming bacteria in stage 1, as per eqn (2):

$$\text{organics} + \text{anaerobic microbes} \rightarrow \text{intermediate organic acids} + CO_2 + H_2O + \text{energy} + \text{more anaerobes} \qquad (2)$$

During stage 2, methane-forming microbes decompose the acids as shown in the eqn (3):

$$\text{organic acid intermediates} + \text{anaerobes} \rightarrow CH_4 + CO_2 + \text{energy} \qquad (3)$$

While eqns (1)–(3) represent simplified versions of the reactions, the following three equations represent the detailed versions of the biochemistry (Tchobanoglous, 1979).

Oxidation (dissimilatory process):

$$\text{COHNS (organic matter)} + O_2 + \text{bacteria} \rightarrow CO_2 + NH_3 + \text{other end products} + \text{energy} \qquad (4)$$

Microbial cells contain about 50% carbon, 20% oxygen, 10–15% nitrogen, 8–10%

hydrogen, 1–3% phosphorus, and 0·5–1·5% sulfur on a dry weight basis. Many trace metals are also present. A commonly used chemical expression for cell material is $C_5H_7NO_2$.

Synthesis (assimilatory process):

$$COHNS + O_2 + \text{bacteria} + \text{energy} \rightarrow C_5H_7NO_2 \text{ (new bacteria)} \quad (5)$$

Endogenous respiration (autoxidation):

$$C_5H_7NO_2 + 5O_2 \rightarrow 5CO_2 + NH_3 + 2H_2O + \text{energy} \quad (6)$$

1.3 Algal Photosynthesis

Photosynthesis is a biological process whereby organisms are able to grow and multiply by using the sun's radiant energy to power the fixation of atmospheric CO_2 and subsequently provide the reducing power to convert the CO_2 to organic compounds. Photosynthesis is usually associated with the green plants; however, certain bacteria as well as algae carry out photosynthesis. In wastewater ponds, the photosynthetic organisms of interest are the algae, cyanobacteria (blue-green algae), and the purple sulfur bacteria (USEPA, 1983).

Photosynthesis may be classified as oxygenic or anoxygenic depending on the source of reducing power used by a particular organism. In oxygenic photosynthesis, water serves as the source of reducing power, with oxygen produced as a by-product. The equation representing oxygenic photosynthesis is:

$$2H_2O \text{ (in presence of sunlight)} \rightarrow O_2 + 4H^+ + 4e^- \quad (7)$$

Oxygenic photosynthesis occurs in green plants, algae, and cyanobacteria. In ponds, the oxygenic photosynthetic algae and cyanobacteria convert carbon dioxide to organic compounds. These compounds and organics present in the incoming waste matter serve as a source of chemical energy for most other living organisms. The by-product oxygen produced in oxygenic photosynthesis allows the aerobic bacteria to aerobically degrade organic waste material.

Anoxygenic photosynthesis produces no oxygen as a by-product, thus occurring in the complete absence of oxygen. The bacteria involved in anoxygenic photosynthesis are mostly anaerobes, with reducing power supplied by reduced inorganic compounds (USEPA, 1983). Many anoxygenic photosynthesis bacteria use reduced sulfur compounds as well as elemental sulfur in anoxygenic photosynthesis:

$$H_2S \rightarrow S_0 + 2H^+ + 2e^- \quad (8)$$

The algal activity in a lagoon system, and the aerobic metabolism of heterotrophic bacteria, can be represented by the following two equations (Smith & Finch, 1983:

$$5CO_2 + 3H_2O + NH_3 \rightarrow C_5H_9O_{2 \cdot 5}N + 5{\cdot}25O_2 \quad (9)$$

$$6(CH_2O)_x + 5O_2 \rightarrow (CH_2O)_x + 5CO_2 + 5H_2O \quad (10)$$

1.4 Symbiotic Relationship

The word symbiosis means living together, in intimate association, of two dissimilar organisms in a mutually beneficial relationship. Aerobic microbes (especially bacteria) and photosynthetic algae exist and multiply in upper layers of a waste stabilization pond's photic zone (Fig. 1): the depth to which sunlight can penetrate is called the photic zone. Algae utilize carbon dioxide from air and wastewater, and produce 'photosynthetic' oxygen. Aerobic bacteria depend on this oxygen to metabolize organics in wastewater and produce CO_2 which in turn is used up by algae. This interdependent relationship is called symbiotic, an adjective of the noun symbiosis.

1.5 Rate of Organic Biodegradation

The rate at which organic matter is decomposed by microbes is normally assumed to be first-order. The concentration of organic substrate (S) decreases with time (t) and the rate of decrease is proportional to the first power of S:

$$-\mathrm{d}S/\mathrm{d}t = KS \tag{11}$$

The constant of proportionality in eqn (11), K, is called the first-order BOD removal coefficient or reaction rate constant, with a unit of reciprocal time. Integrating the differential equation between initial time ($t = 0$) at which the concentration of organic matter was S_0 and final time (t) at which the concentration decreases to S_t, the following relationship can be derived:

$$S_t = S_0 \mathrm{e}^{-Kt} \tag{12}$$

A semi-logarithmic plot of S and t (with ln S on the y-axis) would result in a straight line of slope K. S is usually expressed as 5-day BOD in milligrams of oxygen required to stabilize organics in 1 liter of wastewater. Although the rate of decrease of S is most commonly assumed to be proportional to S, a laboratory-model study (Thirumurthi, 1979) showed that it could be proportional to $S^{1\cdot 1}$. For simplicity as well as ease of mathematical interpretation, almost all designers and researchers assume the kinetics shown in eqns (11) and (12).

The value of K depends, among other things, on temperature (T) of the liquid in which the reaction is taking place. The value of K (K_{20}) is normally expressed at the standard temperature of 20°C at which the BOD tests are conducted in the laboratory. K at any temperature (K_T) can be estimated as follows:

$$K_T = K_{20}\theta^{T-20} \tag{13}$$

The temperature correction factor (θ) for lagoons and ponds is about 1·02 to 1·06, with a recommended usual value of 1·036 (Thirumurthi, 1974). The relationship between K_T and K_{20} delineated in eqn (13) is valid for a temperature range of about 1 to 35°C. At temperatures below this range very little or no biodegradation takes place. Liquid temperatures above this range will be unlikely even in tropical lagoons.

In an optimum or ideal environment (near-neutral pH range of 6 to about 8, high temperatures, presence of adequate nutrients and absence of toxic chemicals), the magnitude of K would be the highest. As the environmental factors deviate more and more from the optimum, the value of K will decrease further and further, slowing the rate of microbial action and the BOD removal efficiency of the lagoon.

2 HYDRAULIC FLOW PATTERNS

The performance of a waste stabilization pond depends not only on the biological/environmental factors which govern K described in the previous section, but also on the pond geometry (shape, size, depth), hydraulic retention time (HRT) and the flow pattern that occurs between the inlet and outlet.

Chemical engineers define the flow patterns as ideal plug flow, dispersed plug flow, completely mixed flow (in a completely stirred tank reactor, CSTR) and partially mixed flow in chemical reactors. Thirumurthi (1969*a*, 1974, 1979) defined the biological wastewater treatment systems as biochemical reactors of first-order BOD removal kinetics and advocated the concept of using the chemical reactor design techniques in environmental engineering. A waste stabilization pond can be designed by understanding the following flow patterns.

2.1 Ideal Plug Flow

In a long rectangular pond liquid travels gently from inlet to outlet without lateral dispersion. Liquid particles travel in parallel paths with an ideal plug flow or piston flow, as shown in Fig. 2, between the sections AA and BB. Ideal plug flow is a theoretical concept which seldom occurs in a real pond.

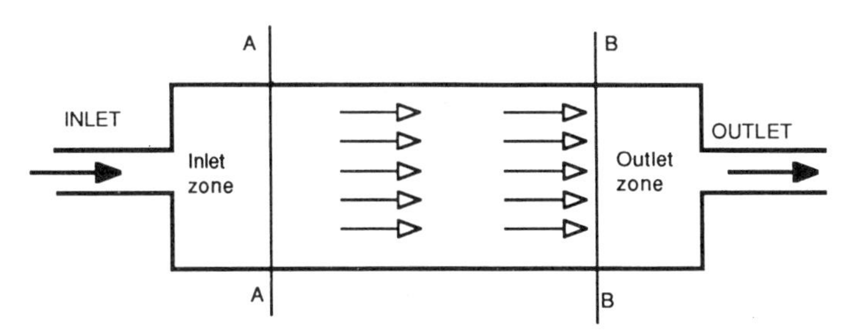

Fig. 2. Ideal plug flow.

2.2 Dispersed Plug Flow

In a long rectangular lagoon a dispersed plug flow not too different from the ideal flow can take place, as shown in Fig. 3. Moderate lateral dispersion is observed between the inlet and outlet zones.

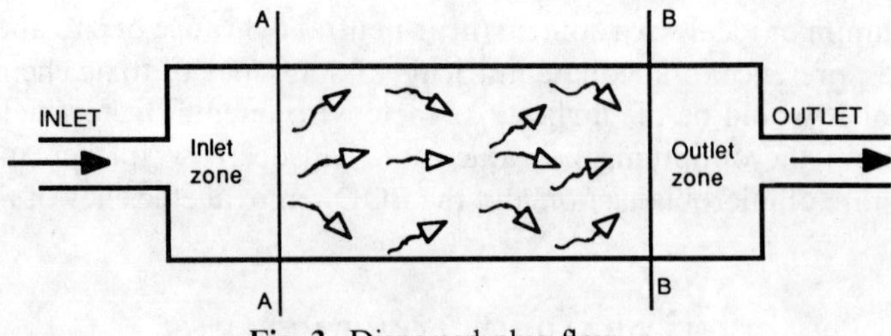

Fig. 3. Dispersed plug flow.

2.3 Tracer Study

Tracer tests conducted in laboratory-model tanks or in the field lagoons can delineate the flow patterns. When a small mass of a tracer (fluorescent dye) is added at initial time ($t = 0$) at the inlet end, it travels toward the outlet. The concentrations of the tracer in the effluent, monitored at various intervals of time, can be plotted as shown in Fig. 4. Unless caution is exercised in maintaining constant temperature and flow rate in the laboratory studies (Thirumurthi, 1969*b*), dependable results cannot be obtained. The curve which depicts the relationship between the effluent tracer concentration and time is called a C-curve.

For an ideal plug flow the C-curve has a very narrow base because most of the tracer, undispersed, exits during a short interval of time. For a dispersed plug flow the curve has a wider base. As the degree of lateral dispersion increases so does the base width. Levenspiel (1965) has developed a technique to quantify the extent of

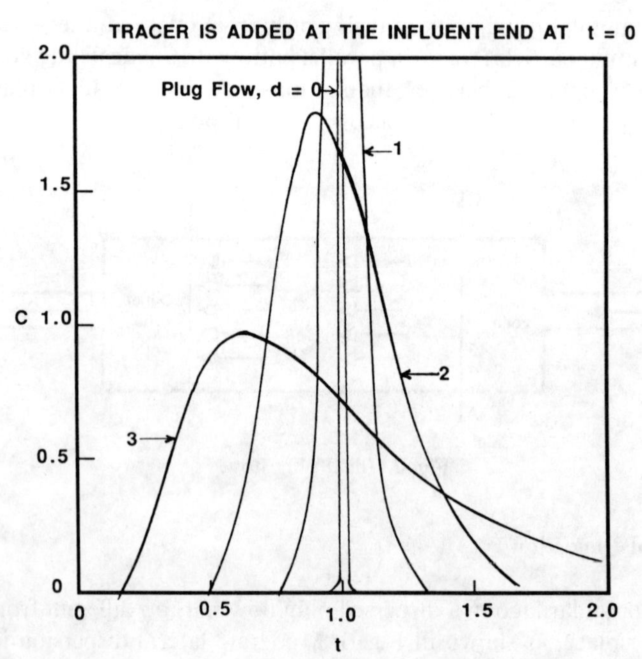

Fig. 4. Typical C-curves for plug flow and dispersed plug flow patterns.

dispersion by introducing a concept of dispersion index (d) which can be defined as follows:

$$d = D/UL \tag{14}$$

where D = axial dispersion coefficient (m^2/h), U = mean flow velocity (m/h) and L = length of liquid travel path (m). For an ideal plug flow, $d = 0$, and for a moderately dispersed plug flow, $d = 0$ to about 0·25. The dispersion index is estimated from the data of the C-curve (Thirumurthi, 1969*b*).

3 KINETICS AND DESIGN

The kinetics of biodegradation and its relationship to design of waste stabilization ponds are highlighted in this section. In an ideal plug flow reactor, tank or lagoon the BOD removal kinetics can be written as:

$$S_e = S_i \, e^{-Kt} \tag{15}$$

where S_i is influent BOD, S_e is effluent BOD, t is theoretical mean hydraulic retention time (HRT) and K is the BOD removal coefficient at liquid temperature. HRT (t) is estimated by dividing the lagoon liquid volume by the average flow rate of wastewater. Equation (15) is applicable when the dispersion index (d) is zero. Therefore, applying eqn (15) to a lagoon is an approximation because no lagoon can represent an ideal plug flow.

Waste stabilization ponds can be more realistically designed by employing the Wehner and Wilhelm (1956) equation which is derived for all hydraulic flow patterns characterized by

$$\infty > d > 0 \tag{16}$$

Thirumurthi (1969*a*, 1974) showed that facultative lagoons are neither precise plug flow reactors nor completely mixed systems. Therefore, either the following Wehner and Wilhelm equation (Levenspiel, 1965) or the Thirumurthi chart (Fig. 1 of Chapter 9) should be used to precisely design a new pond or evaluate an existing pond:

$$\frac{S_e}{S_i} = \frac{4a \exp(1/2d)}{(1+a)^2 \exp(a/2d) - (1-a)^2 \exp(-a/2d)} \tag{17}$$

where $a^2 = 1 + Ktd$. Design engineers who consider eqn (17) too complex may use the following approximate equation which is derived by neglecting the second term in the denominator:

$$\frac{S_e}{S_i} = \frac{4a \exp[(1-a)/2d]}{(1+a)^2} \tag{18}$$

In summary, a waste stabilization pond can be designed approximately by using the ideal plug flow formula, eqn (15), or accurately by eqn (17), or with a moderate degree of approximation with eqn (18). As an example, for given values of $d = 1$,

$K = 0{\cdot}1$/day, and $t = 10$ days, eqn (17) would result in BOD remaining $(C_e/C_i) = 0{\cdot}468$ and eqn (18) in $C_e/C_i = 0{\cdot}458$. The difference is insignificant. The values of K and d are estimated either from monitoring an existing pond or a laboratory-model study or from a literature survey (Thirumurthi, 1974; Mara & Monte, 1987).

3.1 Design Parameters

Successful design and operation of a pond depend on proper understanding of the following parameters normally used:

Theoretical mean hydraulic retention time (HRT) $= t$
= total liquid volume (V) of the pond ÷ average flow rate (Q) of wastewater entering the pond

Actual mean HRT $=$ HRT_A
= effective or useful liquid volume (V_E) of the lagoon ÷ average flow rate (Q)

V_E is usually less than V due to short-circuiting in lagoons. Allan and Jeffreys (1987) reported that about 40% of the volume of a lagoon in Whitehorse, Yukon, Canada, was unused due to short-circuiting. Tracer tests conducted in the Whitehorse lagoon showed that HRT_A was about 60% of t.

Organic load (L_0) entering the lagoon (kg BOD_5/day) = Q (in m^3/day) × influent BOD_5 concentration (in mg/liter of g/m^3)/1000

Surface area organic load (L_A) entering the pond (kg BOD_5/ha.day) = L_0/surface area (A) of the pond (in hectares)

Volumetric organic load (L_V) entering (kg BOD_5/m^3.day) $= L_0/V$

Photosynthetically available radiation (PAR) from the sun is essential for survival and growth of algae which generate the required oxygen to maintain aerobiosis in waste stabilization ponds. The visible spectrum of solar energy (PAR) is measured by different instruments. The quantum radiometer is the most modern and accurate of them all, and it is commonly used by botanists, limnologists (Field & Effler, 1988) and horticulturists to record PAR as photosynthetic photon flux density (PPFD) in the atmosphere, growth chambers and greenhouses. The quantum radiometer measures PAR in units of Einsteins per unit time and area. PAR (waveband 400–700 nm) can be measured either in W/m^2 (which is equivalent to $4{\cdot}6\ \mu mol/m^2$ s for daylight and white fluorescent lamps), or Einsteins/m^2 h (1 Einstein = 1 μmol = $6{\cdot}02 \times 10^{17}$ photons).

Thirumurthi and Andavan (1990) are the first known investigators who used the quantum radiometer for research on waste stabilization ponds in a laboratory-model study which simulated the field conditions in Whitehorse in the Yukon

Territory of Canada. PAR was also recorded in Whitehorse to estimate the duration of available energy at ambient field conditions.

3.2 Design Methods

Several techniques are available for designing the waste stabilization ponds. Some are empirical, and others range from mathematical models to semi-empirical. The empirical methods are based on experience from the existing lagoons. The mathematical models are based on sound scientific principles and interpretation of laboratory-model studies. The semi-empirical techniques involve a combination of the two. As stated by Arceivala *et al.* (1970), 'In spite of its apparent simplicity, the waste stabilization pond is a complex biochemical reactor which defies precise design.' Therefore, applying several methods to meet the given environmental conditions could result in different design parameters.

3.2.1 The Empirical Methods

3.2.1.1 McGARRY AND PESCOD FORMULAE

Based on unfiltered influent sewage and filtered effluents of 134 ponds throughout the world, the following relationships were derived by McGarry and Pescod (1970):

$$L_{max} = 11{\cdot}2(1{\cdot}054)^T \qquad (19)$$

$$L_{DES} = 7{\cdot}5(1{\cdot}054)^T \qquad (20)$$

L stands for volumetric organic load (kg BOD/ha.day), and the subscripts max and DES stand for, respectively, the maximum load that can be applied to a pond before it fails (i.e. becomes fully anaerobic) and design or permissible load; T = liquid temperature in °C.

3.2.1.2 DESIGN GUIDELINES IN THE US

Based on a questionnaire sent to 50 states, Canter and Englande (1970) estimated the mean values used in the warmer southern states as follows: HRT = t = 31 days, L_A = 50 kg/ha.day = approximately 44 lb/acre.day. The corresponding values for the other states are shown in Table 1. The average recommended liquid depth in the

Table 1
The Questionnaire Results of the US Survey

Variables	*Value given in region*		
	North	*Central*	*South*
Number of states	18	17	15
Surface area organic load (L_A) in kg of BOD_5/ha.day (mean)	29	37	50
HRT (t) in days	117	82	31

Source: adapted with permission from Canter and Englande, 1970.
Note: 1 kg/ha.day = 1·12 lb/acre.day.

US is 1·2 m (4 ft), and the freeboard is 0·9 m (3 ft). The majority of the states recommend the use of multiple ponds to enable operational and maintenance flexibility. Hydraulic arrangements of multiple ponds to facilitate either series or parallel operation are desirable (Canter & Englande, 1970). In other words, a given system of lagoons should have provision to be operated either in series or in parallel (Environment Canada, 1987). The distinctions among the three regions shown in Table 1 were made on the following basis: the ponds in the north have extended periods of ice cover in winter, those in the central region have only a short period of ice cover, whereas the lagoons in the south do not have appreciable ice cover during the winter (Benefield & Randall, 1980).

3.2.1.3 ALBERTA GUIDELINES

The province of Alberta, in Canada, employs a combination of multi-cell anaerobic–facultative storage lagoons which discharge either once a year or twice a year. In 1983, out of a total of 360 communities which employ some form of municipal wastewater treatment system, 293 were using lagoon treatment (Beier, 1983). *The Alberta guidelines are recommended for use in cold climates, but not in temperate or warmer climates.* Alaska and certain other northern US states and northern regions of Canada may employ similar design guidelines. Based on extensive monitoring experiences, it was concluded that of five different lagoon configuration groups studied, the system with a four-cell anaerobic lagoon followed by two or more facultative ponds generally produced the best effluent quality. The design is based on retention time in the storage/treatment lagoons. The systems which provided 12 months' storage and treatment produced effluents that met the Alberta requirements. The twice-a-year discharge lagoons are usually drained in fall and spring. Fall drainage repeatedly generated better quality effluents than those associated with spring discharges for both once-a-year and twice-a-year discharge lagoons. The fall-discharge lagoons with detention periods greater than 6–7 months met the Department requirements (Milos & Beier, 1978).

3.2.1.4 INDIAN EMPIRICAL PROCEDURE

In India (Mara, 1976), based on experience of pond operation, the permissible load is related to latitude:

$$L_A = 375 - 6{\cdot}25\ \text{LAT} \tag{21}$$

where LAT = degrees of latitude (range in India: 8–36°N).

3.2.1.5 USEPA AREAL ORGANIC LOAD PROCEDURE

The value of L_A recommended by the United States Environmental Protection Agency (USEPA, 1983) depends on the average winter air temperature of the pond location. In cold climates (average winter air temperature, $T < 0°C$), the recommended range is 11–22 kg/ha.day. In areas where the average winter air temperature is 0–15°C, the range is 22–45 kg/ha.day. However, in warmer climates ($T > 15°C$), the permissible L_A value is 45–90 kg/ha.day.

3.2.2 Mathematical Models

The design formulae derived from scientific principles and interpretation of data from laboratory-model, pilot-scale and actual field lagoons are given in this section.

3.2.2.1 GLOYNA FORMULA

Laboratory-model studies conducted at the University of Texas resulted in the Gloyna equation:

$$V/Q = t = 0{\cdot}035L_u^{35-T}ff' \quad (22)$$

where L_u = ultimate influent BOD or COD (mg/liter)
T = average liquid temperature during the coldest winter month
f = algal toxicity factor (Thirumurthi & Gloyna, 1965) = 1 for municipal wastewater
f' = sulfide oxygen demand factor = 1 when sulfate in influent $<$ 500 mg/liter

Liquid depth is recommended to be 1 m for ideal conditions and warm climates (Gloyna, 1976). In cold climates the depth should be 1·5–2 m.

3.2.2.2 THIRUMURTHI MODEL

The rate of biodegradation is well known to be a first-order relationship, as shown in eqn (11). Any one of three formulae, all based on the first-order kinetics, was suggested by Thirumurthi (1969*b*, 1974) for design of waste stabilization ponds. The plug flow equation (15), the Wehner and Wilhelm equation (17), or its modified approximate equation (18) may be used. Irrespective of which formula is used, the key to the design is estimation of K_S, the 'standard' first-order BOD removal coefficient. This parameter corresponds to the following standard set of environmental and other factors: (1) pond liquid temperature of 20°C, (2) minimum visible solar energy level of 100 Langleys (gm.cal/cm^2)/day, (3) absence of industrial and other toxic chemicals in the influent, (4) absence of benthal (sludge) load in the lower layers, and (5) daily areal organic load (L_A) of 67 kg BOD/ha. When a pond meets all these five standard factors, the biodegradation can be quantified by K_S, which was estimated by Thirumurthi (1974) as 0·056/day. The range of K_S values is from 0·04 to 0·07/day.

When a pond does not meet the standard conditions, a design coefficient, K, corresponding to the actual (annual) average environmental factors should be employed with the following correction factors:

$$K = K_S C_{TE} C_o \quad (23)$$

where C is the correction factor, and the subscripts TE and o, respectively, represent the intended corrections for liquid temperature and organic load:

$$C_{TE} = 1{\cdot}036^{T-20} \quad (24)$$

where T is the expected liquid temperature during the coldest winter month, and 1·036 is the temperature correction factor.

$$C_o = 1 - (0{\cdot}083/K_S)\log_{10}(67/L_A) \quad (25)$$

For most practical purposes, employing the estimated K value in the plug flow equation is adequate. Using eqns (17) and (18) would require an estimation of the dispersion index for a pond, which may not be readily available unless expensive tracer tests are conducted in the field.

3.2.2.3 UNIVERSITY OF CALIFORNIA MODEL

This is one of the most scientific methods available for designing new lagoons or understanding the performance characteristics of existing lagoons. The only disadvantage is that it is not possible to precisely estimate the concentration of dry weight of algae, C_C, which is a key parameter. Oswald *et al.* (1957*b*) and Oswald (1988) of the University of California developed this method from basic principles and experimental evidence. The design is based on the concept of photosynthetic efficiency (F) of algae. F is a fraction which represents the ability of algae to convert solar energy (S = energy in the visible spectrum in cal/cm^2.day) to chemical energy; 3·68 cal energy is required to produce 1 mg of oxygen. The following equation relates F to S:

$$F = C_C h d' / 1000 S t \tag{26}$$

where h = heat of combustion (cal/mg) of dry weight of algae in a lagoon
d' = lagoon depth (cm)
t = mean hydraulic retention time in the lagoon

The next concept is based on the minimum or critical (photosynthetic) efficiency (F_C) required to produce adequate oxygen equal to the BOD load entering the pond so that it will not become anaerobic. In theory, when $F > F_C$, the pond can handle the BOD load. When $F = F_C$, sufficient oxygen would have been produced to stabilize the organic load entering the pond.

$$F_C = 0{\cdot}0094 L_t d'' / S t \tag{27}$$

where L_t = concentration of BOD entering the lagoon (mg/liter)
d'' = liquid depth in inches (as originally expressed by Oswald *et al.*, 1957*b*)
0·0094 = conversion factor employed to render the equation dimensionally acceptable.

The ratio F/F_C is called the oxygenation factor. In theory the oxygenation factor should be equal to 1.

The researchers applied this relationship to pilot-scale and laboratory-model lagoons and proved that when F/F_C is above 1·8 the pond is an overdesign, when it is between 1·2 and 1·8 the design is 'satisfactory', when it is from 0·8 to 1·2 the system is 'fair', from 0·4 to 0·8 is 'poor', and below 0·4 is 'fail'.

In spite of the logical approach used by the researchers, the mathematical model is not as popular as it should be, perhaps because it is not possible to precisely estimate the concentration of dry weight of algae, C_C. An approximate method, however, is available, based on the concentration of the pigment, chlorophyll 'a', in a sample.

An approximate relationship exists between the weight of algae present in the stabilization pond samples and the chlorophyll 'a' content of the algae. The conversion factor used to estimate the algae concentration (C_C) depends on a number of variables such as algal species, light conditions, and nutrient availability. For algae found in wastewater treatment stabilization ponds, the conversion factor would appear to range from about 0·03 to 0·12 with a 'middle-of-the-road' value of about 0·05 mg/liter of algae per μg/liter of chlorophyll 'a' (Stewart & Lewis, 1976). In other words, the mass of chlorophyll 'a' in a sample could range from approximately 0·8% to 3% of dry algal cells (C_C).

4 CONCLUDING REMARKS

Small communities and isolated industries which generate biodegradable liquid wastes should consider the option of treatment by facultative or other types of lagoons. A majority of ponds and lagoons serving municipalities are of the facultative type (Peavy *et al.*, 1985). Developing countries located in warmer climates may find waste stabilization ponds a suitable and economical treatment mechanism (MacLellan, 1984; MacLellan & Thirumurthi, 1987). Such lagoons/ponds are not always feasible due to lack of low-cost land within economical distance. Moreover, potential exists for groundwater pollution (lining the whole lagoon could solve this problem), and nuisances due to occasional odor and insects.

The waste stabilization pond is a complex biochemical reactor which defies precise design (Arceivala *et al.*, 1970). Therefore, application of design equations and mathematical models may not always be precise.

REFERENCES

Allan, R. B. & Jeffreys, Y. (1987). *Performance evaluation of the City of Whitehorse sewage treatment lagoons.* Environment Canada, Conservation and Protection, Environmental Protection, Yukon Branch, Regional Program Report No. 87-17. 45 pp.

Arceivala, S. J., *et al.* (1970). *Waste stabilization ponds: design, construction and operation in India.* A report by Central Public Health Engineering Research Institute, Nagpur, India. 47 pp.

Beier, A. (1983). *A report on lagoon effluent quality evaluation III. 1978/1983.* Municipal Engineering Branch, Alberta Environment, Edmonton, Alberta, Canada. 35 pp.

Benefield, L. D. & Randall, C. W. (1980). *Biological Process Design for Wastewater Treatment.* Teleprint Publishing, Charlottesville, VA.

Canter, L. W. & Englande, A. J. (1970). *J. Water Pollution Control Federation*, **42**, 1842.

Environment Canada (1987). *Cold climate sewage lagoons*, eds A. R. Townshend & H. Knoll. Report EPS 3/NR/1. 159 pp.

Field, S. D. & Effler, S. W. (1988). *Water Resources Bulletin, American Water Resources Association*, **24**(2), 325.

Gloyna, E. F. (1976). In *Advances in Water Quality Improvements: Ponds as a Waste Treatment Alternative.* Water Resources Symposium No. 9, eds E. F. Gloyna & W. Eckenfelder, University of Texas Press, Austin, TX, p. 131.

Government of Canada (1981). *National inventory of municipal waterworks and wastewater systems in Canada.* Minister of Supply and Services, Canadian Information Catalogue No. En-44-10/81, Ottawa, Canada. 50 pp.

Levenspiel, O. (1965). *Chemical Reaction Engineering*. John Wiley, New York, p. 274.
MacLellan, E. J. (1984). *A sanitation improvement development plan for Anse La Raye, St Lucia, Eastern Caribbean*. Master's Project, Technical University of Nova Scotia, Halifax, Nova Scotia, Canada. 85 pp.
MacLellan, E. J. & Thirumurthi, D. (1987). Canadian engineers' solutions to pollution problems in a Caribbean village. Paper presented at a Centennial conference, Canadian Society for Civil Engineering, Montréal, Canada.
Mara, D. (1976). *Sewage Treatment in Hot Climates*. John Wiley & Sons, London.
Mara, D. D. & Monte, M. H. M. D. (eds) (1987). *Waste Stabilization Ponds*. Proceedings of a conference held by International Association of Water Pollution Research and Control in Lisbon, Portugal. Pergamon Press, London.
McGarry, M. G. & Pescod, M. B. (1970). In *Second International Symposium on Wastewater Treatment Lagoons*, ed. R. McKinney. Kansas City, MO, p. 72.
Milos, J. P. & Beier, A. G. (1978). *Sewage lagoon effluent quality evaluation*. Municipal Engineering Branch, Pollution Control Division, Alberta Environment, Edmonton, Alberta, Canada. 37 pp.
Oswald, W. J. (1988). In *Micro-algal Biotechnology*, eds M. A. Borowitzka & L. J. Borowitzka. Cambridge University Press, New York, p. 305.
Oswald, W. J., Golueke, C. G. & Gee, H. K. (1957*a*). *Waste water reclamation through production of algae*. University of California Water Resources Center, Contribution 22.
Oswald, W. J., Gotaas, H. B., ., Golueke, C. G. & Kellen, W. R. (1957*b*). *Sewage and Industrial Waste*, **29**, 437.
Peavy, H. S., Rowe, D. R. & Tchobanoglous, G. (1985). *Environmental Engineering*. McGraw-Hill, New York.
Smith, D. W. & Finch, G. R. (1983). *A critical evaluation of the operation and performance of lagoons in cold climates*. Report submitted to Technical Services Branch, Environmental Protection Programs Directorate, Environmental Protection Service, Ottawa, Canada. 231 pp.
Stewart, M. J. & Lewis, D. H. (1976). *Water & Pollution Control*, **114**, 34.
Tchobanoglous, G. (1979). *Wastewater Engineering: Treatment, Disposal, Reuse*. McGraw-Hill, New York.
Thirumurthi, D. (1969*a*). *Jour. of San Engrg Div. Proceed. of the Amer. Soc. of Civ. Engrs*, **95**(SA2), 311.
Thirumurthi, D. (1969*b*). *Jour. Water Pollution Control Federation*, **41**, R405.
Thirumurthi, D. (1974). *Jour. Water Pollution Control Federation*, **46**(9), 2094.
Thirumurthi, D. (1979). *Amer. Soc. Civil Eng. J. of the San. Eng. Div.*, **105**, 135.
Thirumurthi, D. & Andavan, E. (1990). *Cold climate sewage lagoons for the Yukon: laboratory-model research (Stage I)*. Report to Environment Canada, The Yukon Territorial Government, and Indian and Northern Affairs Canada, Whitehorse, Yukon, Canada. 96 pp.
Thirumurthi, D. & Gloyna, E. F. (1965). *Relative toxicity of organics to* Chlorella pyrenoidosa. Report to US Public Health Service, Contract No. WP-00688-01A1, University of Texas Report No. 4, Austin, TX. 97 pp.
USEPA (1983). *Design Manual: Municipal Wastewater Stabilization Ponds*. Report No. EPA-625/1-83-015, United States Environmental Protection Agency, Washington, DC. 327 pp.
Wehner, J. F. & Wilhelm, R. H. (1956). *Chemical Engineering Science*, **6**, 89.

Chapter 11

APPLICATIONS OF ADAPTED MICROORGANISMS FOR SITE REMEDIATION OF CONTAMINATED SOIL AND GROUND WATER

RALPH J. PORTIER

Aquatic and Industrial Toxicology Laboratory, Institute for Environmental Studies, Louisiana State University, Baton Rouge, Louisiana, USA

CONTENTS

1 INTRODUCTION

Nature has the ability to recycle and purify itself, but in recent years the demand placed on the environment by huge amounts of anthropogenic pollution exceeds its capacity to recover. Bioremediation technologies simply attempt to optimize microorganisms' natural capacity to degrade/recycle by supplying essential inorganic limiting reactants and minimizing abiotic stress. Biodegradation techniques are versatile and can be utilized at various stages of treatment. Applications include removal of contaminants from raw materials prior to processing; treatment of wastes before discharge; treatment of effluent streams; and decontamination of soils, sediments, surface water and groundwater (Portier, 1985).

There are three basic ways that the above can be accomplished:

(1) *Direct release:* Bacteria or their extracellular products may be released directly into the contaminated environment.

(2) *Enhancement of indigenous microbes:* Enhancement of the indigenous population's degradative potential may avoid the aforementioned problems of predation, nutrient competition and subsequent colony inactivation. Enhancement is achieved primarily by supplementing the natural supply of nutrients at the site with additional oxygen, nitrogen, phosphorus, essential vitamins or an organic compound necessary for cometabolism.

(3) *Microbes in contained reactors:* Microorganisms may be used in contained reactors to circumvent the problem of a complex and often unfavorable natural environment. In an enclosed bioreactor parameters like pH/Eh, oxygenation, nutrient concentration, temperature and salinity can be controlled for optimal biodegradation.

The laboratory feasibility/verification methods to be presented in this chapter involve the use of such specialized biological reactors on a laboratory scale linked to chemical cosolvent and mechanical inoculation approaches so as to attack, as preferential substrates, marginally soluble or viscous organics trapped in excavated soil matrices and associated groundwaters.

1.1 Liquid/Solids Contact Reactors

Carcinogenic polycyclic aromatic hydrocarbons (CPAHs), polycyclic aromatic hydrocarbons (PAHs) and associated chlorinated dioxins and furans are known to exist in many sludges, contaminated soils and contaminated slurries of materials having significant hydrocarbon content. Liquid/solids contact reactor technology has been successfully employed in the field to bioremediate soils and sediments having hazardous waste organics content in excess of 1% total organics (Portier & Meyers, 1982). Field deployed versions of these reactors can effectively handle 100–250 m^3/day, depending on type of waste and organics content present in excavated soils. Figure 1 presents a typical LSC reactor deployment in a field situation. Combinations of direct drive mixers and aerators in an enclosed tank or impoundment receive excavated soils and groundwater on a continuous and/or semi-continuous basis from the adjacent hazardous waste impoundment. Many of these systems are deployed within the confines of an actual contaminated site so as to minimize spreading of soils to pristine areas. Figure 2 presents a schematic of a laboratory scale reactor and a summary of the LSC process.

This remediation approach encompasses a two-step treatment process: (1) a roughing cell stage in which contaminated soils are homogenated and inoculated with nutrients and biomass, and (2) a biological treatment step in which significant toxicant mineralization is accomplished. A third step can be mentioned in which the wastewater decanted from the reactor is also polished by means of further biotreatment. Residual settled soils/sediments removed following 'biological contact' are subjected to additional landfarm approaches and/or solidified, and placed into a permanent vault or disposal facility. Soil is inoculated by adapted biomass immobilized on porous diatomaceous earth supports (Celite R630, Manville Corporation). The wastewater polishing step also uses immobilized biomass as the settled sludge fraction remaining in the bottom of the LSC reactor.

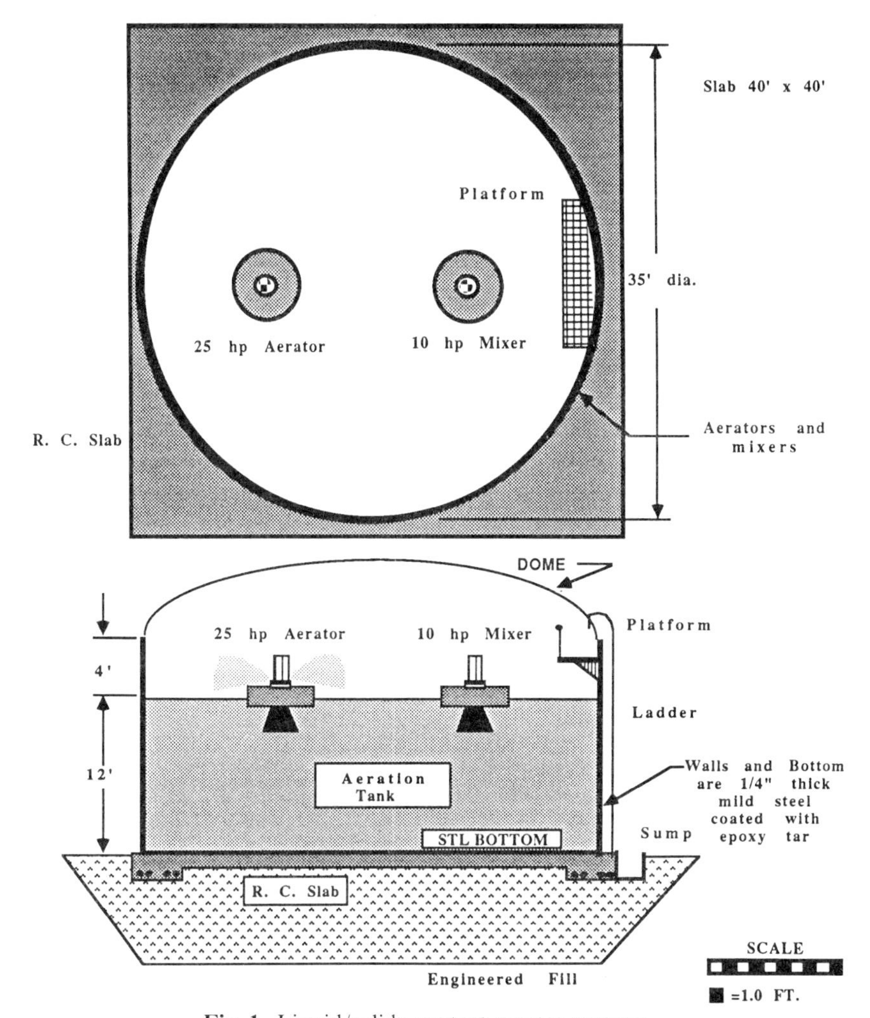

Fig. 1. Liquid/solids contact reactor systems.

Liquid/solids contact reactor technology has been successfully employed in the field to bioremediate soils and sediments having hazardous waste organics content in excess of 1% total organics. For laboratory feasibility tests and field verification studies, data are generated so as to devise a management strategy to maximize the effectiveness of the microbiological component of the process. Specific measurements have included microbial ATP (adenosine triphosphate) for the determination of microbial biomass (Portier *et al.*, 1989), parent compound disappearance, contribution of incident UV light on photolytic decomposition processes and post-treatment residual determinations of PAHs/CPAHs using gas chromatography/mass spectrometry (GC/MS) methods. Each laboratory LSC unit consists of a 2000-ml reaction vessel in which toxicants or substrates are introduced via a peristaltic pump. Temperature is maintained by a heat lamp system regulated by a

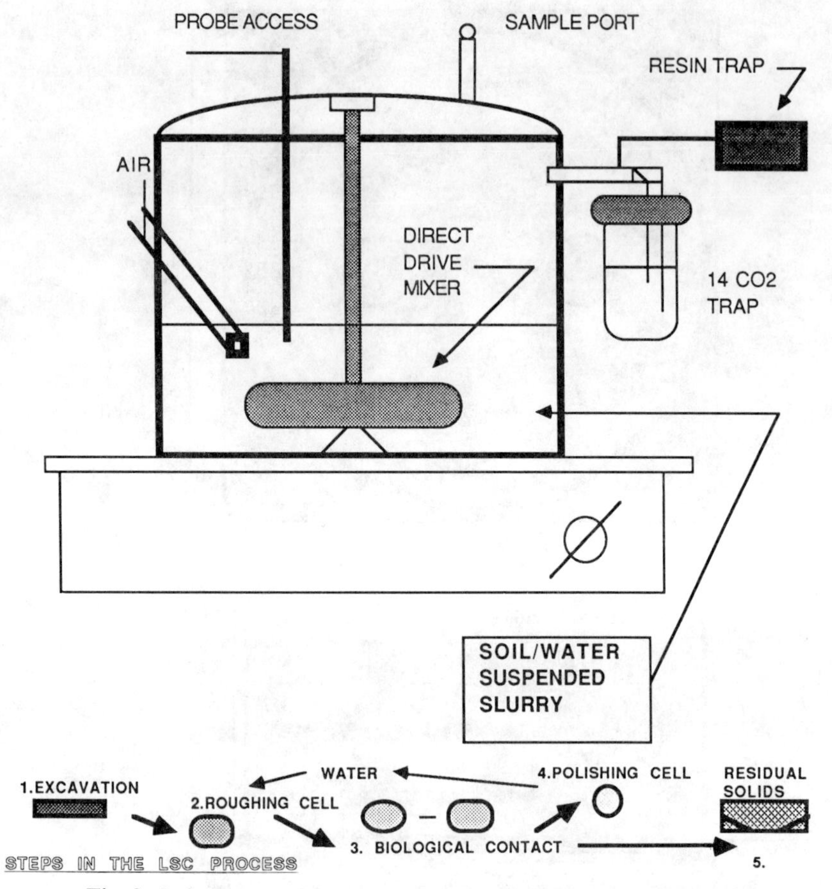

Fig. 2. Laboratory LSC reactor unit for site assessment studies.

proportional indicating temperature controller. pH/Eh of the reaction vessel is maintained by a series of controllers connected to the peristaltic pumps or gas regulators. Samples are withdrawn aseptically from the reaction vessel by means of micropipet or syringe. Samples consist of aqueous slurries grab-sampled from reactor vessels at periodic intervals. Contents of the agitated reactors are presumed to be homogeneous suspensions. However, all calculations of toxicant residuals are determined on a dry weight basis.

2 CREOSOTE AND RELATED CONTAMINANTS: A CASE STUDY

Chlorinated dioxins and the five-ring polynuclear aromatics have long been recognized as major carcinogenic substances associated with creosote- and pentachlorophenol-contaminated sludges and soils. These hazardous waste materials not only have posed a significant threat to groundwater supplies in the southern United States but also have been recognized as a major potential global source of dioxin contamination worldwide via transport mechanisms such as volatilization and particulate transport. Research has been conducted on the feasibility of biologically degrading pentachlorophenol (PCP) groundwater/rinsates and pentachlorophenol/creosote sludge materials using a contact reactor method incorporating the use of acclimated biomass in a highly aerated stirred-tank reactor system. The overall objective of this investigation has been to determine the feasibility of using liquid/solids contact reactors (LSCs) as a site remediation technology for the biotransformation and biodegradation of creosote wastes associated with most Resource Conservation and Recovery Act (RCRA) sites.

3 LIQUID/SOLIDS CONTACT STUDIES: CREOSOTE-CONTAMINATED SOILS

3.1 Experimental Design

Pentachlorophenol-contaminated creosote waste soils were suspended in LSC reactors over a 21-day period. An initial seven-day roughing step provided optimal mixing for creosote, and also provided indications of the fate of CPAHs associated with these wastes. After seven days of high energy contact, the supernatant was transferred to a polishing biological reactor cell, where additional biological treatment was again performed for a 14-day period. Over these time frames GC/MS determinations were made of the primary PAH constituents as well as the chlorinated dioxin and chlorinated furan contaminants. To identify microbial contributions to PAH degradation, sterile LSC reactor soil tests were conducted in parallel with biological treatment using the aforementioned laboratory approach. Antibiotics were used to hinder microbial growth and kinetic response. Comparisons were made between abiotic and biotic tests for targeted compound removal.

3.2 Biotransformation of Creosote Waste Constituents

Figures 3 and 4 provide information on the residual levels of key PAH constituents for all reactors for roughing cell and biological treatment. The roughing step involves the actual resuspension and solubilization of creosote and pentachlorophenol materials over a seven-day period. This key initial step forces the solubilization of the K001 constituents as a result of the addition of surfactant (Triton X100, Sigma) and pH adjustment (to 7·3), resulting in the increased availability of these materials for biological attack. The data presented show high concentrations of fluorene, phenanthrene and fluoranthrene in the initial waste loading. It is important to note that the initial concentrations varied in terms of chemical compound; however, the composite mixed waste represented a 20% loading rate based on solids for all reactors. With combined microbial addition and

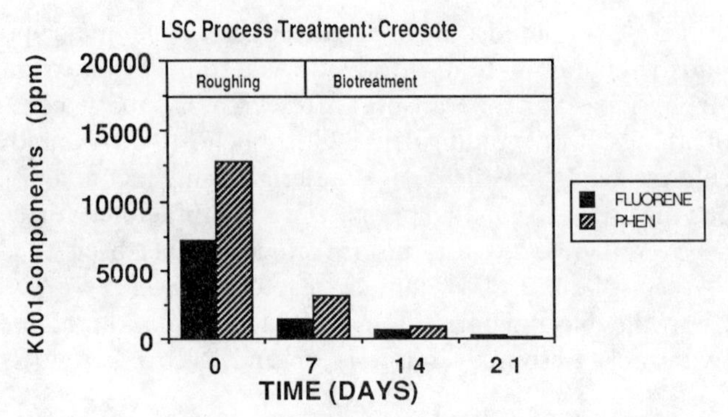

Fig. 3. Biotransformation of fluorene and phenanthrene in a liquid/solids bioreactor over a 21-day treatment period. Data represent mean values of replicate analyses (3) from four replicate reactors ($N=12$) for each sample date.

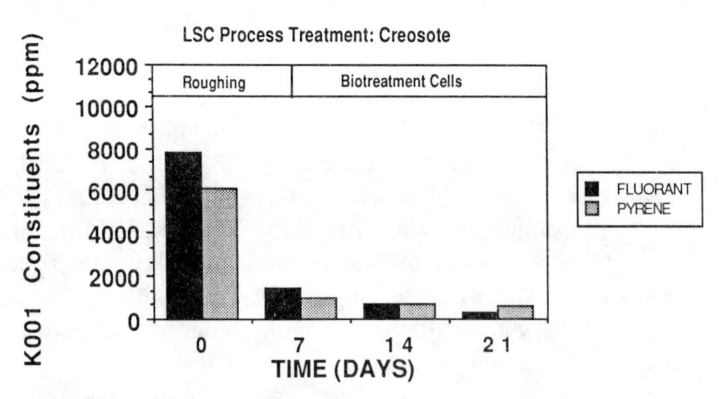

Fig. 4. Biotransformation of fluoranthrene and pyrene in a liquid/solids bioreactor over a 21-day treatment period. Data represent means values of replicate analyses (3) from four replicate reactors ($N=12$) for each sample date.

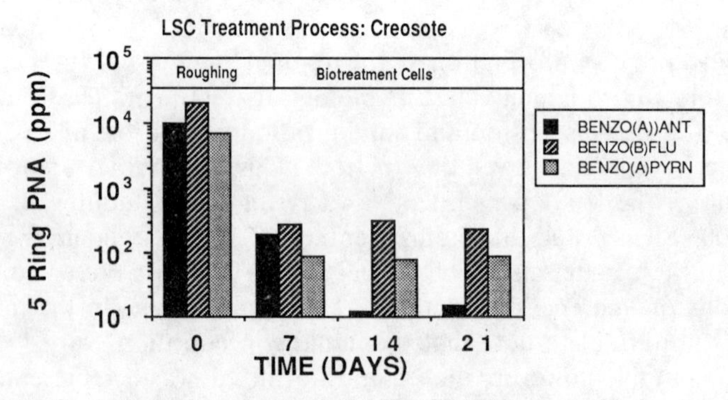

Fig. 5. Biotransformation of carcinogenic PAHs (CPAHs) over a 21-day treatment period.

surfactant addition the residual levels for fluorene, phenanthrene, fluoranthrene and pyrene were greatly reduced after seven days of continuous aeration and agitation.

Residual concentrations from this roughing cell step, prior to transfer to a polishing reactor, averaged in concentration between 500 and 4000 mg/kg soil dry weight residual (ppm), phenanthrene appearing to be the most resistant to the continuous agitation over a seven-day treatment period. Microbial ATP levels exceeded 10^9 cells/ml for continuous treatment. For final biological treatment phenanthrene and fluorene were both significantly degraded to below 100 mg/kg soil dry weight residual within 21 days. Fluoranthrene and pyrene were reduced to levels below 500 mg/kg soil dry weight residual over the same time frame.

3.3 Carcinogenic PAHs

Initial and final concentrations of key five-ring polynuclear aromatics found in creosote/pentachlorophenol waste materials are presented in Fig. 5. Of particular interest is the benzo(b)fluoranthrene and benzo(a)pyrene constituents of these wastes. As in the previous data sets on carcinogenic PAH or CPAH constituents, reactor cell replicates provided significant reduction in the five-ring polynuclear aromatics. Reactor cell replicates, having the highest accumulated biomass in which fungal populations exceeded 10^4 propagules/g soil dry weight, indicated the greatest reductions to <500 mg/kg soil dry weight residual for all constituents. Benzo(b)fluoranthrene was not significantly reduced during this seven-day mixing step. With biological treatment notable reductions were seen for all five-ring polynuclear aromatics (PNAs). Benzo(b)fluoroanthrene was not as significantly reduced as the other CPAHs. Table 1 provides kinetic expressions for LSC biotreatment of highly concentrated carcinogenic PAHs.

Optimal mixing of CPAH waste materials such as are found in creosote can result in significant reductions in all K001 constituents. Optimal reductions in benzo(a)anthracene (BA) as shown in Fig. 5 under optimal mixing, however, required the presence of a fungal isolate, a *Penicillium* sp. designated F101, to further facilitate mineralization from *trans*-BA-8-9 dihydrodiol and *trans*-BA-10-11 dihydrodiol to diol epoxides. Addition of a 10 mg/liter aliquot of salicylic acid at day 14 and the reinoculation by F101 further reduced BA concentrations. Note that the remaining CPAHs did not show any changes in concentration to this supplemental

Table 1
Biokinetic Rates of Carcinogenic PAH Reductions in LSC Reactors

K001 constituent	*Initial concentration (ppm)*[a]	*Rate (mg/kg soil/day)*
Benzo(a)pyrene	9 000 (mean)	$366{\cdot}7 \pm 12{\cdot}6$
Benzo(b)fluoranthrene	13 300 (mean)	$595{\cdot}2 \pm 17{\cdot}9$
Benzo(a)anthracene	11 000 (mean)	$521{\cdot}4 \pm 36{\cdot}8$

[a] Mean values are based on four replicate LSCs for each experimental and control test and three GC sample analyses for each day sampled.

substrate addition. Additional pathway studies for these two CPAHs are under further investigation.

4 PENTACHLOROPHENOL-CONTAMINATED GROUNDWATER

4.1 Pentachlorophenol Mineralization

Radiotracer studies conducted on several isolates have indicated that both bacterial and fungal consortia are preferable for pentachlorophenol (PCP) mineralization in creosote wastes. Figure 6 presents data on $^{14}CO_2$ expiration of ^{14}C-UL-PCP mineralization in creosote LSC reactors at 20% solids loading. Non-labeled PCP at a concentration of 100 mg/kg soil dry weight was present in the reactor. As shown, fungal isolate 3, probably *Penicillium notatum*, which was isolate F101 in CPAH tests, again proved to have a high affinity for creosote wastes and for PCP directly. The bacterial isolate, isolate 2, is an *Acinetobacter* species which later proved to be an important contributor for chlorinated dioxin photolytic product mineralization. The kinetic removal rate for PCP in this creosote test was 55·2 mg/kg dry wt soil/day.

A series of replicate experiments were conducted evaluating both photolytic and photolytic/microbiological methods for biotransformation of pentachlorophenol and its trace chlorinated dioxin contaminant, octachlorodibenzo-*p*-dioxin (OCDD). Abiotic control LSC reactors containing filter-sterilised (0·22 μm filter, Nucleopore™) pentachlorophenol-contaminated waters under filtered aeration (0·45 μm, Nucleopore™) were agitated in the presence and/or absence of ultraviolet light. Experimental reactors were established with all process parameters and, in addition, inoculated with an immobilized *Pseudomonas* species isolated in previous

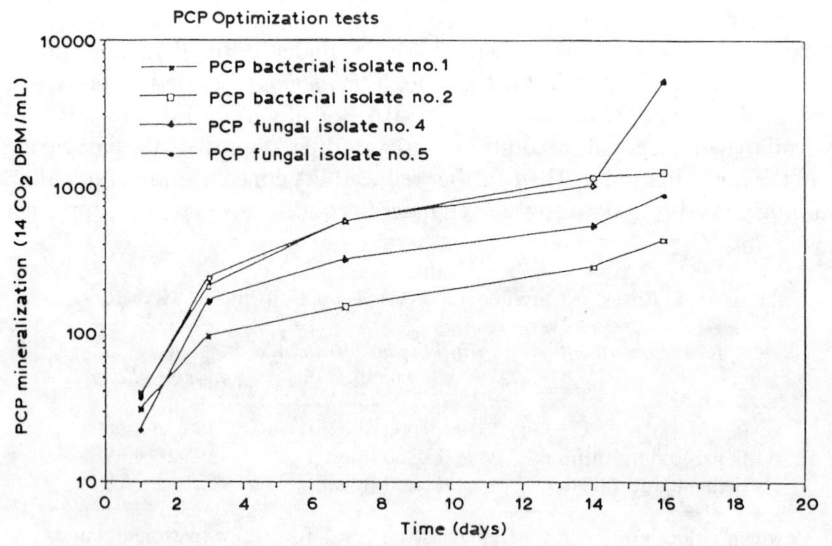

Fig. 6. Mineralization of pentachlorophenol (PCP) by adapted biomass.

investigations (Portier & Fujisaki, 1986). This strain had capabilities of biotransforming pentachlorophenol. Both the control and experimental LSC reactors were maintained at 28°C, agitation being constant at 1200 rpm over a 21-day treatment period. This microbial strain was fixed on a diatomaceous earth support (Manville, Celite R630) using methods previously outlined (Portier & Ahmed, 1988). Biotic control reactors did not receive UV irradiation. GC/MS analyses of groundwater were determined over the 21-day period for both pentachlorophenol parent compound disappearance as well as the fate of the OCDD residuals associated with this waste. At the conclusion of the experiment fugitive emission traps and residual biomass were also evaluated for the presence of both pentachlorophenol and OCDD residuals.

4.2 Chlorinated Dioxin/Furan Mineralization

In research reported previously (Portier & Ahmed, 1988) considerable effort went into the isolation and optimization of microbial strains capable of biotransforming PCP and creosote wastes. Most of the laboratory LSC data generated to date involved the use of bacterial strains. In subsequent investigations an effort was made to further delineate the contributions of fungal strains in mineralization of dioxin/furan congeners. Radiolabeled data sets confirmed chromatography results. Carbon dioxide, $^{14}CO_2$, as shown in Fig. 7, was generated in experimental units from ring-labeled ^{14}C-OCDD in aqueous solution in LSC reactors at 100 mg/liter PCP. No detectable levels above background were noted for sterile controls.

For controls without UV incident light (Fig. 7), minimal CO_2 expiration was noted, indicative perhaps of a cometabolic mineralization due to notable PCP degradation. In the presence of UV light an order-of-magnitude increase in OCDD mineralization was noted. This was particularly apparent with the addition of the fungal isolate F101. Dioxin data sets generated to date indicated an additional

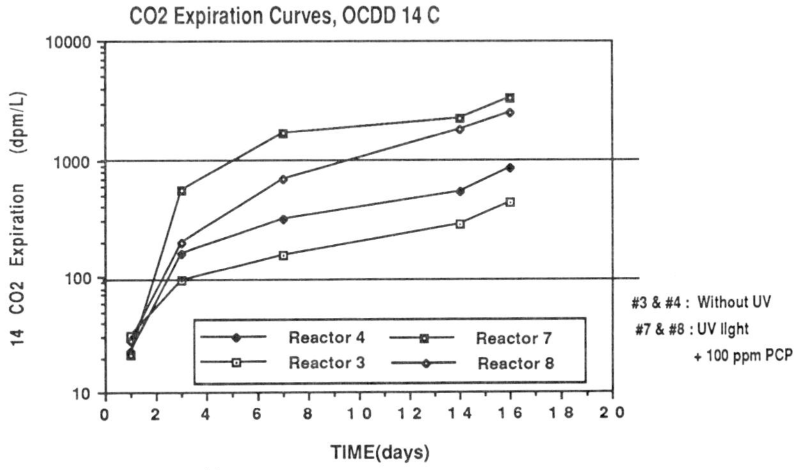

Fig. 7. Mineralization of ^{14}C OCDD in the presence or absence of UV illumination as expressed by $^{14}CO_2$ expiration.

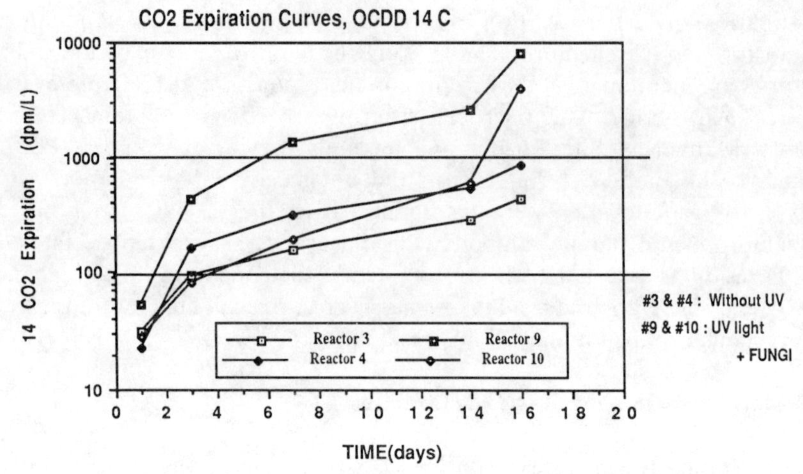

Fig. 8. Mineralization of ^{14}C OCDD in the presence or absence of UV illumination as expressed by $^{14}CO_2$ expiration. Fungi are the dominant microbial population.

increase in $^{14}CO_2$ expiration, approximately 12% of theoretical ^{14}C, in which fungi were the dominant biomass present (Fig. 8).

Chlorinated dioxin congener profiles indicated reductions in octachloro congeners (OCDD) without generation of tetrachlorodibenzo-*p*-dioxins, in particular 2,3,7,8-TCDD. Sequential reduction of such mixed congener assemblages, which include concentration variations as well as UV light incidence, was generated. Microbial populations actively metabolized PCP in the presence or absence of UV light. However, chlorinated dioxin congeners were only reduced in UV light. Sterile controls indicated UV was the primary mechanism for dioxin disappearance (see Fig. 9).

In a repeat of LSC reactor tests 100 ppm PCP at 20% solids creosote loading indicated chlorinated dioxin mineralization over time. Figure 10 presents data on these additional tests under incident UV light. The roughing step played an

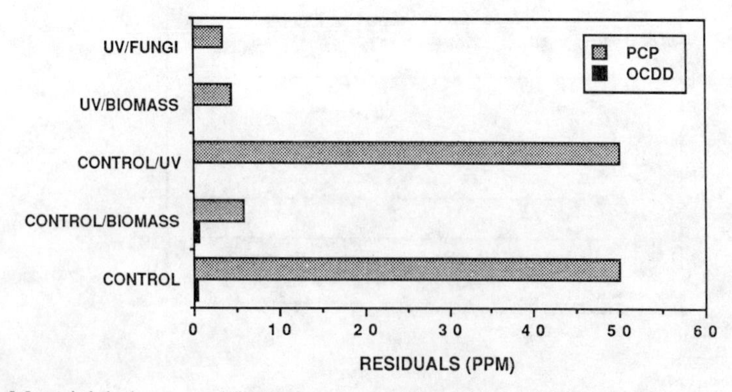

Fig. 9. Material balance residuals for pentachlorophenol (PCP) and its primary dioxin contaminant, octachlorodibenzo-*p*-dioxin (OCDD), in a contaminated leachate.

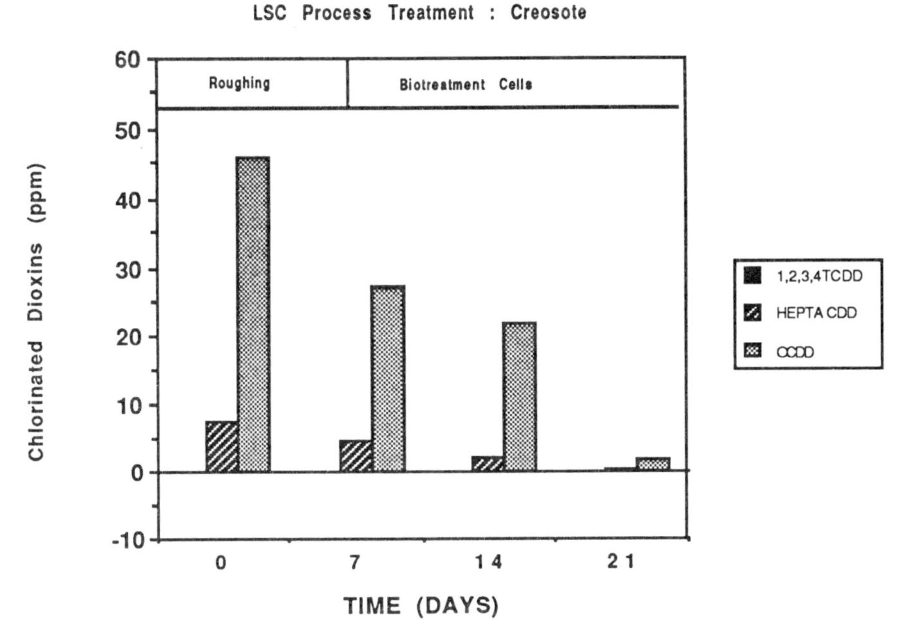

Fig. 10. Reduction of OCDD in creosote LSC reactor tests.

important role in OCDD reductions, namely the optimization of UV incidence on the surface of a well-mixed reactor. Approximately 50% of residual OCDD was reduced during the non-biological mixing step. Additional reductions were noted with continuous mixing in the biotreatment phase, with 92% reductions in concentration noted in 21 days' contact. A residual of 1,2,3,4-TCDD was noted in one reactor only at the conclusion of the study. We are evaluating this possible pathway from photolytic pretreatment.

Evaluation of OCDD mineralization in the presence of dead biomass, namely the *Acinetobacter* sp. noted, indicated less than 0·5% of theoretical ^{14}C was adsorbed to the biomass and 0·1% theoretical ^{14}C was bound to reactor glass surfaces (Fig. 11). Removal of ^{14}C OCDD was apparent primarily through incorporation into viable biomass in the presence of UV light. The residual concentration of PCP also affected assimilation rate.

As reported by Crosby and Wong (1977) the reduction of TCDD, a dioxin congener associated with Agent Orange herbicide formulations, is rapid and complete within a matter of hours for UV light. All dioxins absorb light above 295 nm, so sunlight can serve as a light source. However, three criteria must be met in order for dioxin photolysis to be practical: the wavelengths of incident light must correspond to appreciable absorption by TCDD/HpCDD/OCDD, the light must *penetrate* the medium to contact the dioxin component, and a source of extractable hydrocarbons must be present (although water is not very effective, formulating solvents and related hydrocarbon materials, i.e. PAH/cPAH constituents, would suffice). The LSC reactors provided situations in which these criteria have been met.

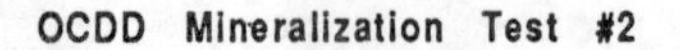

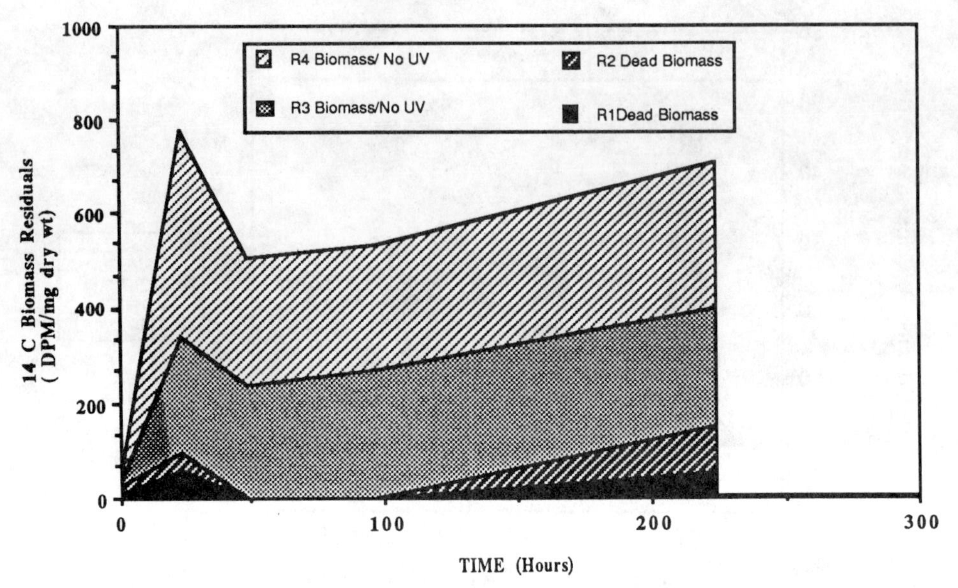

Fig. 11. Assimilation of radiolabeled OCDD by dead biomass and/or a viable microbial consortium.

Finally, Pereira *et al.* (1985) suggest that in the presence of elevated concentrations of dioxins and furans, particularly OCDD and HpCDD, microbial attack may be indicated under aerobic conditions. Specific variations in congener profiles over time indicated that specific microorganisms may be participating in primary or secondary degradative mechanisms for dioxin removal in chlorinated hydrocarbon sludges. Data generated to date in our investigations tend to support these observations.

5 CONCLUSIONS

Liquid/solids contact reactor approaches for site remediation of soils and groundwaters have been used extensively in the remediation of oil exploration production pits in the coastal wetlands of Louisiana and Texas. These systems are currently being deployed for creosote sites in the southeastern United States. Operational costs are site specific, primarily related to toxicant composition and concentration in soils. To date these systems have ranged in cost from $45 to $95 per ton of soils and from $5 to $18 per 10 000 liters of groundwater. Most field applications of this technology have emphasized the ability to operate *in situ*. Thus it has been used as a pretreatment step to traditional landfarming approaches. Since the technology results in soil volume reductions of 20–60% based on organic/water content, the technology has also been used as a pretreatment step to traditional solidification methods. The significant reduction of organic content improves solidification while reducing material costs to achieve a non-leachable composite residual.

Remediation of contaminated soil presents a major challenge to businessmen, scientists and regulators. Remediation of solid and hazardous waste sites containing soils was first required under the Resource Conservation and Recovery Act of 1976 (RCRA) and the Comprehensive Environmental Response, Compensation, and Liability Act of 1980 (CERCLA) (USEPA, 1984). The authority of regulators to order cleanups preferentially through permanent on-site remedies was established under the Hazardous and Solid Waste Amendments of 1984 (HSWA) and the Superfund Amendments and Reauthorization Act of 1986 (SARA). SARA, in subparagraph 121(1)b, says in principal part:

> 'Remedial actions in which treatment permanently and significantly reduces the volume, toxicity or mobility of the hazardous substances, pollutants and contaminants as a principal element are to be preferred over remedial actions not involving such treatment. The off-site transport and disposal of hazardous substances or contaminated materials without such treatment should be the least favored alternative remedial action where practicable treatment technologies are available.'

Bioremediation and thermal treatments are now considered as permanent remedies. Liquid/solids contact (LSC), a methodology using high energy soil slurry reactors, is rapidly becoming the biological counterpart to incineration technology as its remediation equivalent.

ACKNOWLEDGEMENT

The research presented in this manuscript was supported by funding from the Hazardous Waste Research Center, Louisiana State University, Baton Rouge, LA (Cooperative Agreement CR-813888, ORD/EPA).

REFERENCES

Crosby, D. G. & Wong, A. S. (1977). *Science*, **195**, 1337.

Pereira, W. E., Rostad, C. E. & Sisak, M. E. (1985). *Environ. Toxicol. Chem.*, **4**, 629.

Portier, R. J. (1985). In *Validation and Predictability of Laboratory Methods for Assessing the Fate and Effects of Contaminants in Aquatic Ecosystems*, ASTM STP 865, ed. T. P. Boyle. American Society for Testing and Materials, Philadelphia, PA, p. 14.

Portier, R. J. & Ahmed, S. I. (1988). *Marine Technol. Soc. Journal*, **22**, 34.

Portier, R. J. & Fujisaki, K. (1986). *Toxicology Assessment*, **1**, 501.

Portier, R. J. & Meyers, S. P. (1982). *Developments in Industrial Microbiology*, **22**, 459.

Portier, R. J., Reily, L. A., Nelson, J. A., Flynn, B. P. & Bost, R. C. (1989). *Environmental Progress*, **8**, 120.

USEPA (1984). *Permit Guidance Manual on Hazardous Waste Land Treatment Demonstrations*. Report no. EPA/530-SW-84-015, United States Environmental Protection Agency, Washington, DC.

Chapter 12

APPLICATION OF ADAPTED BACTERIAL CULTURES FOR THE DEGRADATION OF XENOBIOTIC COMPOUNDS IN INDUSTRIAL WASTE-WATERS

B. NÖRTEMANN & D. C. HEMPEL

Department of Technical Chemistry and Chemical Engineering, University of Paderborn, FRG

CONTENTS

1 INTRODUCTION AND THEORETICAL BACKGROUND

Xenobiotic compounds are man-made chemicals, but although foreign to the biosphere they need not necessarily constitute an environmental problem (Hutzinger & Veerkamp, 1981; Cook *et al.*, 1983). On the other hand, since most of them have to be chemically inert for their particular application, many xenobiotic compounds have been found to resist biological degradation.

Regarding the release of synthetic chemicals into the biosphere, two types can broadly be distinguished: deliberate area contamination (e.g. by application of herbicides and pesticides), and involuntary emissions (e.g. such as caused by waste-waters, the dumping of wastes and local contamination of waters or soil by accidents). Special attention must be applied to industrial waste-waters which are permanent emission sources of xenobiotic compounds often being recalcitrant or hazardous to the environment.

The proven ability of the biosphere to evolve in a changing world (Monod, 1971; Darwin, 1979; Gibson, 1980; Clarke, 1981; Mortlock, 1982; Mayr, 1983), and the extraordinary evolutionary potential of microorganisms, gave reason to hope that new degradative capabilities for xenobiotic compounds could be developed readily. In recent years, many synthetic chemicals were found to be biodegradable under laboratory conditions by specially enriched, adapted, or constructed micro-organisms. Under in-situ conditions and in communal clarification plants, however, mineralization of recalcitrant xenobiotics is still difficult or not attainable at all for a large number of possible reasons. It should be noted here that synergistic interrelations within microbial communities (Bull & Slater, 1982; Lewis *et al.*, 1984; Hardman, 1987) and threshold concentrations below which growth, adaptation, or induction of catabolic activity will not occur (Boethling & Alexander, 1979; Spain *et al.*, 1980; Alexander, 1985; Schmidt *et al.*, 1985) play an important role in biodegradation. Therefore, even microorganisms harbouring specific catabolic pathways often cannot be established in communal or large industrial clarification plants which collect a variety of waste-water streams.

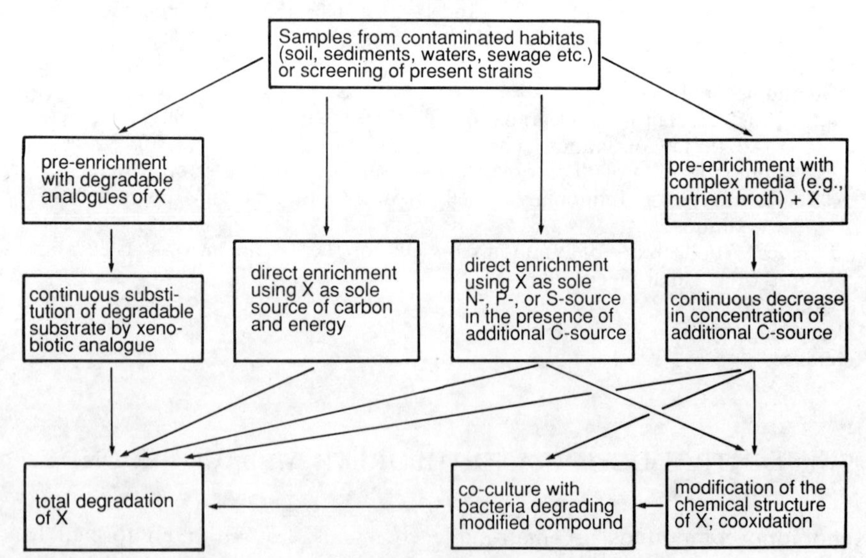

Fig. 1. General methods for the enrichment of bacteria degrading xenobiotic compounds. X, xenobiotic compound; C, carbon; N, nitrogen; P, phosphorus; S, sulphur.

To enable degradation of rather recalcitrant waste-water components, particular treatment systems should be constructed and integrated as a final stage into each production process. This concept, termed 'decentral waste-water treatment', allows the use of specially enriched and adapted microorganisms under selective conditions and suitable substrate concentrations.

Such a decentral waste-water treatment plant must, however, be resistant to adverse parameters such as periods without substrates, periods of high charges or high dilution rates, fluctuations in the feed load and composition of waste-water components, shifts in pH values, high concentrations of salts, heavy metals, toxic organic compounds, etc.

A suitable approach guaranteeing high degradation rates in small treatment plants and the protection of microorganisms against adverse parameters of industrial waste-waters is the use of immobilized cells in specially designed bioreactors.

This chapter describes the interdisciplinary approach by microbiologists, chemists, and engineers towards an efficient solution to special waste-water problems. Experiments were carried out either with aminonaphthalene sulphonic acids or with authentic waste-water from a coal tar refinery plant.

2 ENRICHMENT AND ADAPTATION OF MICROORGANISMS DEGRADING XENOBIOTIC COMPOUNDS SUCH AS NAPHTHALENE SULPHONIC ACIDS

2.1 General Enrichment Techniques

Various methods can be applied in attempts to obtain microorganisms with novel catabolic capabilities (Fig. 1). Present strains may be tested, or new bacteria may be enriched from natural or contaminated habitats. The simplest method would be direct enrichment using the xenobiotic compound in an appropriate concentration as the sole source of carbon and energy. Pre-incubation of soil or water samples with complex nutrients does not seem to be advantageous, since with complex media 'undesired' organisms often grow better or faster than those which are able to degrade xenobiotic compounds.

A direct enrichment technique using the xenobiotic chemical in the presence of a complex substrate such as sugar or nutrient broth could help to isolate bacteria with a cooxidation potential, but dead-end metabolites, possibly more toxic or recalcitrant than the xenobiotic substrate, could be generated by cooxidation processes.

In addition, xenobiotic compounds containing nitrogen, sulphur, or phosphorus can be used as sole N-, S-, or P-source in the presence of an additional carbon source. If bacteria are able to utilize such structural elements, the xenobiotic molecule will be modified and, thereby, could become biodegradable.

Furthermore, structural and biodegradable analogues of recalcitrant compounds can be used as substrates for the enrichment of microorganisms and then progressively replaced by the xenobiotic chemical.

2.2 Enrichment of Bacteria Degrading Naphthalene-1- and Naphthalene-2-sulphonic acid

Naphthalene sulphonic acids and their substituted analogues are multifariously used as intermediates for the large-scale production of azo dyes, wetting agents, dispersants, etc. Azo dyes, for example, are the largest class of dyes with the greatest variety of colours (Johnson *et al.*, 1978; Anliker, 1979). They are usually not degraded under aerobic conditions (Anliker, 1979; Michaels & Lewis, 1985, 1986). Under anaerobic conditions, the azo linkage can be reduced readily to form aromatic amines (Meyer, 1981) such as aminonaphthalene sulphonic acids (see, for example, Fig. 2).

Arylsulphonic acids are very rare among natural compounds (Herbert & Holliman, 1964) and, in general, are known to be non-biodegradable (Bretscher, 1981). On the one hand, the presence of the sulphonic acid group as a structural element of aromatic compounds is necessary for chemical applications (for example, to provide water solubility). On the other hand, it renders the uptake of substrate into cells and electrophilic attack by aryl-dioxygenases more difficult. Brilon *et al.* (1981*a*,*b*) have shown, however, that bacteria mineralizing naphthalene-1- or naphthalene-2-sulphonic acid (1NS, 2NS) can be enriched from naphthalene degrading populations through continuous replacement of the natural substrate by the xenobiotic compound.

Zürrer *et al.* (1987) used sulphur-limited batch enrichment cultures and obtained several bacterial strains which were able to use various aromatic sulphonic acids as sole source of sulphur. None of those substrates served as a carbon source, however, so that naphthols and phenols were accumulated as dead-end products.

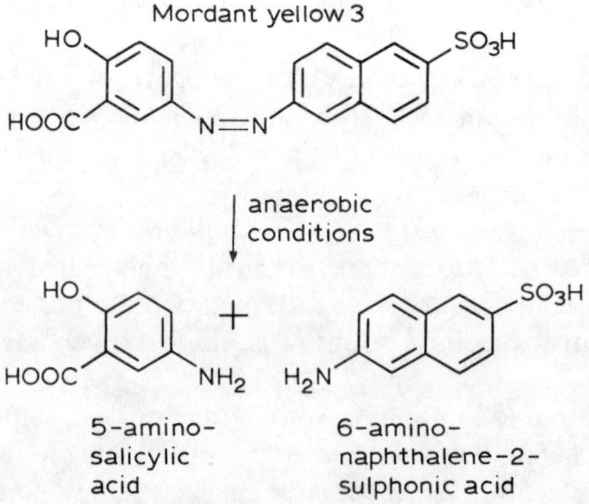

Fig. 2. Reduction of the azo-dye Mordant Yellow 3 by a bacterial mixed culture according to Haug *et al.* (1990).

3 INVESTIGATION OF CATABOLIC PATHWAYS AND THEIR INDUCTION

3.1 Catabolism of Naphthalene-1- and Naphthalene-2-sulphonic Acid (1NS, 2NS)

Extensive studies of the catabolic pathway for 1NS and 2NS were carried out by Brilon *et al.* (1981*b*). Catabolism of 2NS and 1NS by *Pseudomonas testosteroni* A3 and *Pseudomonas* sp. C22 was shown to be initiated by the action of a naphthalene-1,2-dioxygenase with low substrate specificity. Since the C—S bond becomes labile by 1,2-dioxygenation, the SO_3H-group is spontaneously eliminated from the hypothetical intermediate, 1,2-dihydro-1,2-dihydroxynaphthalene sulphonic acid, yielding sulphite and 1,2-dihydroxynaphthalene. The latter is further degraded via the classical naphthalene pathway (Cripps & Watkinson, 1978). The central metabolite, salicylic acid, was able to induce the complete catabolic sequence for 2NS in *P. testosteroni* A3. This is in good agreement with the results obtained by Shamsuzzaman and Barnsley (1974*a,b*) and by Barnsley (1975, 1976) for naphthalene catabolism. Salicylic acid itself is further degraded via catechol (by action of salicylate-1-hydroxylase) in naphthalene degrading bacteria, but via gentisinic acid (by salicylate-5-hydroxylase) in *P. testosteroni* A3 (Brilon *et al.*, 1981*b*).

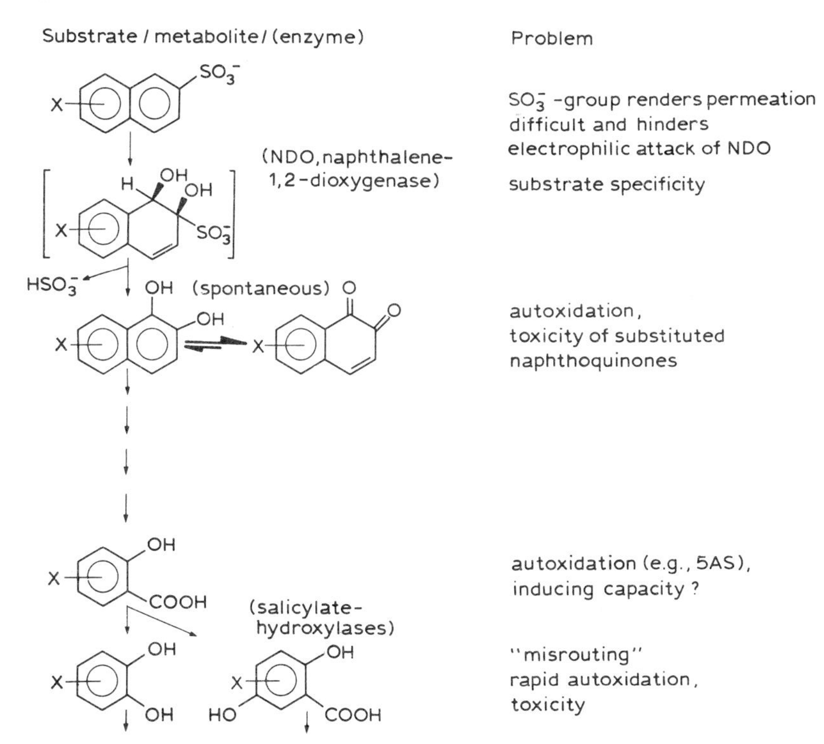

Fig. 3. Critical steps in the degradation of amino- and hydroxynaphthalene sulphonic acids ($X = NH_2$ or OH).

3.2 Total Degradation of Amino- and Hydroxynaphthalene-2-sulphonic Acids by Interspecies Transfer of Salicylic Acids

Although the 2NS or 1NS degrading microorganisms readily cooxidized some isomeric amino- and hydroxynaphthalene-2-sulphonic acids (ANSs and HNSs), no growth was obtained with these compounds. This can be explained by several critical steps in the catabolic pathway (Fig. 3). In particular, the turnover of amino- or hydroxysalicylic acids by the activity of salicylate-1- or salicylate-5-hydroxylase leads to metabolites which are extremely labile and toxic to the cells. Therefore, a 'productive catabolism' of ANSs and HNSs might not involve either of these enzymes.

It was shown that some ANSs and HNSs can be totally degraded by mutual

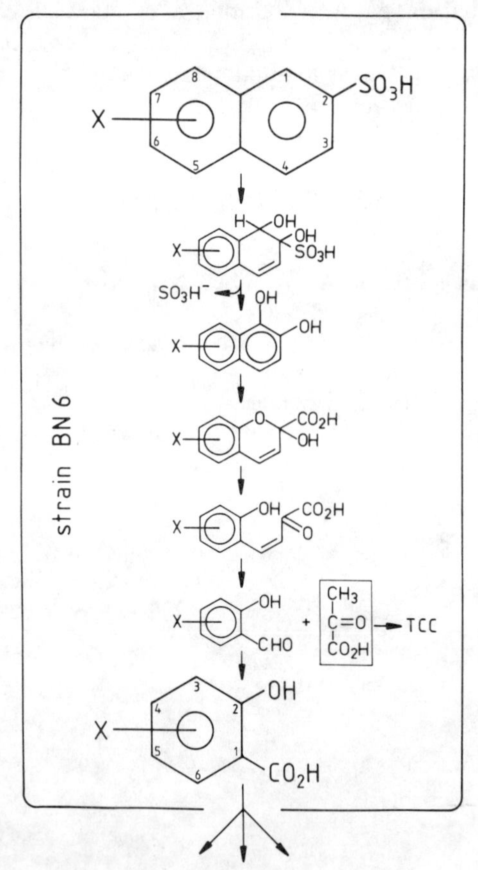

Fig. 4. Total degradation of amino- and hydroxynaphthalene-2-sulphonic acids (X = NH_2 or OH).

interaction within a mixed bacterial culture enriched and isolated from the River Elbe (Nörtemann *et al.*, 1986). Naphthalene-2-sulphonic acids with a NH_2- or OH-group in the position C-5, -6, -7, or -8 are quantitatively converted into the corresponding amino- and hydroxysalicylic acids by one strain, BN6 (Fig. 4). An exception is the oxidation of 5-aminonaphthalene-2-sulphonic acid, which, instead of 6-aminosalicylic acid, generates 5-hydroxyquinoline-2-carboxylic acid as the major product (Nörtemann & Knackmuss, 1988; Nörtemann *et al.*, paper in preparation). Its formation is readily explained by spontaneous ring fission of the hypothetical metabolite 6′-amino-2′-hydroxybenzalpyruvate (Fig. 5).

The enzymes for the catabolism of naphthalene sulphonic acids by strain BN6 are induced by salicylic acid, 3-, 4-, and 5-aminosalicylic acid, and 5- and 6-hydroxysalicylic acid (Nörtemann, 1987). By contrast, structural analogues of naphthalene sulphonic acids which cannot be converted into the corresponding salicylic acids fail to induce turnover activity towards 2NS. Very recently, 3-hydroxy-salicyclic acid was also found to be efficient as an inducer whereas the inducing capacity of 5-hydroxysalicylic acid was only poor (Kuhm, A., 1990, pers. comm.).

The catabolic pathway for the partial conversion of ANSs and HNSs by strain BN6 is identical to that described for 2NS with cells of *Pseudomonas testosteroni* A3 (Brilon *et al.*, 1981*b*). However, in contrast to the latter organism strain BN6 does not metabolize any salicylic acids. Consequently, amino- and hydroxysalicylic acids are not misrouted into unsuitable catabolic pathways but can be totally degraded by other members of the community such as strain BN9 or strain BN11. For example, 6-aminonaphthalene-2-sulphonic acid (6A2NS) is mineralized by a two-species culture of strain BN6 and strain BN9 or BN11 via interspecies transfer of 5-aminosalicylic acid (5AS). This metabolite is completely degraded via 1,2-dioxygenation (Nörtemann, 1987; Stolz *et al.*, paper in preparation).

The mutual interaction between strain BN6 and strain BN9 can clearly be demonstrated by streaking both organisms separately on the same mineral-agar

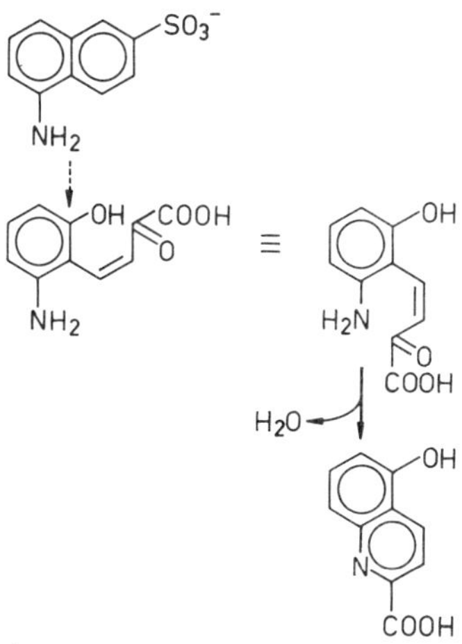

Fig. 5. Conversion of 5-aminonaphthalene-2-sulphonic acid into 5-hydroxyquinoline-2-carboxylic acid.

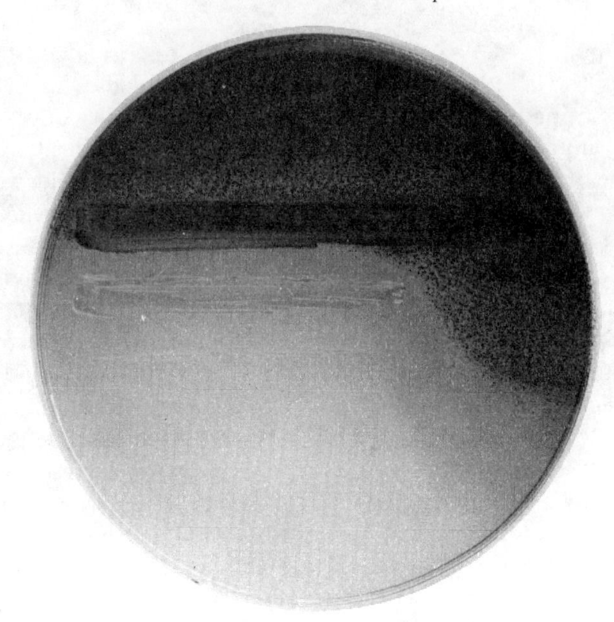

Fig. 6. Degradation of 6A2NS via interspecies transfer of 5-aminosalicylic acid (5AS). Biomass of strain BN6 is visible as a dark streak in the top of the plate. Precipitates result from autoxidation of 5AS. Accumulation of this labile metabolite is avoided in the lower part of the plate, where shorter parallel streaks of the 5AS degrading strain BN9 were inoculated.

plate containing 6A2NS as the sole carbon source (Fig. 6). Strain BN6, represented by the long dark streak in the upper part of the plate, converts 6A2NS into 5AS. This labile metabolite is subject to rapid autoxidation indicated by dark precipitation products. However, accumulation of 5AS is avoided in parts of the plate where parallel streaks of BN9 are inoculated.

Notably, growth of defined two-species communities (strain BN6 plus strain BN9 or BN11) on 6A2NS in liquid cultures was successfully stabilized by contaminants (10–20% of the total cell number) which neither oxidized 6A2NS nor metabolized 5AS. This indicates a cross-feeding of secondary metabolites between the members of the bacterial community. Such synergistic effects are optimized and unproductive loss of metabolites is avoided if the cooperating strains are in close cell-to-cell contact. This can be effected by immobilization of the mixed culture (see Section 4.2).

4 KINETIC DATA AND SUPPORTING METHODS FOR THE DEGRADATION OF NAPHTHALENE SULPHONIC ACIDS

4.1 Kinetic Data for the Degradation of 6-Aminonaphthalene-2-sulphonic Acid (6A2NS) in Continuous Culture

The degradation of 6A2NS by continuous chemostat cultures was investigated by Diekmann *et al.* (1988). From the experimental steady-state data obtained with a

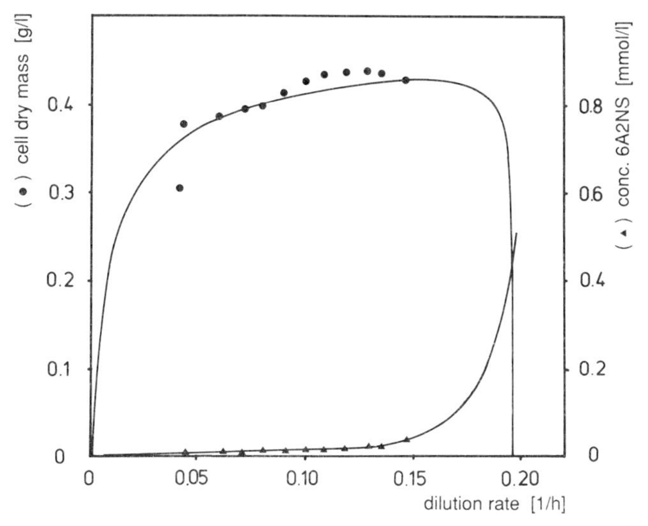

Fig. 7. Mineralization of 6A2NS in continuous cultivation. Inlet concentration of 6A2NS, 630 ppm (2·83 mmol/litre). Data calculated by a structured model on the basis of Monod kinetics are shown by the solid line.

multi-species culture, the kinetic constants for the conversion of 6A2NS into 5AS and for the degradation of 5AS were determined. Conversion of 6A2NS into 5AS by strain BN6 was shown to be rate-limiting for complete mineralization of the substrate. On the basis of Monod kinetics (Monod, 1950), a structured model regarding the interspecies transfer of 5AS was developed (Diekmann *et al.*, 1988). As shown in Fig. 7, the experimental data fit in well with the data calculated by this model.

Surprisingly, after prolonged subcultivation ($\geq$40 months) of the mixed culture on solid or in liquid 6A2NS-media, no stable growth with this substrate could be obtained in continuous cultivation. Only during the first days after the start of continuous flow of substrate was a total degradation of 6A2NS observed. After approximately three to ten generations, growth of the culture came to a halt. A significant increase in the concentration of dissolved organic carbon (DOC) was observed prior to an increase in 6A2NS concentration. Therefore, it can be assumed that growth of the mixed culture was inhibited by a critical metabolite until a complete wash-out of the microorganisms occurred. The reasons for this previously unnoticed behaviour of the bacterial community in continuous cultivation are presently under investigation.

Notably, a stable growth with 6A2NS in continuous cultivation is still readily feasible if cells are immobilized and employed in a fluidized bed reactor.

4.2 Immobilization of Microorganisms on Sand in an Airlift-loop Reactor

Since growth rates of special microorganisms degrading xenobiotic compounds are often low, these organisms should be employed in high concentrations. Moreover, inter-species transfer of labile metabolites such as 5AS should be facilitated and

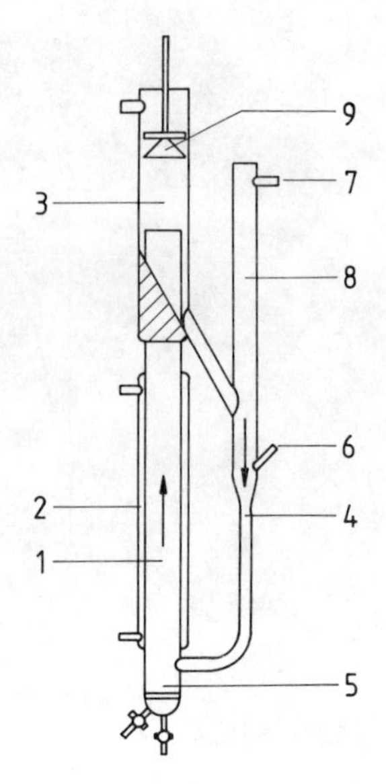

Fig. 8. Airlift-loop reactor. 1, riser; 2, heat exchanger; 3, gas–liquid separator; 4, outer loop (downcomer); 5, aeration plate; 6, feed; 7, liquid outlet; 8, sedimentation tube; 9, foam breaker.

biocatalysts have to be protected against adverse operational parameters. Such requirements can be met by the immobilization of these organisms on support materials. 'Passive' immobilization into polymeric matrices is too expensive and not proper to an aerobic waste-water treatment process, since the particles are disintegrated by high shearing forces. By contrast, biocatalysts resulting from growth of organisms on inert particles are mechanically stable.

Different materials were tested with regard to their applicability as support particles for the immobilization of microorganisms. These experiments were carried out under non-sterile conditions in an airlift-loop reactor (Fig. 8) with fluidized biocatalysts degrading 6A2NS (Gerdes-Kühn *et al.*, 1989). As demonstrated by Lindert *et al.* (1990), the configuration of the airlift-loop reactor ensures, apart from good mixing and fluidization characteristics, that limitations due to mass transfer effects are minimized. Moreover, a good hold-back of the biocatalysts can be obtained if the specific density of the support material is higher than that of water.

Immobilization of 6A2NS-degrading organisms was performed directly within the reactor using 10 g/litre of support material. To avoid a 'submerse' cultivation, i.e. cultivation of unattached or not immobilized cells, the dilution rate ($D = 0{\cdot}25$ litre/h) was set above the maximum growth rate of submerse cultures. For the first three days, the airlift-loop reactor was inoculated continuously with cells from a chemostat. The synthetic waste-waters described by Diekmann *et al.* (1988) contained 630 ppm (2·83 mmol/litre) of 6A2NS as the sole source of carbon and energy.

Significant growth of 6A2NS-degrading organisms on solid particles was obtained only with broken sand (approximately 200 μm in diameter, see Fig. 9) and activated carbon. In both cases, the first colonization was observed in indentations in the surface of the particles after two to three days. Within a few weeks, a thick biofilm and a degradation of $\geq$99% (6A2NS) were obtained.

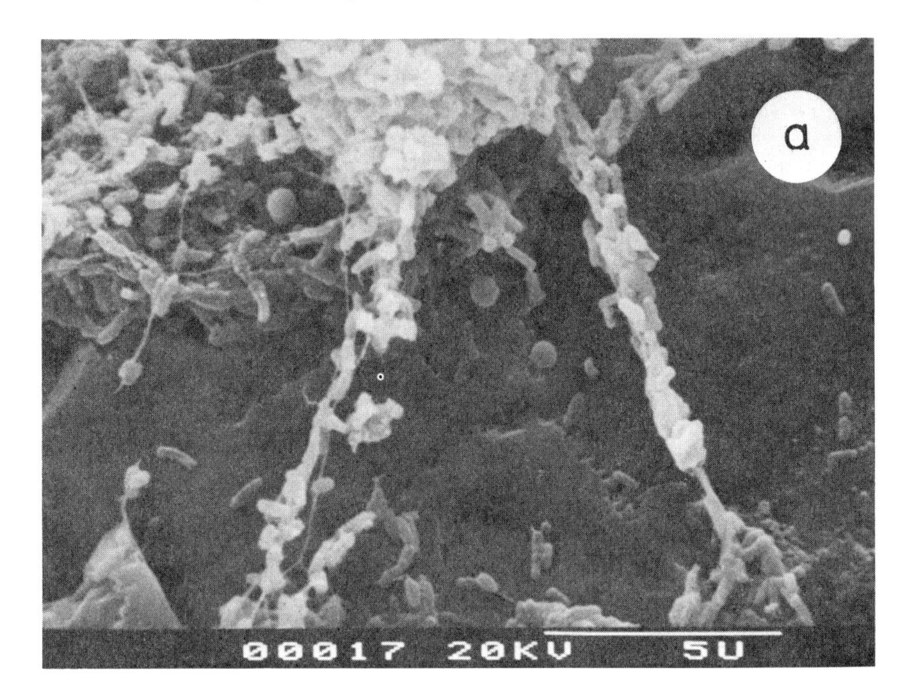

Fig. 9. Biofilm formation. (a) Primary growth of 6A2NS-degrading bacteria on the surface of the sand particle. (b) Accomplished biofilm in detail. EM-photography by G. Masuch, University of Paderborn, FRG.

A drastic increase in biomass concentration on the strength of insufficient shearing forces was observed when activated carbon was used as support material. This could not be avoided even with an increased concentration of activated carbon up to 20 g/litre. On the contrary, a constant biomass concentration of approximately 5 g/litre (cell dry mass) was obtained with sand (20 g/litre) as support material. With free sand particles shearing forces were high enough to remove surplus biomass. Diekmann and Hempel (1989*a*) calculated 6A2NS- and O_2-profiles with regard to the diffusion limitation inside the biofilm. From this, it can be concluded that 6A2NS is completely consumed in a biofilm layer of about 100 μm when the bulk phase concentration of the substrate is 2 mg/litre. A biofilm thickness of 250 μm is conceivable, since dormant cells may occur within the biofilm close to the support material. Their energy requirements for survival could be supplied by primary or secondary metabolites from 6A2NS degradation or, in part, by cell lysis products. With a continued increase in biofilm thickness, however, cells close to the support material will starve and then be subject to lysis. Once the biofilm loses close contact with the sand surface, it is sheared off and removed from the reactor (Wagner & Hempel, 1988).

4.3 Resistance of Immobilized Cells Towards Suboptimal Environmental Conditions

As shown by Diekmann *et al.* (1990), the 6A2NS degradation rate with cells immobilized on sand remained constant ($\geq$99%) over a wide range of dilution rates for more than 20 months (Fig. 10).

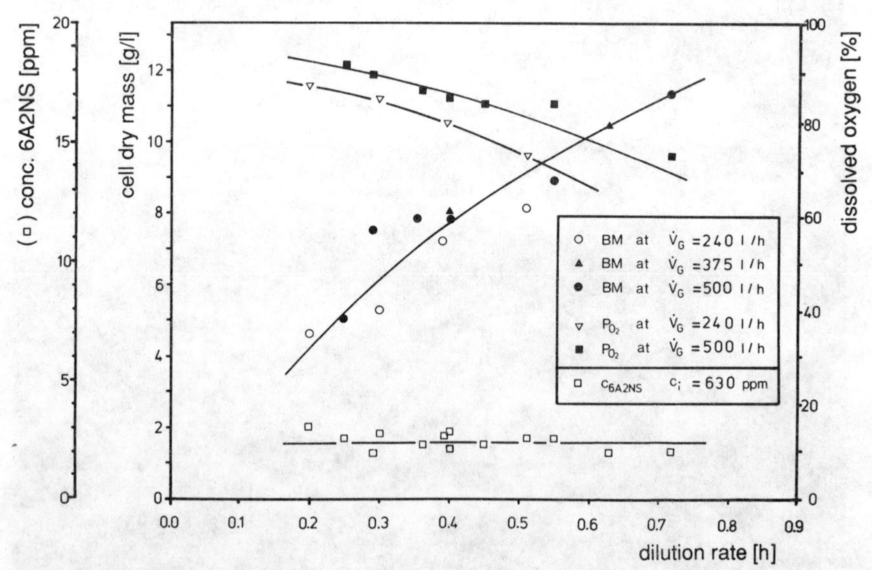

Fig. 10. Degradation of 6A2NS by immobilized cells. Sand (diameter *c.* 200 μm) was used as the support material. Experiments were carried out in an airlift-loop reactor (see Fig. 8). BM, biomass (cell dry weight); $\dot{V}_G$, gas flow rate; P_{O_2}, concentration of dissolved oxygen; $c_{i\ 6A2NS}$, inlet concentration of 6A2NS.

Regardless of the dilution rate, the average growth rate of the immobilized microorganisms was calculated as $\mu = 0{\cdot}02$ litre/h. The DOC (dissolved organic carbon) and COD (chemical oxygen demand) reduction up to 5·56 kg/(m^3 . day) and 17·5 kg/(m^3 . day), respectively, do not represent the maximum values, since the degradation limit was not attained.

When the physico-chemical growth parameters (pH value, temperature, and dissolved oxygen concentration) were suboptimal for 6A2NS-degrading microorganisms, a decreasing growth rate was often compensated by increasing biomass concentrations (e.g. Fig. 11). As a consequence, the 6A2NS degradation rate remained constant even under suboptimal growth conditions. This can be explained mainly by a facilitated supply of the inner cells with substrate when the specific activity of the cells in the outer area of the biofilm decreases.

Immobilized cultures of the 2NS-degrading *Pseudomonas testosteroni* A3 (Brilon *et al.*, 1981*a*,*b*) and of the 6A2NS-mineralizing mixed culture were subjected to pH shock loadings (pH 3 or pH 10) in continuous cultivation for 13 h. Figure 12 shows the 2NS- or 6A2NS-concentrations vs time when the systems were re-adjusted to optimal conditions (pH 7·0 and pH 7·4, respectively). In all cases, immobilized cultures recovered almost immediately so that stable operating conditions were achieved again within less than 48 h. The high concentration of 6A2NS compared to the inlet concentration, observed after the system had been grown at pH 3, is readily explained by the low solubility of 6A2NS at that pH. Crystals of 6A2NS are held back inside the bioreactor, and then redissolved at pH 7.

As shown by Gerdes-Kühn and Hempel (1990), immobilized cells degrading 2NS or 6A2NS are also able to overcome shock loadings with high concentrations of salts (NaCl, Na_2SO_4) and heavy metals such as nickel or cadmium. These findings suggest that, in general, immobilization of special cultures is a suitable method to protect the microorganisms against suboptimal growth conditions which occur in

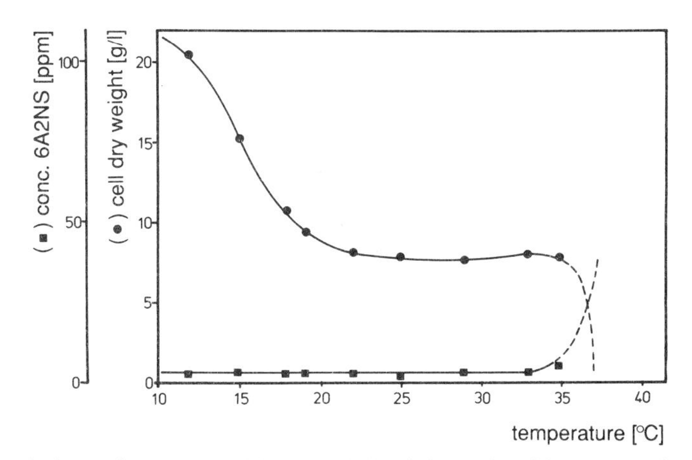

Fig. 11. Degradation of 6A2NS with immobilized bacteria—biomass and substrate concentration vs temperature. Cells of the 6A2NS-degrading mixed culture were immobilized on sand (20 kg/m^3) and employed in an airlift-loop reactor (see Fig. 8). Inlet concentration of 6A2NS, 630 ppm (2·83 mmol/litre); dilution rate, 0·3 litre/h.

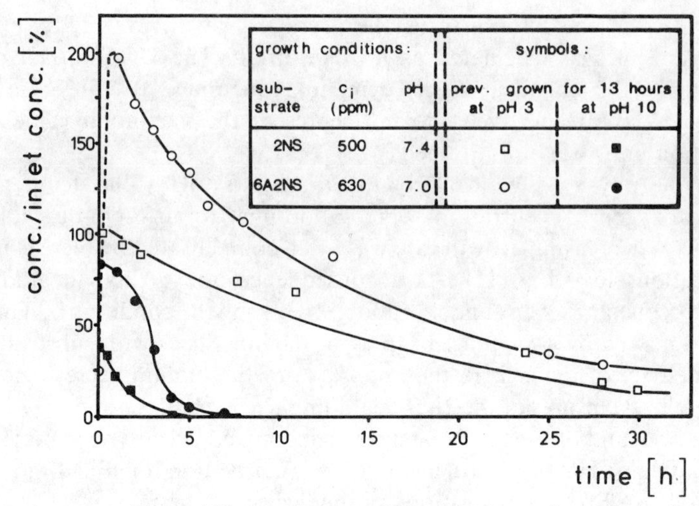

Fig. 12. Convalescence of immobilized cells in continuous culture after pH shock-loadings. Cells of a 2NS- or a 6A2NS-degrading culture were immobilized on sand (20 kg/m^3) and employed in an airlift-loop reactor. Dilution rates were 0·3/h (6A2NS) or 0·55/h (2NS). After 13 hours of continuous cultivation under suboptimal growth conditions (pH 3 or 10, respectively), the pH was re-adjusted to optimal values. c_i, inlet concentration.

practice with waste-water. This may be due to the fact that the outer shells of the biocatalysts guard those organisms located inside the biofilm from serious damage. Therefore, diffusion limitations within biofilm layers can be advantageous for the treatment of waste-waters containing inhibitory or toxic components. It could be demonstrated also for the degradation of naphthalene disulphonic acids (Krull *et al.*, 1990) and a mixture of 6- and 8-aminonaphthalene-2-sulphonic acid (Hattendorf & Hempel, 1990) that growth of special cultures is significantly stabilized by immobilization. Moreover, growth rates of immobilized bacteria are, in spite of high biomass concentrations, in such a low range that the endogenous metabolism is predominant (Diekmann & Hempel, 1989*b*). As a consequence, only a small amount of surplus biomass is produced and thus the costs for deposition are very low.

5 PRACTICAL APPLICABILITY OF IMMOBILIZED CELLS FOR THE TREATMENT OF WASTE-WATER FROM A COAL TAR REFINERY PLANT

Based on the high resistance of immobilized cells towards suboptimal growth conditions, application of specially adapted and immobilized cultures for the degradation of xenobiotic compounds in industrial waste-waters is most promising. Actually, the practical applicability of such systems for the biological treatment of waste-water from a coal tar refinery plant has already been demonstrated (Hüppe *et al.*, 1989, 1990; Höke & Hempel, 1990).

The refining of coal tar is accompanied by the production of heavily polluted

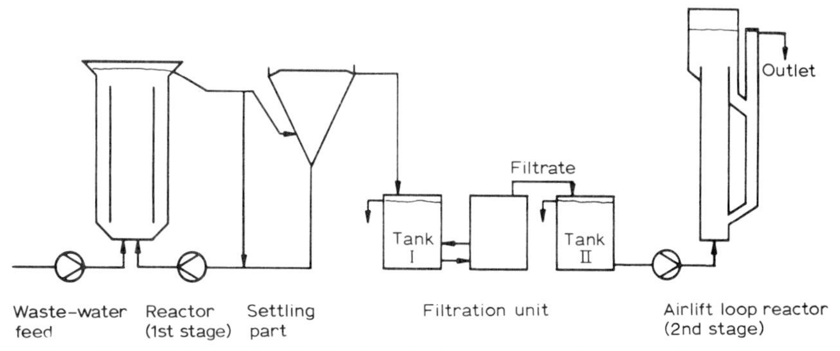

Fig. 13. Two-stage pilot plant for the treatment of waste-water from a coal tar refinery plant.

flows of waste-water. Even after preclarification by various separation and extraction techniques, the waste-water still contains considerable quantities of aromatic compounds. The ratio of biological to chemical oxygen demand varied between 0·25 and 0·35. These low values indicated a high biological recalcitrance of waste-water components. Their composition can be described as 10–30% phenol, mono-, and dimethylphenols, 30–60% neutral aromatic compounds such as naphthalene or acetophenone, and 10–40% nitrogen-containing aromatic bases, i.e. pyridine, mono-, di-, and trimethylpyridines, quinolinc, methylquinolines, and isoquinoline.

From the soil and sludge samples of industrial origin microorganisms were isolated and adapted to growth with the aromatic waste-water components. Preliminary studies were carried out in two-stage bench scale units operating with model and original waste-water (Höke, 1988; Koch *et al.*, 1991). The first stage was run as an activated sludge plant designed to remove the high input concentrations of most phenols and neutral aromatic compounds which are easily degradable. The second stage was designed as a fluidized bed reactor with the special cells immobilized on sand. Based on the promising results of these bench scale investigations, a two-stage pilot plant was designed and run with original waste-water on the industrial terrain (Höke & Hempel, 1990; Hüppe *et al.*, 1990).

The basic arrangement of both pilot plant stages is shown in Fig. 13. The gas throughput in the airlift-loop reactor (second stage) corresponded to a riser superficial gas velocity of 3·5–4·5 cm/s. This efficient fluidization provided circulation of carrier particles and a sufficient supply with oxygen.

Table 1
Basic Data for the Two-stage Pilot Plant

	First stage	*Second stage*
Reactor volume	1 500 litres	160 litres
Carrier concentration	—	20–70 g/litre
Carrier	—	sand (200 μm)
COD_{in}	4 900 mg/litre	1 000 mg/litre
Residence time	25–9·5 h	4–2 h
COD-volume load	5–12 kg/(m^3 day)	5–15 kg/(m^3 day)

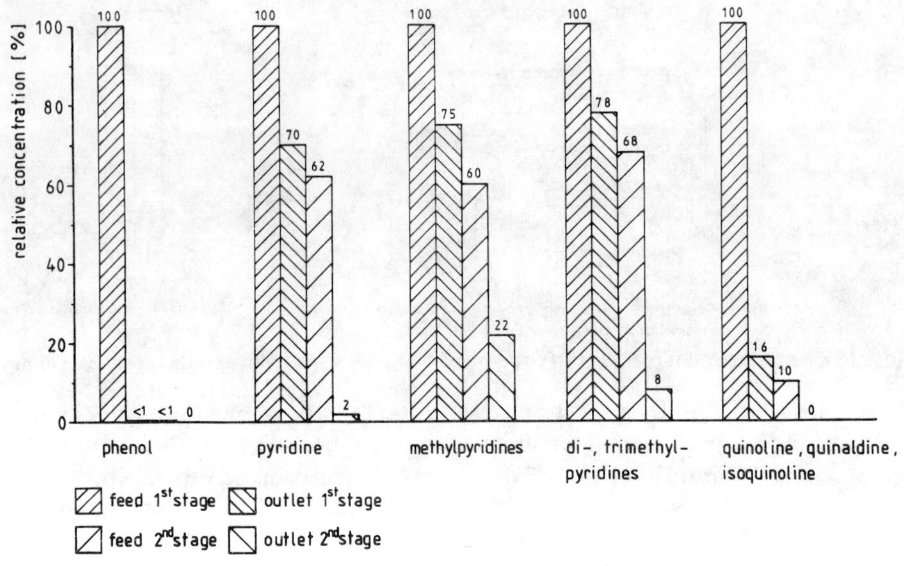

Fig. 14. Degradation of several relevant waste-water components.

The basic data of both pilot plant stages are listed in Table 1. Although the residence times in the second stage were significantly shorter than those in the first stage, COD-volume loads were comparable on the strength of effective biomass retention in the airlift-loop reactor.

As shown by Table 2, the mean input values of 5300 and 1800 mg/litre for COD and DOC, respectively, were mainly reduced by the first stage. Most recalcitrant waste-water components such as pyridine, mono-, di-, and trimethylpyridines were,

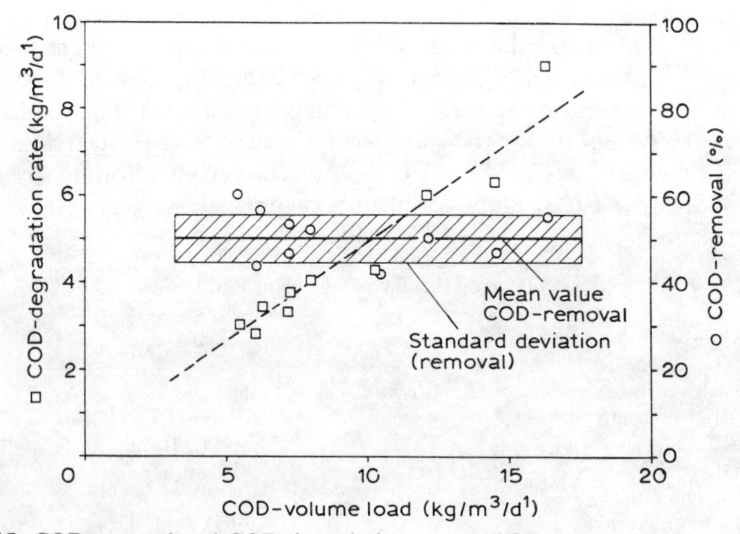

Fig. 15. COD-removal and COD-degradation rate vs COD-volume load (second stage).

Table 2
Average Degradation Values of the Pilot Plant

	First stage	*Second stage*	*Both stages*
COD	75%	50%	88%
DOC	n.d.[a]	56%	n.d.[a]
Phenols	94%	50%	97%
Neutral aromatics	70%	85%	96%
N-containing aromatics	30%	75%	83%

[a] Not determined.

however, only poorly degraded in the first stage, and were chiefly eliminated in the second stage (Fig. 14). The COD-reduction in both stages was approximately 88%.

The COD-removal and COD-degradation rates obtained in the second stage are shown in Fig. 15 as a function of the COD-volume load. Notably, the COD-removal remained constant over a wide range of COD-volume loads. This indicated that the maximum plant capacity in the second stage was not reached, despite a high volume load above 15 kg/(m^3 . day).

6 CONCLUSIONS

Investigations with specially enriched and adapted microorganisms demonstrate their practical applicability for the treatment of industrial waste-waters under suitable process-engineering conditions. Together with an appropriate reactor, immobilization of special cultures on fluidized carrier particles such as sand shows great advantages with regard to aspects of biology and process-engineering. Two important properties are, firstly, high concentrations of active biomass which render small decentral treatment plants possible and, secondly, a high tolerance towards adverse operational conditions such as shock loadings, fluctuations in feed concentrations, pH values, composition of waste-water components, etc.

From this, it can be concluded that many industrial waste-water compounds which are non-degradable in conventional treatment plants could specifically be mineralized by adapted immobilized cultures. Such decentral waste-water treatment plants should be integrated as a final stage into those production processes where more or less continuous flow of effluents containing xenobiotic and recalcitrant compounds occurs.

ACKNOWLEDGEMENTS

The authors gratefully acknowledge the financial support in the form of various grants from the Bundesminister für Forschung und Technologie (BMFT), the Arbeitsgemeinschaft industrieller Forschungsvereinigungen eV (AIF), the Deutsche Forschungsgemeinschaft (DFG), and the Ministerium für Umwelt, Raumordnung und Landwirtschaft des Landes Nordrhein-Westfalen (MURL) (all FRG).

REFERENCES

Alexander, M. (1985). *Environ. Sci. Technol.*, **18**, 106.

Anliker, R. (1979). *Ecotoxicol. Environ. Saf.*, **3**, 59.

Barnsley, E. A. (1975). *J. Gen. Microbiol.*, **88**, 193.

Barnsley, E. A. (1976). *J. Bacteriol.*, **125**, 404.

Boethling, R. S. & Alexander, M. (1979). *Appl. Environ. Microbiol.*, **37**, 1211.

Bretscher, H. (1981). In *Microbial Degradation of Xenobiotics and Recalcitrant Compounds*, eds T. Leisinger, A. M. Cook, R. Hütter & J. Nüesch. Academic Press, London, p. 65.

Brilon, C., Beckmann, W., Hellwig, M. & Knackmuss, H. J. (1981*a*). *Appl. Environ. Microbiol.*, **42**, 39.

Brilon, C., Beckmann, W. & Knackmuss, H. J. (1981*b*). *Appl. Environ. Microbiol.*, **42**, 44.

Bull, A. T. & Slater, J. H. (1982). *Phil. Trans. R. Soc. Lond. B.*, **29**, 575.

Clarke, P. H. (1981). In *Biochemical Evolution*, ed. H. Gutfreund. Cambridge University Press, UK.

Cook, A. M., Grossenbacher, H. & Hütter, R. (1983). *Experientia*, **39**, 1191.

Cripps, R. E. & Watkinson, R. J. (1978). In *Developments in Biodegradation of Hydrocarbons—1*, ed. R. J. Watkinson. Applied Science Publishers, London, p. 113.

Darwin, C. (1979). The Illustrated Origin of Species; abridged and introduced by R. E. Leakey. Faber and Faber, London.

Diekmann, R. & Hempel, D. C. (1989*a*). *Dechema Biotechnology Conferences*, Frankfurt, FRG, Vol. 3. VCH-Verlag, Weinheim, FRG.

Diekmann, R. & Hempel, D. C. (1989*b*). *Appl. Microbiol. Biotechnol.*, **32**, 113.

Diekmann, R., Nörtemann, B., Hempel, D. C. & Knackmuss, H. J. (1988). *Appl. Microbiol. Biotechnol.*, **29**, 85.

Diekmann, R.,Naujoks, M., Gerdes-Kühn, M. & Hempel, D. C. (1990). *Bioproc. Eng.*, **5**, 13.

Gerdes-Kühn, M. & Hempel, D. C. (1990). *Dechema Biotechnology Conferences*, Frankfurt, FRG, Vol. 4. VCH-Verlag, Weinheim, FRG.

Gerdes-Kühn, M., Diekmann, R. & Hempel, D. C. (1989). *Korrespondenz Abwasser*, **7**, 776.

Gibson, D. T. (1980). In *The Handbook of Environmental Chemistry*, ed. O. Hutzinger. Springer-Verlag, Berlin, Vol. 2A, p. 161.

Hardman, D. J. (1987). In *Environmental Biotechnology*, eds C. F. Forster & D. A. J. Wase. Ellis Horwood, Chichester, UK, p. 295.

Hattendorf, C. & Hempel, D. C. (1990). *Dechema Biotechnology Conferences*, Frankfurt, FRG, Vol. 4. VCH-Verlag, Weinheim, FRG.

Haug, W., Schmidt, A., Nörtemann, B., Hempel, D. C., Knackmuss, H. J. & Stolz, A. (1990). In *Biochemical Engineering—Stuttgart*, eds M. Reuss, H. J. Knackmuss, H. Chmiel & E. D. Gilles, Gustav Fischer Verlag, Stuttgart, FRG, p. 409.

Herbert, R. B. & Holliman, F. G. (1964). *Proc. Chem. Soc.*, 19.

Höke, H. (1988). In *Biochemie und Mikrobiologie von Kohle und Kohleinhaltsstoffen*, ed. Studienges. Kohlegewinnung zweite Generation eV, Essen, FRG, 51.

Höke, H. & Hempel, D. C. (1990). Paper presented at *7th Dechema-Fachgespräch Umweltschutz*, Frankfurt, FRG, *GWF-Wasser/Abwasser*, **131**, 660.

Hüppe, P., Höke, H. & Hempel, D. C. (1989). *Dechema Biotechnology Conferences*, Frankfurt, FRG, Vol. 3. VCH-Verlaf, Weinheim, FRG.

Hüppe, P., Höke, H. & Hempel, D. C. (1990). *Chem. Eng. Technol.*, **13**, 73.

Hutzinger, O. & Veerkamp, W. (1981). In *Microbial Degradation of Xenobiotics and Recalcitrant Compounds*, eds T. Leisinger, A. M. Cook, R. Hütter & J. Nüesch. Academic Press, London, p. 3.

Johnson, R. F., Zehnhausern, A. & Zollinger, H. (1978). In *Kirk–Othmer Encyclopedia of Chemical Technology*, 2nd edn, vol. 2, eds H. F. Mark, J. J. McKetta, Jr, D. P. Othmer & A. Standen. John Wiley & Sons, New York.

Koch, B., Ostermann, M., Höke, H. & Hempel, D. C. (1991). *Water Res.*, **25**, 1.

Krull, R., Nörtemann, B., Kuhm, A., Hempel, D. C. & Knackmuss, H. J. (1990). *GWF-Wasser/Abwasser*, **132**, 352.
Lewis, D. L., Hodson, R. E. & Freeman, L. F. (1984). *Appl. Environ. Microbiol.*, **48**, 561.
Lindert, M., Diekmann, R. & Hempel, D. C. (1990). In *Proceedings of the International Biotechnology Conference*, Palmerston, New Zealand.
Mayr, R. (1983). *The Growth of Biological Thought: Diversity, Evolution and Inheritance.* Belknap Press of Harvard University Press, Cambridge, MA.
Meyer, U. (1981). *FEMS Symp.*, **12**, 371.
Michaels, G. B. & Lewis, D. L. (1985). *Environ. Toxicol. Chem.*, **4**, 45.
Michaels, G. B. & Lewis, D. L. (1986). *Environ. Toxicol. Chem.*, **5**, 161.
Monod, J. (1950). *Ann. Inst. Pasteur*, **79**, 390.
Monod, J. (1971). *Chance and Necessity.* Knopf, New York.
Mortlock, R. P. (1982). *Annu. Rev. Microbiol.*, **36**, 259.
Nörtemann, B. (1987). *Bakterieller Abbau von Amino- und Hydroxynaphthalinsulfonsäuren.* PhD thesis, University of Stuttgart, FRG.
Nörtemann, B. & Knackmuss, H. J. (1988). *GWF-Wasser/Abwasser*, **129**, 75.
Nörtemann, B., Baumgarten, J., Rast, H. G. & Knackmuss, H. J. (1986). *Appl. Environ. Microbiol.*, **52**, 1195.
Schmidt, S. K., Alexander, M. & Shuler, M. L. (1985). *J. Theor. Biol.*, **114**, 1.
Shamsuzzaman, K. & Barnsley, E. A. (1974*a*). *J. Gen. Microbiol.*, **83**, 165.
Shamsuzzaman, K. & Barnsley, E. A. (1974*b*). *Biochem. Biophys. Res. Com.*, **60**, 582.
Spain, J. C., Pritchard, P. H. & Bourquin, A. W. (1980). *Appl. Environ. Microbiol.*, **40**, 726.
Wagner, K. & Hempel, D. C. (1988). *Biotechnol. Bioeng.*, **31**, 559.
Zürrer, D., Cook, A. M. & Leisinger, T. (1987). *Appl. Environ. Microbiol.*, **53**, 1459.

Chapter 13

THE RUDAD-PROCESS FOR ENHANCED DEGRADATION OF SOLID ORGANIC WASTE MATERIALS

HUUB J. M. OP DEN CAMP & HUUB J. GIJZEN*

Department of Microbiology, Faculty of Science, University of Nijmegen, Nijmegen, The Netherlands

CONTENTS

* Present address: Applied Microbiology Unit, Department of Botany, University of Dar-es-Salaam, PO Box 35060, Dar-es-Salaam, Tanzania.

1 FROM THE NATURAL SYSTEM TO THE ARTIFICIAL RUMEN REACTOR

1.1 Production and Degradation of Cellulosic Biomass

The cyclic conversion of carbon by means of photosynthesis and microbial degradation is of utmost importance for all life on earth. The gain of this consecutive reduction and oxidation of carbon is the capture of solar energy which is used for growth and maintenance of living cells. Worldwide, photosynthetic fixation of CO_2 is estimated to yield annually up to $1{\cdot}5\text{–}2 \times 10^{11}$ tons of dry plant material (Stephens & Heichel, 1975; Duchesne & Larson, 1989). During the process of CO_2 reduction by solar energy, cellulose (28–50%), hemicellulose (20–30%) and lignin (18–30%) are produced as main constituents of plant biomass (Thompson, 1983). Minor constituents of plant material are pectin, lipids, and protein.

The decomposition of cellulosic biomass is carried out almost exclusively by microbial oxidations under both aerobic and anaerobic conditions (Swift, 1977; Ljungdahl & Eriksson, 1985). Most of the biomass in natural environments is oxidized by aerobic microorganisms. However, a substantial amount is mineralized in anaerobic environments. In this case the organic substrate is oxidized with inorganic electron acceptors such as NO_3^- (nitrate reduction), SO_4^{2-} (sulfate reduction) or CO_2 (methanogenesis). Anaerobic degradation to methane accounts for about 5–10% of the overall mineralization of organic matter (Ehalt, 1976; Vogels, 1979). Methane production occurs in anaerobic habitats such as marshes, aquatic sediments, sewage sludge digestors, paddy-fields, and the digestive tract of many herbivorous animals. The rumen is probably the most investigated methane-producing ecosystem and has been the subject of many reviews (Hungate, 1966, 1975, 1982; Wolin, 1979; Hobson & Wallace, 1982; Hobson, 1988).

1.2 Energy from Waste Materials

A part of the photosynthetic biomass ends up as waste residue from agricultural, industrial and domestic processes, resulting in a deterioration of the environment. Because of the increasing amount of cellulosic wastes and a lack of dumping sites, the costs of current disposal methods such as dumping or incineration are increasing rapidly. On the other hand cellulosic residues represent an abundant and inexpensive renewable resource for the generation of energy and useful products. However, the use of this potential resource requires the conversion of the biomass into a suitable form such as gaseous (CH_4) or liquid (ethanol, methanol) products. In view of the worldwide need for renewable energy sources, present research is focused on the utilization of biomass as an alternative to petroleum for the production of fuel and chemicals (Tsao *et al.*, 1978; Stafford *et al.*, 1980; Lipinsky, 1981; Ladisch *et al.*, 1983). The valorization of cellulosic residues is dependent upon the development of an economically feasible process. A variety of methods are being explored which range from physical treatments, such as pyrolysis, to chemical and biological methods such as acid or enzymic hydrolysis and fermentation. Because of the relatively high moisture content of most biomass, and the high treatment costs of physical and chemical hydrolysis, a biotechnological process is more attractive.

Anaerobic digestion of organic waste may be considered as a rather simple technology for the solution of two concurrent problems: the disposal of solid waste generated by society and the need for new sources of energy. In contrast to other processes aimed at the valorization of organic waste (e.g. methanol, ethanol production), the gaseous end products of methanogenic degradation are spontaneously separated from the liquid, saving the need for an additional expensive separation step (distillation, extraction).

1.3 Anaerobic digestion

Anaerobic decomposition of complex organic biopolymers to methane is brought about by the combined action of a wide range of organisms. According to present knowledge, overall anaerobic digestion is considered to proceed according to the reaction scheme depicted in Fig. 1. The figure is simplified, especially with respect to the acidogenic reactions, in order to emphasize the major metabolic routes. The flow of carbon from polymeric biomass to methane proceeds in several successive stages. However, the degradation process is not a sequence of independent reactions, but is characterized by a complex of mutual interactions between different microbial species (Bryant, 1979; Wolin & Miller, 1987).

1.3.1 Hydrolysis

Since bacteria are unable to take up particulate organic material, the first step in anaerobic degradation consists of the hydrolysis of polymers through the action of extracellular enzymes to produce smaller molecules which can cross the cell barrier. Hydrolytic microorganisms are generally associated with the insoluble substrate (Hobson & Wallace, 1982). Some eukaryotic microorganisms (e.g. rumen ciliates) are known to perform an intracellular hydrolysis of polymers (Coleman, 1983).

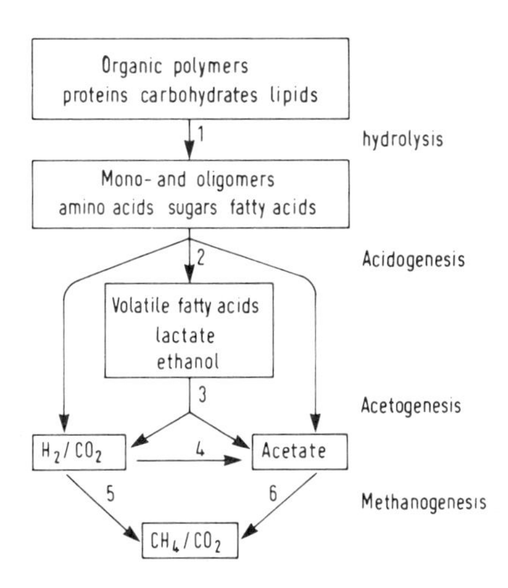

Fig. 1. Anaerobic degradation of polymeric biomass to methane and the microbial groups involved. 1,2, Hydrolytic and fermentative bacteria; 3, hydrogen-producing acetogenic bacteria; 4, hydrogen-consuming acetogenic bacteria; 5, hydrogenotrophic methanogens; 6, acetoclastic methanogens.

Polymers such as starch, glycogen, pectin or protein are readily hydrolyzed. Cellulose, the predominant constituent of biomass, has been shown to be rather resistant to hydrolysis (Noike *et al.*, 1985). The enzymes responsible for the hydrolysis of the β-(1-4)-linked units of D-glucose are referred to as cellulases. Cellulase in fact is a complex of cellulolytic enzymes, composed of exo-glucanases (exo-cellobio-hydrolase), endo-glucanases and cellobiases (β-glucosidase) and their synthesis is regulated by induction and catabolite repression (Brown & Gritzali, 1984; Ljungdahl & Eriksson, 1985). The cellulolytic enzyme systems appear to be almost as diverse as the types of cellulolytic microorganisms.

1.3.2 Acidogenesis

The soluble products of hydrolysis are metabolized intracellularly by a complex consortium of hydrolytic and non-hydrolytic microorganisms. The main end products of acidogenesis by mixed cultures are acetate, propionate, butyrate and H_2/CO_2 (Thauer *et al.*, 1977; McInerney & Bryant, 1981). Furthermore, minor amounts of formate, lactate, valerate, methanol, ethanol, butanediol or acetone may be produced by fermentative microorganisms. Since volatile fatty acids (VFA) are the main products of fermentative organisms, they are usually designated as acidifying or acidogenic microorganisms. The partial pressure of hydrogen plays an important role in controlling the proportions of various fermentation products (Zehnder & Stumm, 1988).

1.3.3 Acetogenesis

The hydrogen-producing acetogenic bacteria are responsible for the oxidation of products generated in the acidogenic phase to substrates suitable for methanogens (Bryant, 1979), and therefore these bacteria form an intermediate metabolic group. The products of acetogenic bacteria are H_2, CO_2 and acetate. The partial pressure of hydrogen also plays an important role in acetogenesis (Zehnder & Stumm, 1988). There often exists a syntrophic association between hydrogen-producing and hydrogen-consuming species, termed 'interspecies hydrogen transfer'.

Interspecies hydrogen transfer has been suggested to occur between sapropelic and rumen ciliates as hydrogen producers and methanogens as hydrogen consumers in endosymbiotic and episymbiotic associations, respectively (Vogels *et al.*, 1988).

1.3.4 Methanogenesis

The final step in the overall anaerobic conversion of organic matter into methane and CO_2 is catalyzed by methanogenic bacteria. Methanogens utilize only a limited number of simple substrates, comprising acetate or the C_1-compounds CO_2/H_2, formate, methanol, methylamines, methylated sulfur compounds and CO (Oremland, 1988).

Both hydrogenotrophic and acetoclastic methanogens are very important in maintaining the progress of anaerobic digestion. It is evident that hydrogenotrophic methanogens fulfill a crucial role as hydrogen scavengers in overall anaerobic digestion (Archer, 1983). Acetoclastic methane formation on the other hand prevents acidification of the anaerobic environment by removal of acetate.

Fermentation of organic matter in the absence of methanogens would consequently result in a rapid acidification and incomplete degradation.

1.4 Rate-limiting Steps

Since anaerobic digestion of organic matter is a multistep process which involves the successive action of metabolically diverse populations of microorganisms, the overall rate of substrate conversion is determined by the kinetic characteristics of the slowest step. Which of the individual steps is rate-limiting is mainly dependent on the composition of the substrate.

During the digestion of refractory insoluble substrates, such as cellulose, the hydrolysis step has been reported to govern the overall degradation rate (Eastman & Ferguson, 1981; Noike *et al.*, 1985). The rate of hydrolysis is determined by both microbial constraints (e.g. generation time, cellulase production) and physical and chemical characteristics of the substrate (e.g. crystallinity of cellulose, degree of association with lignin, surface area/particle size ratio).

If soluble organic components are the main substrates for anaerobic digestion, acetogenesis and methanogenesis have been identified as rate-limiting steps (Ghosh & Pohland, 1974; Kaspar & Wuhrmann, 1978), due to the long generation times of acetogenic bacteria and acetoclastic methanogens (Zehnder & Stumm, 1988). Recent advances both in the technology of anaerobic digestion of soluble waste and in the understanding of the complex process have resulted in drastic improvements of digestor performance (Sahm, 1984). The general strategy to counteract the rate limitation of conventional anaerobic digestors treating wastewater is based on the increase of microbial biomass retention at relatively short hydraulic residence time. Methods to improve microbial biomass retention are all based on the general phenomenon that the bacteria attach readily to surfaces and to one another.

While major advances have been made in the anaerobic treatment of soluble wastes, digestion of solid organic wastes is still mainly performed in low-rate completely stirred tank reactors (CSTR). Immobilization of microbial cells on supports may not be applicable in solid waste treatment, since an optimal contact between the solid substrate and the microorganisms should be maintained. During the digestion of particulate waste materials the microorganisms in fact are immobilized on the substrate particles and therefore increased biomass retention may be realized by selective retainment of the solid substrate, whereas the retention time of rapidly digested soluble waste components may be reduced. This mode of operation has been applied to the digestion of animal manures by means of upflow or horizontal plug flow digestor designs (Callander, 1982; Callander & Barford, 1983).

An alternative means of accelerating the decomposition of cellulosic residues is to make the substrate more accessible to enzymic attack by means of physical, chemical or biological pretreatment. The goal of pretreatments is to break up the lignin–carbohydrate complex in order to increase the degree and the rate of cellulose digestion. Most of the pretreatments have been developed to valorize low quality forages as an animal feed especially for ruminants, but recently they were also

Table 1
Performance Data of Some Natural and Artificial Anaerobic Microbial Ecosystems

Reactor type	*Substrate*	*Loading rate*[a] ($g\ VS.liter^{-1}\ day^{-1}$)	*Retention time* (*days*)	*Conversion* (%)	*Reference*
CSTR[b]	domestic refuse	1–2	20–40	50	Van der Vlugt & Rulkens, 1984
CSTR	pig manure	3–6	10–20	30–40	Van Velsen, 1981
Two-phase	tomato plants	—	14–21	40	Hoffenk, 1983
DRANCO[c]	domestic refuse	20	20–50	50–55	Six & de Baere, 1988
UASB[d]	wastewater	15–18	0·13–0·33	95	Lettinga *et al.*, 1980
Fluidized bed	wastewater	20–60	0·04–0·08	90	Heijnen, 1983
Termite hindgut	lignocellulose	35–70	0·5–1	60	Gijzen, 1987
Rumen	grass	50–100	1–2	40–70	Gijzen, 1987

[a] VS = Volatile Solids.
[b] Completely Stirred Tank Reactor.
[c] Dry Anaerobic Composting.
[d] Upflow Anaerobic Sludge Blanket.

applied to lignocellulosic wastes before anaerobic digestion (Hashimoto, 1986). It is not yet clear whether the additional costs of pretreatment will be recovered in reduced costs of digestion.

A more direct means of improving cellulose degradation would be by the application of microorganisms or microbial communities exhibiting an enhanced cellulolytic activity. In contrast to the inefficient degradation of solid waste materials in current digestors, the anaerobic decomposition of lignocellulosic materials in certain natural microbial ecosystems is known to proceed at high rates. The termite hindgut or the forestomach of ruminants may be considered as high-rate natural digestors which operate at solid residence times of only one or two days (Table 1). These animals and all other herbivores have evolved a symbiotic interrelationship with microorganisms which enable them to convert huge amounts of structural plant polysaccharides that cannot be degraded by animal digestive processes. Herbivorous animals rely on the gut microorganisms to convert these polysaccharides into products which can be utilized for energy and growth. The rate of hydrolysis and subsequent acid formation by intestinal microorganisms may be up to 20–50 times as high as artificial anaerobic systems treating cellulosic waste materials (Table 1). Because of their high cellulolytic activity and relatively short generation times, rumen microorganisms possess an enormous biotechnological potential in the application to anaerobic decomposition of cellulosic residues.

1.5 The Rumen Microbial Ecosystem

The rumen is an open continuous cultivation system of microorganisms in which conditions suitable for microbial growth are provided by the ruminant. The rumen ecosystem is characterized by an almost constant supply of plant material, saliva and water, a constant temperature of 39°C, an almost neutral pH (6–7), a low oxidation–reduction potential and a differential removal of solids and liquids. These conditions favour the growth of a large and complex microbial population which is responsible for the conversion of structural plant fibers. Many aspects of the rumen ecosystem have been reviewed recently by Hobson (1988).

In contrast to the anaerobic degradation pattern depicted in Fig. 1, which represents an overall digestion of organic matter to CH_4 and CO_2, the degradation of plant material in the rumen involves only the stages of hydrolysis and acidogenesis. Consequently, acetate (56–70%), propionate (17–29%), butyrate (9–19%) and minor amounts of longer-chained saturated fatty acids are the predominant fermentation products in the rumen. These products are absorbed from the digestive tract and serve as a major energy source for the ruminant. Hydrogen is an important intermediate in rumen fermentation and is utilized by hydrogenotrophic methanogens for the reduction of CO_2 to CH_4. Fermentation of cellulose and other complex substrates in the rumen proceeds by the interaction of various microbial species. Pure cultures of rumen microorganisms may produce substantial amounts of lactate, ethanol, formate, hydrogen or succinate, products not usually found in high amounts in the rumen (Wolin, 1979; Prins & Clarke, 1980).

Because of the low partial hydrogen pressure created by hydrogenotrophic methanogens, ethanol, lactate and succinate are metabolized in the rumen.

The rumen contains hundreds of different species of microorganisms including bacteria (10^{11}–10^{12} cells ml^{-1}), protozoa (10^5–10^6 cells ml^{-1}) and phycomycete fungi which are firmly attached to the solid substrate during its degradation (Hobson, 1988). Within only a few minutes plant particles ingested by the ruminant are colonized by the microorganisms. The physical coupling of the microorganisms to their substrate enables them to exploit their hydrolytic enzyme activities in an optimal way. Moreover, attachment of microbial cells to the solid digesta provides a means of increasing microbial biomass retention in the rumen, since residence time of solids has been shown to be much longer than that of liquids.

1.6 Outline of this Chapter

Considering the fact that rumen microorganisms exhibit extremely high cellulase activities and relatively short mass doubling times, it is obvious that an application of these organisms to the anaerobic treatment of cellulosic residues might significantly enhance the rate and the economic feasibility of the process. This chapter describes the development of a high-rate artificial rumen reactor and its application to the degradation of several types of waste materials.

2 THE ARTIFICIAL RUMEN REACTOR

A prerequisite for the application of rumen microorganisms in the degradation of solid organic waste is the development of a fermentor with a simple construction and operation. The fermentor design should support the long-term maintenance of rumen microorganisms including the rumen ciliates which can be used as indicator organisms. The fermentation system developed was adapted from Hoover *et al.* (1976) and was operated with differential removal rates of solids and liquid.

2.1 Description of the Fermentation System

Figure 2 shows a schematic diagram of the artificial rumen reactor. The reactor consisted of a double-walled glass vessel of 3 liters (A) with a working volume of 1·5 liters. The temperature was kept at 39°C using a circulating water bath. An adjustable peristaltic pump (D) was used to add rumen buffer solution (according to Rufener *et al.*, 1963) continuously. The products of the rumen fermentation were continuously removed through a filter unit (E, stainless steel wire gauze, 0·3 mm pore size, wrapped in nylon gauze, 30 μm pore size). The total filter area was about 90 cm^2. Mixing of the fermentor contents was achieved by placing it on a rotary shaker (H). The contents were mixed twice per hour for a period of 45 s. Homogeneous reactor contents (residue) were removed immediately after mixing once daily. Biogas production was measured by connecting the fermentor to a 10-liter calibrated Mariotte flask.

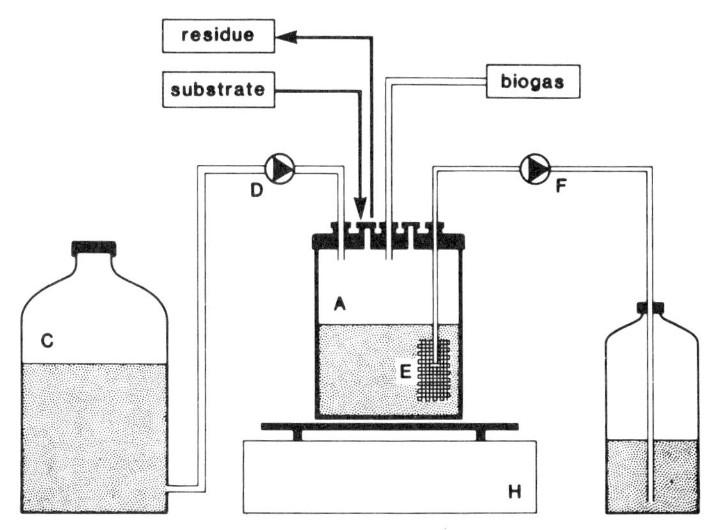

Fig. 2. Schematic diagram of the artificial rumen reactor. A, Acidogenic reactor; C, fermentation medium reservoir; D, fermentation medium supply pump; E, filter, 30 μm pore size; F, filtered effluent removal pump; H, rotary shaker. Adapted from Gijzen *et al.* (1990).

2.2 Experimental Conditions

2.2.1 Inoculation

Rumen fluid from a fistulated sheep was used as inoculum. An amount of 250 ml was added to the fermentor containing prewarmed rumen buffer and a double amount of the daily feeding load. No special precautions were needed to keep anaerobic conditions due to the presence of enough facultative aerobic microorganisms in the inoculum.

2.2.2 Solids (SRT) and Hydraulic Retention Time (HRT)

The SRT and HRT were adjusted to the desired values by means of the adjustable peristaltic pumps D (buffer input) and F (filtered effluent). The HRT was regulated by adjusting the liquid flow through the fermentor, while the SRT was determined by the difference between buffer influent and filter effluent flows. When experiments were performed under 'standard rumen conditions' HRT and SRT were 12 h and 72 h, respectively.

For further details concerning experimental conditions and analytical methods the reader is referred to Gijzen *et al.* (1986).

2.3 Operation of the Artificial Rumen Reactor on Animal Feed

An artificial rumen reactor was started up to test its stability over a long period of time with a grass/grain mixture (4:1, w/w; Hoover *et al.*, 1976) as substrate. The results of this 65-day experimental period clearly demonstrated that under standard rumen conditions a steady state could be maintained with respect to ciliate numbers

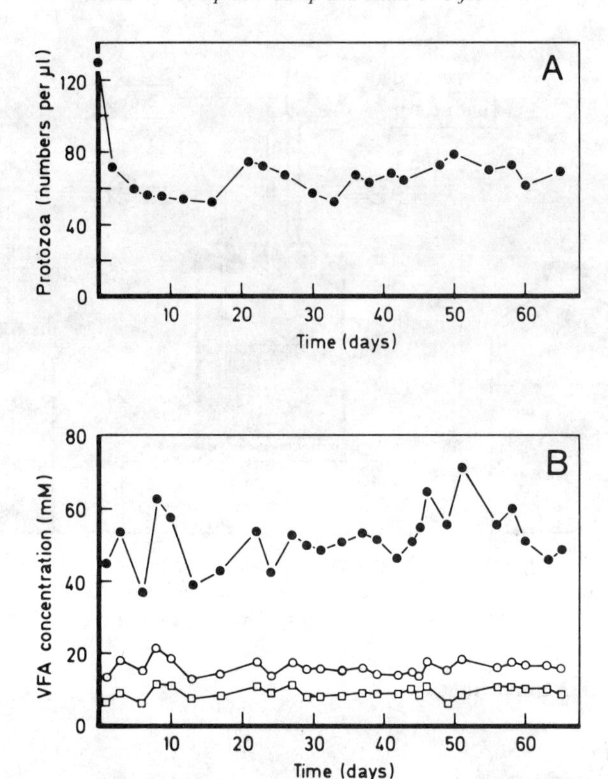

Fig. 3. Variations of protozoal numbers (A) and concentrations of fermentation products (B) during long-term operation. ●, Acetate; ○, propionate; □, butyrate. HRT 12·2 ± 0·6 h; SRT 72·7 ± 4·7 h. Adapted from Gijzen *et al.* (1986).

(Fig. 3A) and fermentation product formation (Fig. 3B). The molar ratio of acetate, propionate and butyrate was 67:21:12. The steady state conditions were reached after an equilibration period of 5–7 days. During two periods (day 7–10 and day 40–43) fiber degradation values of respectively 70% and 67% were determined. Optimal reactor performance was achieved at an HRT of 11–14 h. Above an HRT of 14 h protozoal numbers decreased strongly concomitant with fiber degradation. The high fluid flow rates are necessary to remove the acidic end products in order to keep the artificial rumen within the optimal pH-range. On the other hand, the SRT must be long enough to prevent washout of the rumen microorganisms. An SRT of 72 h was shown to be optimal. It can be concluded that the differential removal rates for liquids and solids are of major importance to maintain the efficiency of the fermentation process. The artificial rumen reactor fed a grass/grain mixture comparable to animal feed resembles well the in-vivo rumen ecosystem (Hungate, 1966, 1975; Hobson, 1988). Microorganisms, degradation efficiency, fermentation products and even the odor are identical. As an example of typical rumen features, Fig. 4 shows ciliates digesting plant fibers (A) and a ciliate covered by episymbiotic methanogenic bacteria (B), both in samples taken from the artificial rumen reactor.

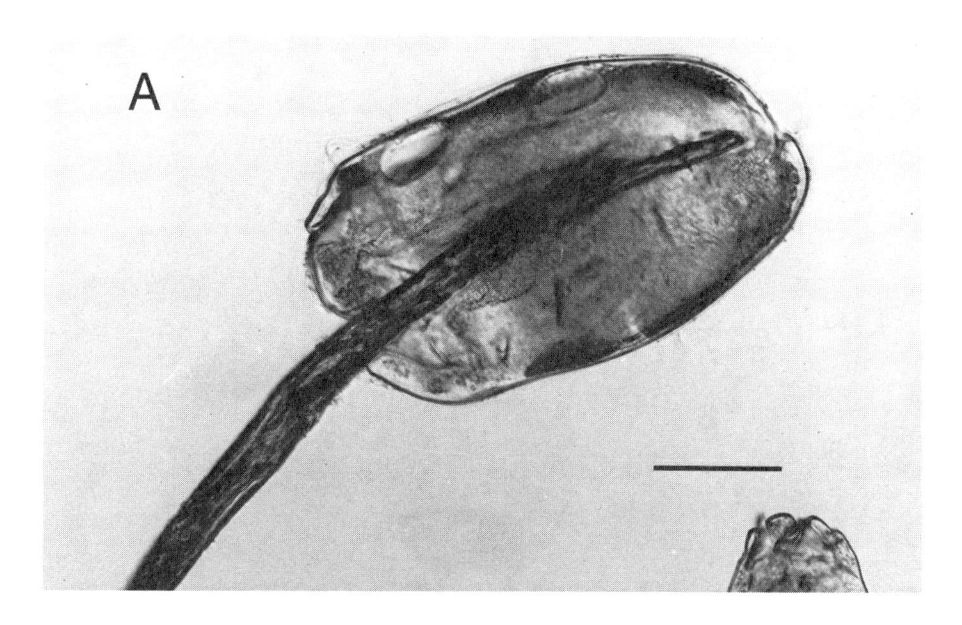

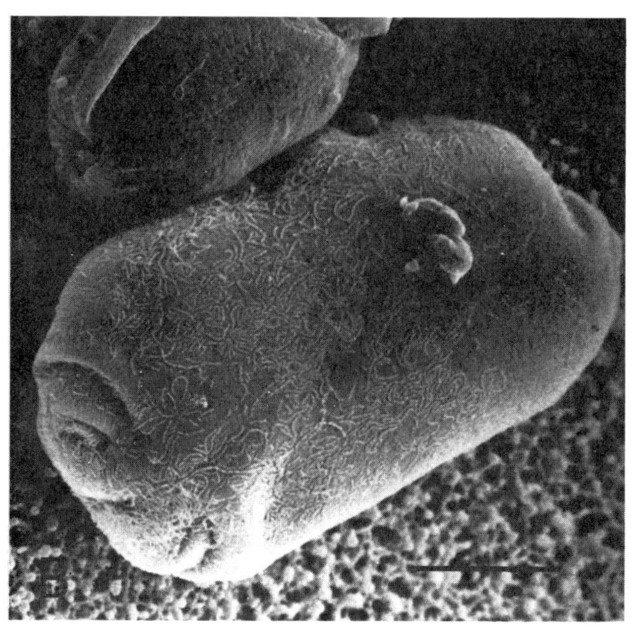

Fig. 4. (A) Micrograph of a ciliate from the artifical rumen reactor digesting fiber. (B) Scanning electron micrograph of a ciliate covered with episymbiotic methanogenic bacteria. Bars represent 50 μm. Adapted from Vogels *et al.* (1988).

Table 2
Chemical Composition of Substrates

Substrate	*TS*[a] (%)	*VS*[b] (% *of TS*)	*NDF*[c] (% *of TS*)	*COD* ($g\,O_2\ g\ TS^{-1}$)	*Cell soluble* (% *of TS*)	*Cellulose* (% *of TS*)	*Hemicellulose* (% *of TS*)	*Lignin* (% *of TS*)
Group A								
VAW[d]	5·7	90·1	34·2	1·15	55·9	22·0	12·7	2·1
Alfalfa	93·9	92·4	47·3	1·08	45·1	23·5	13·5	8·9
Verge grass	44·8	84·6	43·1	nd[h]	41·5	26·0	26·8	4·7
Horticultural waste	21·8	79·7	25·6	0·80	54·1	10·3	8·9	7·8
MSW—OF[e]	38·7	52·1	28·9	1·07	23·2	19·2	7·9	4·4
Onion pulp	7·0	93·0	11·2	1·05	81·8	3·7	3·7	3·7
Group B								
Papermill sludge	39·8	42·6	38·8	0·78	3·8	31·6	1·4	5·8
Papermill pulp	24·0	91·1	90·2	nd	0·9	nd	nd	nd
MSW—RDF[f]	41·4	83·7	68·9	1·10	14·8	61·0	5·8	2·1
Cellulose	93·9	99·7	98·5	1·22	1·2	98·0	0	0
Group C								
Bagasse	94·9	97·1	82·9	1·26	14·2	39·3	27·2	12·2
Coffee pulp	93·3	92·1	56·1	1·25	36·0	27·6	11·4	17·1
Barley straw	88·7	91·0	82·0	1·27	8·9	45·7	18·5	10·2
Rye straw	89·0	97·3	88·6	nd	6·7	48·6	24·5	15·6
Maize stover	94·0	94·0	72·0	nd	22·0	35·2	26·8	10·0
Onion peels	96·0	54·0	27·0	0·68	27·0	nd	nd	nd
SMC[g]	38·0	45·0	30·1	0·79	14·9	16·6	0·0	16·6

[a] Total solids.
[b] Volatile solids.
[c] Neutral detergent fiber.
[d] Vegetable auction waste.
[e] Municipal solid waste—organic fraction.
[f] Cellulose-rich fraction of municipal solid waste, usually referred to as refuse derived fuel.
[g] Spent mushroom compost.
[h] nd = Not determined.

3 IMPROVED ACIDOGENESIS OF SEVERAL WASTE MATERIALS AND NATURAL SUBSTRATES USING THE ARTIFICIAL RUMEN REACTOR

The materials tested thus far as fermentor feed in the artificial rumen reactor can be divided into three groups:

(A) Materials relatively rich in cell solubles ($>40\%$ of the organic matter).
(B) Materials with a high cellulose content ($>40\%$ of the organic matter) and a low amount of cell solubles ($<40\%$ of the organic matter).
(C) Materials with a relatively high lignin content ($>10\%$ of the organic matter) and a low amount of cell solubles ($<40\%$ of the organic matter).

Table 2 shows the chemical composition of the various materials tested. The percentage of inorganic matter differed markedly, but was not used to characterize the materials. Materials from group B and C are both rich in cell wall polymers. The cellulosic (group B) differ from the lignocellulosic materials (group C) mainly because the former are partially or completely delignified in an industrial process.

The results compiled in Table 3 demonstrate that the artificial rumen reactor can be applied for an efficient acidogenesis of various solid organic wastes. Degradation was measured on the basis of neutral detergent fiber (NDF) and/or volatile solids (VS). The relative proportions of individual VFA were markedly affected by the substrate composition. Acidogenesis from materials rich in cell solubles yielded a higher molar proportion of butyrate, while from (ligno)cellulosic materials more propionate was produced. Protozoa were present in high numbers in all experiments, although species distribution was affected by the substrate used (Gijzen *et al.*, 1987).

A low NDF degradation efficiency was obtained with spent mushroom compost (SMC) as substrate. SMC combines a relatively high inorganic content and lignin content. Lignin has been generally recognized to limit fiber degradation. In batch cultures with rumen microorganisms a linear relationship between fiber degradation and lignin content was reported (Op den Camp *et al.*, 1988). When the lignocellulosic substrates were fed to the artificial rumen reactor a loss of lignin was observed to an extent of 32–60%. As shown recently this is due to solubilization rather than degradation (Kivaisi *et al.*, 1990). Lignin–carbohydrate complexes go into solution during the degradation of cell wall polymers, both spontaneously and through the action of rumen microorganisms. The carbohydrate part of the complexes is metabolized by the rumen microorganisms, while the lignin part remains unaffected.

All materials were tested under more or less standard rumen conditions, e.g. SRT 48–110 h and HRT 10–14 h, without further optimization. For municipal solid waste—organic fraction (MSW—OF) (Op den Camp *et al.*, 1989), papermill sludge (Gijzen *et al.*, 1990), MSW—cellulose-rich fraction (MSW—RDF) (Gijzen *et al.*, 1988*b*) and barley straw (Kivaisi, 1990) optimization studies were described. From these results it has become clear that the HRT should not be above 24 h. The relative high fluid flow rate is required to prevent acidification. Furthermore it was observed

Table 3
Performance of the Artificial Rumen Reactor During Steady-state Degradation of Several Substrates

Substrate	*Loading rate* (*g VS liter*$^{-1}$ *day*$^{-1}$)	*Alfalfa added*	*Degradation* (%)		*VFA*[c] *production* (*mmol liter*$^{-1}$ *day*$^{-1}$)	*A*:*P*:*B*[d]	*Biogas* (*liter liter*$^{-1}$ *day*$^{-1}$)	*Ref.*
			NDF[a]	*VS*[b]				
Group A								
VAW	27·0	+	76	92	164	59 18 23	5·7	I
Alfalfa	30·4	−	66	79	171	70 18 12	4·6	I
Verge grass	28·9	−	54	64	138	65 22 13	3·9	I
Horticultural waste	24·4	−	62	73	137	72 15 13	3·5	I
MSW—OF	30·0	−	58	63	134	68 16 16	3·9	II
Onion pulp	13·9	+	81	64	91	54 18 21	4·6	III
Group B								
Papermill sludge	23·0	+	58	nd[e]	127	66 27 7	4·4	IV
Papermill pulp	21·5	+	62	nd	109	69 27 4	2·6	I
MSW—RDF	22·3	+	76	nd	131	69 23 8	3·3	V
Cellulose	18·0	+	70	nd	95	70 24 6	2·4	I
Group C								
Bagasse	18·3	−	64	nd	71	66 25 9	1·9	I
Coffee pulp	17·1	−	72	nd	87	71 21 8	2·2	I
Barley straw	24·3	−	57	nd	113	72 20 8	1·8	VI
Rye straw	26·0	−	45	nd	85	72 22 6	1·6	VI
Maize stover	17·7	−	52	nd	76	74 21 5	1·6	VI
Onion peels	16·6	+	54	63	45	71 24 5	1·7	III
SMC	17·3	+	33	nd	25	61 31 8	0·4	I

For experimental details see the references indicated. Due to a systematic error all SRTs mentioned in these references should be multiplied by a factor of 1·2. For SRT estimation a volume of 1·5 liters instead of 1·8 liters was used. Methane content of the biogas was 30–45%. For substrate abbreviations see Table 2.
References: (I) Gijzen *et al.*, 1987; (II) Op den Camp *et al.*, 1989; (III) Lubberding *et al.*, 1988; (IV) Gijzen *et al.*, 1990; (V) Gijzen *et al.*, 1988*b*; (VI) Kivaisi *et al.*, 1988.
[a] Neutral detergent fiber. [b] Volatile solids. [c] Volatile fatty acids.
[d] Molar ratio of acetate, propionate and butyrate. [e] nd = Not determined.

that an SRT below 72 h resulted in a sharp decrease in degradation efficiency and an SRT longer than 144 h was also suboptimal because of accumulation of recalcitrant substrate components. The loading rate (LR) was limited by the content of total solids in the reactor, since a high TS-content caused mixing problems. The highest LR reported was 40 g VS $liter^{-1}$ day^{-1} for MSW—OF (Op den Camp *et al.*, 1989). In conclusion, it was demonstrated that acidogenic digestion of various organic waste materials could be largely improved by the application of a reactor, simulating rumen conditions and involving rumen microorganisms.

4 THE RUMEN-DERIVED ANAEROBIC DIGESTION (RUDAD) PROCESS

As indicated, rumen microorganisms degrade all substrates listed in Table 2 to volatile fatty acids (VFA) and a relatively small amount of biogas. To achieve a complete conversion to biogas the VFA on turn must be converted to CH_4 and CO_2. The upflow anaerobic sludge blanket (UASB) reactor has found the widest application of the high rate anaerobic wastewater treatment systems (Sahm, 1984; Lettinga & Hulshoff Pol, 1986) and was therefore chosen for conversion of the products of the acidogenic rumen reactor into biogas.

4.1 Description of the Two-stage System

The acidogenesis was performed as outlined in Section 2.1. An UASB-reactor was coupled to the artificial rumen reactor as indicated in Fig. 5. The UASB-reactor (B)

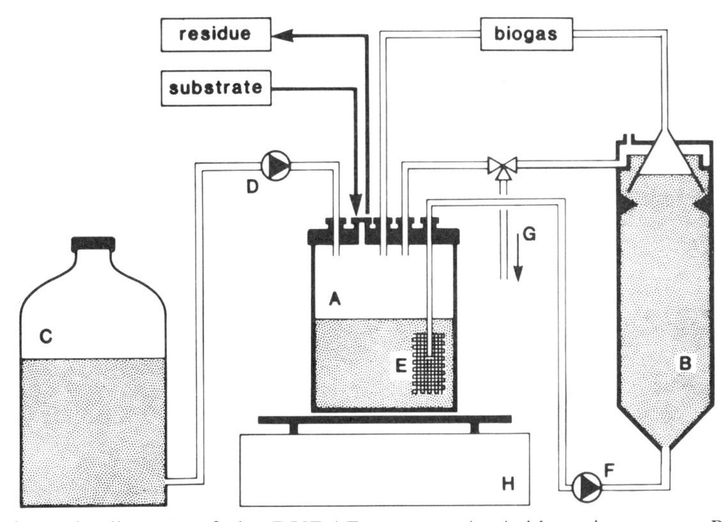

Fig. 5. Schematic diagram of the RUDAD-process. A, Acidogenic reactor; B, UASB-reactor; C, fermentation medium reservoir; D, fermentation medium supply pump; E, filter, 30 μm pore size; F, filtered effluent removal pump; G, UASB-reactor effluent; H, rotary shaker. Adapted from Gijzen *et al.* (1990).

consisted of a glass cylinder (9 cm internal diameter, 28 cm height) with a water jacket, a conical-shaped bottom and a gas collection and settler compartment. The reactor was filled with settled granular sludge from a full-scale potato processing wastewater treatment plant. Liquid effluent from the acidogenic reactor was fed to the UASB-reactor at a rate of about 3 liter day^{-1}, corresponding to an HRT of the acidogenic reactor of 12 h. A big advantage of the two-stage system is the possibility of recirculating the liquid, after passage through the UASB-reactor, to the rumen reactor. In this way a closed fluid circuit is obtained and the buffer supply can be disconnected. The UASB-reactor is in fact used to regenerate buffer capacity of the liquid medium since organic acids produced in the rumen reactor are converted to biogas. The two-phase process described here (Fig.5) is referred to as the 'Rumen Derived Anaerobic Digestion' (RUDAD) process (Gijzen *et al.*, 1988*c*) since it is based on processes and microorganisms of the ruminant.

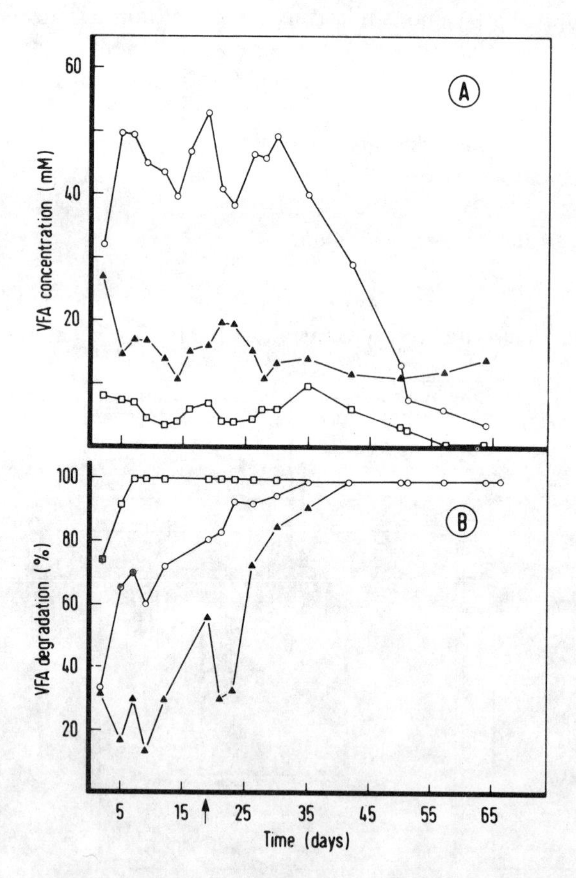

Fig. 6. (A) Effect of prolonged fluid recirculation on the concentration of acetate (○), propionate (▲) and butyrate (□) in the acidogenic reactor; (B) their conversion efficiency in the methanogenic reactor. The arrow indicates the start of fluid recirculation. Adapted from Gijzen *et al.* (1988*c*).

4.2 Stability of the RUDAD-process on Pure Cellulose

Filter paper cellulose appeared to be a good model substrate to test the RUDAD-process (Gijzen *et al.*, 1988*c*). During the experiment no removal of total solid substrate input other than small amounts needed for analysis was necessary. With a closed fluid circuit a situation was created in which cellulose (plus a small amount of alfalfa) was the only input and biogas the only output. Degradation of substrate was virtually 100% with a slight accumulation of alfalfa residues. At a LR of 16·6 g VS $liter^{-1}$ day^{-1} a total methane production of 0·438 liter CH_4 g VS^{-1} added was measured, equivalent to 98% of the theoretical value. The conversion rate for cellulose in the RUDAD-process is significantly higher than those reported for other anaerobic systems (Laube & Martin, 1981; Khan *et al.*, 1983; Petitdemange *et al.*, 1984).

The effect of prolonged fluid recirculation on concentration of VFA in the acidogenic reactor and their conversion in the methanogenic reactor is shown in Fig. 6. Before fluid recirculation it already became clear that the UASB-reactor quickly adapted to the liquid effluent of the acidogenic reactor. After being stable for about 30 days VFA concentrations in the acidogenic reactor dropped most probably because of a distribution of non-flocculent methanogenic bacteria over both reactors. Part of the methanogenic activity was taken over by the acidogenic reactor. This occurred mainly because neither liquid nor solid matter had to be removed and as a consequence all microbial biomass was retained in the system. Concerning the biomass in the acidogenic reactor a striking result was observed (Gijzen *et al.*, 1988*c*). From day 19 to day 68 only a small increase in biomass (from 4 to 7 g protein $liter^{-1}$) was observed, taking into account the high amount (813 g $liter^{-1}$) of cellulose degraded during this period. Apparently, conversion of cellulose to biogas proceeded as a growth-independent process, due either to predation effects of the protozoal and bacterial populations or to uncoupling at a molecular and physiological level.

5 APPLICATIONS OF THE RUDAD-PROCESS

5.1 Anaerobic Digestion of Barley and Rye Straw

Lignification of plant cell walls is considered to be the most important factor limiting digestibility of forage by ruminants (Muntifering *et al.*, 1981). To understand problems and limitations for lignocellulose digestion in the RUDAD-process, Kivaisi *et al.* (1988) studied long-term digestion of barley and rye straw. These experiments provide valuable information which is needed prior to large-scale applications.

The results over a 3-month digestion period using barley and rye straw as fermentor feed are compliled in Fig. 7. The LR and SRT were 17·7 g VS $liter^{-1}$ day^{-1} and 72 h, respectively, throughout the experiment. Period I was used to start, stabilize and couple the acidogenic and methanogenic reactors. During periods II

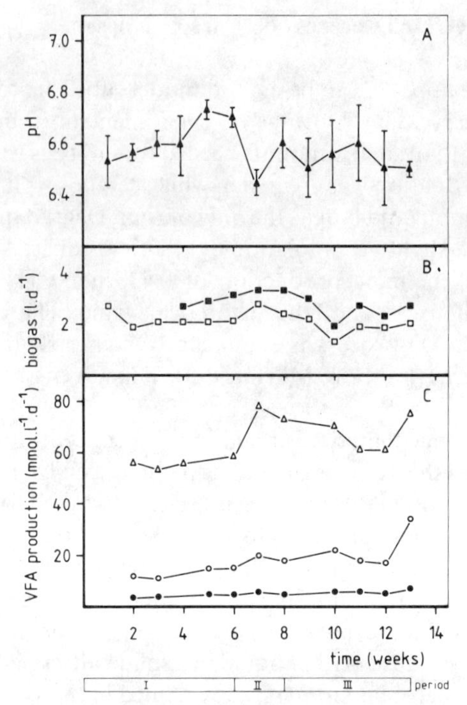

Fig. 7. Performance of the RUDAD-process during long-term digestion of barley straw (Periods I and II) and rye straw (Period III). (A) pH of the acidogenic reactor. (B) Biogas production of acidogenic (□) and methanogenic reactor (■). (C) Production of acetate (△), propionate (○) and butyrate (●) in the acidogenic reactor. Adapted from Kivaisi *et al.* (1988).

and III a stable fiber degradation efficiency of 38% was measured in spite of the use of two different types of straw. The methanogenic reactor converted the products of the acidogenic reactor almost completely into biogas. Rumen ciliates used as indicator organisms were present in the acidogenic fermentor throughout the experimental period. The results demonstrate that the RUDAD-process remains stable for a long time with cereal residues as the sole substrate. After fluctuations in operating conditions, mostly because of technical faults, the system quickly stabilized again without reinoculation.

5.2 Anaerobic Digestion of Papermill Sludge

The pulp and papermill industry produces huge amounts of wastewater and solid wastes. Papermill sludge is produced by papermills using recycled paper as raw material. The major part of papermill sludge is composed of inorganic matter (Table 2). Since this may cause a negative effect on the efficiency and process stability during long-term operation, the digestion of papermill sludge has been tested over a 3-month period (Gijzen *et al.*, 1990) at constant SRT (58 h) and LR (23·8 g VS liter^{-1} day^{-1}). The results obtained are summarized in Fig. 8. Average VFA and

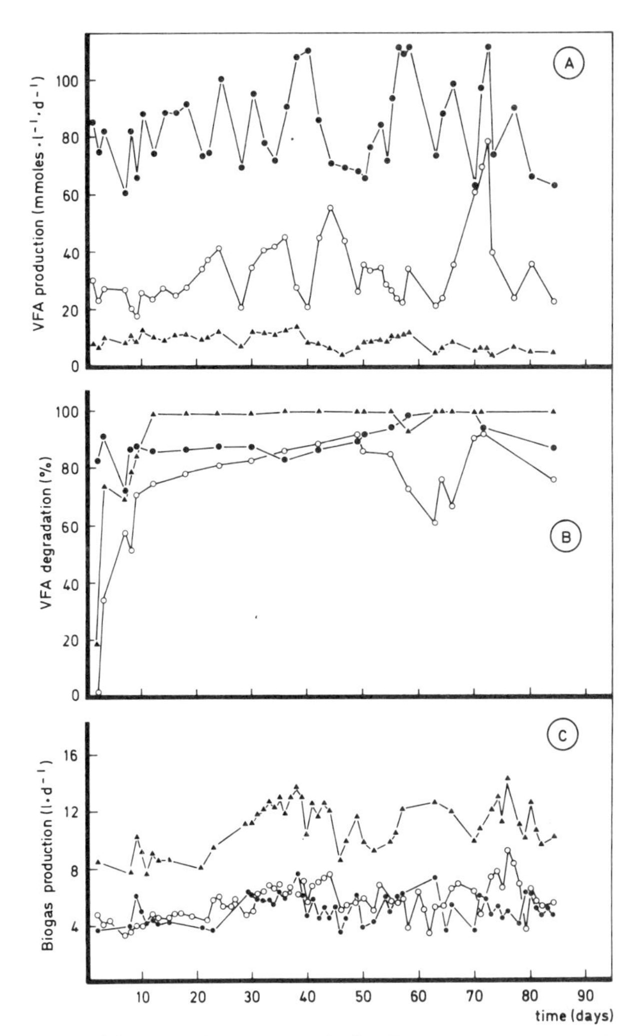

Fig. 8. Performance of the RUDAD-process during long-term digestion of papermill sludge. (A) Production in the acidogenic reactor, and (B) degradation in the methanogenic reactor, of acetate (●), propionate (○) and butyrate (▲). (C) Biogas production from the acidogenic reactor (●), methanogenic reactor (○) and total production (▲). Adapated from Gijzen *et al.* (1990).

biogas production after day 25 was 127 mmol liter^{-1} day^{-1} and 11·4 liter day^{-1}, respectively. A relatively high VS degradation was obtained (50–60%). An increase of the inorganic content of the UASB-sludge from 0·45 to 0·75 g g TS^{-1} was observed during the first 50 days, but the ash content stabilized at the high level. However, further increase will have a negative effect on methanogenesis or even result in washout of the methanogenic biomass. Therefore selective separation of inorganic material from the RUDAD-process by flotation or sedimentation should be introduced to improve reactor performance.

5.3 Anaerobic Digestion of a Cellulose-rich Fraction of MSW

The cellulosic fraction of domestic refuse (RDF) has been studied most extensively (Gijzen *et al.*, 1988*b*; Zwart *et al.*, 1988). After optimization of the acidogenic artificial rumen reactor on this substrate (Gijzen *et al.*, 1988*b*) on 1·5 liter scale the acidogenic phase was scaled-up to 30 liter (20 liter working volume) and coupled to a 12-liter UASB-reactor. In this configuration a long-term digestion trial lasting for more than one year was performed (Zwart *et al.*, 1988). The performance of the

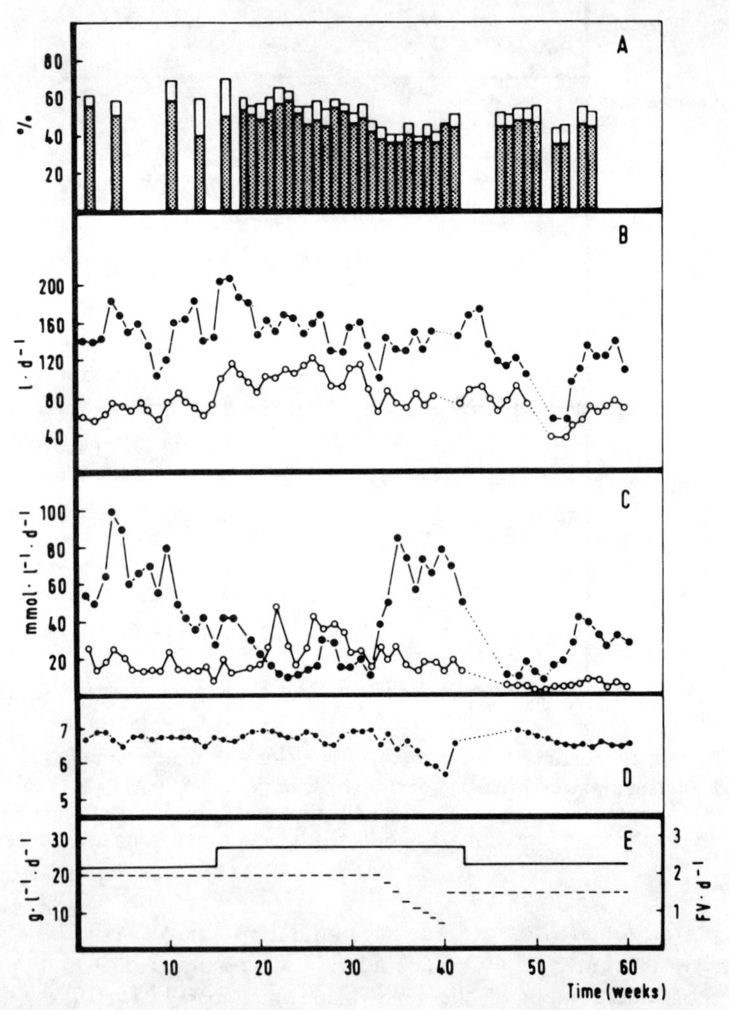

Fig. 9. Performance of the RUDAD-process during long-term digestion of MSW—RDF. (A) Degradation of VS (□) and TS (▦). (B) Biogas production of acidogenic reactor (○) and total system (●). (C) Production of acetate (●) and propionate (○). (D) pH of the acidogenic reactor. (E) Loading rate (——) and fluid removal rate (- - - - -) from the acidogenic reactor. The dotted intervals in the graphs represent periods in which no samples were analyzed. Adapted from Zwart *et al.* (1988).

RUDAD-process during this experiment is illustrated in Fig. 9. An inoculum of rumen fluid of 2·5% by volume was sufficient to start up the acidogenic reactor. Within 3 weeks ciliate numbers reached a level almost comparable to that of the rumen, and no reinoculation was necessary during the rest of the test period. The species composition of the ciliated protozoa was dependent on substrate characteristics (Gijzen *et al.*, 1988*b*). At an LR of 22·5 g VS $liter^{-1}$ day^{-1} stable NDF and VS degradation efficiencies of about 70% were obtained. The COD reduction measured was also 70%. From this pilot study it can be calculated that a large-scale application for RDF will give a methane production of 337 m^3 per tonne VS degraded. Process stability and rapid growth are of course a prerequisite for large-scale applications. Another requirement is the separation of liquid and solids in the acidogenic reactor. Up to the 30-liter reactor the filter systems meet the needs for separation. Filtration on a large scale may not be applicable because of the increasing volume needed for the filter. This problem may be solved by using alternative methods for the separation of liquids and solids (Stanton, 1980).

6 THE HYBRID RUDAD-REACTOR FOR DIGESTION OF PAPERMILL SLUDGE

As indicated in Section 5.2, papermill sludge can be digested effectively by the RUDAD-process. However, this fine particulate waste with a high content of inorganic material can cause problems by clogging of the filter unit of the acidogenic reactor and/or the accumulation of inorganic matter. Especially the filtering problem could be circumvented by combining the acidogenic and methanogenic phase in one reactor (Fig. 10). The methanogenic phase consists of reticulated polyurethane foam, floating on the acidogenic phase, colonized by a bacterial

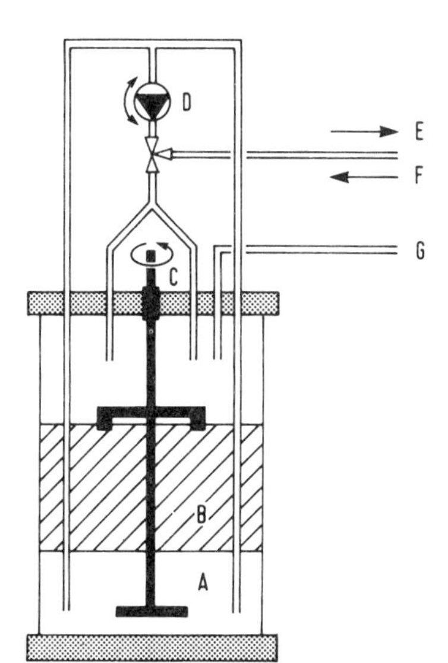

Fig. 10. Schematic diagram of the two-phase reactor. A, Acidogenic phase; B, methanogenic phase; C, stirrer; D, bi-directional sludge pump; E, homogeneous effluent removal; F, feeding and addition of fresh fermentation medium; G, gas outlet. Adapted from Gijzen *et al.* (1988*a*).

population able to convert VFA into biogas. The acidogenic phase of the hybrid reactor was started up first (Gijzen *et al.*, 1988*a*). One day after the inoculation the foam support particles (colonized for 3 weeks on a VFA-based medium and UASB-sludge as inoculum) were brought into the hybrid reactor. The total effective volume was 15 liters. The acidogenic part was stirred and homogeneous slurry was pumped over the layer of foam particles at a rate of 25 liter day^{-1}. Every day 2·5 liters of homogeneous acidogenic phase content was replaced by an equal amount of papermill sludge. The hybrid RUDAD reactor was operated for about 90 days with papermill sludge as a substrate (Caerteling *et al.*, 1987). During this period the LR was increased from 75 to 102 g VS day^{-1}. Rumen ciliates remained present in high numbers and a high conversion of acetate and butyrate was found. Propionate accumulated temporarily, most likely because of slow specific growth rates of propionate-oxidizing populations leading to a longer adaptation time.

In conclusion, the hybrid RUDAD-reactor can be applied successfully for high-rate anaerobic degradation of papermill sludge. The high LR (up to 20 g VS liter^{-1} day^{-1}), short SRT (48 h), simple reactor construction and degradation efficiency of 50–70% make this system attractive for degradation of other particulate solid wastes.

7 THE RUDAD-PROCESS IN COMPARISON TO OTHER SYSTEMS TREATING SOLID ORGANIC WASTE MATERIALS

The results presented in this chapter demonstrate that it is possible to apply a mixed population of rumen microorganisms in a reactor, under conditions similar to those in the rumen, for the degradation of cellulosic waste materials. The rate of acidogenesis of a number of organic waste materials of various origins and compositions appeared to be high in the artificial rumen reactor. The maximum LR obtained so far in this novel reactor amounted to 40 g VS liter^{-1} day^{-1} during the degradation of MSW—OF (Op den Camp *et al.*, 1989). LR applied in the artificial rumen reactors are about 15 to 30 times as high as those reported for conventional reactors treating similar waste materials, such as vegetable wastes (Knol *et al.*, 1978), papermill sludge (Takeshita *et al.*, 1981), rice straw (Lequerica *et al.*, 1984), coffee pulp (Calzada *et al.*, 1981) and MSW (Pfeffer, 1980; Pathe *et al.*, 1982; Van der Vlugt & Rulkens, 1984). In these systems LR of only 0·5–3 g VS liter^{-1} day^{-1} could be achieved because of the long SRT required to obtain satisfactory degradation efficiency.

Because of the limitations of these conventional solid waste digestors, several new reactor designs have been proposed recently. The dry anaerobic composting system (DRANCO) proposed by de Baere and Verstraete (1984) can be operated at high LRs (10–20 kg VS liter^{-1} day^{-1}), but long SRTs of 10–20 days are required to obtain VS degradation efficiencies of 50–95% (Six & de Baere, 1988). The thermophilic conditions applied in this reactor (55°C) require additional energy input, which forms a disadvantage compared to mesophilic systems. Another dry digestion process, referred to as BIOCEL, is operated in a batch mode under mesophilic conditions (ten Brummeler *et al.*, 1988). Because of the extremely long retention times required (4 months) to obtain a VS degradation of about 55% during the treatment of MSW—OF, the overall LR realized was only 3 g VS $liter^{-1}$ day^{-1}. Rivard *et al.* (1990) realised LR as high as 9·5 g VS $liter^{-1}$ day^{-1} during the degradation of MSW in a novel high-solids reactor, whereas Chen *et al.* (1990) obtained LR up to 3·2 g VS $liter^{-1}$ day^{-1} during the digestion of MSW—RDF in a non-mixed solids concentrating reactor. For comparison, LRs obtained with the RUDAD-system digesting MSW—RDF were up to 27 g VS $liter^{-1}$ day^{-1} (Zwart *et al.*, 1988).

The forestomach of ruminants is probably the most efficient anaerobic digestion process for cellulosic materials that is known so far. In this respect it may not be surprising that the RUDAD-process shows markedly better performance compared to other anaerobic digestors which are operated under other conditions and therefore develop different microbial populations. In laboratory-scale models (1·5 liters and 20 liters) the RUDAD-process has shown reliable and stable performance over an extensive period of continuous operation and with a variety of waste materials. The technical and economic feasibility of the RUDAD-process remains to be established on a pilot-plant scale.

Our work has shown that the rumen microbial ecosystem may be of biotechnological significance because of its high cellulolytic and fermentative activities. Besides the cellulolytic organisms, other rumen microorganisms may carry potential biotechnological applications in the production of fuel and chemicals from biomass. Lin *et al.* (1985) suggested a wide range of applications of rumen microorganisms for the production of acetate, propionate, succinate, ethanol, lactate, and methane. Numerous studies report on the metabolic potentials of mono- and co-cultures of rumen microorganisms (Lin *et al.*, 1985; Ohmiya *et al.*, 1986; Wolin & Miller, 1987; Hobson, 1988; Pavlostathis *et al.*, 1990). The products of cellulose fermentation vary depending on the bacterial species used and their interactions. Co-cultures of cellulytic rumen microorganisms with methanogenic bacteria for instance result in the production of more oxidized products such as acetate instead of the more reduced products ethanol or lactate.

It is expected that a further understanding of the metabolism of individual rumen microorganisms and the effect of co-cultures on their metabolism may result in more biotechnological applications. In this respect microorganisms from other high-rate natural ecosystems such as the termite or cockroach hindgut might also prove to carry useful biotechnological potentials.

REFERENCES

Archer, D. B. (1983). *Enzyme Microb. Technol.*, **5**, 162.

Brown, R. D., Jr & Gritzali, M. (1984). In *Genetic Control of Environmental Pollutants*, eds G. S. Omenn & A. Hollaender. Plenum Publishing Corporation, New York, p. 239.

Bryant, M. P. (1979). *J. Anim. Sci.*, **48**, 193.

Caerteling, C. G. M., Gijzen, H. J., Op den Camp, H. J. M., Lubberding, H. J. & Vogels, G. D. (1987). In *Proceedings of the Gasmat Workshop on Granular Anaerobic Sludge;*

Microbiology and Technology, eds G. Lettinga, A. J. B. Zehnder, J. T. C. Gotenhuis & L. W. Hulshoff Pol. PUDOC, Wageningen, The Netherlands, p. 71.
Callander, I. J. (1982). *The development of the tower fermentor for anaerobic digestion.* PhD theses, University of Sydney, Australia.
Callander, I. J. & Barford, J. P. (1983). *Process Biochem.*, **18**, 24.
Calzada, J. F., de Leon, O. R., de Arriola, M. C., de Micheo, F., Rolz, C., de Leon, R. & Menchu, J. F. (1981). *Biotechnol. Lett.*, **3**, 713.
Chen, T. H., Chynoweth, P. & Biljetina, R. (1990). *Appl. Biochem. Biotechnol.*, **24/25**, 533.
Coleman, G. S. (1983). *J. Protozool.*, **30**, 36A.
de Baere, L. & Verstraete, W. (1984). In *Anaerobic Digestion and Carbohydrolysis of Waste*, eds G. L. Ferrero, M. P. Ferranti & H. Naveau. Elsevier Applied Science, London, p. 195.
Duchesne, L. C. & Larson, D. W. (1989). *BioSystems*, **39**, 238.
Eastman, J. A. & Ferguson, J. F. (1981). *J. Water Pollut. Control Fed.*, **53**, 352.
Ehalt, D. H. (1976). In *Microbial Production and Utilization of Gases*, ed. H. G. Schlegel, G. Gottschalk & N. Pfennig, K. G. Goltze, Göttingen, FRG, p. 13.
Ghosh, S. & Pohland, F. G. (1974). *J. Water Pollut. Control Fed.*, **46**, 748.
Gijzen, H. J. (1987). *Anaerobic digestion of cellulosic waste by a rumen derived process.* PhD thesis, University of Nijmegen, The Netherlands.
Gijzen, H. J., Zwart, K. B., van Gelder, P. T. & Vogels, G. D. (1986). *Appl. Microbiol. Biotechnol.*, **25**, 155.
Gijzen, H. J., Lubberding, H. J., Verhagen, F. J., Zwart, K. B. & Vogels, G. D. (1987). *Biol. Wastes*, **22**, 81.
Gijzen, H. J., Schoenmakers, T. J. M., Caerteling, C. G. M. & Vogels, G. D. (1988*a*). *Biotechnol. Lett.*, **10**, 61.
Gijzen, H. J., Zwart, K. B., Teunissen, M. J. & Vogels, G. D. (1988*b*). *Biotech. Bioeng.*, **32**, 749.
Gijzen, H. J., Zwart, K. B., Verhagen, F. J. M. & Vogels, G. D. (1988*c*). *Biotech. Bioeng.*, **31**, 418.
Gijzen, H. J., Derikx, P. J. L. & Vogels, G. D. (1990). *Biol. Wastes*, **32**, 169.
Hashimoto, A. G. (1986). *Biotech. Bioeng.*, **28**, 1857.
Heijnen, J. J. (1983). In *Proceedings of the European Symposium on Anaerobic Wastewater Treatment*, ed. W. J. van der Brink. TNO, Corporate Communication Secretariat, The Netherlands, p. 283.
Hobson, P. N. (1988). *The Rumen Microbial Ecosystem.* Elsevier Applied Science, London & New York.
Hobson, P. N. & Wallace, R. J. (1982). *CRC Microbiol.*, **9**, 165.
Hoffenk, G. (1983). In *Proceedings of the European Symposium on Anaerobic Wastewater Treatment*, ed. W. J. van der Brink. TNO, Corporate Communication Secretariat, The Netherlands, p. 475.
Hoover, W. H., Crooker, B. A. & Sniffen, C. J. (1976). *J. Anim. Sci.*, **93**, 528.
Hungate, R. E. (1966). *The Rumen and its Microbes.* Academic Press, New York.
Hungate, R. E. (1975). *Ann. Rev. Ecol. Syst.*, **6**, 39.
Hungate, R. E. (1982). *Experienta*, **38**, 189.
Kaspar, H. F. & Wuhrmann, K. (1978). *Appl. Environ. Microbiol.*, **36**, 1.
Khan, A. W., Miller, S. S. & Murray, W. D. (1983). *Biotechnol. Bioeng.*, **25**, 1571.
Kivaisi, A. K. (1990). *Anaerobic degradation of cereal residues by a rumen-derived process.* PhD thesis, University of Dar-es-Salaam, Tanzania.
Kivaisi, A. K., Lubberding, H. J., Op den Camp, H. J. M., Robben, A. J. P. M. & Vogels, G. D. (1988). *Med. Fac. Landbouww. Rijksuniv. Gent*, **53**, 1863.
Kivaisi, A. K., Op den Camp, H. J. M., Lubberding, H. J., Boon, J. J. & Vogels, G. D. (1990). *Appl. Microbiol. Biotechnol.*, **33**, 93.
Knol, W., van der Most, M. M. & de Waart, J. (1978). *J. Sci. Fd Agric.*, **29**, 822.
Ladisch, M. R., Lin, K. W., Voloch, M. & Tsao, G. T. (1983). *Enzyme Microb. Technol.*, **5**, 81.
Laube, V. M. & Martin, S. M. (1981). *Appl. Environ. Microbiol.*, **42**, 413.
Lequerica, J. L., Valles, S. & Flors, A. (1984). *Appl. Microbiol. Biotechnol.*, **19**, 70.

Lettinga, G. & Hulshoff Pol, L. W. (1986). *Water Sci. Technol.*, **18**, 99.
Lin, K. W., Patterson, J. A. & Ladisch, M. R. (1985). *Enzyme Microb. Technol.*, **7**, 98.
Lipinsky, E. S. (1981). *Science*, **212**, 1465.
Ljungdahl, L. G. & Eriksson, K. E. (1985). *Adv. Microb. Ecol.*, **8**, 237.
Lubberding, H. J., Gijzen, H. J., Heck, M. & Vogels, G. D. (1988). *Biol. Wastes*, **25**, 61.
McInerney, M. J. & Bryant, M. P. (1981). In *Fuel Gas Production from Biomass*, ed. D. L. Wise. Chemical Rubber Co. Press, Inc., West Palm Beach, FL, p. 26.
Muntifering, R. B., DeGregorio, R. M. & Deetz, L. E. (1981). *Nutr. Rep. Int.*, **24**, 543.
Noike, T., Endo, G., Chang, J. E., Yaguch, J. I. & Matsumoto, J. I. (1985). *Biotech. Bioeng.*, **27**, 1482.
Ohmiya, K., Takeuchi, M., Chen, W., Shimizu, S. & Kawakami, H. (1986). *Appl. Microbiol. Biotechnol.*, **23**, 274.
Op den Camp, H. J. M., Verhagen, F. J. M., Kivaisi, A. K., de Windt, F., Lubberding, H. J., Gijzen, H. J. & Vogels, G. D. (1988). *Appl. Microbiol. Biotechnol.*, **29**, 408.
Op den Camp, H. J. M., Verkley, G. J. M., Gijzen, H. J. & Vogels, G. D. (1989). *Biol. Wastes*, **30**, 309.
Oremland, R. S. (1988). In *Biology of Anaerobic Microorganisms*, ed. A. J. B. Zehnder. John Wiley & Sons, New York, p. 641.
Pathe, P. P., Alone, B. Z., Titus, S. K. & Bhide, A. D. (1982). *Ind. J. Environ. Hlth*, **24**, 8.
Pavlostathis, S. G., Miller, T. L. & Wolin, M. J. (1990). *Appl. Microbiol. Biotechnol.* **33**, 109.
Petitdemange, A., Fond, D., Ravel, G., Petitdemange, H. & Gay, R. (1984). In *Anaerobic Digestion and Carbohydrolysis of Waste*, ed. G. L. Ferrero, M. P. Ferranti & H. Naveau. Elsevier Applied Science, London, p. 223.
Pfeffer, J. T. (1980). In *Anaerobic Digestion 1979*, eds D. A. Stafford, B. I. Wheatly & D. E. Hughes. Applied Science Publishers, London, p. 187.
Prins, R. A. & Clarke, R. T. J. (1980). In *Proceedings of the 5th International Symposium on Rumen Physiology*, ed. Y. Ruckebush & P. Thivend. MPT Press, Lancaster, UK, p. 179.
Rivard, C. J., Himmel, M. E., Vinzant, T. B., Adney, W. S., Wyman, C. E. & Grohmann, K. (1990). *Biotechnol. Lett.*, **12**, 235.
Rufener, W. H., Jr, Nelson, W. O. & Wolin, M. J. (1963). *Appl. Microbiol.*, **11**, 196.
Sahm, H. (1984). In *Advances in Biochemical Engineering Biotechnology*, vol. 29, ed. A. Fiechter. Springer-Verlag, Berlin, p. 83.
Six, W. & de Baere, L. (1988). In *Proceedings of the 5th International Symposium on Anaerobic Digestion*, ed. A. Tilche & A. Rozzi. Monduzzi Editore, Bologna, Italy, p. 793.
Stafford, D. A., Hawkes, D. L. & Horton, R. (1980). *Methane Production from Waste Organic Matter*. CRC Press, Boca Raton, FL.
Stanton, M. J. (1980). In *Anaerobic Digestion*, eds D. A. Stafford, B. I. Wheatly & D. E. Hughes. Applied Science Publishers, London, p. 395.
Stephens, G. R. & Heichel, G. H. (1975). *Biotech. Bioeng. Symp.*, **5**, 27.
Swift, M. J. (1977). *Sci. Prog. Oxford*, **64**, 175.
Takeshita, N., Fujimura, E. & Mimoto, N. (1981). *Pulp and Paper Canada*, **82**, 99.
ten Brummeler, E., Koster, I. W. & Zeevalkink, J. A. (1988). In *Anaerobic Digestion 1988*, eds E. R. Hall & P. N. Hobson. Pergamon Press, Oxford, p. 335.
Thauer, R. K., Jungermann, K. & Decker, K. (1977). *Bacteriol. Rev.*, **41**, 100.
Thompson, N. S. (1983). In *Wood and Agricultural Residues*, ed. E. J. Soltes. Academic Press, New York, p. 101.
Tsao, G. T., Ladisch, M. R., Ladisch, C., Hsu, T. A., Dale, B. & Chou, T. (1978). *Ann. Rep. Ferment. Proc.*, **2**, 1.
Van der Vlugt, A. J. & Rulkens, W. H. (1984). In *Anaerobic Digestion and Carbohydrate Hydrolysis of Waste*, eds G. L. Ferrero, M. P. Ferranti & H. Naveau. Elsevier Applied Science, London, p. 245.
Van Velsen, A. F. M. (1981). *Anaerobic digestion of piggery waste*. PhD thesis, Agricultural University of Wageningen, The Netherlands.
Vogels, G. D. (1979). *Antonie van Leeuwenhoek, J. Microbiol. Serol.*, **45**, 347.

Vogels, G. D., Stumm, C. K., Broers, C. A. M. & Goosen, N. K. (1988). In *Energy Transformations in Cells and Organisms*, eds W. Wieser & E. Gnaiger. Georg Thieme Verlag, Stuttgart, FRG, p. 163.

Wolin, M. J. (1979). *Adv. Microb. Ecol.*, **3**, 49.

Wolin, M. J. & Miller, T. L. (1987). *Geomicrobiol. J.*, **5**, 239.

Zehnder, A. J. B. & Stumm, W. (1988). In *Biology of Anaerobic Microorganisms*, ed. A. J. B. Zehnder. John Wiley & Sons, New York, p. 1.

Zwart, K. B., Gijzen, H. J., Cox, P. & Vogels, G. D. (1988). *Biotech. Bioeng.*, **32**, 719.

Chapter 14

AN INNOVATIVE BIOLOGICAL PROCESS FOR HEAVY METALS REMOVAL FROM MUNICIPAL SLUDGE

R. D. TYAGI & DENIS COUILLARD

Institut National de la Recherche Scientifique (INRS-Eau), Université du Québec, Sainte-Foy, Québec, Canada

CONTENTS

1 INTRODUCTION

Millions of tonnes of sewage sludge are produced worldwide every year (Hayes *et al.*, 1980). The treatment and final disposal cost represents 50% of the overall cost of the wastewater treatment facility. A wide range of processes and options for safe sludge disposal have been used. Landfilling and land application of sewage sludges are the most economical disposal methods (USEPA, 1984; Davis, 1987; Nicholson, 1988). Among these two options, land application of sludge appears to be the most economical solution (Davis, 1986, 1987). However, toxic metals present in the sludge restrict land spreading of sludge (Oliver & Cosgrove, 1974; Cheng *et al.*, 1975; Stoveland *et al.*, 1979; Nelson, 1986; Stephenson & Lester, 1987*a*,*b*). In the USA 50–60% of municipal sludges cannot be applied on agricultural land because the cadmium (Cd) content of the sludge exceeds 25 mg/kg on dry weight basis (Lue Hing *et al.*, 1980). In Ontario, Canada, more than 50% of the sludges exceed the maximum heavy metal limit set for agricultural use (Lue Hing *et al.*, 1980; Wozniak & Huang, 1982; Wong & Henry, 1984*a*; Ste-Yves & Beaulieu, 1988).

The majority of heavy metals being toxic, their high concentration becomes problematic (Sommers & Nelson, 1981; Tjell, 1986; Scheltinga, 1987). Cu, Ni and Zn are phytotoxic if their concentrations increase in soil (Beckett & Davis, 1982; Davis & Carlton-Smith, 1984; Webber, 1986), whereas Cd is zootoxic at low concentration (Lester *et al.*, 1983; Environment Canada, 1985; Nriagu, 1988). Heavy metals can accumulate in plants and can subsequently be transferred to animals through the food chain (Chaney, 1973).

2 PROCESSES OF METAL REMOVAL FROM SLUDGE

To prevent environmental pollution and health risks from heavy metals, the content of heavy metals of sewage sludge must be rendered up to a level recommended by the guidelines of various regional municipalities and local governments (Tyagi & Couillard, 1989). The reduction of heavy metals in sewage sludge can be achieved either by source control of discharge to sewers or by removing metals from sludge.

In source control the major difficulty resides in the identification of the sources. Source control of heavy metals in industry is expensive (Tjell, 1986), and even if the metal content is reduced it may not render the sludge suitable for agricultural use. More than 60% of Cd in sludge comes from industrial sources (Lue Hing *et al.*, 1980). Moreover, even with complete elimination of toxic metals from all industrial discharges to sewers, the problem remains because of the metal content of domestic wastewater and runoff water. Therefore to reduce environmental pollution and health hazards from heavy metals from sewage sludge the metals must be removed before its final disposal.

Since 1975 there have been numerous attempts to remove heavy metals from sewage sludge by acid treatment at pH 1·5–2·0 (Hayes *et al.*, 1980; Jenkins *et al.*, 1981; Wozniak & Huang, 1982; Kiff *et al.*, 1983). The literature on heavy metals removal from municipal sludge by chemical processes has been reviewed by Tyagi and Couillard (1989). This process is expensive since it consumes large amounts of acid (0·5–0·8 g H_2SO_4 per g of dry sludge) (Schonborn & Hartmann, 1978; Jenkins *et al.*, 1981; Wong & Henry, 1984*a*; Tyagi & Couillard, 1987*a*; Tyagi *et al.*, 1988*a*). The chemical process is often found to be ineffective in solubilizing Cu and Pb from anaerobically digested sludge. Recently microbial leaching of heavy metals from municipal sludge has been a subject of study. The microbial leaching process is interesting due to its inherent advantages, such as simplicity, high yield of metal extraction, and less acid and alkali consumption (Couillard & Mercier, 1990).

The objective of the present paper is to provide a state-of-the-art on heavy metals removal from sludge by the microbial leaching process.

3 ENERGY SUBSTRATE FOR LEACHING BACTERIA

The most widely used microorganisms for the leaching process are *Thiobacillus ferrooxidans* and *T. thiooxidans*. Both the organisms, being autotrophs, can derive their metabolic energy from chemical compounds, such as the reduced form of inorganic sulphur and reduced iron. Carbon requirements are fulfilled by CO_2 from the atmosphere. The energy required for the fixation of CO_2 is derived from the oxidation of reduced sulphur compounds to sulphate by either the indirect process or the direct process.

Indirect process

$$2FeSO_4 + 0{\cdot}5O_2 + H_2SO_4 \xrightarrow{T. ferrooxidans} Fe_2(SO_4)_3 + H_2O \quad (1)$$

$$4Fe_2(SO_4)_3 + 2MS + 4H_2O + 2O_2 \longrightarrow 2M^{2+} + 2SO_4^{2-} + 8FeSO_4 + 4H_2SO_4 \quad (2)$$

The second reaction does not require bacteria but proceeds chemically. A cyclic process between reactions (1) and (2) results in more and more metal solubilization. Formation of H_2SO_4 during the process further enhances the overall efficiency.

Direct process

$$MS + 2O_2 \xrightarrow{T. ferrooxidans} M^{2+} + SO_4^{2-} \quad (3)$$

In the direct process metal sulphides are directly attacked by bacterial cells to solubilize metal as metal sulphates. In an anaerobically digested sludge heavy metals exist in the form of sulphides (Tyagi & Couillard, 1989). *T. ferrooxidans* can attack the sulphides to derive the energy for growth and maintenance purposes by the direct process (reaction (2)).

Research has been conducted to remove heavy metals from anaerobically digested sludge without the addition of an external energy source (Wong & Henry, 1984*a*,*b*; Tyagi & Couillard, 1987*a*; Tyagi *et al.*, 1988*a*). However, the reaction time to solubilize heavy metals to a recommended level was found to be very long, 10–15 days. The role of *T. ferrooxidans* and *T. thiooxidans* in the leaching of metals from anaerobically digested sludge without the addition of an external energy source was also investigated in our laboratory (Tyagi & Couillard, 1987*a*; Tyagi *et al.*, 1988*a*). In order to enhance the metal leaching rate in an anaerobically digested sludge and to solubilize heavy metals from the non-digested sludges, an external energy source is required. In non-digested sludges (primary and secondary) the metals do not exist in sulphide form, therefore addition of an energy source is indispensable for the growth of leaching organisms.

Tyagi *et al.* (1991*a*) studied the effect of medium composition on pH change and metal solubilization using anaerobically digested sludge. Addition of N, P, K and Mg did not improve the microbial leaching rates. Obviously sewage sludge is rich in these nutrients. However, addition of ferrous sulphate increased the leaching rate proportionally. The rate of leaching was found to increase until the concentration of $FeSO_4 . 7H_2O$ reached 20 g/litre in the leaching medium. Increase in $FeSO_4 . 7H_2O$ beyond 10 g/litre increased the leaching rate very slowly.

The energy substrates for leaching bacteria other than ferrous sulphate which were evaluated in our laboratory include pyrite mining residue (Co. Aldermac, Abitibi, Canada), spent acid solution (NL Chem, Varrene, Canada) and waste ferrous sulphate (Co. Sidbec-Dosco, Montréal, Canada). It is apparently evident that to minimize the cost of raw material and the operational cost of the process on

Table 1
Composition of Metals in Different Energy Sources

Substrate	*Composition (mg/litre)*						
	Cu	*Zn*	*Pb*	*Ni*	*Cr*	*Cd*	*Mn*
Pyrite	55·12	42·61	10·0	21·34	50·58	0	—
Residue from Aldermac	1 544·6	2 513·0	42·4	4·1	95·6	6·79	—
Spent acid solution	0·0	2·54	0·27	2·71	65·0	0·0	—
Ferrous sulphate residue (Sidbec-Dosco)	2·7	282·4	8·8	29·2	3·8	0·0	1 488
Ferrous sulphate lab. reagent	All metals in trace amount						

an industrial scale it is necessary to adapt the leaching organisms to a less expensive energy substrate. The different substrates tested in our laboratory have the metallic compositions given in Table 1. These energy substrates were tested to remove metals from anaerobic sludge using a 10% inoculum of *T. ferrooxidans* at 25°C in a batch culture.

A concentration of 6–10 g/litre of sludge of $FeSO_4 . 7H_2O$ (commercial laboratory reagent) was established to be the optimum to achieve the maximum rate of metal solubilization. The metal solubilization achieved in 4 days was Zn 85–90%, Cu 70–75%, Cd 65–75%, Ni 70–80%, Cr 10–20% and Pb 0–2%. The optimum quantity of crushed pyrite (53 μm) to achieve maximum efficiency of metal solubilization was 10–16 g/litre, and the following metal solubilizations were observed: Zn 77–80%, Cu 70–85%, Cd 55%, Ni 10–20%, Cr 1–20% and Pb 5–8%. This efficiency of metal solubilization is comparable with the efficiency obtained using $FeSO_4 . 7H_2O$. Thus the crushed pyrite can be a substitute for commercial grade $FeSO_4 . 7H_2O$ as a source of energy for *T. ferrooxidans*. But pyrite has a tendency to decant and settle in the bottom of the reactor, which reduces the availability of pyrite to the suspended bacterial cells. This may cause an important problem during metal leaching on an industrial scale.

The metal solubilization efficiency obtained with a concentration of 13–17 g of mining residue per litre of sludge was Cd 82%, Zn 55% and Cu 46%; Ni, Pb and Cr were slightly solubilized. This is important, in that mining residue contains a very high concentration of heavy metals (Table 1) and is evidently a strong source of sludge contamination.

Poor bacterial growth was observed when ferrous sulphate was substituted with spent acid solution (pickling liquor). Consequently metal solubilization efficiency was very low: Cd 54%, Zn 49%, Cu 43% and Ni 28%. Pb and Cr were solubilized to a very low concentration. Thus spent acid solution served as an inferior substitute to ferrour sulphate because it contaminated the municipal sludge with Cr (from acid solution adsorbed on the sludge solids), and metal solubilization efficiency was comparatively lower.

The metal solubilization efficiency obtained with residual $FeSO_4 . 7H_2O$ (Sidbec-Dosco) was Zn 75–90%, Cu 70–80%, Ni 70–85%, Cd 70–75% and Cr 15–20%. The optimum concentration found for this solubilization was 60 g of $FeSO_4.7H_2O$ residue per litre of sludge. Thus ferrous sulphate residue (Sidbec-Dosco) can substitute for the commercial grade $FeSO_4.7H_2O$ and gives the required metal solubilization efficiency. Metal contamination from this source is also negligible and this residue is easily available at Can$55/t compared to Can$160 kg for the commercial grade ferrous sulphate.

Elemental sulphur (S^0), being a clean and cheaper substrate, has been used for the growth of *T. thiooxidans* to leach metals from anaerobic sludge (Schonborn & Hartmann, 1978). *T. thiooxidans* oxidizes elemental sulphur as follows:

$$2S^0 + 3O_2 + 2H_2O \xrightarrow{T.\ thiooxidans} 2H_2SO_4 \quad (4)$$

The production of H_2SO_4 decreases the pH and contributes towards solubilization of metals. Schonborn and Hartmann (1978) added *T. thiooxidans* and 1% S for the solubilization of heavy metals in the municipal sludge. The pH value decreased from 5·5 to 1·0 in 22–30 days. Sulphur oxidation activity proceeded only after the pH value of the sludge was lowered to 2·5 with an acid addition. Mixed culture of *T. thiooxidans* and *T. ferrooxidans* in the presence of 1% sulphur solubilized more metals than *T. thiooxidans* alone in more than 32 days of incubation. The mechanism for better solubilization by a mixed inoculum was due to the acidification of sludge (initial pH 5·0) by the oxidation of sulphur by *T. thiooxidans* to a pH value of 2·5 that subsequently favoured the growth of *T. ferrooxidans* and oxidized metal sulphides.

4 COMPARISON OF BIOREACTORS

4.1 Batch Process

Bacterial leaching of metals from anaerobically digested sewage sludge has been carried out in a batch reactor, and 8–14 days are necessary to solubilize around 80% of Cu, Zn, Cd and Ni (Wong & Henry, 1984*a*; Tyagi & Couillard, 1987*a*, 1989; Tyagi *et al.*, 1988*a,b*). Although the acid requirement was low (0·15 g of H_2SO_4 per g dry sludge) compared to the acid leaching process, the residence time required to remove metals to a recommended level was too long to be economically viable.

Heavy metals leaching from aerobically digested sludge have been studied in a batch process by Couillard *et al.* (1991). Different options were tested with reference to time periods required for metal solubilization and sludge digestion. These options include sludge digestion followed by metal solubilization, and simultaneous digestion and metal solubilization. In the first option (digestion followed by metal solubilization) the most viable strategy with respect to minimum process time required suggested the inoculation of *T. ferrooxidans* at a time when 36·5% reduction of volatile solids was achieved. In the second option (simultaneous digestion and metal solubilization) the average volatile solids reduction rate was 1·4%/day and 26 days were required to complete metal solubilization as well as 38% volatile solids reduction. According to the USEPA (1984) standard, to complete the sludge digestion, volatile solids should be removed by 38%. In the option of metal solubilization carried out at the end of digestion, 12 days were required for the whole process, i.e. 11 days for digestion and 1 day for metal solubilization. The combined option is therefore 2·6 times slower than the two-steps options (digestion followed by metal solubilization). Also non-digested sludge requires twice as much acid to be acidified to pH 4·0 than does digested sludge (Couillard *et al.*, 1991).

4.2 Continuous Process

Microbial leaching of heavy metals from municipal sludge was studied in a continuously stirred tank reactor (CSTR) with and without sludge recycling at residence times (τ) of 1–4 days (Tyagi *et al.*, 1988*b*, 1989; Couillard & Mercier, 1990). The continuously stirred tank reactor with sludge recycle (CSTRWR) and the CSTR were found to give the same result with respect to metal solubilization when the reactor contents in both cases were fortified with 1 g/litre $FeSO_4 . 7H_2O$. In the CSTR about 62% of Cu and about 77% of Zn were solubilized in 3 days' residence time compared to 50% of Cu and 64% of Zn solubilization in CSTRWR at the same residence time. The addition of an increased amount of soluble iron to the leaching medium resulted in a higher metal solubilization efficiency in the CSTRWR (Couillard & Mercier, 1990). Residence time and pH were the main factors for Zn solubilization, while for Cu solubilization oxidation redox potential (ORP) was also a major factor. The efficiency of metal solubilization increased with residence time (τ), while the rate of solubilization decreased with τ. The CSTR was found to be more efficient than the CSTRWR at $\tau = 2$ days. It was also observed that poor settling persisted at all residence times in the settling tank. The supernatant always contained a solids concentration between 0·8% and 1·4% (Table 2).

Copper recomplexation was observed in the settling tank used to separate the sludge solids (Couillard & Mercier, 1990). Low ORP and high pH contributed to Cu reprecipitation or recomplexation with organic matter in the recycled sludge. Zinc remains unaffected by these variations. About 40% and 50% of Cu were solubilized

Table 2
Total Solids in Reactor, Supernatant and Recycled Sludge

Concentration of $FeSO_4 . 7H_2O$ (g/litre)	*HRT*[a]	*CSTRWR*[a]			*CSTR*[a]
		TS[a] *Reactor (%)*	*TS Recycle (%)*	*TS Supernatant (%)*	*TS Reactor (%)*
1	1	3·1	4·1	0·8	2·8
1	2	3·2	3·2	1·0	2·9
1	3	2·8	3·1	0·9	2·9
1	4	3·2	3·1	1·4	2·9
3	3	3·0	3·3	1·1	2·8

[a] Abbreviations: HRT, hydraulic retention time (days); CSTR, continuously stirred tank reactor; CSTRWR, continuously stirred tank reactor with sludge recycle; TS, total solids.

in the recycled sludge and the CSTRWR, respectively. This lower solubilization in the recycled sludge indicates a reprecipitation (or recomplexation) of Cu in the settling tank. This pattern was not observed for Zn. This reprecipitation or recomplexation of Cu can be explained by low ORP (331 mV) and pH (3·2) in the recycled sludge compared with 510 mV ORP and 2·9 pH in the reactor.

The oxygen concentration in the CSTRWR was lower even if k_La (mass transfer coefficient) values were quite similar (Table 3), and bacteria consumed less oxygen than in the CSTR (oxygen utilization rate, OUR = 0·2, compared to 0·32 in the CSTR). The following explanation could account for this phenomenon. The metals were solubilized in the CSTRWR; the sludge was then transferred to the settling tank, where it stayed between 0·9 and 3·6 days depending on τ (90% of τ). In the settling tank the bacterial cells completely consumed the available oxygen, as a result of which the metabolism slowed down, pH increased and ORP decreased. Although these changes did not interfere much with Zn, they precipitated or recomplexed Cu. About 50% (recycle rate) of this sludge with a low ORP and low oxygen concentration was then returned to the CSTRWR. This precipitated Cu had again to be solubilized. This step produced an additional oxygen demand in the reactor. This led to reduced Cu solubilization in the CSTRWR compared to the CSTR.

The relatively poor efficiency of the CSTRWR is due to the difficulty in operating a rapid solid–liquid separation in the settling tank (Couillard & Mercier, 1990). Consequently a rapid solid–liquid separation technique would be necessary for a CSTRWR to become efficient and for adequate separation. Wong and Henry (1984*b*) found that neither vacuum filtration nor gravity settling could achieve the separation. They found that centrifugation at a centrifugal force of 1100*g* was adequate to produce an 18% solids cake with 95% metal recovery. Filtration could be considered another alternative. This aspect is discussed in the next section.

The pH of the feeding reservoir was maintained at 4·0. Cu and Zn were solubilized

Table 3
Comparison of CSTR and CSTRWR
(τ = 3 days; $FeSO_4 . 7H_2O$ = 3 g/litre)

Parameters	*CSTR*	*CSTRWR*
Cu (% solubilization)	62·2	Reactor: 50·1 Recycled sludge: 39·8
Zn (% solubilization)	77·4	Reactor: 64·1
ORP (mV)	523	Reactor: 501 Recycled sludge: 331
Oxygen (% saturation)	75·6	Reactor: 61 Recycled sludge: 0 to 5
k_La (h^{-1})	7·9	Reactor: 7·6
OUR[a] (mg O_2/litre . min)	0·32	Reactor: 0·20
pH	2·84	Reactor: 2·91 Recycled sludge: 3·15

[a] OUR, oxygen utilization rate.

in this reservoir to 1% and 26% levels, respectively. This difference again can be explained on the basis of ORP (227 mV in the feeding reservoir). Cu solubilization requires a minimum ORP of around 250 mV (Theis & Hayes, 1980).

Major factors influencing bacterial leaching of metals were studied (Tyagi & Couillard, 1987*b*). Cu solubilization is strongly influenced by ORP and sludge solids concentration. This tends to support the hypothesis of Cu solubilization by the indirect mechanism. The pH ranks second with regard to its influence on Cu solubilization, and the residence time has the least effect. In the case of Zn solubilization, however, residence time is the most important factor followed by ORP and pH.

The bacterial process of heavy metal solubilization from sewage sludges has been studied with two types of reactors: a continuously stirred tank reactor (CSTR) and an airlift reactor with sludge recycle (Couillard & Mercier, 1991*a*) at varying hydraulic retention times of 0·5, 0·75, 1·5 and 3 days at 30°C. The volume of the reactor was varied from 2·5 to 30 litres. Sludge feeding, exit, recycling and wasting were carried out hourly or once every two hours depending on residence time. An acclimated strain of *Thiobacillus ferrooxidans* (ATCC 19859) was used during the experiments. It was observed that a 0·5 day residence time was too short to solubilize Zn and Cu to a recommended level. At a residence time of 0·75 day the solubilization efficiencies were similar in both CSTR and the airlift reactors. Zn and Cu were solubilized to a level of 90%. It was concluded that 0·75 day was the shortest mean hydraulic residence time allowing the required metal (Cu and Zn) solubilization. A 20% sludge was recycled to achieve the metal solubilization. This is ten times faster than in batch experiments, in which 8 days were necessary to attain good solubilization (Tyagi & Couillard, 1987*a*). However, Couillard and Mercier (1991*a*) added ferrous sulphate as an energy source, while other workers (Wong & Henry, 1984*a*,*b*; Tyagi & Couillard, 1987*a*) have carried out leaching without the addition of an energy source for bacterial growth. Couillard and Mercier (1991*a*) also found that increase or decrease in residence time from the optimum value of 0·75 day resulted in decreased yield of metal solubilization.

A comparison of CSTR and airlift reactors has also been made by Tyagi *et al.* (1988*b*). By adding 3 g/litre of $FeSO_4 . 7H_2O$ and inoculating the sludge at the natural pH (7·5) of sludge the airlift reactor showed higher rates of metal solubilization and pH decrease. The time required to solubilize metals was much shorter in the airlift reactor than in the CSTR.

Minimum pH and maximum ORP in the airlift reactor were observed at 1·5 days' residence time, while metal solubilization efficiency was higher at $\tau = 0{\cdot}75$ day. In the CSTR the maximum ORP was observed at $\tau = 1{\cdot}5$ days and the minimum pH at $\tau = 0{\cdot}75$ day; the maximum metal solubilization was also observed at $\tau = 0{\cdot}75$ day, at which time Mn was solubilized up to 93% and Ni to 34%. In the CSTR 67% Ni was solubilized at $\tau = 1{\cdot}5$ days, but the maximum Ni solubilization (95%) was reached at $\tau = 3$ days. Cd solubilization varied between 52% and 73% in the CSTR and reached 90% (at $\tau = 1{\cdot}5$ days) in the airlift reactor. It was also observed that the maximum solubilization efficiencies of Cr and Pb were 8% and 7%, respectively, at all residence times (Couillard & Mercier, 1991*a*).

The rate of Cu and Zn solubilization reached a maximum at a residence time of 0·75 day in the CSTR as well as in the airlift reactor. A first-order kinetic relation was proposed without considering the effects of cell mass concentration, and the first-order kinetic constant (k_1) was found to vary with residence time. The highest value of k_1 (1·23) was calculated at a 0·75 day residence time by Couillard and Mercier (1991*a*).

5 SOLID–LIQUID SEPARATION

Once the metals are solubilized, the leached sludge has to be separated from the solubilized metals. The mode of decantation is insufficient to separate the solid from the liquid and also takes a long time, during which metals (especially Cu) get readsorbed on to the solids, thus reducing the overall metal solubilization.

The filterability test on the microbially leached sludge with Percol 757 (polyacrylamide cation) was carried out, according to Degrémont (1978), on the leached sludge obtained from the continuous bioreactor system at residence times of 0·5 and 0·75 day. This method permits one to calculate the coefficient of specific resistance of filtration under 0·5 atm ($r_{0\cdot5}$). The dehydrated sludge and the filtrate thus obtained were analysed for Cu, Zn, Mn, Fe, Cd, Ni, Pb, Cr, Kjeldahl nitrogen (TKN), NH_4^+, P_{total} and $P_{hydrolysable}$. The supernatant was neutralized to pH 10 with 5% lime and 0·25 molar NaOH. The decantability of the metallic sludge was evaluated by measuring the sludge volume index. It was also observed that Cu and Zn were solubilized to 91% and 93%, respectively. However, this was reduced to 78% and 77%, respectively, for Cu and Zn during filtration of sludge (Couillard *et al.*, 1990; Tyagi *et al.*, 1991*a*). The composition of metals in the unleached and leached sludge after filtration is shown in Table 4.

It was found (Tyagi *et al.*, 1991*a*) that the filterability index for anaerobically leached sludge was 38 times lower than for its non-leached counterpart. By adding the polymer (Percol 757) to unleached sludge the filterability index decreased with

Table 4
Composition of Metals in Leached and Unleached Sludge

Metal	*Unleached sludge (mg/kg dry sludge)*	*Leached and filtered sludge (mg/kg dry sludge)*
Cu	2 694	582
Zn	1 491	337
Mn	248	7
Ni	23	16
Cd	16	4
Fe	23 500	60 400
Pb	295	—[a]
Cr	65	—[a]

[a] — Not available.

polymer concentration, while for the leached sludge the filterability index remained unchanged with polymer concentration. The lower filterability index of leached sludge has been attributed to the addition of $FeSO_4 . 7H_2O$ as an energy source for the proliferation of bacterial cells (Couillard *et al.*, 1990; Tyagi *et al.*, 1991*b*). Ferrous sulphate has been used as a flocculant aid (Degrémont, 1978). The filterability index of the leached sludge also varies with the ORP and residence time for which leaching has been carried out in a continuous reactor.

According to Degrémont (1978), if by adjusting the polymer the value of $r_{0\cdot5}$ is less than 8×10^{12} m/kg, a plate filter press can be used to separate the solid from the liquid. However, the equipment used for solid–liquid separation should be corrosion resistant as the pH of the leached sludge is in the range of 2·5 to 3·0. A filter press is more prone to corrosion due to built-in metal parts. According to USEPA (1979), filtration under vacuum is difficult to apply to sludge containing a solids concentration less than 3% (30 kg/m^3). The leached sludge normally contains a solids concentration of less than 3%, because if thickened sludge is used for microbial leaching the metal removal efficiency is decreased (Tyagi *et al.*, 1988*b*). According to Wong and Henry (1984*b*), metals can be separated from leached sludge using a centrifuge and without adding a polymer. Therefore, technically, centrifugation and filtration under vacuum (plate filter) are the possible solutions and in these cases it is also possible to obtain a corrosion-resistant instrument at low pH. However, because of the high cost of a centrifuge against a plate filter press, the latter is recommended for solid–liquid separation after microbial leaching.

6 NEUTRALIZATION OF DECONTAMINATED SLUDGE AND METAL PRECIPITATION

The dehydrated sludge and the supernatant produced during solid–liquid separation have an acid pH (2·5–3·0). Consequently the sludge should be neutralized to pH 7·0 before it can be applied to agricultural land, and metals should be recovered from the supernatant for safe disposal and/or return to the metal industry. Neutralization of sludge and metal precipitation have been studied using $Ca(OH)_2$ and NaOH (Couillard & Mercier, 1991*b*; Tyagi *et al.*, 1991*a*). The amount required to precipitate the metals was found to be the same (0·072 g/g of dry sludge) for $Ca(OH)_2$ or NaOH. However, 0·044 and 0·034 g/g of dry sludge were required to neutralize the leached sludge with $Ca(OH)_2$ and NaOH, respectively.

The amounts of metal sludge produced as a result of precipitation were 0·91 and 0·78 g/litre of filtrate for $Ca(OH)_2$ and NaOH, respectively. The sludge volume index (SVI) of the two metallic sludges produced as a result of precipitation with two different bases was 273–276. According to Tardat-Henry and Beaudry (1984), an SVI of 200 ml/g is considered to be problematic for sludge decantation. Therefore the SVI obtained for the metallic sludge (276–273) suggests that decantation is improbable. This high SVI of metallic sludge is due to the presence of organic suspended solids in the supernatant at concentrations of 630–1200 mg/litre (Table 5).

Table 5
Metal Composition in the Metallic Sludge and the Supernatant

Parameter	*Composition in metal sludge (% dry weight)*	*Composition in supernatant (mg/litre)*
Cu	1–3·8	0·2–0·5
Fe	35·3–46·8	0·0–4·2
Zn	1·1–2·1	0·0
Pb	0·03–0·6	0·1–0·2
Cr	0·002–0·003	0·03–0·06
Cd	0·01	—
Ni	0·02	0·01–0·07
Mn	0·14–0·3	0·0
N	0·13–0·44	—
S	0·07–0·3	—
Organic matter	4·0	630–1 200

The filterability indexes found for metallic sludge were $3{\cdot}6 \times 10^{12}$ and $3{\cdot}1 \times 10^{12}$ m/kg for $Ca(OH)_2$ and NaOH precipitated sludge, respectively. According to Degrémont (1978), this filterability index is very good, so that the metallic sludge can be dehydrated with a plate filter press without adjustment of polymer.

The composition of metals in the metallic sludge is presented in Table 5. The high concentration of iron hinders the recycling of copper. However, selective precipitation techniques can be used to separate iron and copper. According to this method (Vachon *et al.*, 1987) iron is precipitated at pH 6·0 and the other metals at pH 9·5. The precipitate of ferric iron thus obtained can be recycled to an iron industry and the copper obtained can be recycled to a copper smelter.

The supernatant has a very low concentration of metals and therefore can be either used to neutralize the sludge solids (the supernatant has a pH of 10) or returned to the wastewater treatment unit. If the latter is the choice, metals in the supernatant will be diluted 100 times or more and hence will not be toxic to microorganisms in the wastewater treatment plant.

7 FERTILIZER VALUE OF LEACHED SLUDGE

Conservation of nutrient elements during microbial leaching is essential if the sludge is to be used as fertilizer after decontamination. The sludge was analysed before and after microbial leaching for different nutrient elements, and the results are presented in Table 6 (Couillard & Mercier, 1991*b*; Tyagi *et al.*, 1991*a*); 63% ammonia nitrogen was lost whereas total Kjeldahl nitrogen (TKN), P_t and $P_{hydrolysable}$ exhibited little or no loss. Total nitrogen was higher in the leached sludge than in the unleached sludge; this has been attributed to the fixation of atmospheric nitrogen by *T. ferrooxidans* (Mackintosh, 1978). It is to be noted that *T. ferrooxidans* can utilize ammonium ions for its metabolism (Harrison, 1982). The leached sludge has less odour and this could

Table 6
Nutritive Elements of Sludge Before and After Leaching

Parameter	*Unleached sludge (mg)*[a]	*Leached sludge (mg)*	*Loss*
TKN	428	450	No loss
NH_4^+	80	18	63% loss
$NO_3 + NO_2^-$-N	<1	<1	Negligible
P_t[b]	125	125	No loss
$P_{hydrolysable}$[b]	64	64	No loss

[a] Based on per litre of treated sludge.
[b] P_t, Total phosphorus; $P_{hydrolysable}$, hydrolysable phosphorus.

be due to the decrease in ammonia nitrogen. The nitrogen in the supernatant after metal recovery, if returned to the wastewater treatment plant, will be diluted at least 100 times and will not cause adverse effects on the treatment efficiency of the wastewater treatment plant. Total phosphorus and hydrolysable phosphorus were also conserved and no loss was observed. Therefore leached sludge does not lose its nutritive value and is suitable for agricultural land application.

8 ECONOMIC EVALUATION

The costs of metal removal by the microbial leaching process for two different wastewater treatment plant capacities are presented in Table 7. In these calculations the costs of equipment and chemicals are quoted directly from the manufacturers.

Table 7
Cost of Metal Solubilization and Land Application

Item	*Cost (Can$/t dry sludge)*	
	Capacity $20 \times 10^3\ m^3/day$	*Capacity* $388 \times 10^3\ m^3/day$
Metal solubilization process	60·55	17·30
$FeSO_4.7H_2O$	8·27	8·27
H_2SO_4	6·59	6·59
Lime	7·26	7·26
Sludge dehydration (plate filter press)	66·89	28·23
Metal precipitation	10·26	5·02
Dehydration of metal sludge	8·44	1·35
Subtotal	168·26	74·02
Land application	47·36	47·36
Total	215·62	121·38

Table 8
Economic Evaluation of Sludge Disposal Alternatives

Sludge disposal alternative	*Cost of disposal (Can\$/t dry sludge)*		*Remarks*
	Plant capacity 300 m³/day	*Plant capacity 388 × 10³ m³/day*	
(1) Landfill	147·96[a]	191·09	Biomass loss, possibility of groundwater contamination, availability of landfill sites
(2) Solubilization of metals and land application	170·51[a]	146·27	Prevention of soil erosion by organic material, ecologically safe
(3) Incineration (computed from USEPA, 1984)	>500·00	237·60	Biomass loss, air pollution problem, landfill of incinerator ash may cause contamination of groundwater

[a] Does not include the cost of sludge digestion.

Table 8 presents the cost comparison for different sludge disposal alternatives. The cost of application of decontaminated sludge to agricultural land is less than the landfill option by a margin of Can\$22.55/t of dry sludge. Also it is important that the availability of landfill is becoming more and more difficult. The cost of incineration is higher than that of landfill or land application. During metal solubilization iron and copper (major elements in the sludge in Québec) have a high potential for metal recycling, and if this credit could be taken then the overall sludge disposal cost will be further reduced.

The cost of metal removal and land application is much lower for a large (388×10^3 m^3/day) capacity plant than for a small (20×10^3 m^3/day) capacity plant (Table 7). The cost of sludge disposal for large capacity is also lower for other sludge disposal alternatives (Table 8). Sludge digestion is a necessary step for land application and sanitary landfill. For a wastewater treatment plant with a capacity of 388×10^3 m^3/day the digestion cost is US\$55/t $\approx$ Can\$70/t (USEPA, 1984). Including the digestion cost, the net cost of agricultural land application is $121.38 + 70 =$ Can\$191.38/t. If the metallic residue is to be treated the additional cost of Can\$24.0/t of sludge has to be adjusted. The landfill cost is Can\$191.09/t of dry sludge, including digestion, and that for incineration is Can\$273.60/t of dry sludge (Couillard *et al.*, 1990). This shows that incineration is the most expensive process. Individual isolation of sludge metals will further increase the profitability of sludge disposal via metal solubilization followed by land application.

9 CONCLUSIONS

Municipal sludge containing heavy metals is hazardous to health. For safe disposal of municipal sludge heavy metals should be removed. Source control is an expensive way and may not bring down the metal concentration to the required limit.

Chemical methods require a large amount of acid to solubilize metals, and a large amount of alkali is also required to neutralize the sludge at the end of the leaching process. The microbial leaching process is simple, requires less acid and removes metals to a recommended level. Amending the sludge with $FeSO_4 . 7H_2O$ enhances the metal leaching rate. Among the different alternative chemicals tested to replace the $FeSO_4 . 7H_2O$ in order to increase the bacterial growth and metal leaching rate, the residue from Sidbec-Dosco was found to give good results. Different bioreactors studied for metal leaching were presented. In the continuously stirred tank reactor with cell recycle readsorption of Cu to the sludge solids during sludge decantation was observed. Metal solubilization efficiency was found to depend upon the reactor's ORP, pH and residence time.

Different methods of solid–liquid separation, neutralization of sludge solids and precipitation of metals from leachate have been discussed. After microbial leaching the leached sludge does not lose its fertilizer value and is suitable for land application. The microbial leaching process seems to be economically better than incineration and/or sanitary landfills.

ACKNOWLEDGEMENTS

The authors thank the Natural Sciences and Engineering Research Council of Canada (grant A4984), the Ministère de l'Education of the Province of Québec (grants FCAR 90-AS-2713 and FCAR EQ-3029) and the Centre Québécoise de Valorisation de la Biomasse (Québec, Canada) for supporting this research.

REFERENCES

Beckett, P. H. T. & Davis, R. D. (1982). *Water Pollut. Control*, **81**, 112.

Chaney, R. L. (1973). *Crop and food chain effect of toxic elements in sludges and effluents on land.* National Association of State University and Land Grant Colleges, Washington, DC, p. 129.

Cheng, M. H., Patterson, J. W. & Minear, R. A. (1975). *J. Water Pollut. Control Fed.*, **47**, 362.

Couillard, D. & Mercier, G. (1990). *Envir. Pollut.*, **66**, 237.

Couillard, D., Chartier, M. & Mercier, G. (1991). *Bioresource Technology*, **36** (in press).

Couillard, D. & Mercier, G. (1991*a*). *Water Res.*, **25**(2), 237.

Couillard, D. & Mercier, G. (1991*b*). *Can. J. Chem. Eng.*, **69** (in press).

Couillard, D., Mercier, G. & Tyagi, R. D. (1990). In *Proc. 4th Conf. on Toxic Substances*, Montréal, April, 217.

Davis, R. D. (1986). *Experimentia Supplementum*, **50**, 55.

Davis, R. D. (1987). *Water Sci. and Technol.*, **19**(8), 1.

Davis, R. D. & Carlton-Smith, C. H. (1984). *Envir. Pollut.*, **B8**, 163.

Degrémont (1978). *Mémento Technique de l'Eau*, 8th edn. Dégremont, Paris.

Environment Canada (1985). *L'épandage des eaux usées traitées et des boues d'épuration d'origine urbaine.* Direction générale des programmes de protection de l'environnement, Report No. SPE6-EP-84-1, Ottawa, Canada, p. 62.

Harrison, A. P. Jr (1982). *Arch. Microbiol.*, **131**, 68.

Hayes, T. D., Jewell, W. J. & Kabrick, R. M. (1980). In *Proc. 34th Purdue Ind. Waste Conf.*,

Purdue University, West Lafayette, IN. Ann Arbor Science Publishers, Ann Arbor, MI, p. 529.
Jenkins, R. L., Scheybeler, B. J., Smith, M. L., Baird, R., Lo, M. P. & Haung, R. T. (1981). *J. Water Pollut. Control Fed.*, **53**, 25.
Kiff, R. J., Cheng, Y. H. & Brown, S. (1983). In *Proc. of Int. Conf. on Heavy Metals in the Environment*, Heidelberg, FRG. CEP Consultants, Edinburgh, p. 401.
Lester, J. N., Sterrit, R. M. & Kirk, P. W. W. (1983). *Sci. of Total Envir.*, **30**, 45.
Lue Hing, C., Zeng, D. R., Sawyer, B., Guth, E. & Whitebloom, S. (1980). *J. Water Pollut. Control Fed.*, **52**, 2538.
Mackintosh, M. (1978). *J. Gen. Microbiol.*, **105**, 215.
Nelson, P. O. (1986). In *Proc. of Int. Symp. on Metal Speciations, Separations and Recovery*, Chicago, July. Industrial Waste Elimination Research Center, Chicago, Illinois, p. VIII69.
Nicholson, J. P. (1988). *Pollution Engineering*, **20**(3), 14.
Nriagu, J. O. (1988). *Env. Pollut.*, **50**, 139.
Oliver, B. G. & Cosgrove, E. G. (1974). *Water Res.*, **8**, 869.
Scheltinga, H. M. J. (1987). *Water Sci. and Technol.*, **19**(8), 9.
Schonborn, W. & Hartmann, H. (1978). *Europ. J. Applied Microbiol. and Biotechnol.*, **5**, 305.
Sommers, L. E. & Nelson, D. W. (1981). In *Sludge and Its Ultimate Disposal*, eds J. A. Borchardt, W. J. Redman, G. E. Jones & R. T. Sprague. Ann Arbor Science Publishers, Ann Arbor, MI, p. 217.
Stephenson, T. & Lester, J. N. (1987*a*). *Sci. of Total Envir.*, **63**, 199.
Stephenson, T. & Lester, J. N. (1987*b*). *Sci. of Total Envir.*, **63**, 215.
Ste-Yves, A. & Beaulieu, R. (1988). *Caractérisation des boues de 34 stations d'épuration des eaux usées municipales (Jan.–Fev. 1988).* Ministère de l'Environnement du Québec, Direction générale de l'assainissement des eaux, Direction de l'assainissement agricole, Publ. no. 262, Québec, Canada.
Stoveland, S., Astruc, M., Lester, J. M. & Perry, R. (1979). *Sci. of Total Envir.*, **12**, 25.
Tardat-Henry, M. & Beaudry, J. P. (1984). *Chimie des Eaux.* Les éditions Le Griffon d'argile, Ste-Foy, Québec.
Theis, T. L. & Hayes, T. D. (1980). In *Chemistry of Wastewater Technology*, ed. Alan J. Rubin. Ann Arbor Science Publishers, Ann Arbor, MI, p. 403.
Tjell, J. C. (1986). *Trace metal regulations for sludge utilization in agriculture: a critical review.* European Communities, EUR 10361, 348.
Tyagi, R. D. & Couillard, D. (1987*a*). *Process Biochem.*, **22**, 114.
Tyagi, R. D. & Couillard, D. (1987*b*). In *Proc. of Int. Symp. on Small Systems for Water Supply and Wastewater Disposal*, National University of Singapore, Faculty of Engineering, 2–4 July, p. 435.
Tyagi, R. D. & Couillard, D. (1989). In *Library of Environment Control Technology*, ed. P. E. Cheremisinoff. Gulf Publ. Co., Houston, TX, p. 557.
Tyagi, R. D., Couillard, D. & Tran, F. (1988*a*). *Envir. Pollut.*, **50**, 295.
Tyagi, R. D., Couillard, D. & Tran, F. T. (1988*b*). In *Advances in Water Pollution Control: Water Pollution Control in Asia*, eds T. Panswad, C. Polprasert & K. Y. Yamamoto. Pergamon Press, New York, p. 231.
Tyagi, R. D., Tran, F. T. & Couillard, D. (1989). In *Proc. 7th Int. Conf. on Heavy Metals in the Environment*, Geneva, 12–15 September. CEP Consultants, Edinburgh, p. 56.
Tyagi, R. D., Couillard, D. & Tran, F. (1991*a*). *Water Sci. Technol.*, **22**, 229.
Tyagi, R. D., Couillard, D. & Grenier, Y. (1991*b*). *Envir. Pollut.* (in press).
USEPA (1979). *Process design manual for sludge treatment and disposal.* United States Environmental Protection Agency, EPA 625/1-79-011. Municipal Environmental Research Laboratory, Cincinnati, OH.
USEPA (1984). *Environmental regulation and technology: use and disposal of municipal wastewater sludge.* United States Environmental Protection Agency, EPA 625/10-84-003. Intro Agency Sludge Task Force, Washington, DC.

Vachon, D., Siwik, R. S., Schmidt, J. & Wheeland, K. (1987). In *Proc. Séminaire/Atelier sur le Drainage Minier Acide*, Halifax, Nova Scotia. Conservation et Protection Environnement Canada, Ottawa, p. 537.

Webber, M. D. (1986). *Epandage des boues d'épuration sur les terres agricoles—une évaluation.* Environnement Canada et Agriculture Canada, Comité d'Experts sur la Gestion des Sols et de l'Eau, Burlington, Ontario, Canada. 42 pp.

Wong, L. & Henry, J. G. (1984*a*). *Water Sci. Technol.*, **17**, 575.

Wong, L. & Henry, J. G. (1984*b*). In *Proc. 39th Purdue Univ. Ind. Waste Conf.*, Purdue University, West Lafayette, IN. Ann Arbor Science Publishers, Ann Arbor, MI, p. 515.

Wozniak, D. J. & Huang, J. Y. C. (1982). *J. Water Pollut. Control Fed.*, **54**, 1574.

Chapter 15

BIOLOGICAL TREATMENT OF PETROLEUM REFINERY WASTEWATER

R. D. TYAGI

Institut National de la Recherche Scientifique (INRS-Eau), Université du Québec, Sainte-Foy, Québec, Canada

CONTENTS

1 INTRODUCTION

The driving force behind water quality improvement and the trend towards more stringent environmental regulations in general is increasing public concern over the quality of the environment. Public knowledge and concern about trace contaminants have been enhanced by advances in analytical techniques that allow detection of chemicals at very low concentrations.

Crude oil processed in the petroleum refinery industry is a complex mixture of chemical compounds. Integrated refineries include many cracking processes, particularly catalytic cracking, and auxiliary processes which generate wastewaters with high concentrations of hazardous contaminants such as various phenolic compounds.

All refineries are significant water consumers and consequently large wastewater producers. In areas where refining, with its ancillary industries, is practised widely the associated water flow constitutes a major component of the total waste flow treated. But in areas with limited water resources reuse often has to be practised and therefore a good quality effluent is required. Also, in 'water rich' areas, the wastewater has to be efficiently treated for removal of hazardous contaminants before discharge to protect the quality of receiving waters. The pollutants in petroleum refinery wastewaters and their effect upon aquatic organisms were reviewed by Burks (1982). The objective of this paper is to provide pertinent information on the biological treatment of petroleum refinery wastewater.

2 SOURCES OF CHEMICALS FROM PETROLEUM REFINING

Petroleum wastewaters may include wastewater from refineries, petrochemical plants, petroleum tankers, oil, mining, filling stations, etc., containing petroleum and its components (Rhee *et al.*, 1987). Petroleum refineries use water to wash the crude oil stock or fractions derived from the crude oil stock. During the desalting process (the process of washing crude oil with water) the contact water is contaminated with dissolved emulsified organic chemicals from the crude oil stock. In refineries water is also used for stripping undesirable chemicals, such as hydrogen sulphide and ammonia, from overhead gases in thermal and thermocatalytic process units. Condensed water from stripper units is also contaminated with volatile organic chemicals produced during thermal cracking of heavy hydrocarbons. Phenolic compounds occur at high concentration levels in stripper condensate waters. Several other compounds, such as chemicals used within the refinery for controlling scale and corrosion on cooling towers, for reducing frothing in desalters and for miscellaneous other uses—in addition to those originally present in the crude oil—may be picked up as contaminants in the wastewater.

Wastewater generated from indirect sources is also implicated in refinery processes, such as water from storage tanks, ship ballast water, runoff water, etc. The complexity of petroleum refinery wastewater has been reviewed in the literature (Rossini *et al.*, 1952; Bestaugeff, 1967; Burks, 1982; Rhee *et al.*, 1987; Dold, 1989). As a result of different process wastewater streams there exist large differences in parameters between different wastewaters, such as pH, temperature and chemical composition.

3 COMPOSITION OF REFINERY WASTEWATER

Earlier, wastewater composition was classified in terms of oil and grease content, and size of oil droplets in the oil–water mixture. This is the main reason why in the past only primary treatment of the wastewater was carried out. The principal constituents in refinery wastewater are a range of hydrocarbons, both saturated and unsaturated, with a straight chain, a branched chain or a ring structure. In addition to the hydrocarbons, crude oil contains sulphur and nitrogen. Quench water from asphalt oxidation contains significant concentrations of phenolics.

Cyanides in petroleum refineries are generated in fluid catalytic cracking and coke units (Prather & Berkemeyer, 1975; Kunz *et al.*, 1978; Knowlton *et al.*, 1980; Wong & Maroney, 1989*a*). Cracking of organic nitrogen compounds liberates water-soluble cyanides in the form of hydrocyanide (HCN) and other nitrogen compounds (NH_3, thiocyanates). Cyanide production in refinery wastewater increases with nitrogen concentration in crude oil (Prather & Berkemeyer, 1975). Other sources of cyanides may include crude unit fractionator water and gas recovery plant wastewater. The USEPA water quality criterion for cyanide in saltwater is currently 0·001 mg/litre. CN^- forms numerous complex anions with many metals, such as copper, nickel, zinc and iron, that are likely to occur in petroleum wastewater. Cyanide toxicity is primarily attributable to HCN (Dourdoroff, 1980).

Composition of wastewater generated in petroleum refinery operation is defined in three different ways: (1) parameters relating to oil and grease content (i.e. oil droplet size) which provide adequate information on API (American Petroleum Institute) separator and DAF (dissolved air floatation) processes; (2) gross parameters, i.e. biochemical oxygen demand (BOD), chemical oxygen demand (COD) and suspended solids (SS), which are useful in assessing secondary biological treatment processes; and (3) compounds for which specific analyses are adopted to assess levels of toxic trace contaminants (Dold, 1989).

Table 1 indicates the concentration of some pollutants in refinery wastewater after the API separator process. There are other constituents found in refinery wastewater, such as metals (chromium, zinc) and organic compounds (benzene, toluene, ethylbenzene, phenol, 2,4-dimethyl phenol, anthracene, phenanthrene, naphthalene). The quantity and composition of wastewater after primary treatment and before secondary treatment (these treatment configurations are discussed in the next section) will differ significantly between refineries due to several factors, namely refinery process configuration and operating conditions, product lines, and the composition of crude oil feedstock. These individual inorganic and organic compounds are lumped in terms of BOD, COD and SS. Parameters such as these,

Table 1
Composition of Petroleum Refinery Wastewater
(except for pH, all units are mg/litre)

Constituent	*Concentration range*	*Constituent*	*Concentration range*
pH	6–10·8	Ammonia (as N)	0–150
TSS	10–90	Chlorides	10–1 050
VSS	15–60	Sulphide	0–50
COD total	100–3 500	Cyanides total	0·4–18·0
BOD total	5–250	Phenols	10–216
Hydrocarbons	20–250		

although encompassing a range of individual chemical compounds, are nevertheless useful in the design and operation of biological treatment systems. These parameters also enable the assessment, to some degree, of the impact of treated effluents on the receiving waters, such as oxygen profiles in streams and rivers. However, these parameters provide no information on the relative amounts of different compounds.

The US Environmental Protection Agency has compiled a list of priority pollutants (USEPA, 1977). These lists have been used as a basis for identifying sources of pollution, assessing the pollution potential of point sources (including industrial effluents) and formulating regulatory treatment standards. One compound-specific survey of the composition of refinery wastewater has also been carried out by the American Petroleum Institute (API, 1984). Some of these contaminants, which were found at levels of approximately 100 ppb, are chromium, zinc, ethylbenzene, toluene, 2,4-dimethyl phenol, phenol, anthracene, phenanthrene and naphthalene. The end-of-pipe treatment to control the toxicity in refinery wastewater is discussed by Wong and Maroney (1989*b*).

4 REFINERY WASTEWATER TREATMENT CONFIGURATION

Variations in the chemical composition of refinery wastewater and in parameters (pH and temperature) due to different wastewater streams (quench waters from asphalt oxidation, desalter wastewater, etc.) imply that there exists a significant potential for separate treatment of different process streams. Although this approach is practised to some degree, the different wastewater streams are usually combined in a single stream prior to treatment. Treatment schemes commonly used in refineries include (1) primary treatment and (2) secondary treatment.

4.1 Primary Treatment

Prior to primary treatment, hydrogen sulphide and ammonia nitrogen, in refinery wastewater from desulphurization and nitrogen removal processes from crude oil and the hydrotreating process, are removed in a stream called sour water by stripping. Primary treatment consists of gravity oil–water separation by API separators followed by flocculation and the dissolved air flotation (DAF) process. API separators and DAF processes remove free, dispersed and emulsified oils.

Oil droplets with diameters greater than 150 μm are classified as free oil and are separated by the API separator. Dispersed oil containing droplets with diameters in the range of 20 to 150 μm are removed by the DAF process. The oil droplets of less than 5 μm are classified as soluble oil (Rhee *et al.*, 1987). This fraction is not removed by the API separator or DAF. Sludge accumulating at the base of the API separator must be withdrawn and disposed of in an appropriate manner. The insoluble portion in the effluent of primary treatment will principally comprise free and dispersed oil droplets (hydrocarbons), which are not removed in the primary treatment stage. The soluble compounds which have been identified in the effluent from primary treatment are phenolics, short-chain carboxylic acids, naphthenic acids and various substituted aromatics (Sun *et al.*, 1987).

4.2 Secondary Treatment

For the removal of dissolved organics left after primary separation biological treatment is usually necessary. It may be applied either for pretreated refinery wastewater only or as a joint treatment with neighbouring municipal wastewater (El-Abagey & Moursy, 1982). However, the stringent requirements necessitate onsite treatment of refinery wastewater. This could also involve the separate treatment of different streams due to large variations in parameters (temperature, pH, phenol concentration, metal concentration) (Dold, 1989). Phenolic compounds, for example, are inhibitory to biological processes and hence it has been suggested that they should be treated separately.

In addition to the primary and secondary biological treatment processes, several physical methods, such as sand filtration, granular activated carbon adsorption and combustion, have been reported for the treatment of petroleum wastewater (Gloyna & Malina, 1963). Also chemical treatment, such as coagulation in combination with dissolved air flotation, can be used successfully in removing hydrocarbons (Moursy & Abo-El-Ala, 1982). Lie (1985) treated petroleum waste by a combination of physical, chemical and biological treatment. Meiners and Mazewski (1979) treated refinery waste in a combination of biological and physical systems, which showed very good results in terms of high BOD and COD removal.

5 BIODEGRADABILITY AND BIOLOGICAL TREATMENT OF PETROLEUM REFINERY WASTEWATER

Biological treatment is frequently the most economical means of reducing the BOD, removing trace amounts of oil-like materials, improving the overall appearance, and eliminating potential taste and odour problems, Burks (1982) suggested that, although the contaminants in oil refinery wastewater are actually lethal to aquatic organisms, they can be removed by biological treatment. Although all deleterious contaminants are not completely removed, the lethal effects can be eliminated using physical post-treatment processes.

Most organic compounds in waste streams are subjected to degradation by acclimatized microbial populations (Blokker, 1971). Toxicity, temperature, pH, oil content, nutrients and BOD are some of the factors which influence the biological treatment.

Sono and Futaka (1966) found that excessive amounts of petroleum wastes are toxic to biological systems. High temperature may kill bacteria and aquatic organisms while low temperature decreases metabolic activity. Zobbel (1969) showed that microbial oxidation is most rapid when the hydrocarbon molecule is in intimate contact with water and at temperatures ranging from 15 to 35°C. Most biological systems are pH-sensitive, and it is imperative that a pH range between 6·5 and 8·0 be maintained.

Oil retards oxygen transfer, deteriorates settling characteristics of biological sludge and increases the oxygen demand. Nutrients must be present in a BOD:N:20:1 ratio. Microorganisms require available sources of C, N, P, S, Ca, K, Mg, Fe and various trace elements to grow, and these elements are normally available in the wastewater.

Purshell and Miller (1962) showed that aliphatic compounds are oxidized more rapidly than aromatic compounds. Long-chain hydrocarbons are more susceptible to bacterial oxidation than normal-chain homologues. Davies (1967) found that a dozen different genera and more than 50 species of microorganisms have been shown to utilize one or more types of hydrocarbons. Bacteria can attack methane, ether, gasoline, kerosene, tars, paraffin wax, benzene, xylene, mineral oil, lubricating oil and cyclohexane. El-Abagey and Moursy (1982) found that faecal streptococci played a significant role in the biodegradation of hydrocarbons. *Streptococcus faecalis* and *Escherichia coli* can survive in petroleum-contaminated tropical marine waters. A mixed culture is certainly more effective than a pure bacterium in the biodegradation of petroleum refinery wastewater. This can be attributed to the presence of different organic and inorganic compounds in refinery wastewater, hence requiring different types of bacterial strains for their biodegradation.

Davies and Hughes (1968) reported that both agitation and aeration accelerate the biodegradation of oils, because agitation increases the surface area of the oil droplets and also oxygen solubility. This is important because molecular oxygen is essential for the first step in the breakdown of hydrocarbons by oxygenase enzymes. Studies on the biological treatability of petroleum refinery wastewater by activated sludge and aerated lagoons have also been reported by Mahmud and Thanh (1978). Biological treatment can remove simple cyanide, but stable complexed cyanides can pass through the biological treatment system unchanged (Lue-Hing *et al.*, 1975; Wong & Maroney, 1989*a*). Hence the biological treatment system cannot be expected to remove the residual cyanides in the refinery effluent. Other methods for the control of cyanides in petroleum refinery wastewater are discussed by Wong and Maroney (1989*a*).

6 ACTIVATED SLUDGE PROCESS

Biological treatment of refinery wastewater has been reported for many years, with efficient stabilization in the activated sludge process, and continues to be the preferred secondary wastewater treatment method for the removal of soluble biodegradable organic matter. The mode of operation consists of containing the substrate organic matter with a large population of viable microorganisms in an aerobic environment. However, long acclimatization periods for the biota to metabolize the refractory components are required (Dickenson & Giboney, 1970). A detailed process description of an activated sludge process unit is presented elsewhere (Tyagi *et al.*, 1991*b*).

Biotreatability of wastewater from an integrated oil refinery was studied in an onsite pilot plant (Rebhun & Galil, 1987). The wastewater used by the authors was strong, due to low specific water consumption in the refinery, and had significantly high phenol concentrations. The phenols were contributed by the cracking process and special gasoline washeries. The fate of phenols and hydrocarbons and their effect on the biological process, process rates and biofloc properties, and means of overcoming difficulties caused by these pollutants, were studied. Rebhun and Galil (1987) also followed the fate and mechanisms of hydrocarbons removed in the activated sludge process. It was found that 90% of the hydrocarbons were removed by stripping and biodegradation, and about 10% by entrapment in the bioflocs. The magnitude of the fractions removed by stripping varies with hydrocarbon composition and volatility, but at least 50% of the hydrocarbons were always removed by stripping in the aeration tank (Rebhun & Galil, 1987). These authors also showed that phenol inhibited the specific removal rate of COD in a competitive manner and concluded that phenols are the major inhibitors of biological treatment of the refinery wastewater. The practical implication of the inhibitory effects is that the higher the phenols concentration the lower must be the permissible organic loading. A high concentration (40–50 mg/litre) of effluent suspended solids was observed in spite of the secondary clarifier working at a very low surface loading and long residence time. The high effluent SS concentration was attributed to the very poor settleability of the bioflocs (Rebhun & Galil, 1987). The poor settleability of flocs was due to increased hydrocarbon entrapment in the bioflocs. The high effluent suspended solids has two major negative effects:

(1) It adversely affects the quality of the effluent through the high suspended solids itself and by being responsible for a great part of the hydrocarbons' BOD.
(2) The solids washout in the effluent does not enable the MLVSS (mixed liquor volatile suspended solids) to maintain conception concentrations greater than 1500–2000 mg/litre. The low MLVSS affects the size of the reactor required (Tyagi *et al.*, 1986, 1991*b*), since for a given organic loading rate the specific substrate utilization rate, q (which is low for this wastewater), requires a high residence time and a large reactor volume.

Low MLVSS in refinery wastewater treatment have been reported by a number of workers (Dickenson & Giboney, 1970; Mahmud & Thanh, 1978), who also showed high suspended solids in the effluent. Banerje *et al.* (1974) hypothesized that the biofloc is coated by a hydrophobic layer affecting its physical properties and its biochemical performance.

6.1 Start-up Operation

The development of the mixed liquor population during start-up of activated sludge plants can sometimes prove to be a laborious task affected by the seeding method employed. Factors which influence the microbial growth rate and, therefore, the length of time required to attain an acclimatized mixed liquor culture are temperature, organic loading, dissolved oxygen concentration and nutrient levels.

There are several seed origins which can be used to begin development of a viable culture for start-up of an activated sludge plant. These include mixed liquor from another activated sludge plant, naturally occurring microbes in the wastewater to be treated (self-seeding), soil and enriched freeze-dried cultures. With other carbon sources, the use of an activated sludge acclimatized to a petrochemical wastewater was reported by Wilkinson and Hamer (1979). Phenol removal of 99% was obtained at a hydraulic residence time above 3·6 h. A mixed substrate feed containing 100 mg/litre phenol did not affect steady-state operation. The start-up of an activated sludge plant using enriched dried culture is discussed by Reitano (1981). He showed that dried cultures of acclimatized bacteria are a useful tool in fast start-up of activated sludge plants, especially when a similar liquid is not available. The effect of phenol loadings up to 200 mg/litre on the activated sludge operation was also reported.

The use of mixed liquor from another activated sludge plant is a popular method for developing an acclimatized seed. Major problems are normally associated with the different species of microorganisms developed for particular wastewaters. A specific type of substrate will result in the development of a specific microbial population. To minimize acclimatization problems mixed liquor should be obtained from an activated sludge plant treating a wastewater of similar characteristics to that which is to be treated. Typically the use of municipal sludge to develop a stable mixed liquor concentration of 3000–4000 mg/litre suspended solids for a petrochemical wastewater requires 4–8 weeks (Flynn & Stadnik, 1979; Meiners & Mazewski, 1979; Reitano, 1981). Two weeks were required to reach a steady state using inoculum from a municipal activated sludge plant (Tran *et al.*, 1990).

Application of adapted mutant bacterial cultures for the treatment has also been suggested by some researchers (Mahmud & Thanh, 1978; McDowell & Zitrides, 1979).

6.2 Phenol Oxidation Studies

Many workers have reported varying results with respect to biodegradation of phenol in refinery wastewater treatment and in pure culture in synthetic medium. Reynolds *et al.* (1974) estimated that phenol is the controlling toxicant in a significant number of oil refinery wastewater treatments. It was found that high temperature has a significant effect on the toxicity of phenol for the activated sludge process. Lindsay and Prother (1977) also came to similar conclusions. The presence of a high concentration of sulphides along with phenol in the wastewater suppresses the biological oxidation during its treatment. Phenols when in high concentrations (more than 50 mg/litre) have a toxic effect on the microorganisms in the oxidation of organic compounds (Stroud *et al.*, 1963).

Phenols in the range of 15–20 mg/litre are easily amenable to biological oxidation and do not pose any major problems in their handling (Mahmud & Thanh, 1978). But sufficient nutrients must be added to maintain a BOD/N ratio of 20/1 and a

BOD/P ratio of 100/1, as the original wastewater was found to be deficient in nitrogen by these authors.

McKinney (1972) stated that phenol in excess of 50 mg/litre must not be fed to the treatment unit, but this was proved to be invalid as up to 300 mg/litre were treated by Mahmud and Thanh (1978). A phenol concentration of up to 90 mg/litre was removed in the RBC–PUF system (Tran *et al.*, 1990).

While many researchers report that phenol is a controlling toxicant in biological treatment (Reynolds *et al.*, 1974; Christiansen & Sparker, 1982), some researchers report efficient phenol removal at influent concentrations of up to 200 mg/litre after acclimatization but requiring a long reactor residence time to achieve good biofloc settleability (McKinney, 1972; Reitano, 1981). The biodegradability rate of paraffinic and aromatic hydrocarbons was much lower than that of municipal-type organic material, and adverse effects of hydrocarbons on the settling characteristics of the bioflocs were encountered (Groenewold *et al.*, 1982). Nayar and Sylvester (1979) investigated the effect of phenol loads from 1 to 1000 mg/litre on a microbial culture developed with glucose as a substrate. An impulse concentration of up to 100 mg/litre produced no significant effluent TOC (total organic carbon) or MLSS (mixed liquor suspended solids), whereas the MLSS decreased by 39% at the 500 mg/litre phenol level. System washout occurred at 1000 mg/litre phenol.

6.3 Shock Loads of Phenol

Pawlosky and Howell (1973) measured the specific growth rate (μ) as a function of phenol concentration for an acclimatized culture in batch reactors. At phenol concentrations from 50 to 100 mg/litre the specific growth rate varied by 40%, whereas increasing phenol levels from 100 to 1000 mg/litre resulted in a 70% decrease in μ. These studies showed that phenol is rapidly degraded but does exhibit a toxic effect which is dependent on the initial phenol level.

Reitano (1981) revealed that phenol impulses above 200 mg/litre proved to be inhibitory to a refinery mixed liquor. He also showed that increase in feed phenol concentration was followed by a corresponding rise in specific dissolved oxygen uptake rate (SDOUR). Phenol loadings up to 200 mg/litre can be tolerated by refinery activated sludge microorganisms as long as oxygen transfer limitations from increased microbial oxygen uptake do not create anaerobiosis and associated problems such as bulking sludge.

Rebhun and Galil (1987) showed that disruption in the biodegradation (BOD removal) occurred due to the sudden discharge of phenolic wastes (phenol, *para-* and *meta*-cresol, *o*-cresol and xylenols) (15 000 mg/litre of phenols) into the refinery's general wastewater system. The results of such conditions were marked by a sudden steep rise in effluent turbidity and discoloration of biomass (MLVSS) accompanied by a specific strong odour. Bioflocculation was impaired followed by a complete disruption of the whole process.

Adverse effects of a sudden increase in phenol concentration on biological treatment were also reported by Nayar and Sylvester (1979), Reitano (1981), and Rebhun and Galil (1987). The latter reported a steep decline in MLVSS and in substrate utilization. It seems that wastewater containing phenol at relatively high concentrations of up to 100 mg/litre can be treated biologically as long as the feed is steady and continuous. In such cases there is an inhibitory effect of phenols by reducing the process rate, but when operating at the low process rate the process itself is efficient, including good degradation of phenols (Nayar & Sylvester, 1979). The disruption of the process is by sudden discharge—surges of high-phenolic liquids. To prevent such disturbances storage of the suddenly discharged concentrated wastes is necessary and their gradual, low-flow controlled discharge to the general wastewater systems is advisable.

6.4 Problems Encountered in Conventional Activated Sludge Process

The conventional activated sludge process continues to be the most widely used method to remove suspended and dissolved organic compounds in refinery wastewater. This process removes up to 93–95% of BOD, which is satisfactory. Nevertheless, laboratory and pilot-plant as well as plant-scale operation have shown that the activated sludge process is associated with the following problems.

(1) High effluent COD due to the presence of non-biodegradable material (Dold, 1989). The effluent COD is reduced by adding powdered activated carbon (PAC) to the activated sludge process, which significantly improves the COD removal performance (Weber *et al.*, 1987). However, PAC combined with activated sludge cannot be a solution for the sudden high discharges of concentrated phenols. This problem can be solved through interception, storage and regulated gradual discharge. In order to reduce the influent organic matter, partial oxidation using hydrogen peroxide (H_2O_2) as a treatment step prior to activated sludge has been worked out by Bowers *et al.* (1989). It was also found that the by-products of the H_2O_2 oxidation process were less toxic and biodegradability was also improved. Trace organic compounds frequently detected in treated effluents are phenol, benzene, toluene and naphthalene.

(2) Poor sludge settleability results in high effluent suspended solids. Average concentrations of 50 mg/litre are reported regularly, while peaks in excess of 100 mg/litre are not uncommon (Dold, 1989), whereas municipal sludge suspended solids concentrations of 10 mg/litre or below are maintained consistently. The cause of poor settleability has been ascribed to the adsorption of an excess amount of oil in sludge flocs, growth of filamentous organisms (Jenkins *et al.*, 1984) and thus sludge bulking, low food to microorganisms ratio (F/M) at which a refinery wastewater treatment plant generally operates, and the presence of sulphides in the influent wastewater which may lead to the dominance of a specific filamentous organism of the genus *Thiothrix*. Growth of sulphur bacteria *Beggiatoa* has also been found in the activated sludge process and may contribute to the sludge bulking phenomenon. It is recommended that the hydrocarbons should be maintained at a low level by efficient operation of the dissolved air flotation system.

(3) Inhibition due to toxic compounds leading to a very slow biodegradation which necessitates a very long sludge retention time (Tyagi *et al.*, 1986, 1991*b*). This inhibition, as mentioned earlier, is mainly due to the phenol and its derivatives.
(4) The long period of acclimatization and/or start-up that may be necessary to recover from toxic shock loads of phenols and their derivatives, or in other words instability towards shock loads. Also long acclimatization periods for the biota to metabolize the refractory compounds are required (Dickenson & Giboney, 1970). Major process upsets and complete process failure can be encountered even with an acclimatized system. Such disruptions are often associated with sudden discharges of high strength wastewater (phenolics and sulphides). Phenol strength may go as high as 8000 mg/litre. An increase in effluent turbidity has also been observed with sudden discharge of phenol (Rebhun & Galil, 1987).
(5) Washout of biomass in the effluent ascribed to poor bioflocculation and toxic inhibition.
(6) Biological treatment reduces the toxic trace contaminants (USEPA, 1977) to a very low level. However, statistical analysis of data has confirmed that refinery effluent can contribute significant amounts of certain priority pollutants (Dold, 1989).

7 OTHER BIOLOGICAL SYSTEMS TO TREAT REFINERY WASTEWATER

Apart from the activated sludge process, other systems have been used for the treatment of petroleum refinery wastewater. The aerated lagoon process did not provide adequate performance for use as the sole treatment process but gave excellent results as a polishing step for activated sludge effluent (Mahmud & Thanh, 1978). Leising *et al.* (1983) used a trickling filter for refinery wastewater, but this poses practical operating problems, and in particular plugging of the fixed film. This study has shown that improper evacuation of the biomass is the cause of the problem. Hsu (1986) compared a bench-scale SBR (sequencing batch reactor) and a bench-scale conventional activated sludge unit treating petrochemical wastewater. In terms of degradation of BOD and nitrification, the performance of the SBR was slightly superior to that of the conventional activated sludge unit. However, for high strength wastes (with $BOD_5 > 300$ mg/litre) and under organic shock loading conditions, effluent from an SBR has a high solids content as a result of abundant dispersed cell growth. The SBR reacted to a series of phenolic shock loadings. Phenols were degraded from initial concentrations ranging from 250 to 950 mg/litre to $<0{\cdot}1$ mg/litre.

Effective application of the biological treatment technology is likely to reduce markedly the cost of treating refinery wastewaters. A number of successful full-scale biological systems specifically designed for the removal of petroleum hydrocarbons have caught the attention of the concerned industries, regulatory authorities and engineers. However, interest so far has focused on the application of suspended growth biological systems such as the activated sludge process (described previously). In order to overcome the problems encountered in the suspended growth process, a great potential has been recognized in the application of attached growth (fixed-film reactors) biological systems (Tyagi & Vembu, 1990; Vembu & Tyagi, 1990).

8 USE OF RBC IN PETROLEUM REFINERY WASTEWATER TREATMENT

The most commonly used fixed-film process in wastewater treatment is the trickling filter but the conventional rotating biological contactor (RBC) has recently gained more attention (Hynek, 1981; Brenner *et al.*, 1984) for many reasons: a large amount of biomass can be attached at low F/M ratio (Grady & Lim, 1980; Winkler, 1981); RBC can withstand hydraulic and organic surges more effectively (Antonie, 1976); it is an attractive engineering alternative for low-cost wastewater treatment because of its comparatively short process detention time; and process control is simple and energy requirements low (Wu & Smith, 1982). The good resistance can be explained by the relatively low contact time in the bioreactor between biomass and the treated wastewater.

The use of fixed-film technology was studied by Bartoldi *et al.* (1987) in a pilot plant. Their conclusions were that the process could meet the desired specifications regarding effluent quality; the system was found to be tolerant with bacterial versatility for quick recovery time and stability, having low real estate requirements. Knowlton (1977) summarized the information accumulated by Chevron Research Co. regarding the use of the rotating disk in biotreatment of oil refinery wastewater and indicated cases of efficient use.

Godlove *et al.* (1987) have studied the RBC process to determine the design conditions and operating variables for the Kansas City refinery wastewater management programme. The pilot unit demonstrated the applicability of the RBC to wastewater treatment. The full-scale RBC is producing an effluent quality within design conditions, with operational stability and reliability.

The performances of the biological, activated sludge and RBC processes to degrade integrated oil refinery wastewater were compared by Galil and Rebhun (1989), considering the following points: effluent quality in terms of suspended solids and organic matter, especially hydrocarbons and phenols; bioprocess response to disturbances caused by specific pollutants, mainly phenolic compounds; time required for recovery of the bioprocess after disruptions caused by toxic effects; and biosludge production and characteristics.

The effluent suspended solids concentration was higher in the activated sludge process (42 mg/litre) than in the RBC process (7 mg/litre). The total COD was also higher in the activated sludge system than in the RBC process. Good removals of hydrocarbons and phenols were achieved by both systems, as well as high efficiency of ammonia nitrification. The main differences between the systems related to the different biomass qualities accumulated in the bioreactors. While activated sludge could concentrate 1·5 g/litre of VSS (volatile suspended solids), the RBC attached 22·5–80 g/litre of VSS. The organic load on RBC biomass was five times lower than on activated sludge. Although the hydraulic detention time in RBC was about three times lower, the process intensitivity was always substantial. When phenol increased in step from 5 to 30 mg/litre the suspended solids in the activated sludge process effluent rose considerably (from 40 to 80 mg/litre), while no significant change in the suspended solids content could be observed in the RBC effluent.

The two parallel biotreatment systems (activated sludge and RBC) were brought to a high degree of disturbance, until no efficient COD removal took place (Galil & Rebhun, 1989). From this stage activated sludge recovery occurred gradually and continued over a period of 20 days, while the RBC achieved good quality effluent after 4 days. The main reason for the different recovery rates was attributed to the fact that the activated sludge had to rebuild more than 75% of its biomass, while RBC had only to reacclimatize its existing biomass to normal process conditions.

Filion and Murphy (1979) compared activated sludge to RBC and showed that the response of RBC was greater and occurred more rapidly than an activated sludge system operating at similar levels of removal. The biosludge production in RBC was four times lower than in activated sludge, and RBC sludge had better characteristics of thickening and cake filtration. In RBC up to 50% of hydrocarbons were removed by entrapment to the biomass, while only 10% were entrapped in activated sludge flocs (Galil & Rebhun, 1989).

Tanacredi (1980) used a staged, partially submerged rotating biological disk system to determine its performance in the reduction of polynuclear aromatic hydrocarbons attributable to waste crankcase oils (WCCO) in wastewater effluents. The results indicated that such biological systems for the removal of WCCO aromatic hydrocarbons are viable alternatives to the secondary treatment system commonly employed. Varying the flows, loading and recycling may further improve removal efficiencies. Knowlton (1977) covered the current usages of RBC units for the treatment of petroleum wastewaters. Biomass development required 1–7 weeks, in accordance with decreasing process wastewater temperature. The use of RBC for the treatment of refinery wastewater was also discussed by Crame (1976). Chou and Hynek (1981) studied the rotating biological contactor for the treatment of refinery wastewater with and without supplemental aeration. This system was effective in achieving the desired effluent quality and the supplemental aeration was effective in controlling the biomass attached to the media.

Hamoda and Al-Haddad (1987) treated petroleum refinery wastewater in a pilot plant at 0·03, 0·06 and 0·12 m^3/m^2 day hydraulic loading using an aerated submerged fixed-film (ASFF) biological multistage reactor in which a stationary submerged biofilm was developed on the disks. The ASFF system was developed for operation in hot and dry climates (Hamoda & Abd-El-Bary, 1987). It showed that higher removal efficiencies were generally obtained for COD reduction. Removal efficiencies of up to 80% were obtained for COD and ammonia. The results obtained were generally better than those reported by Congram (1976) for the rotating biological contactor system treating a petroleum refinery wastewater.

Alum pretreatment proved to be effective in reducing the levels of some waste constituents prior to biological treatment, thus resulting in improved system performance regarding COD removal. Apparently alum removal of organics present in the colloidal form contributed to such improvements. Removal efficiencies of up to 80% were obtained for COD and ammonia. Phenol concentrations were reduced to undetectable levels. Hydrogen sulphide was eliminated by air stripping. The biologically treated effluent contained only traces of oil while concentrations of metals (copper, cadmium, lead and nickel) were reduced to permissible levels. The effluent suspended solids were reduced to 30 mg/litre. Alum-pretreated wastewater generally resulted in smaller amounts of attached biomass retained in the first stage of the ASFF unit compared with untreated wastewater feed. It was also suggested that staging of the ASFF unit is effective in dampening excessive loadings frequently encountered during refinery effluent treatment. The treatment efficiency of this system was decreased as the loading was increased. Most of the biomass was found to be attached in the first stage (Hamoda & Al-Haddad, 1987).

9 POROUS BIOMASS SUPPORT SYSTEM COUPLED TO RBC

The US Environmental Protection Agency has found the use of polyurethane foam (PUF), which acts as a porous biomass support medium, to be an emerging technology for biological treatment (Lewis *et al.*, 1986). The PUF system is one which provides a structure within which microorganisms can grow, protected from high external shear and within which even weakly adhesive or flocculant organisms can be retained. Park and Mavituna (1986) found that the size distribution of the cell aggregates in relation to the average pore size of the foam matrix is the most important factor affecting the efficiency of immobilization. The hydrodynamics of the bioreactor (i.e. relative movement between the cell aggregates and the foam matrix in which immobilization takes place) also plays a significant role. It is also suggested that by matching the pore size and the aggregate size the amount of entrapment and the depth of the foam into which the aggregates penetrate have a large scope for improvement.

Use of PUF in combination with the conventional RBC system has several advantages over other immobilized techniques: (1) natural entrapment of microbial cells; (2) biologically harmless technique; (3) ease of aseptic manipulation; (4) ease of scale-up of the immobilized system; (5) good mechanical strength and stability of the biomass support particles; (6) extensive capacity to mop up oil and grease (Cooper *et al.*, 1986); (7) enhanced oxygen transfer rate associated with fine bubbles in diffused aeration (Cooper *et al.*, 1986), hence enabling reduction in air usage; and (8) low sludge production.

The PUF system coupled to an RBC unit was studied by our group (Tran *et al.*, 1990; Tyagi *et al.*, 1991*a*). The PUF was attached on disks of an RBC unit comprising a four-stage laboratory cascade. Wastewater was used from a petroleum refinery after treatment by API separator and DAF systems. The temperature was maintained at 25–30°C. The two parallel units of the RBC–PUF system were operated simultaneously at different but constant flow rates, giving hydraulic loadings of 0·01, 0·02, 0·03 and 0·04 m^3/m^2 day at a disk rotation speed of 10 rpm.

For all hydraulic loadings it was found that the removal efficiencies of total chemical oxygen demand (TCOD) and oil were above 80%. NH_3-N and phenol removal were above 95% for all loadings, which is superior to efficiencies achieved by Hamoda and Al-Haddad (1987). The total retention time is 5–7 h (1·25–1·75 h in each stage). Two kinetic models of ammonia nitrogen removal (Monod and Caperon–Meyer kinetics) fitted well with the experimental data. The start-up time in this system was lower (1–2 weeks) (Tyagi *et al.*, 1991*a*) as compared to the conventional RBC (5–8 weeks) to attain a steady state (Meiners & Mazewski, 1979; Chou & Hynek, 1981; Reitano, 1981; Rozzi *et al.*, 1989). A phenol concentration of up to 90 mg/litre was biodegraded in 5–7 h residence time with an effluent concentration lower than 1 mg/litre. Effluent BOD, irrespective of the hydraulic loadings, was found to be below the design criterion (<30 mg/litre) suggested by Chou and Hynek (1981). Increase in flow rate resulted in increased COD reduction rate.

The feed oil concentration varied widely during our experiments (25–200 mg/litre) due to refinery operation, but effluent experimental concentration remained almost constant (3–5 mg/litre), revealing a good absorbance of oil shocks. The effluent suspended solids were lower than 20 mg/litre, which is generally better than those reported by others in the fixed-film process (Hamoda & Al-Haddad, 1987), in the conventional RBC system treating petroleum refinery wastewater (Congram, 1976), as well as in the activated sludge process. Sulphur bacteria *Beggiatoa* were also found in the first-stage and the second-stage disks.

The amount of active biomass in terms of disk biomass index (DBI), the ratio of volatile solids to the total solids of disks, was evaluated. The DBI value remained constant in all stages irrespective of hydraulic and organic loadings. This is better but at the same time contrary to the variations in DBI as observed in the conventional RBC biofilms. An amount of 37–40 g/m^2 of volatile solids (VS) was obtained in the PUF system as compared to 23 g/m^2 in the fixed-film reactor (Hamoda & Al-Haddad, 1987). This comparatively large capacity of the PUF system to retain the biomass and better DBI in different stages is responsible for better biodegradation efficiency and a rapid start-up as compared to the conventional RBC.

10 CONCLUSIONS

In petroleum refineries wastewater streams differ due to different refinery processes having a range of toxic compounds and also differ significantly between refineries. The concentration of these toxic compounds is defined in three ways: oil and grease content, gross parameters and specific compounds analysis.

More stringent environmental requirements necessitate that the quality of effluent should be detected in terms of specific compounds in order to find their consequences on the receiving waters. To minimize the problem of toxicity in the receiving waters different wastewater streams generated in the refinery should be treated separately.

Among different existing biological treatment processes, the activated sludge process continues to be the most widely applied, in spite of several shortcomings. Several other biological processes, especially the fixed-film processes, have been developed and need attention for commercial exploitation. Pertinent information exists with respect to biodegradation and hydraulic organic toxic surges. This information can be used in the design of wastewater works.

The RBC–PUF coupled system is better than the conventional RBC in terms of quick start-up, higher active biomass, higher biodegradation rate, lower effluent suspended solids, and low sludge production.

ACKNOWLEDGEMENT

Sincere thanks are due to the Natural Sciences and Engineering Research Council of Canada (Grant A4984) for supporting this research.

REFERENCES

Antonie, R. L. (1976). *Fixed Biological Surfaces—Wastewater Treatment: RBC*. CRC Press, Cleveland, OH.

API (1984). *Refinery wastewater priority pollutants study—samples analysis and evaluation of data*. Publ. no. 4346, American Petroleum Institute, Washington, DC.

Banerje, S. K., Robson, C. M. & Hyatt, B. S. Jr (1974). In *Proc. 29th Ind. Waste Conf.*, Purdue University, West Lafayette, IN, p. 768.

Bartoldi, A. J., Hillard, G. E. & Blair, J. E. (1987). In *Proc. 42nd Ind. Waste Conf.*, Purdue University, West Lafayette, IN, p. 85.

Bestaugeff, M. A. (1967). In *Fundamental Aspects of Petroleum Geochemistry*, eds B. Nagy & U. Colombo. Elsevier, New York, p. 77.

Blokker, P. C. (1971). Prevention on water pollution from refineries. In *Seminar on Water Pollution by Oil*. The Institute of Petroleum, London, p. 21.

Bowers, A. R., Gaddipati, P., Eckenfelder, W. W. & Monsen, R. M. (1989). *Water Sci. and Technol.*, **21**, 477.

Brenner, R. C., Heidman, J. A., Optaken, J. A. & Petrasek, A. C. Jr (1984). *Design information on rotating biological contactors*. EPA 600/S2-84-106, US Environmental Protection Agency, Washington, DC.

Burks, S. L. (1982). *Envir. Int.*, **7**, 271.

Chou, S. L. & Hynek, R. J. (1981). In *Proc. 35th Ind. Waste Conf.*, Purdue University, West Lafayette, IN, p. 855.

Christiansen, A. J. & Sparker, W. P. (1982). In *Proc. 37th Ind. Waste Conf.*, Purdue University, West Lafayette, IN, p. 567.

Congram, G. E. (1976). *Oil and Gas Journal*, **74**(8), 126, 129, 132.

Cooper, P. E., Walker, I., Crabtree, H. E. & Aldred, R. P. (1986). In *Conf. on Process Eng. Aspects of Immobilized Cell Systems*. Pergamon Press, Manchester, UK, p. 205.

Crame, L. W. (1976). *Hydrocarbon Processing*, **55**(5), 92.

Davies, J. B. (1967). *Petroleum Microbiology*. Elsevier, Amsterdam, London and New York.

Davies, J. A. & Hughes, D. E. (1968). In *Proc. Symp. Biol. Eff. Oil Pollut. Littoral Communities*, ed. J. D. Carthy. Field Studies Council, London, p. 139.
Dickenson, A. G. & Giboney, T. J. (1970). In *Proc. 35th Ind. Waste Conf.*, Purdue University, West Lafayette, IN, p. 294.
Dold, P. L. (1989). *Water Pollut. Res. J. Canada*, **24**, 363.
Dourdoroff, P. (1980). *A Critical Review of Recent Literature on Toxicity of Cyanide to Fish.* American Petroleum Institute, Washington, DC.
El-Abagey, M. M. & Moursy, A. S. (1982). *Envir. Int.*, **7**, 309.
Filion, M. P. & Murphy, K. L. (1979). *J. Water Pollut. Control Fed.*, **51**, 1925.
Flynn, B. P. & Stadnik, J. G. (1979). *J. Water Pollut. Control Fed.*, **51**, 358.
Galil, N. & Rebhun, M. (1989). In *Proc. 44th Ind. Waste Conf.*, Purdue University, West Lafayette, IN, p. 711.
Gloyna, E. F. & Malina, J. F. (1963). *Industrial Water Wastes*, **8**(1), 14.
Godlove, J. W., McCarthy, W. C., Comstock, H. H. & Dunn, R. O. (1987). Kansas City Refinery's wastewater management program using rotating disk technology. Presented at the *Water Pollution Control Fed. Annual Conf.*, Philadelphia, PA.
Grady, C. P. L. Jr & Lim, C. H. (1980). *Biological Wastewater Treatment—Theory and Applications.* Marcel Dekker, New York and Basel.
Groenewold, C. J., Pico, F. R. & Watson, S. K. (1982). *J. Water Pollut. Control Fed.*, **54**, 398.
Hamoda, M. F. & Abd-El-Bary, M. F. (1987). *Water Res.*, **21**, 939.
Hamoda, M. F. & Al-Haddad, A. A. (1987). In *Proc. 19th Symposium on Wastewater Treatment*, 10–11 November, Ministry of Supply and Services Canada, Cat. no. En 44-1411987, p. 283.
Hsu, E. H. (1986). *Envir. Int.*, **5**(2), 71.
Hynek, R. J. (1981). Industrial application of the Bio-surf process. Presented at the *1981 Annual Meeting of the American Petroleum Institute of Chemical Engineers*, New Orleans, LA, November 1981.
Jenkins, D., Richard, M. G. & Daigger, G. T. (1984). *Manual on the Causes and Control of Activated Sludge Bulking and Foaming.* Water Research Commission, PO Box 824, Pretoria 0001, Republic of South Africa.
Knowlton, H. E. (1977). *Hydrocarbon Processing*, **56**(9), 227.
Knowlton, H. E., Coombs, J. & Allen, E. (1980). *Oil and Gas Journal*, **78**(15), 150.
Kunz, R., Casey, J. & Huff, J. (1978). *Hydrocarbon Processing*, **57**(10), 98.
Leising, M., Audoin, L. & Harel, M. (1983). *Marine Rev. Inst., Fr. Pet.*, **38**(5), 655.
Lewis, J. A., Black, G. M., Mavituna, F. & Wilkinson, A. (1986). In *Conf. on the Process Eng. Aspects of Immobilized Systems.* Pergamon Press, Manchester, UK, p. 291.
Lie, L. X. (1985). *Hydrocarbon Processing*, **64**(6), 78.
Lindsay, J. T. & Prother, B. V. (1977). *J. Water Pollut. Control Fed.*, **49**, 1779.
Lue-Hing, C. *et al.* (1975). *Report on cyanide studies—an evaluation of analytical effluent toxicity and compliance of publicly owned treatment works with existing effluent discharge criteria.* Metropolitan Sanitary District of Greater Chicago, Report no. 75-10, Chicago, IL.
Mahmud, Z. & Thanh, C. N. (1978). In *Proc. 33rd Ind. Waste Conf.*, Purdue University, West Lafayette, IN, p. 515.
McDowell, C. S. & Zitrides, G. T. (1979). In *Proc. 34th Ind. Waste Conf.*, Purdue University, West Lafayette, IN, p. 664.
McKinney, R. E. (1972). *Microbiology for Sanitary Engineers.* McGraw-Hill, New York.
Meiners, H. & Mazewski, G. (1979). In *Proc. 34th Ind. Waste Conf.*, Purdue University, West Lafayette, IN, p. 710.
Moursy, A. S. & Abo-El-Ala, S. E. (1982). *Envir. Int.*, **7**(4), 267.
Nayar, S. C. & Sylvester, N. D. (1979). *Water Research*, **13**, 201.
Park, J. M. & Mavituna, F. (1986). Factors affecting the immobilization of plant cells in biomass support particles. In *Conf. on the Process Eng. Aspects of Immobilized Cell Systems*, Pergamon Press, Manchester, UK, p. 115.

Pawlosky, U. & Howell, J. A. (1973). Mixed culture bio-oxidation of phenol—1. Determination of kinetic parameters. *75th National AIChE Meeting*, 3–6 June.
Prather, B. & Berkemeyer, R. (1975). In *Proc. 30th Ind. Waste Conf.*, Purdue University, West Lafayette, IN, p. 306.
Purshell, W. L. & Miller, R. B. (1962). In *Proc. 16th Ind. Waste Conf.*, Purdue University, West Lafayette, IN, p. 292.
Rebhun, M. & Galil, N. (1987). In *Proc. 42nd Ind. Waste Conf.*, Purdue University, West Lafayette, IN, p. 163.
Reitano, A. J. (1981). In *Proc. 36th Ind. Waste Conf.*, Purdue University, West Lafayette, IN, p. 310.
Reynolds, J. H., Middlebrooks, E. J., Asce, F. & Procella, D. B. (1974). *J. Env. Eng. Div. ASCE*, **100**, 557.
Rhee, C. H., Martyn, P. C. & Kremer, J. G. (1987). In *Proc. 42nd Ind. Waste Conf.*, Purdue University, West Lafayette, IN, p. 143.
Rossini, F. D., Mair, B. J. & Streiff, A. J. (1952). *Hydrocarbons from Petroleum*. Reinhold, New York.
Rozzi, A., Passino, R. & Limoni, M. (1989). *Process Biochem.*, **24**, 68.
Sono, K. & Futka, M. (1966). Aspects concerning the treatment of wastewaters from oil refineries in Romania. Presented at *3rd Int. Conf. on Water Poll. Res.*, Munich, FRG, II-11.
Stroud, P. W., Sorg, L. V. & Lamrin, J. C. (1963). In *Proc. 18th Ind. Waste Conf.*, Purdue University, West Lafayette, IN, p. 460.
Sun, P. T., Price, C. L., Raia, J. C. & Baldeas, R. A. (1987). In *Proc. 42nd Ind. Waste Conf.*, Purdue University, West Lafayette, IN, p. 151.
Tanacredi, J. T. (1980). Removal of waste petroleum derived polynuclear aromatic hydrocarbons by rotating biological disks. Paper presented at *First National Symp./Workshop on RBC Technology*, Champain, PA, 4–6 February 1980.
Tran, F. T., Tyagi, R. D. & Chowdhury, A. K. M. M. (1990). *Biological Wastes* (submitted).
Tyagi, R. D. & Vembu, K. (1990). *Wastewater Treatment by Immobilized Cells*. CRC Press, Boca Raton, FL.
Tyagi, R. D., Couillard, D. & Villeneuve, J. P. (1986). *Can. J. Chem. Eng.*, **64**, 97.
Tyagi, R. D., Tran, F. T. & Chowdhury, A. K. M. M. (1991*a*). *Env. Pollut.* (in press).
Tyagi, R. D., Tran, F. T. & Couillard, D. (1991*b*). *Can. J. Chem. Eng.*, **69**, 534.
USEPA (1977). *Development document for effluent limitation guidelines: new source performance standards and pretreatment standards for the petroleum refinery point source category*. EPA 440/1/82/014, EPA Effluent Guidelines Division, US Environmental Protection Agency, Washington, DC.
Vembu, K. & Tyagi, R. D. (1990). In *Wastewater Treatment by Immobilized Cells*, eds R. D. Tyagi & K. Vembu. CRC Press, Boca Raton, FL, p. 253.
Weber, W. J., Jones, B. E. & Katz, L. E. (1987). *Water Sci. Technol.*, **19**, 471.
Wilkinson, T. G. & Hamer, G. (1979). *J. Chem. Technol. Biotechnol.*, **29**, 56.
Winkler, M. (1981). *Biological Treatment of Wastewater*. Ellis Horwood, Chichester, UK.
Wong, J. M. & Maroney, P. M. (1989*a*). In *Proc. 44th Ind. Waste Conf.*, Purdue University, West Lafayette, IN, p. 675.
Wong, J. M. & Maroney, P. M. (1989*b*). In *Proc. 44th Ind. Waste Conf.*, Purdue University, West Lafayette, IN, p. 685.
Wu Y. & Smith, E. (1982). *J. Env. Eng. Div., ASCE*, **108**, 578.
Zobbel, C. E. (1969). In *Proc. API/FWC Conf.*, Publ. no. 4040, American Petroleum Institute, Washington, DC, p. 317.

Chapter 16

PEAT AS AN AGENT IN BIOLOGICAL DEGRADATION: PEAT BIOFILTERS

ANTONIO M. MARTIN

Department of Biochemistry, Memorial University of Newfoundland, St John's, Newfoundland, Canada

CONTENTS

1 INTRODUCTION

As a potential raw material for a variety of uses, peat has been found to possess a unique combination of chemical and physical properties, such as adsorbency and deodorization, which could be employed in environmental protection applications (McLellan & Rock, 1986). With the increasing concern over environmental pollution and the recent demands for safer and more economic pollution-abatement technologies, attention has been focused on those peat properties.

Peat constituents such as lignin bear polar functional groups including acids, alcohols, aldehydes, ethers, ketones, and phenolic hydroxides that can be involved in chemical bonding (Viraraghavan & Ayyaswami, 1987). Humic acids, a constituent responsible for some of peat's particular characteristics, have been

Table 1
Summary of Waste Treatment Processes Using Peat

Pollutant treated	*Remarks*	*Reference*
Cd	Usually $<3\,\mu g$/litre from laboratory columns	Chaney & Hundemann (1979)
Ag, Cd, Cr, Cu, Fe, Hg, Ni, Pb, Sb, Zn; BOD; colour	Peat on a moving screen belt	Lalancette & Coupal (1972); Coupal & Lalancette (1976)
Cd, Cr, Cu, Fe, Ni, Pb, Zn; colour; COD, BOD	Dyehouse effluent using Hussong/Couplan unit	Leslie (1974)
Cu	Adsorption low if Na present	Dissanayake & Weerasooriya (1981)
Ca, Mg	P, N, COD removals low	Schwartz (1968)
Cd, Cu, Pb, Zn	Ca in peat increased removal	Wolf *et al.* (1977)
Cd, Cr, Ni, Pb, Zn	Laboratory tests	Zhipei *et al.* (1984)
Ti, Zr	Practical for metal recovery	Parkash & Brown (1976)
Alkyl benzene sulphonate; beef extract (COD)	Excellent on ABS; COD removal low	Tinh *et al.* (1971)
Dieldrin	Peat superior to soils	Eye (1968)
Dieldrin	Peat columns removed 88%	Brown *et al.* (1979)
Textile dyes (acid or basic)	Best on basic dyes	Dufort & Ruel (1972)
Acid blue dye	2-hour contact time	Poots *et al.* (1976)
Dyes (acid, basic, direct, disposed)	Economic analysis favoured peat vs carbon	McKay *et al.* (1978)
Dimethylamine, NH_3, H_2S	Odours adsorbed, but carbon better	Soniassy (1974)
BOD, P, N, fats	Pretreatment of slaughterhouse waste	Silvo (1972)
Oil	90–95% adsorbed at oil spill	D'Hennezel & Coupal (1972)
Oil	Artificially dried peat	Asplund *et al.* (1976)
Phenol	87% adsorbed at 40 mg/litre concentration	Mueller (1972)
BOD, N, P, TSS	Filtration of secondary wastewater	Farnham & Brown (1972); Brown & Farnham (1976)
Coliforms, N, P	Tertiary treatment	Osborne (1975)
Coliforms, N, P	Peat–sand filter	Nichols & Boelter (1982)
Coliforms, BOD, N, P	Septic tank effluent after sand filter	Stanlick (1976)
Coliforms, BOD, N, P	Septic tank effluent	Rock *et al.* (1984); Brooks *et al.* (1984)
BOD, N, P	Dairy, cattle, pig wastes	Barton *et al.* (1984)
Ca, Fe, K, Mg, Mn, P, Pb, Zn, BOD	Landfill leachate	Lidkea (1974); Cameron (1978)

Source: McLellan & Rock, 1986; reprinted with permission of the *International Peat Journal*, International Peat Society.

identified by many workers as providing the carboxylic acid (COOH) groups implicated in the removal of metal ions from solution by the formation of chelate rings involving adjacent COOH and phenolic (OH) groups. Less predominantly, two adjacent COOH groups could also be involved. Also, the formation of complexes of metal ions with the same types of functional groups could be accomplished without the formation of chelate rings (Boyd *et al.*, 1981). Several investigators have shown the effectiveness of peat in the removal of heavy metals, some of the more recent studies being reported by Gosset *et al.* (1986).

The use of peat has been reported in the treatment of industrial and municipal wastewaters. Mueller (1972) discussed the pollution abatement potential of peat derived from its capacity to adsorb organic and inorganic matter and its properties as a filter material. D'Hennezel and Coupal (1972), among other researchers, have studied the affinity of peat moss for oil. Several studies have been published on the use of peat in the removal of dyes, and for the reduction of oxygen demands from slaughterhouse, dairy and animal wastewaters and septic tank effluents, which have been previously reviewed (Ruel *et al.*, 1977; McLellan & Rock, 1986; Viraraghavan & Ayyaswami, 1987). Table 1 presents a summary of the reported uses of peat in waste treatment processes.

Recently, the need to dispose of toxic wastes has generated renewed interest in the properties of peat which could be employed towards those aims. For example, Cloutier *et al.* (1985) have conducted research on the peat adsorption of the herbicide 2,4-D (dichloro-phenoxy acetic acid). Treatment of gaseous effluents has also been highlighted as one of the possible applications for which peat properties are suitable (Furusawa *et al.*, 1984).

This chapter will review and discuss the potential of peat utilization in pollution control with emphasis on the biodegradative processes that occur in the presence of peat. The characteristics of peat as support for microbial populations will be highlighted. A new theory will be presented, suggesting that peat could play the role of an active agent for biological degradation. This role could complement the present use of peat in waste treatment, which is mostly based on its physical and chemical properties.

2 PEAT IN WASTEWATER TREATMENT

Tinh *et al.* (1971) reported studies using peat moss as an adsorbing agent for beef extract and alkyl benzene sulphonate solution. Contact time, particle size and concentration of pollutant and adsorbent were among the parameters studied. The authors reported a 27% reduction in the chemical oxygen demand (COD) of the beef extract and between 72 and 95% reduction in the COD of the alkyl benzene sulphonate.

During 1973, the US Forest Service began operation of a tertiary waste treatment facility at a campground near Marcell, Minnesota, using peat as a filtration material (Osborne, 1975). This facility was designed following previous studies, including those by Farnham and Brown (1972) in which peat and peat–soil mixtures significantly reduced the phosphorus and organic compound level of wastewaters. Later, it was reported that there were approximately 12 peat-filter tertiary waste treatment systems being operated by the US Department of Agriculture (USDA) Forest Service in the northeastern United States, satisfactorily solving wastewater problems for selected situations (Parrott & Boelter, 1979).

Poots and McKay (1980) studied the filtration characteristics of spruce wood and peat using air and a water and dye solution as flow media. The use of peat for on-site wastewater treatment was studied by Rock *et al.* (1984) and Brooks *et al.* (1984). Viraraghavan and Ayyaswami (1987) reviewed the use of peat in industrial, municipal and on-site wastewater treatment systems. The same authors reported that peat was effective in adsorbing 30–50% of soluble biological oxygen demand (BOD), COD and organic carbon from the septic tank effluent (Ayyaswami & Viraraghavan, 1985).

Animal wastes and animal processing wastewaters have been treated with peat, showing that treatment of slaughterhouse wastewaters reduced appreciably the organic matter, BOD and COD (Silvo, 1972; Gravelle & Landreville, 1980). Barton *et al.* (1984) also obtained good results with a peat filter system for dairy and other animal wastes.

In addition to the abovementioned chemical and physical characteristics, which enable peat to adsorb and bond waste materials, the biological properties of peat should be taken into consideration when analysing its effects in pollution control. It has been recognized for a long time that some of the properties of peat are appropriate for maintaining microorganisms in viable condition. Lochhead and Thexton (1947*a*,*b*) found in comparative tests of various powdered materials that peat was superior to other preparations for maintaining viable test bacteria. Jaouich (1975), in his studies on nitrate reduction in peat, reported the role of microorganisms in the denitrification process and also the isolation of representative strains of denitrifying *Pseudomonas* species found in peat. Nichols and Boelter (1982), in their studies on the treatment of secondary sewage effluent with a peat–sand filter bed, found that microbial immobilization in the peat contributed to nitrogen and phosphorus removal.

2.1 Peat Sorptive Properties

The removal by peat of heavy metals such as antimony, copper, cadmium, lead, mercury, nickel, uranium, zinc and zirconium has been reported (Viraraghavan & Ayyaswami, 1987).

Parkash and Brown (1976) found that adsorption on peat and low-rank subbituminous coal could be a practical way for recovering zirconium and titanium from aqueous solutions such as oil sands tailings. The authors recovered almost four times more zirconium than titanium, which resulted in a simple and cheap initial separation method. The equilibria as well as the rates of adsorption and desorption of the ions Ca^{2+}, Cd^{2+}, Cu^{2+}, Pb^{2+} and Zn^{2+} were determined in experiments with HCl-washed peat (Bunzl *et al.*, 1976). Lalancette and Coupal (1972) studied the removal of mercury from wastewater using peat, and Coupal and Lalancette (1976) reported good removals of copper, chromium, lead, nickel and zinc from wastewaters. Takamatsu and Yoshida (1978) studied the stability constants of metal–humic acid complexes and Boyd *et al.* (1981) demonstrated that copper and iron ions formed complexes with the carboxylate groups of humic acids. Chaney and Hundemann (1979) used peat moss columns to remove cadmium from simulated wastewaters.

The enhanced peat-cation exchange capacities after treatment with acid were studied by Smith *et al.* (1977) and Bloom and McBride (1979). The role of the humic substance components of peat in its sorption properties was also reported by Bunzl (1974*a*,*b*) and Wolf *et al.* (1977). A study of the physical properties of humic substances from several sources, including peat, and of their metal complexes, was presented by Sipos *et al.* (1978). Meisel *et al.* (1979) studied the ion exchange and redox capacity of the peat humic substances. Extensive work has been conducted on the ion exchange properties of peat (Tummavuori & Aho, 1980*a*,*b*) and on the effect of experimental conditions on those properties (Aho & Tummavuori, 1984).

More recent studies have included the works of Dissanayake and Weerasooriya (1981) concerning the effects of pH on metal ion adsorption by peat, of Kadlec and Rathbun (1983) on copper sorption on peat, of Zhipei *et al.* (1984) on the removal of several metals from wastewaters using Chinese peats, and of Rock *et al.* (1985) on the removal of Cu, Fe, Pb and Zn from landfill leachate.

One of the most interesting applications of peat filters has been in the removal of colour from wastewaters. Coloured wastewaters are the result of undegraded materials being present in the effluent. Unwanted colour in streams is one source of pollution in itself.

It is well known that many dyes, particularly those used in the textile industry, are fairly stable to light and oxidizing agents and resistant to aerobic digestion. Most commercial systems use activated carbon as the adsorbent agent. Peat has been suggested as a potentially more economical method (McKay, 1980). Leslie (1974) presented a design for a plant to treat dye house effluent with peat. The adsorption of Telon Blue (Acid Blue 25) and Astrazone Blue (Basic Blue 69) on peat has been studied by Poots *et al.* (1976, 1978, respectively). In both cases, the adsorption parameters for the Langmuir and Freundich isotherms were determined and the effects of contact time, initial dye concentration and peat particle size were studied.

2.2 Peatlands and Wastewater Treatment

The use of peat bogs to treat contaminated waters has been suggested (Tilton *et al.*, 1976). Tilton and Kadlec (1979) studied the application of secondary effluent to, and the removal of nutrients by, a natural peatland. They reported very high removal of dissolved P and (nitrate plus nitrite)-N (more than 90%) and ammonium-N (more than 70%). Nichols (1983) reviewed the potential of wetlands to remove nutrients from wastewaters, emphasizing the 'phosphorus dynamics', 'nitrogen dynamics' and the 'vegetation dynamics' in purifying wastewater. Hall (1987) presented the idea that 'natural wastewater bioengineering offers a way to reduce pollution while beautifying the environment', suggesting the creation of artificial marshes that could treat wastewater in approximately one-third the area required for traditional lagoons, thanks to the properties of natural systems to extract heavy metals and other toxic compounds. It can be expected that, in his analysis, the author took into consideration the properties of peat within the context of the natural environment proper for wastewater treatment.

2.3 Air and Gas Purification

The removal of low concentrations of pollutant gases from air streams is a difficult and expensive process. The existing techniques, which include water scrubbing, adsorption by activated charcoal and burning, involve the input of energy and water, and a high cost in initial capital and maintenance operations. In addition, the water used to wash the gases becomes polluted itself (Bohn, 1975). To overcome these problems, the same author demonstrated the potential of soil or compost air filters for the removal of pollutants and the control of odours in gases. He defined 'filter' as the mechanism to remove gases, aerosols and particulate matter from air.

Air sterilization and disinfection is also a requirement in many industries and in institutions such as hospitals. Sterilization and disinfection of air can be achieved by chemicals, heating, irradiation, scrubbing and filtration (including the use of activated carbon). It has been noticed that the use of filtration to clean and sterilize air and other gases is the most efficient, economic and versatile of all air-cleaning and sterilizing systems (Firman, 1970). Air filters range from roughing filters for the removal of large airborne particles to ultra-efficient filters that can be applied to virus-sized particles.

Mueller (1972) pointed out the high carbon content, the large surface area to weight ratio and the adsorptive properties of dried peat moss. He also mentioned the potential of peat to adsorb odours, by virtue of its physical and chemical structure, in the same way that activated carbon is being used by industry. Soniassy (1974)

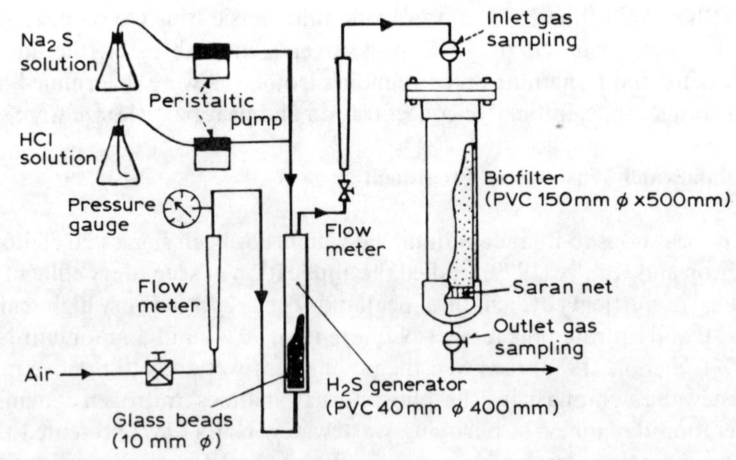

Fig. 1. Schematic diagram of a laboratory-scale deodorising system (from Furusawa *et al.*, 1984, reprinted with permission from The Society of Fermentation Technology, Japan).

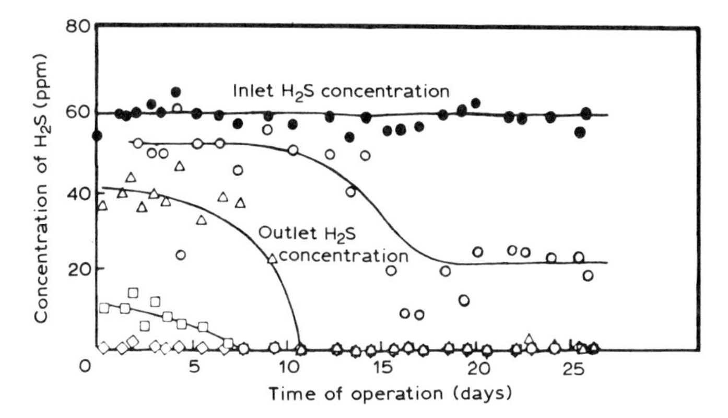

Fig. 2. Time course of H_2O removal at inlet H_2S concentration of about 60 ppm under various air feed rates (from Furusawa *et al.*, 1984, reprinted with permission from the Society of Fermentation Technology, Japan). Outlet H_2S concentrations (N m^3/day kg-dry peat): ○, 36; △, 18; □, 9·1; ◇, 3·6. Temperature: 13–19°C.

reported inconclusive experiments carried out to determine the feasibility of using peat moss as an adsorbent for the odorous gases dimethylamine, ammonia and hydrogen sulphide.

Pomeroy (1982) discussed the important role of microorganisms in the treatment of odorous air. He compared the problem of purifying polluted air as similar to the problem of purifying polluted water. At the beginning, efforts were confined to chemical and physical methods. Then, the importance of biological methods in pollution control was realized. Smith *et al.* (1973) showed the importance of microbial action in the capability of soil to degrade SO_2, H_2S, CH_3SH, CO, C_2H_2 and C_2H_4.

Furusawa *et al.* (1984) used a packed bed of fibrous peat as a deodorizing material to remove H_2S from air in a laboratory-scale column (Fig. 1). The authors obtained an efficient removal that was mostly due to biological oxidation by indigenous microorganisms in the peat (Fig. 2). Wada *et al.* (1986) determined the characteristics of the H_2S oxidizing bacteria inhabiting a peat biofilter.

3 FILTRATION AND BIOFILTRATION

The use of carbon as adsorbent dates back many centuries. Carbon adsorbents have been subjected to much research to develop new techniques and applications. Low concentrations of toxic or undesirable products in wastewaters are difficult to remove by conventional treatments, and some compounds have been found to be resistant to biological degradation. Carbon is activated by processes that include dehydration, carbonization and activation (burning of tars and pore enlargement), which require treatment at high temperatures. Activated carbon has been applied extensively to the removal of contaminants from wastewaters and gases due to its large surface area. The surface area of activated carbon, in the range of approximately 500 to 1400 m^2/g, has been a critical factor in its adsorptive properties (Cheremisinoff & Morresi, 1978). The chemical nature of carbon varies with the type of carbon but, for the most part, activated carbon surfaces are nonpolar, which makes the adsorption of inorganic electrolytes difficult. The utilization of activated carbon in a water or a wastewater treatment facility can be in the form of a granule or a powder.

The use of activated carbon in filtration includes the need for regeneration when the carbon reaches its adsorption saturation point. The reuse of the material is required to improve the economics of its use, although its regeneration implies additional operating costs. For example, drying of the activated carbon and volatilization and oxidation of organic adsorbents require furnace temperatures usually in the range of approximately 870 to 980°C (Cheremisinoff & Morresi, 1978).

3.1 Activated Carbon and Biological Processes

The biological regeneration of activated carbon was presented by Rodman *et al.* (1978). The objectives were the study of the ways in which activated carbon could be utilized in wastewater treatment without the need for thermal regeneration. Hutton (1978) reported on combined activated carbon–biological processes applied to the activated sludge process and to others, including aerated lagoons, contact stabilization processes and rotating disc contactors. The author indicated many technical advantages achieved by adding powdered carbon to a submerged culture biological process for waste treatment. Thus, activated carbon, as one preferred filtering medium, could be used for the removal of contaminants from streams and effluents. However, its effectiveness has limits. For example, Netzer and Hughes (1984) reported that the adsorption of either cobalt, copper or lead was hindered by the presence of the other metals, apparently due to competition for adsorption sites. They also reported significant differences in the ability of different commercially available activated carbons to adsorb metals from aqueous solutions. Subden *et al.* (1986) reported the use of 'charcoal' (activated carbon) in the wine industry to decolorize must and wines and to remove organic molecules associated with unwanted flavours and odours. They also observed that the activated carbon reduced copper and iron concentrations, increased the sodium concentration and produced small changes in calcium, magnesium and potassium concentrations.

An anaerobic activated carbon filter was demonstrated to be an effective process for the degradation of dihydroxyphenol catechol, which is a constituent of some industrial wastewaters. The process was also effective in removing total organic carbon and reducing COD with efficiencies higher than 81 and 95%, respectively (Suidan *et al.*, 1980). Wilson (1981) reported growing interest in the systematic exploitation of microbial activity on granular activated carbon (identified as 'GAC') for the removal of organics from wastewaters. The effectiveness of a fluidized-bed GAC biological process in the treatment of refinery sour water stripped bottoms and the role of reactor attachment media was studied by Gardner *et al.* (1988). The fluidized-bed GAC anaerobic reactor has been reported to be an effective process for the continuous long-term treatment of wastewaters containing biodegradable and nonbiodegradable toxic compounds (Pfeffer & Suidan, 1989). The authors observed that toxic materials that are neither biodegradable nor adsorbed by activated carbon may inhibit the biofilm activity. Therefore, successful operation of the process required carbon replacement that will keep concentrations of the toxic adsorbable compounds below their toxic threshold levels. Still, if the toxic substances are adsorbed by the carbon surface, there would not be any inhibition of the anaerobic biofilm.

3.2 Peat as a Filtering Agent

The characteristics of peat which make it a potential filtering agent have been presented elesewhere in this chapter. However, the final decision on developing commercial peat filters will depend on how their properties compare with those of existing filter materials, and on the economics of their use.

In comparing peat with one of the most accepted media for filtration and wastewater treatments, i.e. activated carbon, it has been noted above that the surface of the latter is, for the most part, nonpolar. This makes carbon a good adsorbent for organics, but the adsorption of inorganic electrolytes is more difficult (Netzer & Hughes, 1984). Moreover, the major drawback attributed to the use of activated carbon is its expense (McLellan & Rock, 1986).

4 BIOSORPTION

It has been proposed, after studying them, that the biological processes for the removal of metals ions from solution can be divided into three general categories: (a) biosorption (adsorption) of metal ions onto the surface of microorganisms, (b) transformation of metal ions by microorganisms, and (c) intracellular uptake of metal ions (Darnall *et al.*, 1986). It is recognized that process (b) is most suitable for mining operations, while the advantages of developing biological processes such as (a) and (c) for water purification and wastewater treatment operations are clear. Also, it has been indicated that processes (b) and (c) require living organisms, while process (a) may occur under conditions that would normally be toxic to them (Darnall *et al.*, 1986).

The development of new biosorbent material from microbial biomass is a rapidly-developing new area of research. It has been reported that the physico-chemical process of metal biosorption is based on adsorption, ion exchange, complexation and/or microprecipitation by the cell walls of living and nonliving microbial biomass, and that the process is somewhat rapid and can be reversible (Kuyucak & Volesky, 1988). Tsezos and Volesky (1981) screened samples of waste microbial biomass from fermentation processes and biological treatment plants for biosorbent properties relating to thorium and uranium in aqueous solutions. They found that the biomass of *Rhizopus arrhizus* exhibited the highest biosorptive uptake capacity for those metals, higher than ion exchange resins and activated carbon. A study of the mechanism of uranium biosorption by the abovementioned organisms has been presented (Tsezos & Volesky, 1982). Treen-Sears *et al.* (1984)

studied the propagation of *Rhizopus javanicus* biosorbent biomass. The authors mentioned that members of the order *Mucorales* have been reported to sequester ions from metal solutions by concentrating them in the cell walls. It appears that chitosan, which is a major structural polysaccharide in the cell walls of this order, has an important role in the uptake of thorium and uranium. Darnall *et al.* (1986) observed that the pH dependence of the binding of Ag^{+}, Au^{3+} and Hg^{2+} to the walls of the alga *Chlorella vulgaris* is different from that of the binding of other metal ions. Kuyucak and Volesky (1988) reported that algal biomass of *Sargassum natans* and *Ascophyllum nodosum* performed better than ion-exchange resins in removing cobalt and gold from solutions. Working with *Saccharomyces cerevisiae* and *Rhizopus arrhizus* biomass, they also observed that nonliving biomass exhibits a higher capacity for metal uptake than living biomass. Other findings reported by the same authors included the effect of pH on the metal-uptake capacity and the fact that the equilibrium adsorption isotherms were independent of the initial concentration of the metal in the solution.

In general, nonliving microbial biomass biosorption has been suggested as the basis for a new technology for environmental control and the recovery of valuable heavy metals, nuclear fuel and radioactive elements. Biosorbents can be cheap and efficient, highly selective, and reconditioned for reuse (Volesky, 1987).

It is interesting to note, at this point, that peat's complex composition includes the existence of dead microbial biomass. It is probable that some of the adsorptive characteristics exhibited by the peat fibres are of the same kind as those properties found in dead microbial cells.

5 MICROBIAL DEGRADATION

When an organic chemical is introduced into an ecosystem, it may be subjected to enzymatic and/or nonenzymatic reactions caused by the inhabitants of the environment. Few abiotic mechanisms for chemical changes totally convert organic compounds to inorganic products. Although plants and animals can metabolize a variety of chemicals, the degradations produced as a consequence of their metabolic activities are generally small when compared to the transformations effected by heterotrophic bacteria and fungi present in the same habitat. Therefore, the complete biodegradation or mineralization of organic molecules will be, in practically all cases, a consequence of microbial activity (Alexander, 1980).

Biodegradation by microorganisms is generally a growth-associated process, in which the carbon in the substrate is used by the microbial populations. In this process, the energy required for the biosynthetic reactions is released, and the by-products of the reactions are converted to cell constituents. Consequently, the microbial population increases in number and biomass.

Because of the diversity of compounds that are frequently present in industrial waste streams and others with a wide range of organic compounds, the biodegradation processes occurring probably involve the activities of several microbial species. These species interact by competing for nutrients, excreting growth factors, destroying antimicrobial chemicals, and also producing toxic compounds (Murakami & Alexander, 1989).

It has been noted that many waste chemicals can occur at concentrations that inhibit microbial populations that produce the biodegradation of other chemicals present. In addition, certain innocuous chemicals could be converted enzymatically to products that are hazardous. The generation of toxicants from innocuous precursors is often called 'activation' (Alexander, 1980).

The rate of degradation of foreign compounds under natural conditions is usually low. Many environmental factors could limit the activity of microorganisms, such as temperature, pH, surface tension, oxygen availability, salinity (in sea waters) and others. In general, it is to be expected that the concentration of a given degrading microorganism is low in the surroundings of an area invaded by foreign materials. For example, except near land, the ratio of oil oxidizers to the total bacteria population in unpolluted environments is low, probably ranging from 1:100 to 1:10 000. To overcome this natural low concentration of hydrocarbon-oxidizing microorganisms, seeding of oil slicks and other methods of waste oil disposal by microbial technology have been suggested (Berwick & Stafford, 1985). Still, besides the need to solve several important problems to make this method successful, such as finding which specific cultures are effective under varying environmental conditions, nutrient limitations to the microbial populations need to be overcome. Horowitz and Atlas (1977) reported enhanced biodegradation of hydrocarbons with bacterial inoculation and nutrient supplementation. The latter is accomplished by adding the necessary supplements to the seeding, particularly N and P sources (Bertrand *et al.*, 1983). These growth limitations are a result of the low nutrient levels of the waters, and in turn limit the growth of the microbial population (Atlas, 1981). In this case, the nutrients would have to be in a form that would adhere to the oil–water interface where the organisms are found (Schwartz & Leathen, 1976). Miget (1973) also suggested the possibility of periodic application of water-soluble encapsulated nutrient salts.

There are inherent disadvantages in the utilization of several soaking materials to clean oil spills on water, such as diatomaceous earth, sawdust, natural rubber latex granules, perlite, straw and the like, because a large portion of the adsorptive capacity of the various materials is consumed by saturation with water. In addition, some of the conventional sorbents have tendencies to release the oils already adsorbed and replace them with water, and then sink. Work conducted in laboratories at Memorial University of Newfoundland, Canada, have shown that when peat moss fibres are dried under specific conditions, the colloidal-like structure of peat is affected and the broken chemical bonds alter the hydrophilic characteristics of peat. The resulting dry peat has selective adsorption capacity with oil being readily adsorbed (and absorbed, given the spongy characteristics of the peat fibres). This phenomenon is the basis of a commercial process developed in Newfoundland, Canada, to produce an oil sorbent for aquatic environments (Canadian Patent no. 1160201, 10 January 1984).

Based on the abovementioned properties of peat and some peat products, a research project has been developed at Memorial University of Newfoundland on a combined soaking–biological method for cleaning up oil spills. The process is based on the attachment of oil-degrading microbial populations onto peat particles. While the peat particles adsorb and absorb the hydrocarbons, specialized microbial populations immobilized in the peat fibres will degrade the oil. In addition to acting as a support for the microbial seeding of the oil spill, the peat should provide the microbial population with the additional nutrients required for its metabolic activities. Those nutrients could be either components of the peat that have been shown to be metabolizable by microorganisms (as will be discussed in another part of this chapter), or nutrient supplements that have been added to the peat along with the microorganisms.

The use of natural products to immobilize whole cell adapted microorganisms for biodegradation of toxic products has been studied. Portier (1986, 1987*a*) suggested that polysaccharide materials may constitute inexpensive support materials and demonstrated the use of chitin and chitosan for handling mixtures of chlorinated phenols and ethanes.

Organic chemicals that persist for long periods in natural ecosystems because of the inability of microorganisms to degrade them, or because of their inability to degrade them rapidly, are known as recalcitrant molecules. Alexander (1980) discussed the compounds in which they appear (recalcitrant compounds), and the inherent problems with these molecules such as biomagnification and long-distance migration. The author pointed out that the lack of rapid microbial attack on a chemical and its persistence can be due to chemical or environmental causes. If a molecule is completely refractory to microbial degradation, it is probably because it possesses a chemical structure characteristic of absolutely recalcitrant compounds such as synthetic polymers. Alternatively, some recalcitrant molecules could be acted on enzymatically, but the degrading microorganisms do not proliferate and the rate of transformation does not increase as a function of time. As a consequence, the compound is likely to remain in the environment. This is the case with cometabolic reactions. Grady (1985) presented a comprehensive review of the microbiological basis of biodegradation and its measurement. In acknowledging the use of the term 'recalcitrance', the author preferred to use the term 'persistence' to recognize that a chemical has failed to undergo biodegradation under a specified set of conditions. Therefore, the chemical could be biodegradable but still persist in the environment. These definitions were introduced by Bull (1980), who reserved the term 'recalcitrant' for chemicals that are inherently resistant to any degradation. Therefore, much important research work could be devoted to the recognition of persistent compounds and to finding the appropriate methods for their degradation.

5.1 Cometabolism

With many chemicals, including some of the most notorious pollutants such as DDT, aldrin, and many chlorinated and nonchlorinated compounds, it has been found that a different microbial conversion takes place. Some compounds have been transformed biologically in nonsterile environments, but no microbial populations responsible for using those compounds as substrates have been reported. Still, sterilized samples of the same environment resulted in no transformations. This phenomenon has been termed 'cometabolism' or 'cooxidation', implying that the microbial populations are growing in another substrate while performing the cometabolic transformation. Consequently, the introduction of the compound into the environment does not cause the populations responsible for the transformation to increase in number or biomass, reflecting their inability to use the compound for biosynthetic purposes. Because of this, a compound subjected to cometabolic degradation is converted very slowly (Alexander, 1980).

The phenomenon of cooxidation was reported, initially, by Leadbetter and Foster (1959), and was used to describe a process in which a microorganism oxidized a substance without being able to use, for supporting its growth, the energy released from the oxidation (Horvath, 1972). Jensen (1963) suggested the term 'cometabolism' for this process, expanding it to cover dehalogenation reactions by microorganisms. Generally, both terms have been used interchangeably and they are considered synonymous. Alexander (1980) commented on the objections voiced to the use of the term 'cometabolism' based on the fact that it does not describe a new metabolic phenomenon. However, he justifies its use because of the important environmental repercussions of this kind of biotransformation process. Horvath (1972), in reviewing the existing literature on cometabolism, pointed out that the interpretation of experimental results which indicate the presence of cometabolism requires considerable care. Alexander (1979) discussed the possible role of cometabolism in nature, suggesting that a probable explanation for this phenomenon is the existence of an enzyme of broad substrate specificity within the microbial active population in which the initial enzymatic reactions result in products that do not serve as substrates for the other enzymes of the microbial cells. Baker and Woods (1977) reported the isolation of bacteria capable of degrading the ixodicide amitraz without the utilization of the ixodicide as a substrate or energy source. The authors reported that yeast extract (or a component of the yeast extract) was required for the microbial process, to act as what the authors identified as 'cometabolite'.

Most cases of cometabolic reactions have been found among toxic and recalcitrant chemical compounds such as pesticides and other industrial chemicals. Chou and Bohonos (1979) studied mixed-culture biodegradation of methyl parathion, benzothiophene, dibenzothiophene, benzoquinoline and isoquinoline and the induction of degradative cometabolic and enzymatic reactions with chemicals structurally related to those compounds. Francis *et al.* (1978) reported that a *Pseudomonas* sp. able to grow on several nonchlorinated and mono-*p*-chloro-substituted analogues of DDT as a sole carbon source degraded bis(*p*-chlorophenyl)methane and 1,1-bis(*p*-cholorophenyl)ethane only in the presence of diphenylethane. However, major metabolites produced in the reactions were not metabolized further. In studying the microbial metabolism of 2,4,5-trichlorophenoxyacetic acid (2,4,5-T) in soil, soil suspension and axenic culture, Rosenberg and Alexander (1980) suggested that 2,4,5-T was acted on by cometabolism by the isolated microbial population. They also indicated that the usually long persistence of 2,3,4-T in soil could be the result of the inability of a small cometabolizing microbial population to replicate by using the pesticide as a carbon source. But, if there is a product of the cometabolic reaction that can be used as a carbon source, that product would have served as a substrate for microbial growth and it would not be found in natural conditions in appreciable concentrations. Venkataramani and Ahlert (1985) proposed a substrate utilization model to quantify the role of cometabolism coupled with cellular maintenance.

A negative result of cometabolism has been pointed out by Alexander (1980), who suggested that the generation of toxicants from nontoxic precursors (the activation process previously mentioned) could also occur during cometabolism. Still, the author also indicated that cometabolism of a toxicant could result in detoxification, referring to the publications by several researchers which stated that certain synthetic compounds that had been apparently cometabolized were actually transformed into products that are not of ecological concern.

Microorganisms require energy to maintain viability, and to obtain sufficient energy for their essential functions they must carry out oxidations. If the available concentration of a chemical is very low, a species able to metabolize that chemical may not reproduce or even survive. Therefore, some compounds available at low concentration persist in the environment. Alexander (1980) suggested that this hypothesis (which could account for the longevity of compounds present at low concentrations as water-soluble materials) could also apply to chemicals with low solubilities or to those that are not emulsified enough to have a rapid penetration into the microbial cell. Therefore, a substrate will persist in an environment if it does not enter the cell at the required rate that will permit the organism to get energy, both for growth and maintenance, rapidly enough.

If cometabolism occurs, a potential way of using it for the degradation of recalcitrant molecules is by providing the microbial population responsible for the cooxidation with sufficient nutrient sources for their basic metabolic reactions. In case the cometabolic phenomenon is brought about by the accumulation of an intermediary metabolite(s) that inhibits the total reaction, a continuous or flow system in a biological filter could be the solution.

As mentioned before, Portier *et al.* (1987*a*,*b*) presented technologies for the detoxification of contaminated groundwaters utilizing packed bed biological reactors. They claimed that their methods are based on the principle that natural populations of microorganisms may adapt to degrade molecules considered as 'refractory' or difficult to decompose by biological means. The same authors have conducted the evaluation of packed bed immobilized reactors for the continuous biodegradation of contaminated groundwaters and industrial effluents (Portier *et al.*, 1987*c*, 1988). Their research is presented in more detail in Chapter 11.

6 PEAT AS A NUTRIENT SOURCE FOR MICROORGANISMS

The commercial utilization of peat is of interest to countries that have significant peat reserves (Table 2), and the possibilities of using peat as a raw material in the chemical and biochemical industries have been reported (Fuchsman, 1980). Peat is

Table 2
World Peat Resources

Country	*Biological peatland (ha × 10⁶)*	*Peat production (t × 10³)*		
		Fuel peat	*Moss peat*	*Total*
Canada	170·0	—	488	488
USSR	150·0	80 000	120 000	200 000
USA	40·0	—	800	800
Indonesia	26·0	—	—	—
Finland	10·0	3 100	500	3 600
Sweden	7·0	—	270	270
People's Republic of China	3·5	800	1 300	2 100
Norway	3·0	1	83	84
Malaysia	2·4	—	—	—
United Kingdom	1·6	50	500	550
Poland	1·4	—	280	280
Republic of Ireland	1·2	5 570	380	5 950
FRG	1·1	250	2 000	2 250
Total	417·2	89 771	126 601	216 372

Source: Kivinen & Pakarinen, 1980, as presented by Viraraghavan & Ayyaswami, 1987; reprinted with permission of the Organizer and Publisher of the *Proceedings of the 6th International Peat Congress*, Duluth, MN, 1980.

an organic material composed of carbohydrates, minerals and a group of substances identified as humic acids, among other components. The complex mixture of organic compounds in peat and its relatively low price make this material a potential source of nutrients for fermentation processes (Martin, 1984). LeDuy (1979, 1981) presented the basics for the microbial bioconversion of peat. The process of extracting nutrients from peat to be employed in submerged fermentations for the growth of various microorganisms with potential commercial value has been studied for a long time. Figure 3 is a schematic diagram of a process for the hydrolysis of peat and the production of peat extract to be employed in

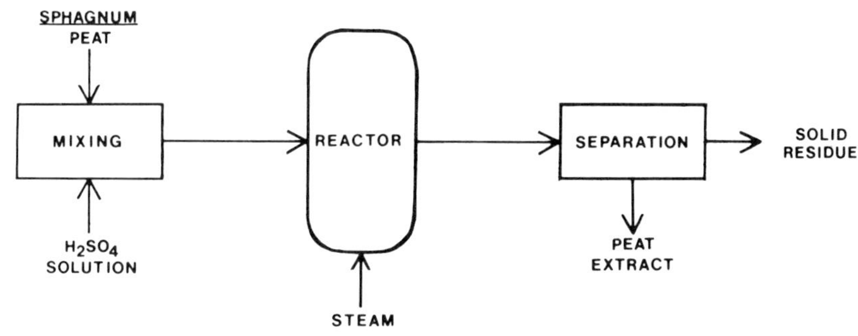

Fig. 3. Process flow for the acid hydrolysis of peat and the production of peat extract (from Martin & Manu-Tawiah, 1989, reprinted with permission of the Society of Chemical Industry, London).

Table 3
Chemical Composition of *Sphagnum* Peat and Peat Extract

Component	*Raw peat (% of dry weight)*[a]	*Peat extract (g/litre)*[a]
pH[b]	4·85 ± 0·11	0·99 ± 0·15
Moisture[c]	80·50 ± 1·30	—
Total solids	19·50 ± 0·50	62·26 ± 1·59
Dissolved solids	—[d]	49·41 ± 0·77
TCH	—[d]	32·75 ± 1·23
Total reducing sugars	—[d]	16·47 ± 0·51
Total lipids	2·52 ± 0·17	0·99 ± 0·03
Total nitrogen	0·80 ± 0·08	0·60 ± 0·01
Ash	2·60 ± 0·58	4·51 ± 0·01

Source: Martin & Manu-Tawiah, 1989; reprinted with permission of the Society of Chemical Industry, London.
[a] Mean values of three determinations of three replicate samples ± standard deviations.
[b] No units.
[c] Percent of wet weight.
[d] Not determined.

fermentation processes. A review of the literature (Fuchsman, 1980; Martin, 1983) shows that the fermentation of peat extract (also called peat hydrolysate) has been conducted extensively in the Soviet Union and to a lesser extent in Canada, Ireland and the United States.

Table 3 shows the chemical composition of *Sphagnum* peat and of the extract produced by acid hydrolysis, and Table 4 presents their amino acid contents. The major monosaccharide composition of the peat extract, which is an important piece of information from the point of view of its potential as fermentation substrate, is reported in Table 5. It is important to specify that the quantitative contents of the components listed in these tables are a function of the values of the process parameters utilized in the extraction process. Peat hydrolysates have been used as the main medium in the submerged culture of yeasts (Quierzy *et al.*, 1979) and fungi (Boa & Le Duy, 1982; Martin & White, 1986). Other recent works on the fermentation of peat hydrolysate have been aimed at the production of microbial products with potential industrial application such as pullulan (LeDuy & Boa, 1983). The production of submerged mushroom mycelium has also received attention (Martin, 1986; Manu-Tawiah & Martin, 1987).

The main objective of the acid treatment of peat at high temperature is to hydrolyse the carbohydrates in the peat. The hydrolysation process will also yield a wide range of other substances, many of them with nutrient characteristics (Martin & Manu-Tawiah, 1989). The potential of enzymatic hydrolysis of peat to yield nutrients for microbial processes has been also reported (Fuchsman, 1980).

The abovementioned information, together with many other reports found in the scientific literature concerning the properties of peat for use in the maintenance,

Table 4
Amino Acid Composition of *Sphagnum* Peat and Peat Extract

Amino acid	*Raw peat (mg/g of dry weight)*[a]	*Peat extract (mg/litre)*[a]
Alanine	1·82 ± 0·12	2·65 ± 0·18
Arginine	0·58 ± 0·02	1·22 ± 0·05
Aspartic acid	1·93 ± 0·07	4·37 ± 0·18
Cysteic acid	0·73 ± 0·01	4·64 ± 0·18
Glutamic acid	1·93 ± 0·15	2·45 ± 0·21
Glycine	1·91 ± 0·12	2·95 ± 0·12
Histidine	0·62 ± 0·07	0·79 ± 0·14
Isoleucine	0·62 ± 0·07	0·84 ± 0·08
Leucine	1·31 ± 0·11	1·71 ± 0·13
Lysine	0·86 ± 0·13	1·22 ± 0·08
Methionine	Trace	0·42 ± 0·03
Phenylalanine	0·86 ± 0·04	0·78 ± 0·03
Proline	1·06 ± 0·00	2·24 ± 0·12
Serine	1·54 ± 0·21	1·93 ± 0·24
Threonine	1·54 ± 0·02	2·52 ± 0·11
Tryptophane	Trace	0·00
Tyrosine	0·74 ± 0·01	0·86 ± 0·06
Valine	1·24 ± 0·04	1·68 ± 0·15

Source: Martin & Manu-Tawiah, 1989; reprinted with permission of the Society of Chemical Industry, London.
[a] Mean values of three determinations of three replicate samples ± standard deviations.

Table 5
Major Monosaccharide Composition of *Sphagnum* Peat Extract

Monosaccharide	*Percentage of total reducing sugars*[a]
Arabinose	2·48 ± 1·11
Galactose	19·07 ± 1·67
Glucose	38·20 ± 1·31
Mannose	16·46 ± 1·83
Rhamnose	6·96 ± 1·62
Xylose	12·03 ± 1·15

Source: Martin & Manu-Tawiah, 1989; reprinted with permission of the Society of Chemical Industry, London.
[a] Mean values of three determinations of three replicate samples ± standard deviations.

incubation and growth of microbial populations, has pointed out the properties of peat that allow its use as a support and substrate source for microorganisms.

7 CONCLUSIONS

In spite of the ability of nature to recycle waste materials and to purify itself, the increasing demand placed on the earth's environment during the last decades (and in the foreseeable future) by large amounts of wastes and pollutants could threaten the capacity of natural systems to recover and to maintain proper environmental conditions for mankind. For many obvious reasons, already explained in this chapter and elsewhere in this book, biodegradation techniques are the ones which, in the long term, could provide the tools to fight pollution and preserve the environment. The biodegradation techniques are multiple, they are versatile and they can be used at various stages of waste treatment.

7.1 Bioremediation and the Potential of Peat Biofilters

Bioremediation technologies are those that attempt to optimize the natural microbial capacity to degrade pollutants by providing proper conditions for the microbial population, including essential nutrients, some of which could otherwise limit their growth. Also, waste treatment processes for recalcitrant compounds, which have been found to be biodegradable by cometabolic reactions ('persistent' compounds) could then be enhanced by providing sources of nutrients required by the degrading microbial population.

Portier *et al.* (1987*c*) reported that adapted microorganisms present in a biological reactor selectively utilized the pesticide contaminant of a wastewater stream when appropriate amounts of nitrogen and phosphorus were provided to obtain complete degradation. An interesting conclusion was drawn by Portier *et al.* (1988), who indicated that recalcitrant chemicals can be biotransformed, at the molecular level, when these materials are in close proximity to adapted microbial populations on biodegradable surfaces. Polychlorinated biphenyls (PCBs) were degraded anaerobically by cometabolic/cooxidative microbial processes in which the aminopolysaccharide chitin was identified as providing an amenable substrate for degrading microbial populations and also an effective adsorbing surface for water-soluble PCB isomers (Portier & Fujisaki, 1988).

The convenience of having a natural material acting as support for degrading microbial populations and added nutrients, and if possible as a source of nutrients itself, is obvious. Peat possesses those properties, besides possessing sorbent properties that could contribute to its employment in waste treatment systems. The use of peat as an agent in combating pollutants in liquid and gaseous streams and its role in supporting microbial degrading populations make it an ideal biological filter material for waste treatment systems. Research on this subject is being conducted and it is predicted that new technologies will evolve based on these principles.

ACKNOWLEDGEMENT

The assistance of Mr Paul Bemister, Department of Biochemistry, Memorial University of Newfoundland, in proofreading this manuscript, is appreciated.

REFERENCES

Aho, M. & Tummavuori, J. (1984). *Suo*, **35**, 47.
Alexander, M. A. (1979). In *Proc. Workshop Microbial Degradation of Pollutants in Marine Environments*, eds A. W. Bourquin & P. H. Pritchard. US Environmental Protection Agency, Gulf Breeze, FL, p. 67.
Alexander, M. (1980). *Science*, **211**, 132.
Asplund, D., Ekman, E. & Thun, R. (1976). In *Proc. Fifth Int. Peat Cong.*, Poznan, Poland, **1**, 358.
Atlas, R. M. (1981). *Microbiol. Rev.*, **45**, 180.
Ayyaswami, A. & Viraraghavan, T. (1985). In *Proc. Can. Soc. Civ. Eng. Ann. Conf.*, Saskatoon, SK, Canada, p. 343.
Baker, P. B. & Woods, D. R. (1977). *J. Appl. Bacteriol.*, **42**, 187.
Barton, P., Buggy, M., Deane, S., Kelly, J. & Lyon, H. J. (1984). In *Proc. Seventh Int. Peat Cong.*, Irish National Peat Committee, Dublin, **2**, 148.
Bertrand, J. C., Rambeloarisoa, E., Rontani, J. F., Giusti, G. & Mattei, G. (1983). *Biotechnol. Letters*, **8**, 567.
Berwick, P. G. & Stafford, D. A. (1985). *Proc. Biochem.*, **20**, 175.
Bloom, P. R. & McBride, M. B. (1979). *Soil Sci. Soc. Am. J.*, **43**, 687.
Boa, J. M. & LeDuy, A. (1982). *Can. J. Chem. Eng.*, **60**, 532.
Bohn, H. L. (1975). *Air Pollut. Control Assoc. J.*, **25**, 953.
Boyd, S. A., Sommers, L. E. & Nelson, D. W. (1981). *Soil Sci. Soc. Am. J.*, **45**, 1241.
Brooks, J. L., Rock, C. A. & Struchtemeyer, R. A. (1984). *J. Environ. Qual.*, **13**, 524.
Brown, J. L. & Farnham, R. S. (1976). In *Proc. Fifth Int. Peat Cong.*, Poznan, Poland, **1**, 349.
Brown, L., Bellinger, G. & Day, J. P. (1979). *J. Inst. Water Engin. Scient.*, **33**(5), 478.
Bull, A. T. (1980). In *Contemporary Microbial Ecology*, eds J. N. Hedger, M. J. Lathane, J. M. Lynch & J. H. Slater. Academic Press, London, p. 107.
Bunzl, K. (1974*a*). *J. Soil Sci.*, **25**, 343.
Bunzl, K. (1974*b*). *J. Soil Sci.*, **25**, 517.
Bunzl, K., Schmidt, W. & Sansoni, B. (1976). *J. Soil Sci.*, **27**, 32.
Cameron, R. D. (1978). *Can. J. Civil Eng.*, **5**, 83.
Chaney, R. L. & Hundemann, P. T. (1979). *Water Pollut. Control Fed. J.*, **51**, 17.
Cheremisinoff, P. N. & Morresi, A. C. (1978). In *Carbon Adsorption Handbook*, eds P. N. Cheremisinoff & F. Ellerbusch. Ann Arbor Science Publishers, Ann Arbor, MI, p. 1.
Chou, T.-W. & Bohonos, N. (1979). In *Proc. Workshop Microbial Degradation of Pollutants in Marine Environments*, eds A. W. Bourquin & P. H. Pritchard. US Environmental Protection Agency, Gulf Breeze, FL, p. 76.
Cloutier, J.-N., LeDuy, A. & Ramalho, R. S. (1985). *Can. J. Chem. Eng.*, **63**, 250.
Coupal, B. & Lalancette, J. M. (1976). *Water Res.*, **10**, 1071.
Darnall, D. W., Greene, B., Henzl, M. T., Hosea, J. M., McPherson, R. A., Sneddon, J. & Alexander, M. D. (1986). *Environ. Sci. Technol.*, **20**, 206.
D'Hennezel, F. & Coupal, B. (1972). *Can. Mining Metallurgical Bull.*, **65**, 51.
Dissanayake, C. B. & Weerasooriya, V. R. (1981). *Int. J. Environ. Studies*, **17**, 233.
Dufort, J. & Ruel, M. (1972). *Proc. Fourth Int. Peat Cong.*, Helsinki, **4**, 299.
Eye, J. D. (1968). *Water Pollut. Control Fed. J.*, **40**(8), Part 2, R316.

Farnham, R. S. & Brown, J. L. (1972). In *Proc. Fourth Int. Peat Cong.*, Otaniemi, Finland, **4**, 271.
Firman, J. E. (1970). *Process Biochem.*, **5**, 21.
Francis, A. J., Spanggord, R. J., Ouchi, G. I. & Bohonos, N. (1978). *Appl. Environ. Microbiol.*, **35**, 364.
Fuchsman, C. H. (1980). In *Peat, Industrial Chemistry and Technology*. Academic Press, New York.
Furusawa, N., Togashi, I., Hirai, M., Shoda, M. & Kubota, H. (1984). *J. Ferment. Technol.*, **62**, 589.
Gardner, D. A., Suidan, M. T. & Kobayashi, H. A. (1988). *Water Pollut. Control Fed. J.*, **60**, 505.
Gosset, T., Trancart, J.-L. & Thevenot, D. R. (1986). *Water Res.*, **20**, 21.
Grady, C. P. L., Jr (1985). *Biotechnol. Bioeng.*, **27**, 660.
Gravelle, D. V. & Landreville, S. (1980). *Can. J. Chem. Eng.*, **58**, 235.
Hall, R. (1987). *Can. Res./Biotechnol. Can.*, April, 34.
Horowitz, A. & Atlas, R. M. (1977). *Appl. Environ. Microbiol.*, **33**, 1252.
Horvath, R. S. (1972). *Bacteriol. Rev.*, **36**, 146.
Hutton, D. G. (1978). In *Carbon Adsorption Handbook*, ed. P. N. Cheremisinoff & F. Ellerbusch. Ann Arbor Science Publishers, Ann Arbor, MI, p. 389.
Jaouich, B. A. (1975). *Nitrate reduction in peat.* PhD thesis, University of Minnesota, Minneapolis, MN.
Jensen, H. L. (1963). *Acta Agr. Scand.*, **13**, 404.
Kadlec, R. H. & Rathbun, M. A. (1983). In *Proc. Int. Symp. Peat Utilization*, eds C. H. Fuchsman & S. A. Spigarelli. Bemidji State University, Bemidji, MN, p. 351.
Kivinen, E. & Pakarinen, P. (1980). In *Proc. Sixth Int. Cong.*, Duluth, MN, p. 52.
Kuyucak, N. & Volesky, B. (1988). *Biotechnol. Letters*, **10**, 137.
Lalancette, J. M. & Coupal, B. (1972). In *Proc. Fourth Int. Peat Cong.*, Otaniemi, Finland, **4**, 213.
Leadbetter, E. R. & Foster, J. W. (1959). *Arch. Biochem. Biophys.*, **82**, 491.
LeDuy, A. (1979). *Proc. Biochem.*, **15**, 5.
LeDuy, A. (1981). In *Proc. Peat as an Energy Alternative Symp.*, Institute of Gas Technology, Chicago, IL, p. 479.
LeDuy, A. & Boa, J. M. (1983). *Can. J. Microbiol.*, **29**, 143.
Leslie, M. E. (1974). *Amer. Dyestuff Reporter*, August, 15.
Lidkea, T. R. (1974). *Treatment of sanitary landfill leachate with peat.* MSc thesis, University of British Columbia, Vancouver, BC, Canada.
Lochhead, A. G. & Thexton, R. H. (1947*a*). *Can. J. Res.*, **25**, 1.
Lochhead, A. G. & Thexton, R. H. (1947*b*). *Can. J. Res.*, **25**, 14.
Manu-Tawiah, W. & Martin, A. M. (1987). *Appl. Biochem. Biotechnol.*, **14**, 221.
Martin, A. M. (1983). In *Proc. Int. Symp. Peat Utilization*, ed. C. H. Fuchsman & S. A. Spigarelli. Bemidji State University, Bemidji, MN, p. 301.
Martin, A. M. (1984). *J. Chem. Technol. Biotechnol.*, **34B**, 70.
Martin, A. M. (1986). *J. Food Proc. Eng.*, **8**, 81.
Martin, A. M. & Manu-Tawiah, W. (1989). *J. Chem. Technol. Biotechnol.*, **45**, 171.
Martin, A. M. & White, M. D. (1986). *Appl. Microbiol. Biotechnol.*, **24**, 84.
McKay, G. (1980). *Water Services*, **84**, 357.
McKay, G., Otterburn, M. S. & Sweeney, A. G. (1978). *JSDC*, August, 357.
McLellan, J. K. & Rock, C. A. (1986). *Int. Peat J.*, **1**, 1.
Meisel, J., Lakatos, B. & Mady, G. (1979). *Acta Agro. Acad. Sci. Hung.*, **28**, 75.
Miget, G. (1973). In *The Microbial Degradation of Oil Pollutants*, Center for Wetland Resources, Louisiana State University, Baton Rouge, LA, p. 291.
Mueller, J. C. (1972). In *Proc. Symp. Peat Moss in Canada*, University of Sherbrooke, Sherbrooke, Québec, Canada, p. 274.
Murakami, Y. & Alexander, M. (1989). *Biotechnol. Bioeng.*, **33**, 832.

Netzer, A. & Hughes, D. E. (1984). *Water Res.*, **18**, 927.
Nichols, D. S. (1983). *Water Pollut. Control Fed. J.*, **55**, 495.
Nichols, D. S. & Boelter, D. H. (1982). *J. Environ. Qual.*, **11**, 86.
Osborne, J. M. (1975). *J. Soil Water Conserv.*, **30**, 235.
Parkash, S. & Brown, R. A. S. (1976). *CIM Bull.*, **69**, 59.
Parrott, H. A. & Boelter, D. H. (1979). In *Utilization of Municipal Sewage Effluent and Sludge on Forest and Disturbed Land*, ed. W. E. Sopper & S. N. Kerr, Pennsylvania State University Press, University Park, PA, p. 115.
Pfeffer, J. T. & Suidan, M. T. (1989). *Biotechnol. Bioeng.*, **33**, 139.
Pomeroy, R. D. (1982). *Water Pollut. Control. Fed. J.*, **54**, 1541.
Poots, V. J. P. & McKay, G. (1980). *Sci. Proc. Royal Dublin Soc., Series A*, **6**(15), 409.
Poots, V. J. P., McKay, G. & Healy, J. J. (1976). *Water Res.*, **10**, 1061.
Poots, V. J. P., McKay, G. & Healy, J. J. (1978). *Sci. Proc. Royal Dublin Soc., Series A*, **6**(6), 61.
Portier, R. J. (1986). In *Immobilisation of Ions by Bio-Sorption*, ed. H. Eccles. Ellis Horwood, Chichester, UK, p. 229.
Portier, R. J. & Fujisaki, K. (1988). In *Aquatic Toxicology and Hazard Assessment*, Vol. 10, ASTM STP 971, eds W. J. Adams, G. A. Chapman & W. G. Landis, American Society for Testing Materials, Philadelphia, PA, p. 517.
Portier, R. J., Fujisaki, K., Reilly, L. A. & Henry, C. B. (1987*a*). *Detoxification of contaminated groundwaters using a marine polysaccharide/diatomaceous earth packed bed biological reactor*. Institute for Environmental Studies, Louisiana State University, Baton Rouge, LA, p. 1709.
Portier, R. J., Fujisaki, K., Reilly, L. A. & McMilin, D. J. (1987*b*). *Detoxification of rinsates from aerial pesticide applications using a marine polysaccharide/diatomaceous earth packed bed biological reactor*. Institute for Environmental Studies, Louisiana State University, Baton Rouge, LA, p. 1713.
Portier, R. J., Christiansen, J. A., Wilkerson, J. M., Flynn, B. P. & Bost, R. C. (1987*c*). Evaluation of a packed bed immobilized microbe reactor for the continuous biodegradation of pesticide-contaminated ground water. Presented at the *AICHE Annual Meeting*, New York, 16–20 November.
Portier, R. J., Friday, D. D., Christianson, J. A., Nelson, J. F. & Eaton, D. L. (1988). Evaluation of a packed bed immobilized microbe bioreactor for the continuous biodegradation of contaminated ground waters and industry effluents: case studies. Presented at the *18th Intersociety Conference on Environmental Systems*, San Francisco, CA, 11–13 July.
Quierzy, P., Therien, N. & LeDuy, A. (1979). *Biotechnol. Bioeng.*, **21**, 1175.
Rock, C. A., Brooks, J. L., Bradeen, S. A. & Struchtemeyer, R. A. (1984). *J. Environ. Qual.*, **13**, 518.
Rock, C. A., Fiola, J. W., Greer, T. F. & Woodward, F. E. (1985). *J. New England Water Pollut. Control Assoc.*, **19**, 32.
Rodman, C. A., Shunney, E. L. & Perrotti, A. E. (1978). In *Carbon Adsorption Handbook*, eds P. N. Cheremisinoff & F. Ellerbusch. Ann Arbor Science Publishers, Ann Arbor, MI, p. 483.
Rosenberg, A. & Alexander, M. (1980). *J. Agric. Food Chem.*, **28**, 297.
Ruel, M., Chornet, E., Coupal, B., Aitcin, P. & Cossette, M. (1977). In *Muskeg and the Northern Environment in Canada*, eds N. W. Radforth & C. O. Brawner. University of Toronto Press, Toronto, ON, Canada, p. 221.
Schwartz, W. A. (1968). Report on evaluation of the waste treatment potential of peat. US Department of the Interior, FWPCA (unpublished manuscript).
Schwartz, R. D. & Leathen, W. W. (1976). In *Industrial Microbiology*, eds B. M. Miller & W. Litsky. McGraw-Hill, New York, p. 384.
Silvo, O. E. J. (1972). In *Proc. Fourth Int. Peat Cong.*, Otaniemi, Finland, **4**, 311.
Sipos, S, Sipos, E., Dekany, I., Deer, A., Meisel, J. & Lakatos, B. (1978). *Acta Agro. Acad. Sci. Hung.*, **27**, 31.

Smith, K. A., Bremner, J. M. Tabatabai, M. A. (1973). *Soil Sci.*, **116**, 313.
Smith, E. F., MacCarthy, P., Yu, T. C. & Mark, H. B., Jr (1977). *Water Pollut. Control Fed. J.*, **49**, 633.
Soniassy, R. N. (1974). *CIM Bull.*, **67**, 95.
Stanlick, H. T. (1976). Treatment of septic tank effluent using underground peat filters. US Forest Service, Milwaukee, WI (unpublished manuscript).
Subden, R. E., Akhtar, M. & Cunningham, J. D. (1986). *Can. Inst. Food Sci. Technol. J.*, **19**, 145.
Suidan, M. T., Cross, W. H. & Fong, M. (1980). *Prog. Water Technol.*, **12**, 203.
Takamatsu, T. & Yoshida, T. (1978). *Soil Sci.*, **125**, 377.
Tilton, D. L. & Kadlec, R. H. (1979). *J. Environ. Qual.*, **8**, 328.
Tilton, D. L., Kadlec, R. H. & Richardson, C. J. (eds) (1976). *National Symp. Freshwater Wetlands and Sewage Effluent Disposal*, University of Michigan, Ann Arbor, MI.
Tinh, V. Q., Leblanc, R., Janssens, J. M. & Ruel, M. (1971). *Can. Mining Metallurgical Bull.*, **64**, 99.
Treen-Sears, M. E., Martin, S. M. & Volesky, B. (1984). *Appl. Environ. Microbiol.*, **48**, 137.
Tsezos, M. & Volesky, B. (1981). *Biotechnol. Bioeng.*, **22**, 583.
Tsezos, M. & Volesky, B. (1982). *Biotechnol. Bioeng.*, **24**, 385.
Tummavuori, J. & Aho, M. (1980*a*). *Suo*, **31**, 45.
Tummavuori, J. & Aho, M. (1980*b*). *Suo*, **31**, 79.
Venkataramani, E. S. & Ahlert, R. C. (1985). *Biotechnol. Bioeng.*, **27**, 1306.
Viraraghavan, T. & Ayyaswami, A. (1987). *Can. J. Civ. Eng.*, **14**, 230.
Volesky, B. (1987). *TIBTECH*, **5**, 96.
Wada, A., Shoda, M., Kubota, H., Kobayashi, T., Katayama-Fujimura, Y. & Kuraishi, H. (1986). *J. Ferm. Technol.*, **64**, 161.
Wilson, J. (1981). *Proc. Biochem.*, **16**, 9.
Wolf, A., Bunzl, K., Dietl, F. & Schmidt, W. F. (1977). *Chemosphere*, **5**, 207.
Zhipei, Z., Junlu, Y., Zenghui, W. & Piya, C. (1984). In *Proc. Seventh Int. Peat Cong.*, Irish National Peat Committee, Dublin, **3**, 147.

Chapter 17

BIODEGRADATION OF PROCESS INDUSTRY WASTEWATER. CASE PROBLEM: SUGARCANE INDUSTRY

CARMEN DURÁN DE BAZÚA

Departamento de Alimentos y Biotecnología, División de Ingeniería, Facultad de Química, Universidad Nacional Autónoma de México, México

ADALBERTO NOYOLA

Coordinación de Ingeniería Ambiental, Instituto de Ingeniería, Universidad Nacional Autónoma de México, México

HÉCTOR POGGI

Departamento de Biotecnología y Bioingeniería, Centro de Investigación y Estudios Avanzados, Instituto Politécnico Nacional, México

&

LUIS EDUARDO ZEDILLO

Instituto para el Mejoramiento de la Producción de Azúcar, Grupo de Países de Latinoamérica y del Caribe Exportadores de Azúcar, Tuxpan, México

CONTENTS

1 INTRODUCTION

Industrialization in Third World countries has become a necessity, especially because of the very low prices associated with raw materials and very high prices associated with manufactured products.

However, due to the high prices of the technological packages and know-how, obsolescence in the acquired technology is more the rule than the exception, when installation of new plants in these countries is concerned. Therefore ecological concepts of clean technologies are difficult to meet, and generation of wastes or losses of raw materials and products in the effluents is a common problem associated with most industrial branches.

In Mexico, in particular, the estimated data on wastewater generation for 1990 were 184 m^3/s (105 of municipal discharges and 79 of industrial liquid effluents), and for the year 2000 are expected to be 207 m^3/s, with an organic load, measured as biochemical oxygen demand, of 2·4 million metric tonnes (36% belonging to the municipal sector and 64% to the industrial one). These figures give the magnitude of the enormous challenge that research–education, governmental, private and social sectors of Mexican society confront, not only for the provision of water services but for the treatment and disposal of these wastewaters.

There are 39 industrial branches considered as the main consumers of the hydric resource, and nine of them (sugarcane, chemical, cellulose and paper, oil and petrochemical, softdrinks and alcohol beverages, textiles, metalurgical, electrical and food industries) produce the highest amounts of wastewaters (82% of the total amount of industrial liquid effluents). The first two branches alone, sugarcane and chemical industries, generate 59·8% of the total amount for the industrial sector (Sedue, 1990).

In such a context, priorities for research are clearly indicated by these data. Therefore this presents the results of a research project, obtained by a multidisciplinary and interinstitutional group, taking as an example of the biological degradation of a specific waste the sugarcane industry's most polluting liquid effluent, vinasses.

2 PROBLEMATICS OF THE SUGARCANE INDUSTRY IN DEVELOPING COUNTRIES

Sugarcane is one of the main agricultural crops in Mexico. Its only product is refined sugar. However, in the last 20 years its market has been a fluctuating one, mainly due to the challenge given by the artificial edulcorants (sweeteners). This situation has adversely affected the producing countries, most of them belonging to the Third World economies that heavily depend on raw materials exports.

In Mexico the average annual production of sugarcane is around 40 million metric tonnes, with a yield of 3·5 million tonnes of refined or partially refined sugar. Clearly the remaining byproducts represent considerable amounts of reusable organic material that presently are mostly taken as wastes. In the next paragraphs some remarks on the point are made.

2.1 Wastes Generated in Sugarcane Factories and Their Utilization

Roughly one metric tonne of sugarcane renders the following products:

350 kg wet bagasse	35%
100 kg sugar	10%
60 kg straw and leaves	6%
40 kg final molasses	4%
40 kg cachasses	4%
100 kg cane heads	10%
310 kg evaporated water	31%

The bagasse is the cellulosic residue of the cane after juice extraction, and it is mainly constituted by fiber, moisture and soluble organics (65% fiber, 25% pith cells and 10% water solubles). In Mexico its use is constrained to produce steam in the sugarcane factories by its burning. This combustion process is quite inefficient and generates considerable amounts of unrecovered soot that heavily pollutes the sugarcane factories' neighboring areas.

There are only two types of plants in Mexico that have another use for bagasse, one type to produce cellulose (about ten factories) and another type for pressed building boards (only two plants). There was one plant to produce furfural, an important chemical product, but it is presently closed down.

The straws and leaves and the cane heads are given to the cattle as raw or ensilaged feed. However, its recollection represents associated costs that, on occasions, are higher than its value as a feedlot. Therefore these byproducts are often burnt in the fields without rendering any benefit and creating air pollution problems.

Molasses is the final crystallization liquor obtained during sugar crystals recovery. It has been used over centuries as an energy source for animal feedlots as it also improves its palatability. In Mexico about 50% of the produced molasses are used in cattle feeds.

Molasses is also extensively used as raw material for ethyl alcohol production, for industrial and human consumption, by fermentation using *Saccharomyces cerevisiae*. Other minor uses, mainly employing biotechnological systems, are the production of citric acid, bread and beer yeasts, lysine, methionine, etc.

As some sugarcane factories carry out the fermentation of the molasses *in situ*, the byproducts of the fermentation and distillation operations are usually considered as part of the cane sugar factory too.

For the fermentation, the main byproduct is the spent yeast. When recovered it is used as a supplement in feedlots, but there are very few sugarcane factories–distilleries in Mexico that have the equipment to separate the yeast before distillation of the must or fermented molasses liquor. Therefore these suspended organics are generally carried on to the distillation operation.

For the distillation step, the main byproduct is the stillage or liquid bottoms of the columns. In Mexico and other Latin American countries these liquid effluents are known as vinasses or de-alcoholized musts. Their amounts are quite considerable, since for each volume unit of ethyl alcohol produced ten volume units of vinasses are generated. As mentioned above, these effluents usually carry considerable quantities of suspended or colloidal matter (spent yeast, molasses impurities, etc.) as well as of dissolved organics. This organic matter creates severe pollution problems when vinasses are dumped either in water sources or in the soil.

However, its use is quite restricted due to these organic matter concentrations, both dissolved and suspended. Irrigation is one use for it but not a very common one, at least in Mexico. Some work has been carried out by the Mexican Institute for the Improvement of Sugar Production (Instituto para el Mejoramiento de la Producción de Azúcar, IMPA) to dewater it and use it as animal feed, but the energy costs associated with this process make this option financially unviable (Zedillo, 1990).

In Puerto Rico, in a rum factory, a huge installation for anaerobic treatment of vinasses has been in operation for almost nine years and the biogas generated has been used as a secondary fuel (Szendrey, 1982, 1984, 1986).

Finally, cachasses is the resulting sludge coming out from the milk of lime clarification of the cane juice. It mainly contains organic material from the cane itself, and inorganics from the lime suspension and other impurities. Its main use in Mexico is as a soil improver, due to the presence of phosphorus and organic matter.

To summarize these data, Table 1 presents the Mexican figures for these

Table 1
Cane Sugar Processing Byproducts (Zedillo, 1990)

Year	*Cane*	*Sugar*	*Bagasse*	*Molasses*	*Cachasses*	*Ethanol*	*Vinasses*
	($\times 10^6$ *metric tonnes*)					($\times 10^6\ m^3$)	
1980	31·3	2·60	11·03	1·27	1·32	0·09	0·96
1981	28·7	2·37	10·10	1·12	1·20	0·10	1·18
1982	31·8	2·68	10·77	1·30	1·33	0·09	0·92
1983	32·5	2·89	11·47	1·28	1·36	0·11	1·22
1984	34·7	3·05	11·92	1·36	1·46	0·11	1·25
1985	35·7	3·23	12·06	1·39	1·50	0·11	1·24
1986	40·8	3·69	14·13	1·59	1·70	0·12	1·40
1987	41·4	3·74	13·82	1·53	1·37	0·073	0·81
1988	37·2	3·59	12·95	1·38	1·83	0·070	0·77
1989	35·5	3·47	11·62	1·32	1·17	0·068	0·76
1990	34·9	3·17	15·71	1·32	1·15	0·071	0·78

byproduct quantities. It is clear that an important economic asset to restore the economics of the sugarcane as a source of money, particularly of foreign currency is to use in a very efficient way the generated byproducts of its processing.

In the next paragraphs the biotechnological approach to recover these wastes, which really should be considered as byproducts, is presented.

2.2 Biotechnological Approach to Sugarcane Wastes

2.2.1 Bagasse

The first waste in order of importance, due to its amount, is the bagasse. As shown in Table 1, it reaches about 14 to 16 million tons, that is almost the annual production of maize in Mexico.

To have an idea of its use in Mexico, Table 2 presents some data on the industrial use of bagasse. This total amount represents only 2% of the annual generation of this byproduct. The rest is burnt in the sugarcane factories to generate the steam mainly used in the evaporators and crystallizers, and in the distillation columns to enrich ethanol up to 96% in volume. By comparison, regarding the use of bagasse as a fuel in the sugarcane/ethanol factories in Third World countries, in India an average of 3 kg of steam are required to obtain 1 liter of ethanol and they are produced with 2 kg of bagasse (Juneja, 1990).

Even this use would not be undesirable, due to the fact that some Third World countries do not have inexpensive fuel sources, if it were performed in a proper way with an energy-saving scope, and soot were recovered and further employed in the sugar refining process.

Another obvious use of bagasse is to produce cellulose and/or paper, after taking proper care of the treatment of the wastewaters generated during its processing, either with soda or with kraft cooking liquor, since most of the sugarcane producing countries have not enough forest resources and must import considerable amounts

Table 2
Industrial Use of Sugarcane Bagasse (Zedillo, 1986)

Industry	*Annual capacity (metric tonnes)*	*Product*
Celox, SA	—	Bleched cellulose
Celulosa y Fibra Nales, SA de CV	—	Bleached cellulose
Mexicana de Papel Periódico	100 000	Newsprint
Compañía Industrial San Cristóbal	76 000	Pulp, paper
Fábrica de Celulosa El Pilar, SA	22 500	Bleached pulp
Kimberley Clark de México, SA	60 000	Pulp
Productora de Papel, SA	18 000	Pulp
Fibrasin	40 000	Pressed board
Productos Furánicos Mexicanos[a]	18 000[a]	Furfural
Total use of bagasse	316 500	
(percent)	0·2	

[a] Presently not in operation.

of cellulose or wastepaper for their domestic paper, carton and corrugated items production. It remains only to find appropriate technologies to remove a substantial proportion of the pith prior to the pulping operation as well as to overcome the ever-present problems of processing wastewater pollution. This problem is being approached by biotechnological processes, mainly through the use of anaerobic bacteria.

Other uses would be to make pressed building boards, door cores and even mulch, again with emphasis on countries with poor forest resources.

Finally, bagasse may be saccharified and fermented through biotechnological processes to produce either liquid fuels such as alcohol or protein-rich microbial mass. Naturally, as raw material costs are a predominant portion of the total alcohol and microbial protein costs (that is, fermentable raw material), the use of acid saccharification increases the cost of the process, making it at present economically unfeasible (Bailey & Ollis, 1986).

2.2.2 Molasses

This byproduct is widely employed and does not really represent a so-called waste. Its primary use, as already explained, is as part of inexpensive feedlots, mainly for cattle. It is also a raw material in biotechnological enterprises as an inexpensive carbon source, since it contains about 14% invert sugar, a mixture of glucose and fructose, compared with about 1% in beet molasses.

The most extensive fermentation use of molasses is for producing ethanol, and the wastes are really generated during the unit operations involved with its separation. Data on this issue were presented in Table 1, and later, since this is the subject of this chapter, they will be further expanded.

2.2.3 Cachasses

As mentioned above, cachasses has a specific use as a soil improver. The main cost associated with its use is its transportation to the sugarcane fields, where it is applied. Also the social work bound to its dissemination in the fields, especially when landowners are not convinced of its potential benefit as soil improver, may be a crucial issue for its disposal.

It is clear from this panorama that biotechnology may play an important role to make an adequate utilization of these wastes, and the example that will be presented here reinforces this statement.

2.3 Case Problem: Vinasses

Mexican ethyl alcohol annual production reached, in the last five years, 115 million liters, and around 1500 million liters of vinasses were generated (Zedillo, 1990).

As already mentioned, research has been conducted in several countries trying to find effective methods to handle vinasses, since its discharge to the environment, mainly to water bodies and soil, creates severe pollution problems (Szendrey, 1982, 1984, 1986; Craveiro *et al.*, 1986; Durán *et al.*, 1988).

Biological treatment appears to be the most promising technique, since most of the vinasses' organic matter is biodegradable. Both anaerobic and aerobic systems are suitable, but due to the high concentration of dissolved organics anaerobic bacteria are more efficient. However, the anaerobic systems alone are, normally, not capable of dealing with a complete removal of organics. If these systems are followed by an aerobic process, a polishing effect is obtained on the dissolved organic matter and the colloidal and suspended solids. As the economics of the wastewater treatment plants is a decisive factor, the use of the treatment plants' byproducts may represent an added money value to offset the operating and maintenance costs. These byproducts would be biogas for the anaerobic processes and microbial biomass for the aerobic ones. Biogas may be used as an alternative energy source and biomass as a potential feed supplement for farm animals and aquaculture (Obayashi *et al.*, 1981; Szendrey, 1984; Sanna *et al.*, 1987).

This combination anaerobic–aerobic process has been successfully used for the laboratory and pilot-plant treatment of the Mexican corn industry wastewaters, which have a carbonaceous composition similar to vinasses. Biogas yield and methane composition, and biomass yield and protein composition, were quite satisfactory, especially with reactors using biofilm systems that were found to be the most suitable for carbonaceous wastewater treatment (Durán, 1983, 1987*a*,*b*, 1988; González-Martínez *et al.*, 1985; Montesinos & Durán, 1986; Luna-Pabello *et al.*, 1990; Pedroza-Islas & Durán-de-Bazúa, 1990).

In this work vinasses were biologically treated both in laboratory-scale experiments and in pilot-plant tests. The biological systems studied were, for the anaerobic type of microorganisms, a fluidized bed anaerobic reactor, upflow sludge blanket reactors and three packed anaerobic reactors (two were arranged in an upflow mode and one in a downflow mode), and, for the aerobic type of microorganisms, rotating biological reactors. The organics removal efficiencies and the biogas production for the different anaerobic systems, and the biomass yield for both aerobic systems, were studied.

3 APPLICATION OF AEROBIC AND ANAEROBIC PROCESSES

Vinasses from a molasses/alcohol distillery were directly used as the influent for the laboratory and pilot-plant equipment. Two sets of experiments were carried out. In the first set, the equipment was run with vinasses fed in a parallel mode (Fig. 1). For the second set, vinasses were fed only to the anaerobic equipment, and effluents to the aerobic laboratory and pilot-plant equipment.

Vinasses and reactor effluent characterization were carried out with the following parameters: temperature, pH, alkalinity, turbidity, suspended and dissolved solids, and organic matter (measured as chemical oxygen demand, COD, total and dissolved or soluble, COD_t and COD_s, respectively, and as biochemical oxygen demand, BOD).

For the anaerobic systems, redox potential (ORP), volatile acids, biogas composition, sulfides and sulfates analyses were also performed. For the aerobic systems, biomass composition and productivity were monitored. Biomass

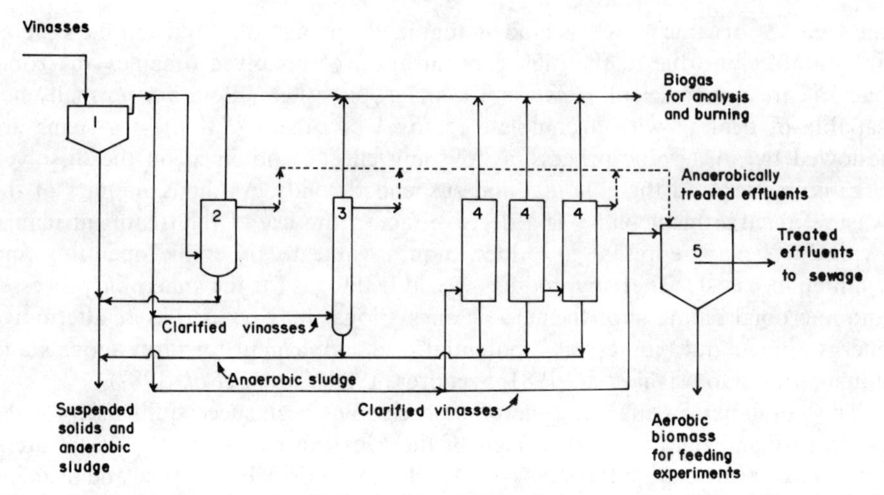

Fig. 1. Mexico's pilot plant for biological treatment of vinasses: 1, sedimentation tank; 2, UASB reactor; 3, fluidized bed anaerobic reactor; 4, anaerobic packed bed reactors; 5, RBR aerobic unit.

composition was evaluated from the solids obtained by centrifuging mixed liquor of the biodisc reactor (Durán, 1987*b*).

Standard methods were used for all determinations (AOAC, 1970; APHA, 1985), with the exception of the volatile organic acids, determined by the methodology of Dilalo and Albertson (1961), and the BOD, which was measured using a Voith Sapromat respirometer following the instructions of the manufacturer. Redox potential (ORP) was measured with a platinum electrode and Ag/AgCl. Reported values were corrected to the hydrogen electrode (Eh) by addition of 216 mV to the meter readings.

Biogas production was measured by water displacement, and corrected to standard temperature and pressure conditions. Gas quality was determined by the syringe method (Poggi *et al.*, 1987), and gas chromatography with 'Porapak' Q and molecular sieve columns connected in series (Noyola *et al.*, 1988). Biomass productivity was obtained following the method developed by Durán (1983).

It is important to mention that, as the sugarcane industry is a seasonal operating process, during the idle period vinasses were stored, and physical and chemical changes occurred during this period. Consequently some experiments were carried out with fresh vinasses and others with stored vinasses.

3.1 Anaerobic Systems

3.1.1 Upflow Anaerobic Fluidized Bed Reactor

The fluidized bed anaerobic reactor at pilot-plant scale was a tall, narrow column 0·3 m in diameter by 3·85 m in height, with a total working volume of 0·3 m^3. It has an expansion chamber at the top to control suspended solids and biomass losses. The granular media support was spent ion-exchange resin of 700 μm diameter. It contained 0·08 m^3 of quiescent medium, and its expansion was controlled up to 0·12 m^3. It was an isothermic reactor, and heating was provided by two electrical resistances with manual control. Temperature was maintained at 30–37°C. Continuous recirculation was performed by a centrifugal pump, influent entering into the bottom of the reactor. Biogas was collected from the top in a water-filled drum (Durán *et al.*, 1988).

To start it up the anaerobic reactor was filled with tap water, the recirculation pump was started, and the spent resin was charged from the top. The reactor was closed and 0·2 m^3 of a heavy inoculum were fed. The inoculum was screened cow manure (30 g/liter), pig excreta (30 g/liter), molasses (5 g/liter), deep soil (30 g/liter) and tap water to complete 1 liter. A semi-synthetic influent was intermittently introduced to the reactor, 0·01 m^3 twice a day. This influent contained molasses (4 g/liter), glacial acetic acid (0·5 g/liter), sodium carbonate (1 g/liter), sodium bicarbonate (1 g/liter), soil mesh 30–40 (0·3 g/liter), potassium phosphate (0·2 g/liter), ammonium chloride (1 g/liter) and tap water to complete 1 liter. Temperature was fixed to 32°C. The maturation period was 20 days. During this time frequent pH adjustments were necessary, using a sodium carbonate solution, in order to maintain a neutral pH (Poggi & Durán-de-Bazúa, 1988).

After the first 20 days an adaptation period was started to, first, establish a methanogenic regime and, secondly, change the influent to 100% vinasses. This process is described elsewhere (Durán *et al.*, 1988).

The operation was performed using three different hydraulic residence times (4, 3 and 2 days, based on operating fluidized bed volume). Samples were taken twice a week to determine organics degradation using COD as evaluation parameter. Each operation was considered to have reached pseudo-steady state when variations in COD of samples did not exceed $\pm 10\%$. At that time all control parameters were daily determined during five consecutive days to obtain statistically significant data (pH, alkalinity, turbidity, suspended and dissolved solids, dissolved organic matter as COD and BOD, volatile organic acids, biogas composition and sulfates). Fresh vinasses were used for the 4-day residence time, and stored vinasses were employed for the other two periods.

3.1.2 Upflow Anaerobic Sludge Blanket Reactors

During two production seasons a 0·120-m^3 reactor, known as UASB-1, was employed. It consisted of a polyethylene container with a diameter of 0·45 m and a height of 0·77 m. Influent vinasses were continuously fed with a peristaltic pump. It was inoculated with anaerobically-adapted activated sludge as described elsewhere (Noyola & Briones, 1988). Characteristics of this sludge were the following: sludge volume index of 50 ml/g, flocculent granules with a most probable number (MPN) of $1{\cdot}5 \times 10^9$ per g of volatile suspended solids (VSS) for acidogenic bacteria, $1{\cdot}4 \times 10^7$ per g VSS for hydrogenotrophic methanogenic bacteria and 4×10^5 per g VSS for acetoclastic methanogens. Operating conditions for these experiments are presented in Table 3.

A parallel laboratory-scale study was carried out using two identical UASB reactors (0·08 m diameter, 0·45 m height, 0·0023 m^3), known as UASB(a) and (b). The

Table 3
Operating Conditions for an Upflow Anaerobic Sludge Blanket Reactor (UASB-1) at Bench Scale (120 Liters Working Volume), Two Upflow Anaerobic Packed Bed Reactors (AF-1 and AF-2) and One Downflow Anaerobic Packed Bed Reactor (AF-3) (270 Liters Working Volume Each) for Vinasses Treatment

Step	*Vinasses*	*HRT*[a] *(days)*	*Organic load (kg COD/m^3 day)*	*Temperature (°C)*	*Observations*
Start-up	Diluted (1:1)	5	9	24	Without pH control
1	Raw	2·5	36	29	Without pH control
2	Raw stored[b]	2·5	29	30	Without pH control
3	Stored,[b] diluted (1:1)	2·5	22	27	With pH control

[a] HRT = Hydraulic residence time.
[b] Vinasses were stored in steel tanks under ambient conditions (around 30°C).

main objective of this part of the study was the assessment of the extent of inhibition to the methanogenic activity due to H_2S and K concentration in vinasses during treatment. Both reactors were seeded with the same inoculum as the 0·12-m^3 bench-scale reactor, and fed under the same conditions during starting-up. Table 4 shows the operating conditions for both reactors. For these specific experiments vinasses were diluted to 20% v/v.

An absorption device for H_2S was added to UASB(a) reactor after the second pseudo-steady-state period. It consisted of a peristaltic pump (Masterflex), a gas-washing flask (Fisher Milligan), filled with NaOH 2N, and phenolphthaleine. The biogas was pumped from the top of the reactor through the absorption flask and returned to the bottom of the UASB(a). In order to have a more reliable comparison basis, the same modification was performed on the UASB(b), with the exception of the gas-washing flask.

Finally, a 5-m^3 UASB reactor, known as UASB-2, also made of polyethylene (1·8 m diameter and 2·45 m height), has been started this season (December 1989).

Table 4
Operating Conditions for Upflow Anaerobic Sludge Blanket Reactors at Laboratory Scale (2·3 Liters Working Volume) for Vinasses Treatment

Step	*Reactor*	*HRT (days)*	*Organic load (kg COD/m^3 day)*	*Temperature (°C)*	*Observations*
1	UASB(a)[a]	6·6	2·8	35	No pH control
1	UASB(b)[a]	7·1	2·1	35	No pH control
2	UASB(a)	3·5	5·8	35	No pH control
2	UASB(b)	3·2	6·2	35	No pH control
3	UASB(a)	3·2	5·7	35	H_2S absorption, no pH control

[a] Starting-up period with pH control: addition of $NaHCO_3$ during 40 days, UASB(a), and 22 days, UASB(b) (vinasses were diluted to 20% v/v).

The main problem so far has been the inoculum, since the size of the reactor and the unavailability of enough anaerobically-adapted activated sludge has made its proper performance somewhat slow. However, available data seem to indicate that it will work in a similar manner to the bench-scale reactor. In this chapter no results for this piece of equipment of the pilot plant will be presented.

3.1.3 Upflow and Downflow Packed Bed Reactors

Three identical reactors were made of steel pipe (0·5 m diameter and 2 m height). The packing material was plastic pall-rings of 0·09 m (3·5 in). Each reactor has an empty volume of 0·294 m^3 and a void volume of 0·272 m^3, with a surface to volume ratio of 59 m^2/m^3. There were two upflow reactors, known as AF-1 and AF-2, and one downflow reactor, known as AF-3.

The reactor AF-1 was seeded with septic tank sludge, AF-3 was inoculated with a one-month-old mixture of cow manure and molasses, and AF-2 with a mixture of both inocula. The septic tank sludge had a MPN of 4×10^8 per g VSS for acidogenic bacteria, $4{\cdot}2 \times 10^8$ per g VSS for hydrogenotrophic methanogens and 2×10^7 per g VSS for acetoclastic methanogens. No MPN determinations were made on the cow manure mixture or the mixture of inocula.

The operating conditions for these three reactors are also presented in Table 3.

3.2 Aerobic Systems

3.2.1 Rotating Biological Reactors

The laboratory rotating biological reactor construction and operation has been described elsewhere (Luna-Pabello *et al.*, 1990; Pedroza-Islas & Durán de Bazúa, 1990). Its characteristics and the operating conditions are shown in Table 5. For simplicity this reactor will be called RBR-50. Fresh vinasses were used for the whole experiment.

Table 5
Characteristics and Operation of the Laboratory and Pilot-Plant Biodisc Reactors

Characteristics	*Laboratory RBR-50*	*Pilot plant RBR-3000*	*Units*
Chambers	10	4	
Discs per chamber	5	18	
Discs' diameter	0·3	2	m
Discs' thickness	5	10	mm
Discs' total surface area	7	450	m^2
Total working volume	0·05	3	m^3
Area:volume ratio	140	150	m^2/m^3
Rotational speed	22	2	min^{-1}
Peripheral velocity	0·346	0·209	m/s
Feed rate	0·001	0·060	m^3/h
Hydraulic retention time	50	50	h
Organic load (COD)	5	60	g/liter

In order to adapt the microbial communities already growing on the discs' surface, a gradual change in influent was carried out. It was started using 900 ml fresh vinasses and 1500 ml maize processing wastewaters, with tap water to 25 liters. This feed was used for 15 days. For the following 7 days a feed liquid made with 1200 ml vinasses and 90 ml maize wastewaters, with tap water to 20 liters, was pumped to the reactor. After the 23rd day a water-diluted vinasses influent, with an organic matter concentration of around 5000 mg/liter, measured as COD, was fed. This organic load was chosen theoretically considering that this concentration would be the expected one after the anaerobic treatment. It is clear that dilution would not really simulate chemical composition of anaerobically-treated vinasses, but this experiment may indicate if the biodiscs system works properly. A second experiment is contemplated using the effluent of the anaerobic fluidized bed reactor as the influent for this reactor.

The pilot-plant biodisc reactor has also been described elsewhere, since it was installed in a pilot plant used for degradation of corn wastewater (Montesinos & Durán, 1986). Table 5 also presents its characteristics and operating conditions, and this reactor will be called RBR-3000. For this system fresh vinasses were used.

The reactors were started using a 50–50 vinasses–tap water influent in order to promote the film growth on the discs' surface. The maturation period was about eight weeks, and at that time the biofilm was about 200 μm thick. Vinasses direct from the primary settling tank of the pilot plant were then fed to the reactor (Fig. 1). This was done because the anaerobic reactor was being operated at this time in parallel with the RBR in order to generate microbial biomass and to corroborate the performance of aerobic systems with highly polluted wastewaters.

Both aerobic systems were brought to pseudo-steady state under the same considerations as the anaerobic reactor. When a pseudo-steady state was reached, samples were taken daily and analyzed for five consecutive days, and averages of data are presented as the results for these experiments.

4 RESULTS AND DISCUSSION

Vinasses characterization is shown in Table 6. Mean data for both seasons, including standard deviation, are presented elsewhere (Moreno *et al.*, 1990). Undiluted vinasses were used for the pilot-plant equipment, both fresh and stored. For the first set of aerobic laboratory experiments, fresh vinasses dilution was made with tap water. For the first season of operation of the pilot plant, at the beginning of the experiments, fresh vinasses pH was increased by lime addition during distillery plant operations. However, as the experiments were running, this lime addition was stopped, and, for the case of the UASB-1 and the packed bed reactors, fresh vinasses used did not contain lime. Stored vinasses, used for the anaerobic experiments, were not neutralized but stored as they were produced. Finally, all vinasses used in the second season of operation were not neutralized with lime but used as they came from the distillation equipment.

Table 6
Vinasses Analyses (Fresh, Stored and Diluted Lots)
(Bazúa *et al.*, 1990; Moreno *et al.*, 1990)

Parameter	*Vinasses*		
	Fresh	*Stored*	*Diluted*
Temperature (°C)	25–35	25–30	15–20
pH	4·2–7·0	4·5–5·5	4·8–5·5
Alkalinity (g $CaCO_3$/liter)	5·8	9·0	2·95
Turbidity (NTU)	30 000	30 000	3 100
Total solids (g/liter)	63–90	47–70	13–17
Total volatile solids (g/liter)	65·1	—	—
Total fixed solids (g/liter)	24·5	—	—
Volatile suspended solids (g/liter)	2·6	—	—
Fixed suspended solids (g/liter)	1·0	—	—
BOD (mg O_2/liter)	31 500[a]	27 500[a]	1 960[b]
COD_t (g O_2/liter)	69–128	64–120	4·9
Kjeldahl nitrogen (g N/liter)	1·2–1·6	1·0–1·3	—
Ammonia-nitrogen (g N/liter)	0·11–0·15	0·5	—
Sulfate ions (g/liter)	3·1–5·8	2·8–3·5	—
Sulfide ions (g/liter)	1·2	—	—
Potassium ion (g/liter)	8·1	—	—
Sodium ion (g/liter)	0·14	—	—

[a] Dilution of 1/100.
[b] Dilution of 1/5.

4.1 Anaerobic Systems

4.1.1 Upflow Anaerobic Fluidized Bed Reactor

Results obtained for the pilot-plant anaerobic experiments are presented in Table 7. During all hydraulic residence times studied, when pseudo-steady state was reached, conditions maintained a methanogenic regime without any acidogenic disturbance or other problem. Results obtained in the anaerobic reactor proved that the system can be considered as a high-intensity process. High loading rates at short retention times were reached (up to 34 kg COD/m^3 day at a residence time as low as 2 days, based on operating fluidized bed volume), maintaining a purification efficiency up to 70% total COD removed, and a total gas production of 7 m^3 (STP)/m^3 day with 70–80% methane content.

The increase of the partial alkalinity of the effluent and the volatile organic acids removal, and the increase in volumetric biogas production with respect to loading rate (Table 8), suggest that the process may be even further accelerated, reducing overall hydraulic retention time below 2 days.

Results obtained for loading rate and methane production are comparable to or somewhat better than the ones obtained by Craveiro *et al.* (1986), with wastewaters formed from diluted vinasses treated in a pilot-plant UASB reactor (Table 9). Considering its organic load, removal rates obtained here were quite acceptable (up to 70%).

Table 7
Vinasses Treatment Results for the Anaerobic Fluidized Bed Reactor at Pseudo-Steady State (Bazúa *et al.*, 1990)

	Influent flowrate (liter/day)		
	30	*40*	*60*
Recirculation (liter/day)	6·7	5·8	6·0
Upward velocity (m/h)	5·7	4·9	5·1
Hydraulic residence time (days)			
(1)[a]	10·0	7·5	5·0
(2)[b]	4	3	2
Volumetric organic loading rate (kg COD/m^3 day)			
(1)[a]	7·9	7·8	13·5
(2)[b]	19·7	19·3	33·9
Chemical oxygen demand (mg/liter)			
Influent	78 941	58 820	67 731
Effluent	26 032	17 660	22 802
% Removal efficiency	65·3	70·0	66·3
pH			
Influent	5·05	5·17	5·33
Effluent	7·30	7·35	7·34
Total alkalinity (mg $CaCO_3$/liter)			
Influent	8 840	9 391	10 500
Effluent	10 387	10 910	11 000
Partial alkalinity (mg $CaCO_3$/liter)[c]			
Influent	0	0	0
Effluent	5 477	5 974	5 640
Intermediate alkalinity (mg $CaCO_3$/liter)[d]			
Influent	8 840	9 591	10 500
Effluent	4 910	4 933	5 359
Intermediate/partial alkalinity effluent	0·90	0·82	0·95
Volatile organic acids (mg acetic acid/liter)			
Influent	—[f]	18 063	19 646
Effluent	—	7 712	5 126
Biogas production (m^3(TP)/day)[e]	0·58	0·66	0·85
Biogas yield (m^3(TP)/kg COD removed)	0·40	0·40	0·32
Volumetric biogas production (m^3(TP)/m^3 day)			
(1)[a]	1·91	2·21	2·83
(2)[b]	4·83	5·52	7·10
Biogas composition			
CO_2 (%)	—	21·80	28·00
CH_4 + others (%)	—	78·20	72·00
Flammability	Good	Good	Good

[a] Based on total working reactor volume.
[b] Based on operating fluidized bed volume.
[c] Alkalinity measured to end-point pH = 5·8.
[d] Alkalinity measured between pH = 5·8 and 4·3.
[e] Values corrected to 0°C, 1 atm, and for humidity.
[f] — Not determined.

Table 8
Results Obtained in the Pilot-Plant Scale Study Using UASB and AF Reactors (Noyola *et al.*, 1989)[a]

Step	pH	TS[b]	TVS[c]	TFS[d]	VSS[e]	COD_t[f]	Removal COD_t	COD_s[g]	Removal COD_s	SO_4^{2-}[h]	S^{2-}[i]
1 Infl.	5·0	88·0	58·1	29·9	—	90·3	—	—	—	—	—
UASB-1	5·8	66·6	35·4	31·2	—	39·7	56	—	—	—	—
AF-1	5·4	75·0	35·6	39·4	—	42·4	53	—	—	—	—
AF-2	5·6	65·7	28·4	37·3	—	40·6	55	—	—	—	—
AF-3	5·6	64·0	28·3	35·7	—	38·8	57	—	—	—	—
2 Infl.	5·2	51·8	27·8	24·0	2·7	72·7	—	—	—	6·1	1·2
UASB-1	5·4	49·5	27·8	21·7	2·0	47·9	34	—	—	7·2	1·1
AF-1	5·4	44·9	23·8	21·1	4·3	51·7	29	—	—	6·8	1·1
AF-2	5·4	49·8	28·8	21·0	2·8	51·0	30	—	—	6·5	1·1
AF-3	5·4	46·5	26·0	20·5	0·8	48·6	33	—	—	6·9	1·1
3 Infl.	5·3	38·6	22·5	16·1	—	56·8	—	36·2	—	—	—
UASB-1	5·6	42·6	24·4	18·2	—	30·0	47	—	—	—	—
AF-1	6·2	43·7	23·6	20·1	—	35·8	37	21·5	40	—	—
AF-2	6·2	44·1	23·8	20·3	—	34·7	39	20·0	45	—	—
AF-3	5·8	44·5	22·7	19·8	—	37·7	34	17·7	51	—	—

[a] Values in g/liter except pH (dimensionless).
[b] TS = Total solids.
[c] TVS = Total volatile solids.
[d] TFS = Total fixed solids.
[e] VSS = Volatile suspended solids.
[f] COD_t = Chemical oxygen demand (total).
[g] COD_s = Chemical oxygen demand (soluble).
[h] SO_4^{2-} = Sulfate ions.
[i] SO^{2-} = Sulfide ions.

Table 9
Anaerobic Treatment of Vinasses with Upflow Anaerobic Sludge Blanket Reactors (UASB), Anaerobic Packed Bed Reactors (AF), and a Fluidized Bed Reactor

Vinasses type[a]	*Reactor type*	*Temperature (°C)*	*HRT*[b] *(days)*	*Reactor volume* (m^3)	*Organic load* $(kg\ COD/m^3\ day)$	*COD removal (%)*	*Reference*
Alcohol (M)	AF downflow	29	2·5	0·120	12	57	Carrondo *et al.* (1983)
Alcohol (M)	UASB	40	2·3	0·110	23·3	71	Sánchez Riera *et al.* (1985)
Alcohol (M)	UASB	28	2·5	0·120	36	56	This work
Alcohol (M)	AF	28	2·4	0·280	36	57	This work
Alcohol (M)	Fluidized bed	35	2·0	0·300	33·9	66–70	This work
Rum (M)	Anaerobic contact	35	5·5	0·030	11	75	Roth and Lentz (1977)
Rum (M)	AF downflow	37	4–2·7	10·000	11·4–20·4	72–62	Bories *et al.* (1988)
Rum (M)	AF downflow	37	—	1 700	—	—	Bories *et al.* (1988)
Rum (M)	AF	37	9	13·000	7	50–70	Szendrey (1983)
Rum (M)	AF	37	—	13 000	—	—	Szendrey (1983)
Alcohol (CJ)	AF	35	6	0·040	3·4	90	Russo *et al.* (1985)
Alcohol (CJ)	UASB	31	2·4	11·000	13	83	Craveiro *et al.* (1986)
Rum (CJ)	Upflow floc tower	35	4·9	0·011	5·3	86	Callender and Barford (1983)
Rum (CJ)	UASB	35	1·2	0·010	25·4	90	Cail and Barford (1985*a*)
Rum (CJ)	Upflow floc tower	35	1·2	0·010	22·5	90	Cail and Barford (1985*b*)

[a] (M) = From molasses fermentation; (CJ) = from sugarcane juice fermentation.
[b] HRT = Hydraulic residence time.

The methane-rich biogas obtained after removal of hydrogen sulfide may be used in substitution of bagasse in the steam boilers, leaving bagasse for cellulose production. A previous energy balance made with these biogas productivity data gave a production of 6×10^8 MJ per year, equivalent to 14 400 m^3 of 'combustoleum' (Bazúa *et al.*, 1990).

4.1.2 Upflow Anaerobic Sludge Blanket Reactors and Upflow and Downflow Packed Bed Reactors

The use of different inocula permitted the confirmation of the great importance of this aspect of the process. In Mexico there is no granular sludge, and therefore a replacement must be found. According to the starting-up results, the best inocula were the anaerobically-adapted activated sludge and the septic tank sludge. The MPN of methanogenic bacteria were somewhat similar, but the adapted activated sludge had better physical characteristics (incipient granulation and lower sludge volume index) than the septic tank sludge.

One month after starting-up UASB-1 and AF-1 reached 65% COD removal efficiencies, while AF-2 and AF-3 reached only 41% and 29% removal, respectively. The last reactor was seeded again with septic tank sludge, and one month afterwards the four reactors had similar COD removal efficiencies (between 55% and 65%).

The results obtained during the pseudo-steady-state period are presented in Table 8. It may be observed that COD removal efficiencies of step 1 are the highest, around 56%. The COD removal efficiency decreases significantly for steps 2 (around 32%) and 3 (between 34% and 47%). The step 1 results are comparable with those reported in the literature using the same type of reactors, as may be noticed in Table 9.

With vinasses from molasses the best results were obtained by Roth and Lentz (1977), with a conservative organic load and hydraulic residence time. A particularity of their work is that they added nitrogen and phosphorus to a COD:N:P ratio of 200:5:1, as well as yeast extract.

According to Table 9, the results obtained during step 1 of this work compare well with those using vinasses from molasses, and present the highest organic load applied. From the same table it may be said the vinasses from cane juice contain a higher amount of biodegradable matter than those from molasses. This may be explained by the fact that molasses is a byproduct of sugarcane production and contains lesser amounts of sucrose and reducing sugars, and a higher content of complex organics and salts, than cane juice.

The relatively good results of step 1 that were obtained under severe pH conditions, as shown in Table 8, seem to indicate that these microbial communities have the ability of creating a microenvironment around them. The fact that methanogenesis took place may be taken as a sound proof of it, since pH measurements inside the sludge bed showed that sludge pH was 0·2 to 0·4 units higher than effluent pH (Noyola & Briones, 1988).

The use of stored vinasses seems to be detrimental for the anaerobic reactors' performance (Table 8). In step 2 COD removal efficiencies markedly decreased. To corroborate this experiment, in step 3, with a lower organic load and pH control,

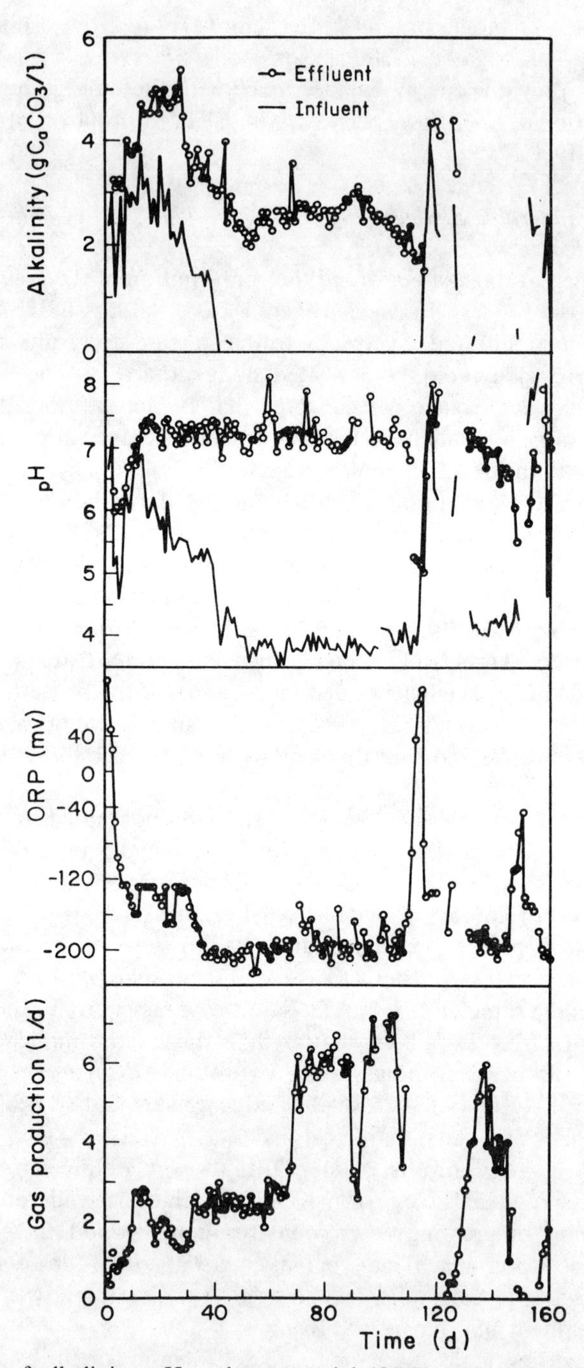

Fig. 2. Evolution of alkalinity, pH, redox potential (ORP) and gas production during UASB(a) start-up and two steady-state periods.

these efficiencies increased but never reached the values obtained in step 1. Apparently stored vinasses organic matter is less biodegradable or becomes more inhibitory than organics in fresh vinasses. It is presumed that, during storage (at a temperature about 30°C), fermentation uncontrolled processes consume most of the biodegradable organics generating inhibitory metabolites for acidogenic, acetoclastic and methanogenic bacteria. In practice, these results suggest that vinasses should be immediately treated, avoiding storage periods.

The high sulfate and potassium concentrations in raw vinasses (Table 6) may also be responsible for the limitation on COD removal. The inhibition appears even at low organic loads, as was noticed during the starting-up period (initial organic load of 9 kg COD/m^3 day, 50% diluted vinasses and COD removal of 65%). A four-fold increase in organic load with undiluted vinasses (step 1) reduced the efficiency only to 55%, which could be expected even for highly biodegradable substrates.

Due to the low initial pH in the reactors, the H_2S gas would be the predominant form of reduced sulfur, which is the most inhibitory one. According to Speece (1983), H_2S concentrations around 150 mg/liter are highly inhibitory for methanogens. Koster *et al.* (1986) found that for pH between 6·4 and 7·2 an H_2S concentration of 250 mg/liter decreased methanogenic activity to 50%. In the present study the H_2S concentration was much higher (Table 8, step 2). In fact these results suggest that even the sulfate-reducing bacteria were inhibited, probably due to the stored vinasses characteristics, as mentioned above.

Another reason for the limited COD removal may be the high potassium content. McCarty (1964) states that K concentrations between 2·5 and 4·5 g/liter are moderately inhibitory and up to 12 g/liter are highly inhibitory. Vinasses used in this study had around 10 g/liter.

These inhibitory effects were confirmed with respirometric tests. It was noticed that biochemical oxygen demand (BOD) values increased with the dilution factor. This behavior is typical for inhibitory or toxic wastewaters. A parallel observation is that BOD tests may mean nothing for vinasses characterization and assessment of treatment efficiency, unless the dilution factor is reported.

As previously said, laboratory UASB reactors were operated in order to confirm the inhibitory effect of H_2S and K. The reactors were fed with diluted vinasses (20% raw vinasses, 80% tap water) in order to lower the potassium concentration from around 8·1 to 1·6 g/liter (Table 6). This concentration may be considered slightly inhibitory (McCarty, 1964). Therefore inhibition effects, if found, might be attributed to H_2S.

As is observed from Table 4, both reactors were started up under similar conditions (step 1). The only difference was that for UASB(a) the pH control was maintained for 40 days, whereas for UASB(b) it lasted 22 days only. The stabilization of effluent COD was accomplished in 40 and 37 days, respectively. The evolution of some physico-chemical parameters during the UASB(a) starting-up is shown in Fig. 2.

It may be noticed that a sharp decrease in ORP and a rapid increase of effluent pH and alkalinity occur. The effluent COD removal evolves in a slower manner. The ORP values passed for a 20-day stable period (ORP values around −135 mV), and

Table 10
Results Obtained in the Anaerobic Laboratory-Scale Study (Moreno *et al.*, 1990)

Parameter	*1 UASB(a)*	*1 UASB(b)*	*2 UASB(a)*	*2 UASB(b)*	*3 UASB(a)*
COD_i (g/liter)	17·5	14·9	20·0	20·0	18·4
COD_e (g/liter)	5·1	9·0	9·8	9·5	6·8
COD_{rem} (%)	71	40	51	52	63
SO_{4i}^{2-} (mg/liter)	1 087	980	1 177	1 042	1 042
SO_{4e}^{2-} (mg/liter)	181	206	269	154	138
SO_i^{2-} (mg/liter)	74	18	36	88	88
SO_e^{2-} (mg/liter)	208	260	155	119	104
CH_4 yield (m^3/kg COD rem.)	0·28	0·5	0·3	0·3	0·3
CH_4 in gas (%)	68	69	71	70	91
Load (kg COD/m^3 day)	2·8	2·1	5·8	6·2	5·7

Note: Vinasses characteristics change from sample to sample, since they come from a sugarcane factory. For 1 UASB(a) the influent came from a different lot than the rest, which were run in a parallel manner (1b and 2a, 2b and 3a). Subscripts i and e indicate influent and effluent characteristics, and subcript rem indicates removed.

then they decreased to −204 mV and remained stable. The observed pattern may be due to the sulfate-reducing bacteria adaptation, followed by the starting-up of the methanogenic bacteria activity. It is worth noting that the ORP value of −204 mV is higher than the reported value for the methanogenic reaction of −320 mV (Zehnder, 1978).

After 40 days no alkalinity was added through the influent and the reactor was able to maintain a high alkalinity value and a neutral pH, even with influent pH around 4.

The results obtained for the pseudo-steady-state periods are shown in Table 10. A direct comparison may be done with pseudo-steady states 2 UASB(a) and 3 UASB(b), as they were run with the same batch of wastewaters. It may be seen that the absorption of hydrogen sulfide increased the COD removal efficiencies from 52% to 63% and the methane content from 71% to 91% (the CO_2 was also retained).

Nevertheless, the sulfide content in the effluent did not decrease significantly in UASB(a). This fact suggests that the absorption unit was not so efficient in removing H_2S, and the expected increase in COD removal rate was partially achieved. Anyway gas absorption gave a very good stability to the UASB(a) reactor. The changes in the batch of vinasses did not affect it, while the UASB(b) reactor was perturbed to some extent.

Even if there was still some reduced sulfur in the effluent the pH of the UASB(a) was 7·2. Under these conditions the predominant form would be HS^-, which is less toxic to the methanogenic bacteria. It seems that limited COD removal efficiency is not only due to high H_2S and K concentrations. Some other factors may contribute to the limitation in vinasses biodegradability. A combination effect of various ions, the unavailability of micronutrients (mainly the metallic ones) due to K displacement and/or sulfide precipitation, and the possible high content of recalcitrant organic matter may all play a part.

Table 11
Most Probable Number (MPN) of Various Metabolic Groups of Anaerobic Bacteria Found in the UASB(a) Sludge per Gram of Volatile Suspended Solids (Moreno *et al.*, 1990)

Group of bacteria	*Inoculum* $\times 10^{-6}$	*110-day-old sludge* $\times 10^{-10}$
Acid formers	4·0	49·0
Propionate users	5·2	4·9
Butyrate users	1·7	4·9
Acetotrophic methanogens	17·0	2·2
Hydrogenotrophic methanogens	0·54	75·0
Sulfate reducers	n.d.[a]	5·7

[a] n.d. = Not determined.

Finally, Table 11 presents the MPN of organism determinations for the inoculum sludge and a sample taken from UASB(a) 110 days after starting-up. The values confirm the rather good inoculum quality of the anaerobically-adapted activated sludge.

4.2 Aerobic Systems

4.2.1 Rotating Biological Reactors

For the laboratory reactor (Table 12), where organic matter is not as concentrated as in the pilot-plant reactor, removal efficiencies are somewhat higher. However, the differences are not as considerable as expected, taking into account the extreme contrast between the two influent concentrations.

Dissolved organic matter load in the pilot-plant bioreactor appreciably decreases, considering that aerobic systems are not as effective for high organic load removals (Durán, 1983). Biodegradability differences are easily evaluated with BOD and COD data (46% versus 64% for the pilot-plant experiments and 95% versus 68% for the laboratory-scale experiments). Therefore vinasses possess a certain amount of recalcitrant or non-biodegradable organic compounds that should be taken into account for the programmed experiment using anaerobically-treated vinasses as the influent for this type of reactor.

Table 12
Data Obtained in the Aerobic Biodisc Reactors (Bazúa *et al.*, 1990)

Parameter	*Effluent*		*Percent removal*	
	RBR-50	*RBR-3000*	*RBR-50*	*RBR-3000*
Temperature (°C)	12–15	20–25		
pH	8·0–8·4	6·0–7·1		
BOD (mg O_2/liter)	85	13 450[a]	95	57
COD (mg O_2/liter)	1 550	37 000	68	46

[a] Dilution of 1/100.

Table 13
Proximal Analysis of RBR Biomass

Composition		*Percent (dry basis)*	
		In parallel	*In series*
Crude protein (factor = 6·25)		27·70	27·74
Ashes		26·60	26·62
Crude fiber		0·20	4·01
Ether extract		14·40	14·43
Essential amino acids[a] *(g/100 g protein)*	*FAO Standard*[b]	*In parallel*	*In series*
Lysine	4·00	4·43	4·80
Threonine	2·80	4·22	4·10
Isoleucine	4·20	4·20	4·50
Leucine	4·80	6·76	6·60
Valine	4·20	5·44	5·50
Methionine[c]	2·20	1·81	1·90
Phenylalanine	2·80	3·74	3·70

[a] Analyses carried out by CIMMYT's Protein Quality Laboratories, El Batán, México.
[b] In *Protein from Hydrocarbons*, ed. H. Gounelle de Pontanel, Academic Press, New York, 1972.
[c] Partially destroyed during hydrolysis.

The analysis of the biomass obtained from the aerobic experiments is presented in Table 13. It may be seen that an appreciable amount of nitrogen is found in the biomass, and therefore it may represent a possible protein source for animal feeds (fish, pigs, poultry, even cattle). Also the ash content is considerable due to the lime used to increase the pH of fresh vinasses after distillation. Further experiments will be carried out avoiding the lime additions to evaluate ash in biomass, since experiments with corn-processing wastewaters biomass using laboratory Wistar rats demonstrated that an excess of calcium salts in the diet may create gastric problems (Valderrama *et al.*, 1988).

For the series experiment (anaerobically-treated effluents were sent to the pilot-plant biodisc reactor) no statistically significant differences were found in the quality of the biomass, either in overall bromatological composition or in amino acids content (Table 13). Therefore the biomass has the same possibility of being used for animal feeds.

During the pilot-plant experiments, run in parallel, the system productivity based on biomass production per COD consumption was evaluated. It was assumed that biomass produced per mass unit of removed organic matter was not time-dependent. Data obtained for a 2-day hydraulic residence time were (Bazúa *et al.*, 1990)

$$(69 - 37\ \text{g O}_2/\text{liter})(3000\ \text{liters})/(2\ \text{days}) = 48\,000\ \text{g removed COD/day}$$

Generated biomass in 400 cm^2 average disc area in the eight considered discs was 30 g (wet basis) in 4 days. Therefore productivity was 187·5 g wet biomass/m^2 day, and if total disc area is 454 m^2 the calculated yield is 1·76 kg biomass per kg COD:

$$Y = (0{\cdot}1875\ \text{kg biomass/m}^2\ \text{day})(454\ \text{m}^2)/(48\ \text{kg COD/day}) = 1{\cdot}76\ \text{kg/kg}$$

This yield would represent a considerable amount of microbial protein if all vinasses were aerobically treated. For 1500 million liters generated in 1989

$$(1{\cdot}76\ \text{kg wet biomass/kg COD})(32\ \text{kg COD removed/m}^3\ \text{vinasses}) \times (1{\cdot}5 \times 10^6\ \text{m}^3\text{/year}) = 84{\cdot}3744 \times 10^6\ \text{kg wet biomass/year}$$

Assuming average water contents in biomass to be around 80–90%, and protein around 28%, an annual production of non-conventional protein source of 8000–17 000 tonnes would be obtained.

If soy protein is used to compare with the non-conventional microbial protein obtained for feeds, with a market price of 3750 Mexican pesos (MP) per kg soymeal with 389 g protein per kg soymeal, a value of 9640 MP/kg protein would be obtained.

Following the methodology proposed by Peters and Timmerhaus, already used for maize wastewaters (Durán, 1983, 1987*b*), for the pilot-plant equipment that may treat 1·5 m^3/day and with a capital cost of 600 000 Mexican pesos per cubic meter of effluents per year, the total capital investment (TCI) would be

$$(600\,000\ \text{MP})(1{\cdot}5 \times 10^6) = 0{\cdot}9 \times 10^{12}\ \text{MP}$$

Using this figure to evaluate the annual gross earnings (GE) from biomass sale, a value of 70 000–160 000 million Mexican pesos would be obtained.

The turnover ratio (TR), defined as gross earnings (GE) divided into the total capital investment (TCI), would be

$$\text{TR} = \text{GE/TCI} = (0{\cdot}16 \times 10^{12})/(0{\cdot}9 \times 10^{12}) = 0{\cdot}18$$

This is a very similar value to the one obtained for the aerobic treatment of corn-processing wastewaters (TR = 0·193).

Consequently, although the system is not as good as a chemical enterprise with turnover ratios from 0·2 to 1·0, gross earnings may guarantee the payment of the wastewater treatment plant operation costs, with a gross reduction on the vinasses dissolved organic load of at least 50%.

5 CONCLUSIONS

Sugarcane is a renewable resource that has a very promising potential. Sucrochemistry has never been truly applied but it is hoped that biotechnology, and of course biotechnologists, will have the ability to use microorganisms and microbial products to enhance the use of sugarcane byproducts for a better future of Third World countries, its main producers.

The following conclusions may be drawn from this general overview:

(1) As mentioned at the beginning of this chapter, the first waste in order of importance of the sugarcane industry, due to its amount, is the bagasse. Biotechnology can be employed to convert it to good quality cellulose for the paper industry, to produce low-cost saccharified products for liquid fuels production, and even to clean the wastewaters generated during its processing.

(2) Cachasses, as already stated, has a specific use as a soil improver. The main cost associated with it is its transportation to the sugarcane fields, where it is applied. So far no biotechnological use has been devised, perhaps because its disposal causes no major problem, but no doubt something may be found in the future.

(3) Molasses, a widely employed byproduct, does not really represent a so-called waste. As already explained, it is a substantial part of inexpensive cattle feedlots. It is also a raw material in biotechnological enterprises as an inexpensive carbon source, since it contains about 14% invert sugar, a mixture of glucose and fructose. Its main biotechnological use, as mentioned, is for producing ethyl alcohol, for which, in order to reach a high concentration in process separation units, it generates the commonly known stillages or vinasses or musts. Its biotechnological treatment, namely using anaerobic and aerobic bacteria, has given the following results:

—The anaerobic treatment of vinasses from sugarcane molasses presented a limitation in COD removal efficiencies above 70% for the raw wastewater. More research is needed in order to increase the COD removal of the process.

—Its anaerobic treatment may be accomplished at very high organic loads, up to 36 kg COD/m^3 day, and under severe pH conditions, even below 6. It was observed that a pH gradient was present in the sludge bed of a UASB reactor, increasing its pH by 0·2–0·4 units.

—Anaerobically-adapted activated sludge is a good source of inoculum when no granular sludge is available for seeding anaerobic reactors.

—Biogas productivity, measured with the data obtained in the anaerobic fluidized bed reactor, for the amount of vinasses produced in Mexico in 1989 may represent the generation of 6×10^8 MJ which compared with the calorific power of 'combustoleum' would give an energy equivalent to 14 000 m^3 'combustoleum' (Bazúa *et al.*, 1990).

—For the aerobic treatment of vinasses, both fresh and already anaerobically pretreated, microbial biomass produced has a good chemical value, measured by its protein's amino acids. A preliminary economic balance, considering this biomass as a protein equivalent to soymeal, would give an added value to the aerobic treatment.

ACKNOWLEDGMENTS

The authors acknowledge the manufacturing company AZUCAR, SA, especially its authorities and coworkers in Delegación Huastecas and Ingenio Alianza Popular, as well as Instituto para el Mejoramiento de la Producción Azucarera, for financial and logistic support for this research study. Especially, Mr Gilberto Arriaga's always encouraging support to the execution of this work is gratefully acknowledged. Experimental and analytical work, as well as supervision of operations, was carried out by Mr Roberto Briones, Mr Rafael Hernández, and the undergraduate students (now practicing engineers) Juana María Castro, Arturo Contreras, Gustavo García-Díaz, Miguel Angel García-Rocha, Rosaura Villegas and Héctor Alejandro Zámano under the authors' supervision. Their invaluable support is greatly acknowledged. The academic support of the Universidad Autónoma de San Luis Potosí, Facultad de Ciencias Químicas Graduate Department, through its staff (Dr Pedro Medellín, Mr Miguel Angel Cooper, Ms María de los Angeles Cabrero and Ms Catalina Alfaro), is gratefully acknowledged.

REFERENCES

AOAC (1970). *Official Methods of Analysis*, 11th edn. Association of Official Analytical Chemists, Washington, DC.

APHA (1985). *Standard Methods for the Examination of Water and Wastewater*, 16th edn. American Public Health Association, Washington, DC.

Bailey, J. E. & Ollis, D. F. (1986). *Biochemical Engineering Fundamentals*. McGraw-Hill, New York, p. 831.

Bazúa, C. D. de, Cabrero, M. A. & Poggi, H. M. (1990). *Bioresource Technology*, **35**, 87–93.

Bories, A., Raynal, J. & Bazile, F. (1988). *Biol. Wastes*, **23**, 261.

Cail, R. G. & Barford, J. P. (1985*a*). *Biotechnol. Letters*, **7**(7), 493.

Cail, R. G. & Barford, J. P. (1985*b*). *Agric. Wastes*, **14**, 292.

Callender, F. J. & Barford, J. P. (1983). *Biotechnol. Letters*, **5**, 755.

Carrondo, M. J. T., Silva, J. M. C., Figueira, M. I. I., Ganho, R. M. B. & Oliveira, J. F. S. (1983). *Water Sci. Technol.*, **8–9**(15), 117.

Craveiro, A. M., Soares, H. M. & Schmidell, W. (1986). *Water Sci. Technol.*, **18**(12), 123.

Dilallo, R. & Albertson, O. E. (1961). Volatile fatty acids by direct titration. *J. WPCF*, **33**, 356–65.

Durán, C. (1983). *Problemas sanitarios en la industria de alimentos. Problema tipo: Tratamiento de los efluentes de la industria del maíz en México*. Tesis doctoral, Pub. Facultad de Química, UNAM, México.

Durán, C. (1987*a*). In *Global Bioconversions*, Vol. II, ed. D. L. Wise. CRC Press, Boca Raton, FL, p. 75.

Durán, C. (1987*b*). *Fortschritt-Berichte*, Series 15: Environmental Techniques, No. 51. VDI Press, Düsseldorf, FRG.

Durán, C. (1988). *Recycling of corn processing effluents/Reaprovechamiento de efluentes de la industria del maíz*. UNAM-UNEP-BMFT-Conacyt-Miconsa Final Project Report. Azteca, México DF.

Durán, C., Medellin, P., Noyola, A., Poggi, H. & Zedillo, L. E. (1988). *Tecnol. Ciencia Ed. (IMIQ, Méx.)*, **3**(2), 33.

González-Martínez, S., Pedroza-de-Brenes, R., Durán-de-Bazúa, C. & Norouzian, M. (1985). In *Proceedings of the 1985 Spec. Conf. Environm. Eng.*, ed. J. C. O'Shaughnessy. ASCE, Northeastern University Press, Boston, MA, p. 606.

Gounelle de Pontanel, H. (1972). *Protein from Hydrocarbons*. Academic Press, New York.

Juneja, J. S. (1990). Use of biomass as fuel for the rural industries and village applications.

Presented at *International Symposium on Application and Management of Energy in Agriculture—The Role of Biomass Fuels*, New Delhi, 21–24 May.

Koster, I. W., Rinzema, A., de Vegt, A. L. & Lettinga, G. (1986). *Water Res.*, **20**(12), 1561.

Licht, F. O. (1986). World Sugar Balance. Cited in Zedillo, L. E., *Aprovechamiento de los subproductos de la caña de azúcar en México. Situación actual y perspectivas.* IMPA, México DF. 39 pp.

Luna-Pabello, V. M., Mayén, R., Olvera-Viascan, V., Saavedra, J. & Durán de Bazúa, C. (1990). *Biol. Wastes*, **32**, 82.

McCarty, P. L. (1964). *Pub. Works* 95: (9), 107; (10), 123; (11), 91; (12), 95.

Montesinos, M. A. & Durán, C. (1986). Estudio dinámico de un reactor biológico rotatorio en la producción de proteína microbiana. In *Recycling of Corn Processing Wastes, XVII Congresso Nacional Ciencia y Tecnologia Alimentos*, Cholula, Puebla, México, 7–8 March.

Moreno, G., Rodríguez, F., Jiménez, C. & Noyola, A. (1990). In *Memorias VII Congresso Nacional SMISAAC*, ed. A. Noyola. SMISAAC, Oaxaca, Oax., México, p. C172.

Noyola, A. & Briones, R. (1988). Tratamiento anaerobio de vinazas a escala piloto: Inoculación y arranque de reactores tipo lecho de lodos y filtro anaerobio. In *Memorias VI Congreso Nacional SMISAAC*, ed. A. Noyola. SMISAAC, Querétaro, Qro., México, Section III. 5 pp.

Noyola, A., Capdeville, B. & Roques, H. (1988). *Water Res.*, **22**(12), 1582.

Noyola, A., Briones, R. & Jiménez, C. (1989). *Tratamiento anaerobio de vinazas a nivel planta piloto con dos tipos de reactores anaerobios.* Research Report Nos 8335 and 9341. Institute of Engineering, UNAM, México. 154 pp.

Obayashi, A. W., Stensel, H. D. & Kominek, E. (1981). *CEP*, April, 68.

Pedroza-Islas, R. & Durán de Bazúa, C. (1990). *Biol. Wastes*, **32**, 17.

Poggi, H. M. & Durán-de-Bazúa, C. (1988). Preprints of papers presented at *3rd Chemical Congress of North America and 195th ACS Meeting*, Vol. 28, No. 1, p. 496. Am. Chem. Soc., Ann Arbor, MI.

Poggi, H. M., Hernández, R. & Rindernecht, N. (1987). *Manual de técnicas de laboratorio.* CINVESTAV-IPN, México DF.

Roth, L. A. & Lentz, C. P. (1977). *Can. Inst. Food Sci. Technol. J.*, **10**, 105.

Russo, C., Sant'Anna, G. L. & De Carvalho Pereira, S. E. (1985). *Agric. Wastes*, **14**, 301.

Sánchez Riera, F., Córdoba, P. & Siñeriz, F. (1985). *Biotechnol. and Bioeng.*, **27**, 1710.

Sanna, P., Camilli, M. & Degen, L. (1987). In *Global Bioconversions*, Vol. II, ed. D. L. Wise. CRC Press, Boca Raton, FL, p. 49.

Sedue (1990). *Programa Nacional para la Protección del Medio Ambiente 1990–1994.* Secretaría de Desarrollo Urbano y Ecología, México DF, p. 22.

Speece, R. E. (1983). *Environ. Sci. Technol.*, **9**(7), 416A.

Szendrey, L. M. (1982). *Ind. Wastes*, **28**, 31.

Szendrey, L. M. (1983). Start-up and operation of the Bacardi corporation anaerobic filter. In *Proceedings of Third Int. Conf. Anaerobic Digestion*, Boston, MA. 13 pp.

Szendrey, L. M. (1984). *Environm. Progress*, **3**(4), 222.

Szendrey, L. M. (1986). In *Biomass Energy Development*, ed. W. H. Smith. Plenum Press, New York, p. 517.

Valderrama, S. B., Pedroza, R., Nieto, Z. & Durán, C. (1988). *Informe CONACYT-UNAM.* NEX-01-88, Facultad de Química, UNAM, México DF.

Zedillo, L. E. (1986). *Aprovechamiento de los subproductos de la caña de azúcar en México. Situación actual y perspectivas.* Instituto para el Mejoramiento de la Producción Azucarera (IMPA), Azúcar, SA, México DF.

Zedillo, L. E. (1987). *ATAM*, **1**, 13.

Zedillo, L. E. (1990). Internal Report, Instituto para el Mejoramiento de la Producción Azucarera (IMPA), Azúcar, SA, México DF.

Zehnder, A. J. B. (1978). In *Water Pollution Microbiology*, ed. R. Mitchell. John Wiley & Sons, New York, p. 349.

Chapter 18

BIODEGRADATION OF TEXTILE WASTEWATERS

NIKOS ATHANASOPOULOS

Perivallontiki-Energiaki Ltd, 106B Lontou Street, Patras, Greece

CONTENTS

1 INTRODUCTION

The production process for textile articles consists of a number of interlinked unit operations in which many different raw materials are used to produce a great variety of products. In order to satisfactorily discuss the pollution abatement characteristics of the industry, an operational disaggregation of its activities must be undertaken. The most realistic method of categorizing industrial activity and pollution abatement is that developed by the United States Environmental Protection Agency (USEPA). By it the major sources of waste from textile

Table 1
Established Wastewater Characteristics for the Textile Categories of Activity

Parameters	*USEPA Categories*						
	1	*2*	*3*	*4*	*5*	*6*	*7*
BOD_5/COD	0·20	0·29	0·35	0·54	0·35	0·30	0·31
BOD_5	6 000	300	350	650	350	300	250
TSS	8 000	130	200	300	300	120	75
COD	30 000	1 040	1 000	1 200	1 000	1 000	800
Oil and grease	5 500	—	—	14	53	—	—
Total chrome	0·05	4	0·014	0·04	0·05	0·42	0·27
Phenol	1·50	0·5		0·04	0·24	0·13	0·12
Sulphide	0·20	0·1	8·0	3·0	0·20	0·14	0·09
Colour (ADMI)	2 000	1 000		325	400	600	600
pH	8·0	7	10	10	8	8	11
Temperature (°F)	82	144	70	99	102	67	100
Water use (US gal/lb)	4·3	40	1·5	13·5	18	8·3	18

Explanatory notes:
1—Raw wool scouring
2—Yarn and fabric manufacture
3—Wool finishing
4—Woven fabric finishing
5—Knitted fabric finishing
6—Carpet manufacture
7—Stock and yarn dyeing and finishing

operations can be attributed to the following seven categories of activity:

(1) Raw wool scouring
(2) Yarn and fabric manufacture
(3) Wool finishing
(4) Woven fabric finishing
(5) Knitted fabric finishing
(6) Carpet manufacture
(7) Stock and yarn dyeing, and finishing

A detailed description of each of these categories is given by many authors (Cooper, 1978; USEPA, 1978; OECD, 1981). The wastewater characteristics established by USEPA for each category are given in Table 1. These characteristics were established after a comparison of results of the studies carried out for the American Textile Manufacturers' Institute, Inc., and the National Commission in Water Quality (USEPA, 1978). The extent to which waste must be treated to remove pollutants will depend on the discharge requirements set by regulatory bodies. These vary from country to country and from one enforcement agency to another.

To date the best available technology economically achievable for textile wastewater treatment involves the preliminary screening, primary settling, coagulation, secondary biological treatment, chlorination and advanced treatment techniques such as multi-media filtration and/or activated carbon following biological treatment.

Biological treatment technologies in most cases give the highest degree of confidence from an engineering and economic practicability point of view.

2 BIODEGRADABILITY OF TEXTILE CHEMICALS

A list of chemicals found in textile wastewater and their biodegradability is given in Table 2. Some explanatory notes are also included. The chemicals are listed according to the unit operation and the fibre; the category is also referenced (OECD, 1976). Category 8 is the commission finishing, which may include any of the unit operations of the other seven categories established by USEPA. Typical process or waste characteristics are not applicable to this category.

2.1 Dyes

In a study (Fledge, 1970) it was shown that the anthraquinone disperse dyes examined were partially degraded when subjected to aeration under conditions similar to those provided by an activated sludge aeration basin. Metabolites resulting from degradation were derivatives of the original dyes. The anthraquinone ring system, or backbone of the original dye molecule, had not been broken. These dyes in a practical sense were not degraded. Also the reactive dyes of the vinyl sulphone class studied were not degraded in a laboratory treatment simulating an activated sludge aeration basin. The reactive dyes were degraded when subjected to conditions simulating those employed in a conventional anaerobic digester but the disperse dyes were resistant to anaerobic degradation.

With any of the soluble reactive or direct dyes studied the most interesting result discovered was that no significant colour loss was observed after 30-day BOD studies (Porter & Snider, 1976). The disperse and vat dyes which are pigment dispersions rather than true solutions showed an average of 15% colour loss after 30 days in the incubated BOD bottle.

The elimination of dyestuffs in the activated sludge process is caused by the adsorption on the sludge (Hitz *et al.*, 1978). The adsorption capacity of sludge depends mainly on the chemical structure of the dyestuffs and less on the specific properties of the sludge. The substantive, disperse and basic dyestuffs had a high adsorption capacity while the reactive dyestuffs and a very low one. In the examined acid dyestuffs the adsorption capacity decreased with the increasing solubility. The adsorption capacity is also influenced by other chemicals in the wastewater.

Facultative anaerobic bacteria, which were able to degrade azo dyes, were isolated from draining ditches at dyestuff factories (Idaka & Ogawa, 1978). Concentrations up to 100 mg/litre were tested on shaking and static cultures. In shaking cultures about 50% of the dye was eliminated with elimination occurring rapidly. In the static culture elimination was more complete but occurred at a slower rate.

Under anaerobic sludge digestion conditions adsorbed dyestuffs are in general susceptible to degradation (Brown & Laboureur, 1983).

Table 2
List of Waterborne Pollutants Arising from Textile Processes
(reprinted with permission of OECD)

Operational subprocess	*Operational fibre*	*Substances contributing to residual pollution load*			*Biodegradability of organic fraction*
		Inorganic		*Organic*	
		Cations	*Anions*		
	Desizing				
	Cotton	Na^+	Cl^-	Enzymes	A
	Linen	Ca^{2+}	SO_4^{2-}	Nonionic surfactants	A
	Viscose	NH_4^+(*)		Starch	B
				Modified starches	B
				Carboxymethyl cellulose	S–NB
4				Hemicelluloses	A
				Fats, waxes and oils	S–NB
	Silk	Na^+		Gelatine	A
	Acetates	NH_4^+		Polyvinylalcohol	A
	Synthetics		CO_3^{2-}	Enzymes	A
			PO_4^{3-}	Starch	B
				Carboxymethyl cellulose	S–NB
				Polymeric sizes	NB
				Fats, waxes and oils	S–NB
	Degumming				
3	Silk	Na^+	CO_3^{2-}	Sencin (silk gum)	S–NB
		NH_4^+	PO_4^{3-}	Soap	A
	Scouring				
	Cotton	Na^+	CO_3^{2-}	Soaps	A
			PO_4^{3-}	Anionic surfactants	A
				Nonionic surfactants	A
				Hemicellulose	A
				Cotton waxes	NB
				Pectic matter	A
				Starch and sizes	A
4, 5, 6, 7, 8				Fats	S–NB
				Glycerol	B
	Viscose	Na^+	CO_3^{2-}	Soaps, anionic and	B
	Acetate		PO_4^{3-}	nonionic detergents	B
				Sizes	B
				Fats, waxes and oils	S–NB
	Synthetics	Na^+	Co_3^{2-}	Soaps, anionic and	A
			PO^{3-}	nonionic detergents	A
				Sizes	B
				Fats, waxes and oils	S–NB
				Antistatic agents	NB
				Petroleum spirit	A
3, 6, 7, 8	Wool	Na^+	CO_3^{2-}	Soaps	A
	(yarns and fabrics)	NH_4^+	PO_4^{3-}	Anionic detergents	A

Table 2—*contd.*

Operational subprocess	*Operational fibre*	*Substances contributing to residual pollution load*			*Biodegradability of organic fraction*
		Inorganic		*Organic*	
		Cations	*Anions*		
3, 6, 7, 8	Wool (yarns and fabrics)	Na^+	CO_3^{2-}	Soaps	A
		NH_4^+	PO_4^{3-}	Nonionic detergents	A
1	Wool (loose fibre)	Na^+	CO_3^{2-}	Wool, grease or wax	S–NB
		Ca^{2+}	Cl^-	Suint	A
		K^+	PO_4^{3-}	Soaps	A
		NH_4^+		Nitrogenous matter	U
				Anionic surfactants	A
				Formate, acetate	B
4	**Creping**				
	Cotton	Na^+	CO_3^{2-}	Phenolic wetting agents	NB
	Viscose		BO_3^{3-}	Anionic detergent	A
	Silk	NH_4^+			
3	**Crabbing**				
	Potting				
	Roll-boiling				
	Wool	Na^+	CO_3^{2-}	Oils	U
		NH_4^+	S^{2-}		
	Carbonizing				
	Wool	Al^{3+}	SO_4^{2-}	Surfactants	A
		Mg^{2+}	Cl^-	Wool grease	S–NB
			CO_3^{2-}	Suint	A
	Milling	Na^+	CO_3^{2-}	Wool grease	S–NB
			SO_4^{2-}	Suint	A
				Soaps	A
				Acetate, formate	B
	Oiling				
	Softening				
	Wool	Na^+	Br^-	Cetylpyridinium bromide	NB
				Soaps, nonionic and anionic detergents	A
				Oleine oil	A
				Mineral oil	S–NB
				Polypropylene oxides	A
5, 6, 7, 8	Synthetics			Fatty acid condensation products	U
				Quaternary ammonium compounds	NB
				Nonionic surfactants	A

(*continued*)

Table 2—*contd.*

Operational subprocess	*Operational fibre*	*Substances contributing to residual pollution load*			*Biodegradability of organic fraction*
		Inorganic		*Organic*	
		Cations	*Anions*		
	Cotton	Na^+	SO_4^{2-}	Vegetable, animal and mineral oils	U
		NH_4^+	Cl^-		
				Soaps and fatty acids	A
4, 6, 7, 8				Sulphonated oils and alcohols	U
				Quaternary ammonium compounds	NB
				Glycerol/polyglycols	A/NB
				Acetate	B
	Mercerizing				
	Cotton	Na^+	CO_3^{2-}	Cresols	A
	Linen	NH_4^+(*)	SO_4^{2-}	Cyclohexanol	A
				Alcohol sulphates	A
				Anionic surfactants	A
4, 7, 8	**Bleaching**				
	Cotton	Na^+	ClO^-	Formate	B
	Linen		Cl^-		
	Viscose		O_2^{2-}(*)		
	Jute	Na^+	SiO_3^{2-}		
		NH_4^+	Cl^-		
			F^-		
3, 8	Wool	Na^+	O_2^{2-}(*)	Oxalate	B
	Synthetics		PO_4^{3-}		
4, 5, 7, 8	Acetates		SiO_3^{2-}		
			F^-		
	Shrink resist				
		Na^+	ClO^-	Formaldehyde	A
			Cl^-	Formate	B
3, 7, 8				Proteolytic enzymes	A
			H_2SO_5	Melamine resins	NB
				Polyamide cationic resin	U
	Waterproofing				
	Cotton	Al^{3+}	Cl^-	Gelatine stearate	B
	Linen	Na^+	SO_4^{2-}	Paraffin wax	NB
	Jute	Zr^{4+}		Dispersing agents	U
	Wool	K^+		Acetate	B
4, 7, 8				Formate	B
				Stearamidemethyl pyridinium chloride	NB
				Silicone resins	NB
				Fluoroacrylic esters	U
				Melamine resins	NB
				Titanates	NB

Table 2—*contd.*

Operational subprocess	*Operational fibre*	*Substances contributing to residual pollution load*			*Biodegradability of organic fraction*
		Inorganic		*Organic*	
		Cations	*Anions*		
	Mothproofing				
3, 6, 7, 8	Wool	Na^+	F^-	Pentachlorophenyl laurate	NB
		K^+	SiF_6^{2-}		
		Al^{3+}		Formate	B
				Chlorinated compounds such as mitin and dieldrin	NB
4, 7, 8	**Rot and mildew proofing**				
	Cotton			Salicylanilide	A
	Jute	Cu^{2+}	SO_4^{2-}	Dihydroxy–dichlorophenyl	NB
		NH_4^+(*)	CO_3^{2-}		
		Zn^{2+}		8-Hydroxy-quinoline	NB
		Na^+		Naphthenic acid	A
		Al^{3+}		Petroleum solvents	A
				Pentachlorophenyl laurate	NB
				Formate	B
				Sulphonated oils	A
				'Eulans'	NB
	Fireproofing				
	Cotton	NH_4^+	PO_4^{3-}	Chlorinated rubber	NB
		Na^+	BO_3^{3-}	Synthetic resin binders	U
		Sb^{3+}	Cl^-	Tetrabishydroxymethyl-phosphonium chloride	U
		Ti^{4+}	NO_3^-		
			Br^-	Melamine resin	NB
				Thiourea resin	
3, 6, 7, 8	Wool		F^-		
4, 7, 8	**Loading**				
	Silk	Sn^{4+}	Cl^-	soaps	B
		Na^+	PO_4^{3-}	Sulphated alcohols	A
			SO_4^{2-}	Sulphonated oils	A
			SiO_3^{2-}		
	Surface coating				
	Cotton	NH_4^+(*)	SO_4^{2-}	Urea	B
	Viscose	Mg^{2+}	Cl^-	Formaldehyde	A
	Linen	Zn^{2+}	NO_3^-(*)	Methylol ureas	A
		Na^+	CO_3^{2-}	Melamine resin and precondensates	NB
				Substituted ureas and precondensates	U
				Phenol and PF resin	NB
				Polyvinyl chloride	NB
				Buna and chlorinated rubber	NB
				Polyurethane	NB
				Polystyrene	NB

(*continued*)

Table 2—*contd.*

Operational subprocess	*Operational fibre*	*Substances contributing to residual pollution load*			*Biodegradability of organic fraction*
		Inorganic		*Organic*	
		Cations	*Anions*		
				Polyethylene	NB
				Polypropylene	NB
4, 7, 8				Polyacrylics	NB
				Polyalkyls	NB
				Starch and modified starches	B
	Finishing (with resins)				
	Cotton	NH_4^+	SO_4^{2-}	Formaldehyde	A
		Mg^{2+}	Cl^-	Urea	B
		Zn^{2+}	NO_3^-(*)	Substituted ureas	A
		Na^+	Co_3^{2-}	Methylol ureas	A
				Substituted urea(s) precondensates	U
				UF resin	NB
				MF resin	NB
				DMEU resin	NB
				DMPU resin	NB
	Dyeing				
	Cotton	Na^+	Cl^-	Residual dyestuffs (S) (IS)	NB
	Viscose	Cr^{3+}	SO_4^{2-}		
	Linen	Cu^{2+}	CO_3^{2-}	Sulphated oils	A
4, 5, 7, 8		Sb^{3+}	F^-	Anionic dispersing agents	A
		K^+	NO_2^-(*)		
		NH_4^+	S^{2-}(*)	Formaldehyde	A
			$S_2O_3^{2-}$	β-Naphthol	A
			SO_3^{2-}	Cationic fixing agents	NB
			CO_4^{2-}	Tannic acid	A
			$H_2BO_4^-$	Acetate	B
			O_2^{2-}	Formate	B
				Tartrate	B
				Soaps	A
				Anionic surfactants	A
				Nonionic surfactants	A
				Urea	B
				Amides of naphthoic acids, etc.	B
				Nitro and chloro amines	S–NB
				Soluble oils	S–NB
	Wool	Na^+	SO_4^{2-}	Residual dyestuffs	NB
		NH_4^+(*)	SO_3^{2-}	Acetate	B
		Cr^{3+}	CO_3^{2-}	Formate	B
3, 7, 8		K^+	Cl^-	Sulphonated oils	A
		Sb^{3+}	$S_2O_4^{2-}$	Dispersing agents	U
		Al^{3+}		Lactate	B
		Cu^{2+}		Tartrate	B

Table 2—*contd.*

Operational subprocess	*Operational fibre*	*Substances contributing to residual pollution load*			*Biodegradability of organic fraction*
		Inorganic		*Organic*	
		Cations	*Anions*		
	Silk	Na^+	SO_4^{2-}	Residual dyestuffs	NB
		Al^{3+}	CO_3^{2-}	Sericin (S) (IS)	S–NB
		Ca^{2+}	NO_3^-(*)	Acetate	B
		Cr^{3+}	Cl^-	Soaps	A
		Fe^{3+}	SiO_3^{2-}	Tannic acid	A
		Cu^{2+}	$Fe(CN)_6^{4-}$	Vegetable oils	A
			$S_2O_4^{2-}$	Sulphonated oils	U
4, 5, 7, 8			NO_2^-(*)	Anionic detergents	A
	Jute	Na^+	Cl^-	Residual dyestuffs (S) (IS)	NB
			SO_4^{2-}		
			S^{2-}(*)	Acetate	B
			SO_3^{2-}(*)		
	Coir	Na^+	Cl^-	Residual dyestuffs (S)	NB
	(coconut fibre)	Al^{3+}	SO_4^{2-}	Acetate	B
	Acetates	Na^+	Cl^-	Residual dyestuffs (IS)	NB
				Soluble oils	S–NB
				Sulphated alcohols	A
				Urea	B
	Synthetics				
	Polyamide	Na^+	Cl^-	Residual dyestuffs (S) (IS)	NB
			CO_3^{2-}	Acetate	B
				Formate	B
				Sulphonated oils	A
				Polyamide oligeins	U
	Polyacrylic	Na^+	SO_4^{2-}	Residual dyestuffs (S)	NB
		Cu^{2+}		Phenolic compounds	A
		NH_4^+(*)		Aromatic amines	A
				Surfactants	A
				Levelling and retarding agents	U
				Formate, acetate	B
4, 5, 7, 8				Thiourea dioxide	A
	Polyester	Na^+	Cl^-	Residual dyestuffs (IS)	NB
		NH_4^+	$S_4O_6^{2-}$	Antistatic agents	NB
			SO_3^{2-}	Mineral oils	S–NB
			ClO^-	Nonionic surfactants	A
			NO_3^-(*)	Anionic surfactants	A
				Acetate, formate	B
				Soaps	A
				Dispersing agents	A
				Solvents	A
				Dye 'carriers'	S–NB
				Ethylene oxide condensation	U
				EDTA	NB

(*continued*)

Table 2—*contd.*

Operational subprocess	*Operational fibre*	*Substances contributing to residual pollution load*			*Biodegradability of organic fraction*
		Inorganic		*Organic*	
		Cations	*Anions*		
	Printing				
4, 5, 7, 8	Cotton	Na^+	$S_2O_4^{2-}$	Dyestuffs (S) (IS)	NB
				Formaldehyde	A
	Linen	Cr^{3+}	CO_3^{2-}	Starch and modified starches	B
	Viscose		NO_2^-	Vegetable gums	NB
				Alginates	A
				Carboxymethyl cellulose	NB
				Anionic surfactants	A
				Nonionic surfactants	A
				Acetate	B
				Urea	B
				Petroleum solvent	A
4, 5, 7, 8	Acetates	Na^+	CO_3^{2-}	Dyestuffs (S) (IS)	NB
	Synthetics	Cr^{3+}	NO_2^-	Formaldehyde	A
		NH_4^+	$S_2O_4^{2-}$	Acetate	B
			PO_4^{3-}	Starch and modified starches	A
				Vegetable gums	NB
				Anionic surfactants	A
				Nonionic surfactants	A
				Urea	B
				Alginates	A
				Petroleum solvents	A
	Glass fibre			Dyestuffs (IS)	NB
				Petroleum solvents	A
3, 7, 8	Wool	Na^+	ClO_3^-	Glycol, thiodiglycol	B
		NH_4^+		Tartrate, acetate	B
				Oxalate	B
				Urea	B
				Alginates	A
				Vegetable gums	NB
5, 7, 8	Antistatic treatment of synthetics			Polyglycols	S–NB

Explanatory notes:
(S) = soluble; (IS) = insoluble; B = readily biodegradable; A = biodegradable after acclimatization; NB = non-biodegradable; S–NB = slowly degradable—effectively non-degradable during normal biological treatment; U = unknown—in some cases the description of the material involved is too imprecise for the substance to be classified; (*) some radicals, such as ammonium nitrite, sulphide, thiosulphate, tetrathional and sulphite can be oxidized by micro-organisms to nitrate and sulphate, and to this extent may be considered to be biodegradable. Under appropriate conditions, nitrate can be reduced to nitrogen gas or assimilated by plant life and thus removed from the water. Also peroxides can be catalytically decomposed by enzymes produced by micro-organisms.

In an extensive work on biodegradation of dyestuffs (Pagga & Brown, 1986) it has been shown that dyestuffs are most unlikely to show any significant biodegradation in aerobic tests and the substantial colour removal observed was attributed to the elimination of the dyes by adsorption of the intact dye molecules. It has not been possible to demonstrate unequivocally if aerobic biodegradation processes have occurred in the chromophore components of the dyes and if, for some dyestuffs, it is possible that relatively minor biodegradation process changes occur to render the dye more amenable to removal.

It was concluded (Shaul *et al.*, 1987) that CI Acid Red 1 appeared to be very nominally adsorbed but not biodegradable; CI Acid Red 337 and Acid Yellow 151 appeared to be moderately adsorbed but not biodegraded; CI Acid Blue 113 and Acid Red 151 appeared to be strongly adsorbed and possibly partially biodegraded; and CI Acid Orange 7 and CI Acid Red 88 appeared to be moderately adsorbed and significantly biodegraded. The above dyes are azo dyes and could constitute submission of toxic substances through their degradation products, such as aromatic amines.

2.2 Surface Active Agents

A series of research studies on biodegradation of surface active agents has been reviewed (Swisher, 1970; Schmid *et al.*, 1976). For the biodegradation of surface active agents their hydrophobic ends play a fundamental role. Less fundamental for biodegradation but fundamental for their water solubility, when water is the transportation medium for bacteria, are the hydrophilic groups of a surface active agent ($—SO_3H$, ethylenoxides). It was shown (Swisher, 1970) that the hydrophilic ends of molecules are settled on the enzymes where the oxidation at the end of the chain begins to take place.

For the biodegradation of chain molecules there are two ways: ω-oxidation and β-oxidation. The ω-oxidation begins on the hydrophobic end (higher than C_6); the β-oxidation, in the presence of coenzyme A, begins on the carboxyl group. These two ways are shown in Fig. 1. For example, paraffin is oxidized through ω-oxidation. The end products carbonic acids or dicarbonic acids, when both ends of the chain are oxidized, are further attacked through β-oxidation and produce acetate and a carbonic acid with longer chain, which is further attacked by β-oxidation. The hydrogen produced during the β-oxidation under aerobic conditions gives water and under anaerobic conditions gives methane or hydrogen sulphide.

A third possibility is the biochemical opening of the ring of aromatic molecules to β-ketoadipic acid and further through β-oxidation to simpler compounds.

Anionic surface active agents are moderately bactericidal to Gram-positive bacteria but only slightly to Gram-negative bacteria. Cationic surface active agents are more active than anionic against all bacteria because they are attached by the net negative charge of the bacteria surface. In general, the bactericidal action of surface active agents occurs by preferential adsorption of the agent on the bacterial surface. When the agent concentration builds up, the membrane is disrupted, thus allowing agents to combine with bacteria lipids (Woodworth, 1975). Marked inhibition of

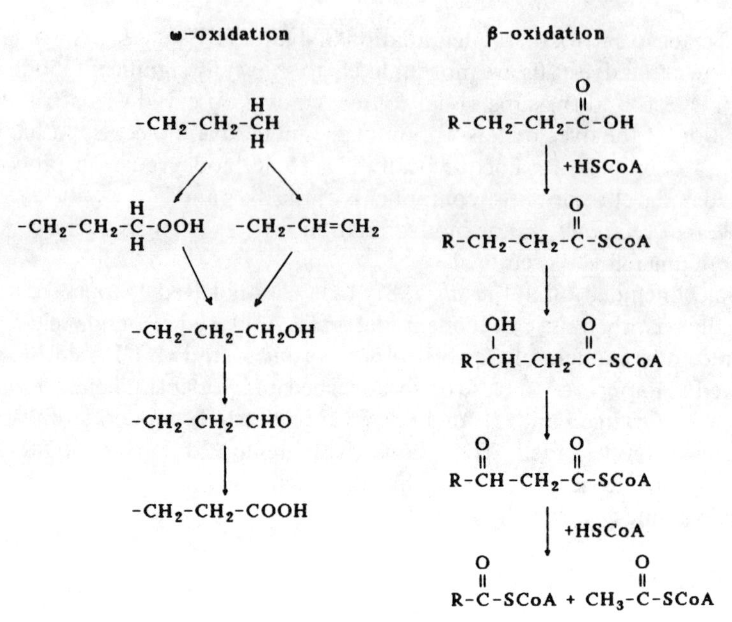

Fig. 1. Biodegradation of chain molecules through ω- and β-oxidation.

bacteria growth occurs when the content of ABS approaches 750 mg/litre, although the ratio of ABS to the total solids will have an effect on the concentration at which inhibition occurs. Antibacterial action by cationic agents can occur at 1 mg/litre but adsorbing macromolecules can prevent this action from occurring until concentrations reach 100–250 mg/litre. The inhibitory action of the anionic and cationic agents is influenced greatly by their pH. Cationic agents, which account for the 4–5% of the total surfactants used in the textile industry (Fisher, 1978), exhibit their maximum activity in the alkaline pH range, whereas the anionic agents are more active in the acid range. Nonionic agents are not particularly inhibitory to bacteria (Kremer *et al.*, 1982). Cationic agents react with anionic agents and decrease their toxicity to bacteria.

Nonionic surfactants are used more extensively in the textile industry than any other type of surfactant. The two principal nonionic surfactants used are alcohol ethoxylates (AE) having essentially linear alkyl chains and alkylphenol ethoxylates (APE) having highly branched alkyl chains. These surfactants are used in such textile applications as wetting, scouring, dye levelling, emulsification, fibre lubrication and more recently in foam finishing. Biodegradation of surfactants is important because they interfere with effective oxygen transfer in biological treatment systems, they are characterized by relatively high aquatic toxicity and contribute to final BOD_5 or COD of treated effluents.

After many years of laboratory studies (Kravetz, 1983) it was concluded that alcohol ethoxylates biodegrade faster and more extensively than alkylphenol ethoxylates. Under realistic summer field conditions both AE and APE appear to undergo adequate primary biodegradation while AE undergoes more extensive ultimate biodegradation than APE. Biodegradation of AE also permits lower levels of biodegradation intermediates to enter receiving waters compared to APE. Under conditions of plant stress, such as cold winter conditions or plant upset, the primary biodegradability of APE, but not AE, decreases significantly and can permit undegraded APE to enter receiving waters and cause foaming.

Even though literature indicates that surfactants are more easily biodegradable than ever, they are still very difficult to degrade. For example, the surfactant alkylphenol ethoxylate is removed in activated sludge treatment and it was thought to be biodegradable. This surfactant is resistant to attack by bacteria found in activated sludge but it is effectively adsorbed on the sludge. The results shown by the Organization for Economic Cooperation and Development (OECD) confirmatory test (Pagga & Brown, 1986) include both biodegradability and elimination due to adsorption on the sludge. Thus most of the surface active agents used today are actually eliminated (and not biodegraded) by more than 80%.

2.3 Sizes

PVA is degraded well after acclimatization (Porter & Snider, 1976; USEPA, 1978). The hydrolysis of polyvinyl acetate to yield water-soluble alcohol produces PVA. The hydrolysis is not complete, and several forms are offered for commercial use which contain a small percentage of the original acetate groups remaining in the polymer. As small amounts of substituent groups (acetate and acrylonitrile) on starch and cellulose have been used for many years to make these polymers resistant to rotting and decay, PVA resists biodegradation when it contains small amounts of acetate groups in the polymer. The number and location of these residual groups may vary from batch to batch, causing the biodegradability to vary considerably. A portion of PVA is removed by sedimentation in the biological system. Starch is readily biodegradable. Polyacrylates and CMC are partially removed by sedimentation in activated sludge. CMC degrades slowly with acclimatization. Polyacrylic and polyester sizes are recalcitrant to biodegradation.

2.4 Finishing Agents

Many different chemicals are used in treating textile fabric to impart desirable properties to the final garment. The resin finishes have extremely low BOD values compared to their COD values. This may be caused by the presence of toxic chemicals or a metal catalyst used in the finish formulation (Porter & Snider, 1976). Most finishing agents are eliminated by adsorption or sedimentation in the biological treatment plants and not by biodegradation.

2.5 Grease and Oils

Oil and grease may present problems at the wastewater treatment plant because they adhere to particulate matter, causing it to float. Small quantities can be assimilated with little effect by the effluent treatment process, particularly if emulsified. Not all oils are equally troublesome. Vegetable oils and natural grease tend to be fairly readily degradable. The reduction of grease from wool scouring wastewater by various biological treatments is HRT-dependent (Christoe *et al.*, 1976). At an HRT of 25 h the removal of grease was 60%.

The modern trend is towards lubricants based on petroleum and these materials are very much less easily degraded. Generally pretreatment for removal of these substances will not be required unless floating oil or grease is obviously present in the effluent.

The naphthenic and paraffinic mineral oils are biodegradable with acclimatization. The silicon oil (dimethyl polysiloxane) is not biodegradable. Waxes degrade slowly and the fatty esters are biodegradable.

2.6 Complexing Agents

EDTA (ethylene diamine tetraacetate), which is the most widely used complexing agent, is very stable against oxidation and is not biodegradable. Due to its low molecular weight it is not adsorbed on activated sludge and escapes with the effluent of the treatment plant.

3 ACCLIMATIZATION OF BACTERIA TO CHEMICALS

Acclimatization describes the bacteria 'getting used to' a particular waste. When a new compound is introduced to a living biomass, the population of microorganisms may not be able to biodegrade the material initially. As time passes, however, the biomass may become active towards the material. This process, of course, is the acclimatization of sludge. The acclimatization phenomenon occurs due to one or more characteristics of mixed bacteria cultures (Straley, 1984):

(1) Bacteria are conservative. Although a bacterium may possess the genetic information necessary to produce enzymes to degrade a particular new compound, it will not spend the energy and actually synthesize the necessary enzymes unless the compound is present. This lagging response to a new compound is called 'induction'.
(2) Population dynamics affect the removal of a new compound in mixed cultures. If a particular strain of bacteria acclimatizes, and the co-existing populations do not acclimatize the compound, the strain has a competitive advantage. Food is available to a strain which can increase its relative predominance in the total population. Sometimes the strain can be a minor part of the population prior to the introduction of a new compound, and gradually grows into a much larger proportion. This corresponds generally to a decrease in the effluent COD/BOD ratio. This gross effect is often referred to as the acclimatization of the biomass.
(3) A bacteria strain capable of degrading the new compound may not be present as a component of the existing biomass. When this happens no

acclimatization can take place in the system. This phenomenon accounts for the situation where a particular new component will be degraded in one plant and yet will pass through another plant untouched, even though conditions may be similar.

In textile finishing and dyeing wastewater treatment many new compounds are only present occasionally, due to the batch nature of the process. A biomass has become acclimatized to a new compound and then the compound disappears from the influent wastewater due to a process change. The acclimatization population will lose its activity toward the novel compound because the daughter cells will not exhibit activity toward the compound, if it was not present during cell division. Thus the number of bacteria having activity towards the compound is rapidly reduced due to a dilution effect.

A biomass can lose acclimatization during plant shutdowns. In this case different selection pressures are brought to bear on the biomass than are present during normal operation. During plant shutdowns the food is cut off. The bacterial population crashes and only those bacteria that are predisposed toward survival are left. Spore-forming bacteria such as *Bacillus* leave a fair seed behind, but bacteria such as *Pseudomonas*—very important in the breakdown of aromatic compounds—die off to very low levels. Another problem that can upset the acclimatization of the plant is that of shocks due to pH, high load of BOD or toxic conditions.

3.1 A Case Study of Acclimatization: Piraiki–Patraiki's Wastewater Treatment Plant

Piraiki–Patraiki is the largest textile company in Greece. According to the USEPA classification of industrial activity and pollution abatement the company is characterized by the following categories: yarn and fabric manufacture; woven fabric finishing; and knit fabric finishing.

Simply, it can be said that the company includes all unit operations related to cotton and cotton polyester blend products. It discharges to Patraikos Gulf, which is almost a closed sea without strong water streams. It was necessary to build such a wastewater treatment plant that could meet the discharge limitations of this area (BOD_5 40 mg/litre, COD 150 mg/litre, SS 40 mg/litre).

The systematic research procedure carried out to establish the wastewater treatment plant design characteristics is described elsewhere (Athanasopoulos, 1990). The results of bench-scale activated sludge units clearly demonstrated that for all the various waste samples the COD removal rate coefficient, k, was roughly equal whereas the non-biodegradable fraction of COD, z, was different. This coefficient was measured according to Eckenfelder's model:

$$(S_o - S_e)/Xt = k(S_e - z)$$

where S_o = influent total COD (mg/litre), S_e = effluent dissolved COD (mg/litre), X = MLVSS (mg/litre) and t = detention time (days), and the results are shown in Fig. 2.

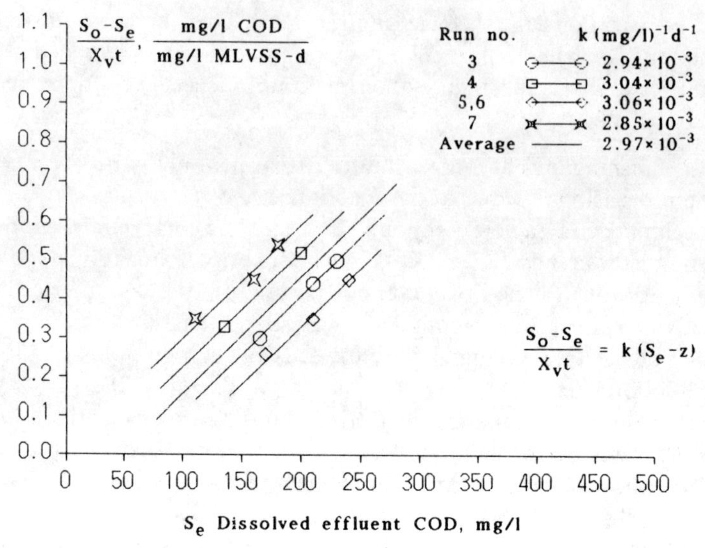

Fig. 2. COD removal rate coefficient calculation for Piraiki–Patraiki's wastewater treatment, in bench scale activated sludge units. Each run number represents an experimental run which lasted two or three weeks and was carried out with the same feed (Athanasopoulos, 1990).

The wastewater characteristics during the research period (1981) are shown in Table 3. BOD_5/COD ratio was 0·48. The theoretical concentration of COD in the effluents, as a function of aeration time, was calculated according to the kinetic parameters developed after the analysis of experimental results and is presented in Fig. 3.

The characteristics of wastewater and activated sludge treatment plant after four years of operation (1988) are shown in Table 4. The BOD_5 and COD values have

Table 3
Average Composition of Piraiki–Patraiki's Raw Wastewater During the Research Period (1981)

Parameter	*Average value*	*Standard deviation*	*Range of 95% confidence level*
COD total	825 ppm	0·18	534–1 116
COD dissolved	748 ppm	0·17	499–997
BOD_5 total	396 ppm	0·18	256–536
BOF_5 dissolved	346 ppm	0·17	231–461
TSS	68 ppm	0·60	0–140
pH	11·4	0·02	11·0–11·8
P-alkalinity	8·7 meq/litre	0·32	3·2–14·1
M-alkalinity	15·7 meq/litre	0·27	7·4–24·0
Colour	820 Co-Pt	0·33	290–1 350
	1:8 dilution	0·34	1:3–1:14
Conductivity	2 300 mhos/cm	0·24	1 200–3 380
Cr^{6+}	0·06 ppm	0·56	
S^{2-}	0·01–6 ppm		

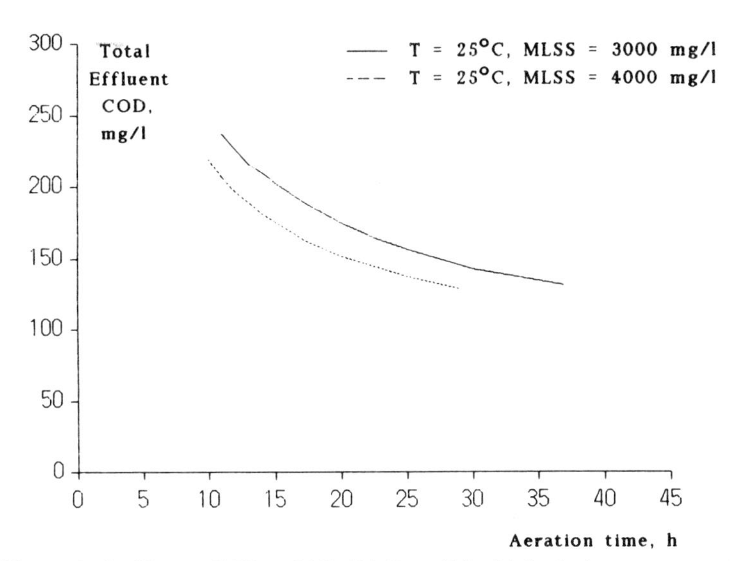

Fig. 3. Theoretical effluent COD of Piraiki–Patraiki's biological treatment plant as a function of aeration time.

Table 4
Characteristics of Piraiki–Patraiki's Wastewater and Its Activated Sludge Treatment Plant After Four Years of Operation (1988)

Parameter	*Average value*	*Number of measurements*
Wastewater		
Flow rate (m^3/day)	13 219	225
pH	>10·5	
Total COD (mg/litre)	908	86
Total BOD_5 (mg/litre)	450	25
TSS (mg/litre)	65·4	70
NH_3–N (mg/litre)	5·0	202
PO_4–P (mg/litre)	3·6	203
Temperature	35–45	225
Colour, dilution	1:8	178
Activated sludge treatment plant		
Aeration tank volume (m^3)	12 000	
Total effluent COD (mg/litre)	80·1	86
Total effluent BOD_5 (mg/litre)	19	25
Total effluent SS (mg/litre)	15	98
Effluent colour, dilution	1:4	178
MLSS (mg/litre)	2 797	176
MLVSS (mg/litre)	2 505	176
Dissolved oxygen (mg/litre)	3·65	209
Temperature (°C)	25·5	208
SVI	350	176
Consumptionof 75% H_2SO_4 (kg/day)	4 023	225
Consumption of electricity (kWh/day)	13 552	225
Consumption of liquid cationic polyelectrolyte (kg/day)	32·3	225
Sludge production (22·5%) (kg/day)	8 162	116

increased due to water conservation policy and countercurrent systems applied (Athanasopoulos & Karadimitris, 1989). BOD_5/COD ratio is now 0·49. Actual COD reduction is 91·2% instead of 81·2% expected from the laboratory research, as shown in Fig. 3. COD removal is explained by the long acclimatization of biomass to various chemicals. The COD removal rate coefficient of effluents has been increased three times compared to that found in laboratory research.

4 CARRIER BIOLOGY

Carrier biology has been described recently (Oehme, 1986). Suspended materials have two significant advantages: owing to their granular form and consequently better ratio between surface and volume, suspended materials have a larger specific surface, and this offers more space for colonization (this is the external geometrical factor). To this we can add the enormous inner surface of 'active' carrier substances, e.g. activated carbon (this can be regarded as the inner geometrical factor). When we use certain types of such activated carbon as the carrier material in a biological treatment system as activated sludge, we can obtain a significant synergistic acceleration of the reaction because the surface of the activated carbon has affinity for oxygen. The biological degradation of the pollutants takes place on the millions of square metres of inner surface of the carrier, where large amounts of oxygen activated by chemisorption are available. During this process the carrier material is not consumed because the reaction is controlled, not by its adsorption capacity but by the continuously regenerated oxygen-activated surfaces.

The bacterial biocenosis that is formed on the carrier material (carrier biology) profits from the oxygen excess and also from the substrate offered, because both of these are available in a 'concentrated' form due to temporary adsorption (Fig. 4). As

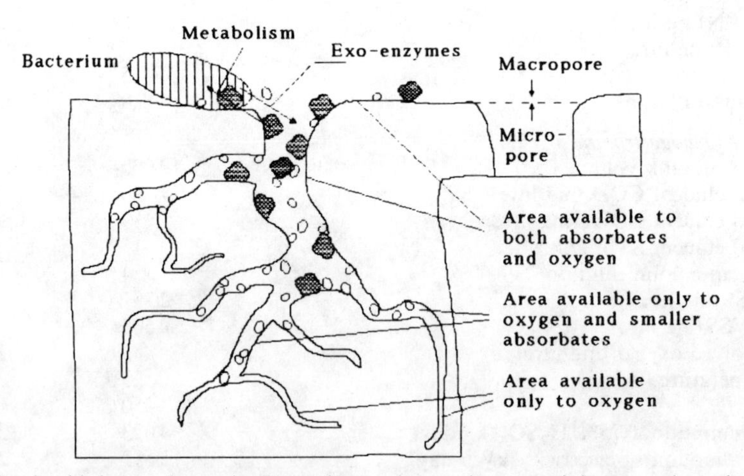

Fig. 4. Carrier biology schematic. Bacteria on the carrier material, e.g. activated carbon, profit from the oxygen and substrate excess, because both of these are available in a concentrated form due to temporary adsorption (Courtesy: Melliand Textilberichte GmbH).

a result of the immobilization, even special bacteria with a normally slow growth rate are accumulated and thus kept available and active for pollutants that are difficult to degrade. The substances which are difficult to degrade are held back for a longer time on the carrier material by temporary adsorption and they are also exposed longer to the action of the special bacteria that also settle on the carrier material. The ferments and enzymes that have to be developed for degrading complicated organic pollutants are formed more rapidly on the carrier, and they remain there as fixed and available exoenzymes.

The acceleration and the extent of the degradation of the organic pollutants dissolved in the wastewater are closely connected to the structure and texture of the carrier materials used.

Various mechanisms have been proposed to interpret the effect of activated carbon, and these mechanisms have been reviewed by Sublette *et al.* (1982):

—adsorption of toxic or inhibitory substances either fed or produced from cell activity;
—substrate concentration increase in the superficial microenvironment due to adsorption;
—carbon bioregeneration: the adsorbed molecules of organics with low biodegradability can be broken by extracellular enzymes (easily adsorbable) into smaller molecules that desorb and are utilized by the microorganisms present in the carbon micropores;
—the superficial macropores of activated carbon act as protected zones for prey-bacteria from predator-protozoa.

The removal of an organic reactive azo dyestuff particularly resistant to biodegradation was studied in an activated sludge pilot plant on the basis of the PACT process by Speccia *et al.* (1988). They showed that carbon bioregeneration, existing only in the presence of adsorbent solids, is probably the fundamental mechanism governing biological dye removal, and the carbon physical adsorption capacity seemed also to play a fundamental role.

Reactive dyes and other water-soluble dyes, as well as biologically more resistant surfactants and the complexing agent EDTA, which cannot be eliminated satisfactorily in a conventional mechanical–biological water treatment system, can be biologically degraded with the aid of a subsequent activated carbon solid-bed filter covered with biological growth (Croissant *et al.*, 1983). This can be carried out in the same manner on industrial effluents. The degradation of these compounds is based on biological respiration.

5 BIOLOGICAL TREATMENT METHODS

Biological treatment encompasses basically aerobic and anaerobic treatment. Biological treatment may be preceded by primary treatment, typically sediment-ation. Pretreatment of textile wastes before biological treatment could include any or all of the following: screening, sedimentation, equalization, neutralization, chrome reduction, coagulation or any of the other physical–chemical treatments. The performance of biological treatment in COD removal depends on the BOD_5/COD ratio. As shown in Table 1, the mean value for BOD_5/COD ratios for all categories is about 0·35, which is unfavourable for complete COD elimination. This ratio should change to at least 0·6 by replacing the difficult to biodegrade organic chemicals with other more biodegradable ones (König, 1989).

5.1 Aerobic Treatment

The biological treatment methods applicable to textile wastewater follow in order of increasing detention time: (1) trickling filters; (2) activated sludge; (3) rotating biological disks; (4) extended aeration; (5) lagoons; and (6) aquatic plants. Details of these methods are described in the literature (Cooper, 1978; USEPA, 1978).

Applying the AS–PACT process (activated sludge–powder activated carbon treatment) by dosing powder activated carbon in the aeration tank, it is possible to have a very effective COD removal. The main result is the fact that certain effects of the addition of PAC only become visible after many months of uninterrupted dosing. Once an adapted culture capable of regenerating the activated carbon is developed, the economic utilization depends on the number of times the adsorbed molecule is degraded and replaced by another molecule at the same site. The quality of the treated effluent from a Belgian textile factory applying the AS–PACT method was as good as the quality of the receiving surface water (Bettens, 1988). Belgian experience from another textile plant shows that addition of PAC reduces toxicity, enhances the treatability and augments the removal of soluble dyestuffs. Acclimatization of activated sludge to various chemicals enhances the biodegradation, as shown in Fig. 3 and Table 4.

Aquatic macrophyte-based treatment systems offer a promising low-cost method for removing contaminants from textile wastewater. The vascular plants cultured in such treatment systems perform several functions, including assimilating and storing contaminants, transporting O_2 to the root zone and providing a substrate for microbial activity. Among the various types of aquatic treatment systems, pond systems containing floating macrophytes such as the water hyacinth are most commonly utilized for wastewater treatment in tropical and subtropical regions, whereas in temperate regions emergent plants cultured in artificial wetlands appear to be more appropriate (Reddy & DeBusk, 1987).

In a study (Trivedy & Gudekar, 1987) with cotton finishing wastewater it was shown that water hyacinth treatment results in high BOD_5, COD and solids removal at a retention time of 3–4 days. Its main advantage over the other methods of treatment, besides BOD_5 and COD removal, is the high solids removal (suspended and dissolved). No other system gives such a high reduction of dissolved solids.

A cotton finishing wastewater combined with sewage (flow rate 100 m^3/day) was studied in a series of two hyacinth tanks of 30 days' retention time (Kulkarni *et al.*, 1982). The COD removal was higher than 80%.

5.2 Anaerobic Treatment

Anaerobic lagoons and facultative ponds are used in treating textile wastewater and are operating successfully mainly in rural areas.

Sulzer has applied anaerobic treatment to indigo dyeing wastewater, after indigo dyestuff recovery by reverse osmosis, in a four-day retention time anaerobic digester. Cotton fabric desizing and scouring wastewater is very well treated anaerobically in an upflow anaerobic filter (UAF) reactor. The COD removal was 60–90% for a COD loading of up to 2·75 kg/m^3 day (Athanasopoulos, 1986).

Cotton textile finishing wastewater with characteristics described in Table 4 was successfully treated in a UAF reactor. The COD removal was 50–90% for a COD loading of up to 1 kg/m^3 day (Athanasopoulos & Karadimitris, 1988).

The same wastewater was treated in an anaerobic expanded bed reactor of 2 m^3 volume. The COD removal was 50–87% for a COD loading of up to 0·63 kg/m^3 day (Athanasopoulos, 1991).

It seems that the acidification phase is the rate-limiting step and needs further research in order to use high loaded reactors. The achieved COD loading of the reactors was almost equal to the loading of the aeration tank in the activated sludge treatment. The investment cost for the installation of a UAF reactor with such a loading is very high, to be justified by the conservation of energy, nutrients and neutralization chemicals by using the flue gas of biogas burning.

The anaerobic treatment of wool scouring wastewater was studied in a single-stage anaerobic contact process reactor of 4·5 m^3 volume (Seyfried *et al.*, 1984). Up to an HRT of 10 days and a COD loading of 4·2 kg/m^3 day COD and grease were removable up to approximately 55% under stable operating conditions. The addition of polymers significantly improved the sludge settling factors and recirculation. This allowed an increase in load of up to 5 days HRT or 8·1 kg/m^3 day COD.

6 CONCLUSIONS

The aerobic biological method of treatment of textile wastewater is so far the most commonly used method all over the world.

The BOD_5/COD ratio for all categories of textile industry is unfavourable for complete COD elimination. The trend in future must be the replacing of difficult to biodegrade organic chemicals with other more biodegradable ones so that the BOD_5/COD ratio will become higher than 0·5.

Acclimatization is very important in textile wastewater biodegradation. Due to the batch nature of the textile production process it is necessary to implement high SRT in the biological reactor. Shutdowns of the biological reactor should be avoided by installing wastewater buffering or storage tanks.

Approaches relating to improvement of the metabolic diversity and affinity of the aerobic microbial community should be followed. Such an approach is the addition of PAC in the activated sludge basin or the application of the ‘carrier biology’.

Aquatic macrophyte-based treatment systems offer a promising low-cost method for removing solids (suspended and dissolved) besides BOD_5 and COD.

Anaerobic digestion is assumed to be more sensitive to toxicants than its aerobic counterpart. Though not a misconception, this assumption currently requires re-evaluation. It seems that anaerobic digestion could be applied, in many cases, as a pretreatment of textile wastewater. Further research is needed to use high loaded reactors so that the method will outcompete the aerobic counterpart.

REFERENCES

Athanasopoulos, N. (1986). *Biotech. Lett.*, **8**, 377.
Athansopoulos, N. (1990). *Melliand Textilberichte*, **8**, 619.
Athanasopoulos, N. (1991). *Bioresource Technology* (in press).
Athanasopoulos, N. & Karadimitris, T. (1988). *Biotech. Lett.*, **10**, 443.
Athanasopoulos, N. & Karadimitris, T. (1989). *Melliand Textilberichte*, **2**, 133.
Bettens, L. (1988). *Textile Month*, **10**, 49.
Brown, D. & Laboureur, P. (1983). *Chemosphere*, **12**, 397.
Christoe, J. R., Anderson, C. A. & Wood, G. F. (1976). *J. WPCF*, **48**(4), 729.
Cooper, S. G. (1978). *The Textile Industry, Environmental Control and Energy Conservation.* Noyes Data Co., Park Ridge, NJ, p. 21.
Croissant, B., Efferehn, K. & Fahne, D. (1983). *Melliand Textilberichte*, **64**(9), 686.
Fisher, K. (1978). *Melliand Textilberichte*, **59**(8), 659.
Fledge, R. K. (1970). *Determination of Degraded Dyes and Auxiliary Chemicals in Effluents from Textile Dyeing Processes.* ERC-0270, Georgia Institute of Technology, Atlanta, GA.
Hitz, H. R., Huber, W. & Reed, R. H. (1978). *J. Soc. Dyers Colorists*, **94**(2), 71.
Idaka, E. & Ogawa, T. (1978). *J. Soc. Dyers Colorists*, **94**, 91.
König, K. (1989). *Melliand Textilberichte*, **70**(3), 209.
Kravetz, L. (1983). *AATCC*, **15**(4), 57.
Kremer, F., Broomfield, B. & Fradkin, L. (1982). In *Proc. 37th Industrial Waste Conference, Purdue University,* Ann Arbor Science Publishers, Ann Arbor, MI. p. 157.
Kulkarni, M. V., Agnihotri, S. B. & Jain, S. M. (1982). *Colourage*, **5**, 7.
OECD (1976). *Programme on the Control of Specific Water Pollutants, Industrial Branch Strategies—Textile Finishing Industry.* Water Management Group, Organisation for Economic Control and Development, Paris, pp. 124–129.
OECD (1981). *Emission Control in the Textile Industry.* Organisation for Economic Control and Development, Paris, pp. 32–41.
Oehme, C. (1986). *Melliand Textilberichte*, **67**(8), 582.
Pagga, N. & Brown, D. (1986). *Chemosphere*, **15**(4), 479.
Porter, J. J. & Snider, E. H. (1976). *J. WPCF*, **48**(9), 2198.
Reddy, K. R. & DeBusk, T. A. (1987). *Wat. Sci. Tech.*, **19**(10), 61.
Schmid, R. D., Fischer, W. K., Gerike, P. & Gode, P. (1976). *Waschmittelchemie.* Alfred Huthig Verlag GmbH, Heidelberg, FRG.
Seyfried, C. F., Bode, H., Rosenwinkel, K. H., Saake, M. & Spies, P. (1984). *Anaerobic Treatment of Organic High Polluted Industrial Wastes—Comparison of Different Anaerobic Processes.* Institut für Siedlungswasserwirtschaft und Abfalltechnik, Universität Hannover, FRG.
Shaul, G. M., Dempsey, C. R., Dostal, K. A. & Lieberman, R. J. (1987). Fate of azo dyes in the activated sludge process. In *Proc. 41st Industrial Waste Conference, Purdue University,* Ann Arbor Science Publishers, Ann Arbor, MI.
Speccia, V., Ruggeri, B. & Gianetto, A. (1988). *Chem. Eng. Comm.*, **68**, 99.

Straley, J. P. (1984). *American Dyestuff Reporter*, **73**(9), 46.
Sublette, K. L., Snider, E. H. & Sylvester, N. D. (1982). *Water Research*, **16**, 1075.
Swisher, R. D. (1970). *Surfactant Science Series, Vol. III. Surfactant Biodegradation.* Marcel Dekker, New York.
Trivedy, P. K. & Gudekar, N. R. (1987). *Wat. Sci. Tech.*, **19**(10), 103.
USEPA (1978). *Textile Processing Industry*, EPA-625/778-002. United States Environmental Protection Agency, Washington, DC, p. 3.1.
Woodworth, G. M. (1975). *New Zealand J. Sci.*, **18**, 131.

INDEX